D1154161

Beyond Foraging and Collecting

Evolutionary Change in Hunter-Gatherer Settlement Systems

FUNDAMENTAL ISSUES IN ARCHAEOLOGY

BEYOND FORAGING AND COLLECTING
Evolutionary Change in Hunter-Gatherer Settlement Systems
Edited by Ben Fitzhugh and Junko Habu

EMERGENCE AND CHANGE IN EARLY URBAN SOCIETIES
Edited by Linda Manzanilla

FOUNDATIONS OF SOCIAL INEQUALITY
Edited by T. Douglas Price and Gary M. Feinman

FROM LEADERS TO RULERS
Edited by Jonathan Haas

LANDSCAPES OF POWER, LANDSCAPES OF CONFLICT
State Formation in the South Scandinavian Iron Age
Tina L. Thurston

LEADERSHIP STRATEGIES, ECONOMIC ACTIVITY, AND INTERREGIONAL INTERACTION
Social Complexity in Northeast China
Gideon Shelach

LIFE IN NEOLITHIC FARMING COMMUNITIES
Social Organization, Identity, and Differentiation
Edited by Ian Kuijt

A Continuation Order Plan is available for this series. A continuation order will bring delivery of each new volume immediately upon publication. Volumes are billed only upon actual shipment. For further information please contact the publisher.

Beyond Foraging and Collecting

Evolutionary Change in Hunter-Gatherer Settlement Systems

Edited by

BEN FITZHUGH

University of Washington
Seattle, Washington

and

JUNKO HABU

University of California
Berkeley, California

Kluwer Academic / Plenum Publishers
New York Boston Dordrecht London Moscow

ISBN: 0-306-46753-4

233 Spring Street, New York, New York 10013

http://www.wkap.nl/

10 9 8 7 6 5 4 3 2 1

A C.I.P. record for this book is available from the Library of Congress

Printed in the United States of America

Contributors

Mark Aldenderfer, Department of Anthropology, University of California, Santa Barbara, California 93106

Kenneth M. Ames, Department of Anthropology, Portland State University, Portland, Oregon 97207

Ofer Bar-Yosef, Peabody Museum, Harvard University, Cambridge, Massachusetts 02138

Lewis R. Binford, Department of Anthropology, Southern Methodist University, Dallas, Texas 75275

Aubrey Cannon, Department of Anthropology, McMaster University, Hamilton, Ontario, L8S 4L9 CANADA

Lynn E. Fisher, Sociology/Anthropology Program, University of Illinois at Springfield, Springfield, Illinois 62794

Ben Fitzhugh, Department of Anthropology, University of Washington, Seattle, Washington 98195

Junko Habu, Department of Anthropology, University of California, Berkeley, California 94720

Amber L. Johnson, Department of Anthropology, Southern Methodist University, Dallas, Texas 75275

Laura Lee Junker, Department of Anthropology, University of Illinois at Chicago, Chicago, Illinois 60607

Renato Kipnis, Laboratório de Estudos de Evolutivos Humanos, Instituto de Biociências, Universidade de São Paulo, São Paulo, SP 05508–900 Brasil

T. Douglas Price, Department of Anthropology, University of Wisconsin-Madison, Madison Wisconsin 53706

James M. Savelle, Department of Anthropology, McGill University, Montreal, Quebec H3A 2T7 CANADA

David W. Zeanah, Department of Anthropology, California State University, Sacramento, California 95819

Foreword

Lewis R. Binford and Amber L. Johnson

The organizers of this volume have brought together authors who have worked on local sequences, much as traditional archaeologists tended to do, however, with the modern goal of addressing evolutionary change in hunter-gatherer systems over long time spans. Given this ambitious goal they wisely chose to ask the authors to build their treatments around a focal question, the utility of the forager–collector continuum (Binford 1980) for research on archaeological sequences. Needless to say, Binford was flattered by their choice and understandably read the papers with a great deal of interest. When he was asked to write the foreword to this provocative book he expected to learn new things and in this he has not been disappointed.

The common organizing questions addressed among the contributors to this volume are simply, how useful is the forager–collector continuum for explanatory research on sequences, and what else might we need to know to explain evolutionary change in hunter-gatherer adaptations? Most sequences document systems change, in some sense. Though we don't necessarily know how much synchronous systemic variability there might have been relative to the documented sequence, most authors have tried to address the problem of within systems variability. In this sense, most are operating with sophistication not seen among traditional culture historians.

The primary problem for archaeologists of the generation prior to Binford was how to date archaeological materials. Techniques of seriation were developed to figure out how to use artifacts and cultural materials to

LEWIS R. BINFORD AND AMBER L. JOHNSON • Department of Anthropology, Southern Methodist University, Dallas Texas 75275.

produce a temporal sequence for a site or region (see O'Brien and Lyman 2000:245–396 for a current reactionary discussion of this point of view). Nevertheless, the traditional strategy of grouping similar assemblages into the same chronological unit meant that synchronous internal variation in the system could be inappropriately assigned chronological significance. Sequences were not necessarily sequences.

In the 1950s, the advent of radiocarbon dating made it possible to date many materials directly, therefore reducing the relevance of artifact seriation. Not surprisingly, the goals of research for some of the generation of archaeologists who were students in the 1950s changed from simply producing a sequence to studying variability in the archaeological record at various scales—hopefully referable to within systems postures or functional variants and between distinct states of systems. It was in the context of dealing with variability among hunter-gatherer cultural systems that the forager–collector distinction was advanced.

When "Willow Smoke and Dog's Tails" and other related papers were written, the dominant view in anthropology was that the Kalahari San were representative of Paleolithic hunter-gatherers anywhere around the globe under any environmental conditions. The ethnography of the San was used to interpret Paleolithic archaeology in Europe (Clarke 1976), a near-glacial environment! A prominent part of the dominant argument was that there was no important diversity among hunter-gatherers that was not related to either colonization, acculturation, or other "corrupting influences" (See Schrire 1984a, b; Wilmsen and Denbow 1990) as imagined earlier by Service (1962, 1966)

Binford (1980) was trying to emphasize that there were legitimate hunter-gatherers that were not organized like the San and that the Nunamiut was one such example. The forager–collector continuum was an attempt to dimensionalize the recognized variability so that it could hopefully be measured in some meaningful sense. In turn, it was suggested that such variability was conditioned by the relative spatial congruence among necessary resources in different habitats. The latter had as much to do with the distribution and availability of exploited products, that were not foods, as with foods themselves, although this point has often been missed. The argument was simply that, other things being equal, the more challenges the members of a cultural system faced, the more things they would need to survive, the higher the probability of incongruence in the distribution of necessary resources, and the greater the likelihood that some things would be acquired through logistical mobility. Different strategies for positioning people relative to resources were expected to occur in (1) places where all or most necessary resources were available in the immediate proximity of a residential site, whereas in (2) places where there was some considerable distance between access locations for necessary resources, more distant ones would be procured using logistical principles of organization. This was not an argument about patch choice, but about positioning with respect to a variety of necessary resources or patches. The contrast between the Nunamiut and the Kalahari cases was used to propose some factors, other than colonization or acculturation, that might condition variability among hunter-gatherer systems.

Though generally recognizing the utility of the forager–collector continuum, several contributors to this volume have realized that all other things are not equal among groups identified as hunter-gatherers, that is, there are dimensions of variability other than congruence of resources that must be controlled to anticipate archaeological variability accurately. Among those recognized here are transport technology, access to market trading opportunities, and demography or population packing.

At about the same time the forager–collector continuum was introduced to the literature, Binford spelled out the packing model in his book *In Pursuit of the Past* (1983a:210–213). This was presented as a contrast to then current arguments about population pressure. Unlike the forager–collector continuum which was initially viewed as appropriate to studying variability among hunter-gatherer systems in different environments, but which was not stressed as resulting in a linear trajectory, packing did address a mechanism for evolutionary change within archaeological sequences. It was only after Eder (1984) published a critical argument based on his experience with the Batek in the Philippines that Binford seriously considered the relationship between logistical strategies and sedentism. His original thought had been focused on mobile peoples. The link between mobile and sedentary peoples relative to foraging versus collecting makes it possible to think about collecting replacing foraging in some settings, but this was never fully spelled out by Binford as an evolutionary argument (see, however, Binford 2001).

Other responses to the original presentation of the forager–collector continuum were less productive. There wasn't anything new in Woodburn (1980) or Bettinger and Baumhoff (1982), except the point of view they brought to the empirical ideas. Wiessner (1982), however, argued that social considerations had to be taken into account when discussing mobility. This seems an appropriate place to point out that Binford never answered Polly Wiessner's critique because he thought it was an example of working at the wrong scale. In the Kalahari, families move with respect to social networks in and out of places where they have kin. Neither the places nor the compositions of groups in a given place were stable. Wiessner was still working with a band model, discussing a socially defined unit with set membership and mobility in those terms. Binford (1983b) had stopped thinking about bands, having recognized that among some mobile collectors and perhaps most foragers, the unit that made decisions about when and where to move was the family. It was the pileup of such decisions by a number of families that resulted in camps coalescing and disappearing. Thus, the camp could not be thought of as a band. Kin-based social networks are important in all small-scale societies, and kinship relationships are a constant matrix in which decisions are made. This does not mean, however, that they are the determinants of mobility patterning such that differences between residential versus logistical forms of mobility could be understood in "social terms." Wiessner overlooked the relevant question, which is, Why do camps coalesce, dissolve, or organizationally restructure while kin presence is a constant? I suggest that this occurs with respect to variation in environmental and demographic variables. There is little evidence suggesting that this viewpoint is nonproductive.

The above developments of the early 1980s era focused upon variability arising from processes internal to systems, but they occurred in an intellectual context where there was great concern for identifying the "correct" locus of cause, for example, social versus environmental causes or mental versus behavioral causes standing behind properties of the archaeological record. The latter concern might be best thought of as a search for the "reasons" behind variability in the archaeological record, much as historians seek "reasons" for events (See Hexter 1971). In spite of the drawbacks, these approaches marked a contrast to the earlier (pre-carbon-14) emphasis upon sequence building and comparison for purposes of historical reconstruction.

Although the issue of variability was addressed, the early 1980s also witnessed a turn to intellectual chaos. Proponents of dueling paradigms were seeking to capture the field and the minds of young students, upon whom the future of the field depended. This infighting lasted thru the early 1990s.

The organizers of this book should be strongly commended for putting together a collection of papers which were not designed to illustrate the opined superiority of one paradigm over another; instead they have asked a well-chosen group of archaeological specialists, who represent a large segment of the globe, to use the archaeological record as a source of problems when viewed relative to the collector-forager range of mobility organization. The goal is an assessment of its utility and directly addresses the issue of explaining sequential variability, or change through time, in their respective regions. The latter, of course, has the goal of moving "beyond foraging–collecting" toward more sophisticated forms of understanding. The results are fascinating and can be viewed as a sample of the analytical challenges arising in contemporary archaeology and a cross section of the intellectual approaches common to contemporary archaeology. Seen in this light, this book can be viewed as a most useful anthology for use as a comparative reader in general courses treating the state of the art in contemporary archaeology.

One of the most striking features of the material synthesized by the contributors is the large size of the data sets used and the fact that such large-scale archaeological coverage is generally a product of many different field workers. This is certainly a new characteristic when compared to one field worker, one sequence, one site, types of products common in the pre-carbon-14 era of sequence production. Several features of change within archaeology contribute to the contrast noted here. The expansion of the field both in terms of overall funds available for research and the numbers of persons engaged in archaeological research ensures that there is a fast accumulation of large quantities of data. This data must be addressed analytically because it does not speak for itself. Archaeologists are just now beginning to address this issue and, thus far, few new techniques appropriate to large-scale comparative research have been forthcoming. This is apparent from the analytical approaches employed by many of the authors. For instance, one of the unfortunate uses of the forager–collector continuum has been as a taxonomy to assign a label to sites and/or systems which was, in most cases, operationally defined to use the particular data available. This taxonomic approach to understanding is present in a large number of the papers that seek to evaluate the forager–collector continuum. Such an approach normally results in the logical fallacy of confirming the consequent or finding in the data what was a priori dictated by definition to accommodate the particular data used. The use of simple site size as an indicator of camps versus residential locations could be an example. This type of potential failure to link data and concepts such that data get "to talk back to our ideas" must be solved if archaeology is to realize the learning opportunities that the analysis of large data sets actually holds out to contemporary researchers.

A legacy of the "period of intellectual chaos" between the early 1980s and the 1990s is best referred to as the top-down approach to accommodative argument. One advocates the paradigm of their choice and then accommodates the data to fit a priori beliefs. This problem is certainly manifest in a few of the papers in this book. Such an approach ensures that the data are prevented from ever "talking back" to the ideas with which the researcher started. Needless to say, no learning is possible when such approaches are employed.

An additional characteristic of contemporary archaeology that is well illustrated in some of the papers in this book is the tendency to "interpret" one's data by simply applying an accommodative argument to the patterning observed and thereby preserving one's beliefs as to the causes, reasons behind, or events in the past believed responsible for the human behavior manifest thru archaeological patterns. This mind-numbing approach is well illustrated by the frequent appeals to conservatism as an explanation for the perceived lack of change or long periods of stability characteristic of reported sequences. This kind of word "magic" is not an explanation: it is the descriptive translation of properties in one domain of observation to descriptive terms, which are perhaps appropriate to another domain of observation.

Despite any problems with individual papers, the volume as a whole is a valuable source of exactly the kind of comparative data that are necessary to move beyond the analytical traps so common in contemporary archaeology. For example, though it may be difficult to figure out how to explain a marked contrast in the rates of culture change within any given sequence, by comparing variability in rates of culture change across sequences, one recognizes patterns that can be clues to general evolutionary processes. This strategy has been used productively to ask what conditions variability in the duration of cultural adaptations both on regional (Johnson 1997) and global scales of analysis (Raish 1992). There is enormous explanatory potential if we can determine the dimensions that condition variability in the stability of cultural adaptations. Such explanatory potential must be developed through theory building that is really possible only when similar patterning can be demonstrated across a substantial set of cases synthesized on roughly the same scale. This volume provides just such a set of cases, that can be used as the foundation for the comparative analysis necessary to fuel theory building in archaeology.

Lewis R. Binford and Amber L. Johnson
Dallas, TX

REFERENCES

Bettinger, R. L., 1999, From Traveler to Processor: Regional Trajectories of Hunter-Gatherer Sedentism in the Inyo-Mono Region, California. In *Settlement Pattern Studies in the Americas: Fifty Years since Virú*, edited by B. R. Billman and G. M. Feinman, pp. 39–55. Smithsonian Institution Press, Washington, DC.

Bettinger, R. L., and Baumhoff, M. A., 1982, The Numic Spread: Great Basin Cultures in Competition. *American Antiquity* 47:485–503.

Binford, L. R., 1980, Willow Smoke and Dog's Tails: Hunter-Gatherer Settlement Systems and Archaeological Site Formation. *American Antiquity* 45(1): 1–17.

Binford, L. R., 1983a, *In Pursuit of the Past: Decoding the Archaeological Record*. Thames and Hudson, New York.

Binford, L. R., 1983b, Long Term Land Use Patterns: Some Implications for Archaeology. In *Lulu Linear Punctated: Essays in Honor of George Irving Quimby*, edited by R. C. Dunnell and D. K. Grayson, pp. 27–54. Anthropological Papers, Museum of Anthropology, University of Michigan No. 72.

Binford, L. R., 2001, *Constructing Frames of Reference; Using Ethnographic and Environmental Data Sets.* University of California Press, Berkeley.

Clarke, D., 1976, Mesolithic Europe: The Economic Basis. In *Problems in Economic and Social Archaeology*, edited by G. G. Sieveking, I. H. Longworth, and K. E. Wilson, pp. 449–481. Duckworth, London.

Eder, J. F., 1984, The Impact of Subsistence Change on Mobility and Settlement Pattern in a Tropical Forest Foraging Economy: Some Implications for Archeology. *American Anthropologist* 86:837–853.

Hexter, J. H., 1971, Chapter 1: The Cases of the Muddy Pants, the Dead Mr. Sweet, and the Convergence of Particles, or Explanation Why and Prediction in History. In *The History Primer*, edited by J. H. Hexter, pp. 21–42. Basic Books, New York.

Johnson, A. L., 1997, Explaining Variability in the Pace and Pattern of Cultural Evolution in the North American Southwest: An Exercise in Theory Building. Ph.D. Dissertation, Southern Methodist University, University Microfilms, Ann Arbor.

O'Brien, M. J. and Lyman, R. L., 2000, *Applying Evolutionary Archaeology: A Systemic Approach.* Plenum, New York.

Raish, C., 1992, *Domestic Animals and Stability in Pre-State Farming Societies.* BAR International Series 579, Tempus Repartum, Oxford.

Schrire, C., 1984a, An Inquiry into the Evolutionary Status and Apparent Identity of San Hunter-Gatherers, *Human Ecology* 8(1): 9–31.

Schrire, C., 1984b, Wild Surmises on Savage Thoughts, in: *Past and Present in Hunter-Gatherer Studies*, edited by C. Schrire, pp. 1–26. Academic Press, Orlando.

Service, E. R., 1962, *Primitive Social Organization.* Random House, New York.

Service, E. R., 1966, *The Hunters.* Prentice-Hall, Englewood Cliffs, NJ.

Wiessner, P., 1982, Beyond Willow Smoke and Dogs' Tails: A Comment on Binford's Analysis of Hunter-Gatherer Settlement Systems. *American Antiquity* 47(1): 171–178.

Wilmsen, E. N., and Denbow, J. R., 1990, Paradigmatic History of San-Speaking Peoples and Current Attempts at Revision. *Current Anthropology* 31:489–524.

Woodburn, J., 1980, Hunters and Gatherers Today and Reconstruction of the Past. In *Soviet and Western Anthropology*, edited by E. Gellner, pp. 95–117. Columbia University Press, New York.

Contents

PART II. MICROEVOLUTIONARY APPROACHES TO LONG-TERM HUNTER-GATHERER SETTLEMENT CHANGE

PART III. BEYOND ECOLOGICAL APPROACHES TO HUNTER-GATHERER SETTLEMENT CHANGE

Chapter **1**

Introduction

Beyond Foraging and Collecting: Evolutionary Change in Hunter-Gatherer Settlement Systems

JUNKO HABU AND BEN FITZHUGH

THE FORAGER/COLLECTOR MODEL

Twenty years ago, Lewis Binford published an article that revolutionized the study of hunter-gatherer settlement and land use. The article, *Willow Smoke and Dogs' Tails: Hunter-Gatherer Settlement Systems and Archaeological Site Formation* (Binford 1980), made the simple but elegant argument that seasonal or short-term hunter-gatherer mobility should be patterned in predictable ways with respect to spatial and temporal variation in resource availability. In the model, Binford distinguished residential mobility (the movement of all members of a residential base from one locality to another) from logistical mobility (the movement of specially organized task groups on temporary excursions from a residential base). Based on these distinctions, Binford identified two basic subsistence-settlement systems: forager systems that are characterized by low logistical mobility and high residential mobility and collector systems that have high logistical mobility and low residential mobility. According to Binford, the former systems are responses to environments where the distribution of important resources is spatially and/or temporally (seasonally) homogeneous, whereas the latter

JUNKO HABU • Department of Anthropology, University of California, Berkeley, California 94720 **BEN FITZHUGH** • Department of Anthropology, University of Washington, Seattle, Washington 98195

systems are adapted to environments where the distributions of critical resources are spatially or temporally uneven.

Binford's (1980) distinction between residentially mobile foragers and logistically mobile collectors has contributed significantly to our understanding of hunter-gatherer settlement systems and is probably the most influential source of hunter-gatherer settlement theory. Unlike many other models of hunter-gatherer mobility, Binford's forager/collector model "stresses the strategies behind the observed patterns, rather than the empirical patterns themselves" (Thomas 1983: 11). In other words, the primary objective of the model was to explain hunter-gatherer variability, rather than to create another set of normative generalizations about hunter-gatherer behavior. As a result, even though the forager/collector model was an informal model based on ethnographic examples of the G/wi San (Silberbauer 1972) and Nunamiut (Binford 1978), the model is applicable to a wide range of archaeological and ethnographic cases from various parts of the world.

Furthermore, the fact that the model specified the material consequences of hunter-gatherer behavior in terms of site types and intersite variability in associated tool assemblages (Binford 1980, 1982; see also Binford 1978) made this model extremely attractive to many archaeologists who were eager to find middle-range theories to bridge the gap between archaeological data and past people's behavior. Examples of the applications of this model to archaeological and ethnographic hunter-gatherer data include Schalk (1981), Thomas (1981), Kelly (1983), Savelle (1987), Savelle and McCartney (1988), Bang Anderson (1996), and Cowan (1999).

One dimension that has rarely been systematically discussed in the archaeological literature is the relevance of the forager/collector model in the study of long-term changes in hunter-gatherer subsistence-settlement systems. Because the model was based on short-term ethnographic observations, the primary focus was placed on the annual cycles of subsistence activities and resulting settlement pattern changes. The exception is Binford's 1983 article, which was entitled *Long-Term Land-Use Patterning: Some Implications for Archaeology*. Based on his interviews with elderly Nunamiut men, Binford defined an annual range as the area where people lived, hunted, fished, and collected during an annual cycle. According to his article, each Nunamiut group typically moved its annual range to a new area every nine years or so, and they came back to the same annual range after approximately 40 years. Although these observations are extremely insightful, the shift of annual range discussed in Binford's (1983) article did not lead to overall system changes, nor did it reveal changes during periods of several hundred to more than a thousand years. In other words, "the archaeology of the *longue durée*" (Ames 1991) in relation to the forager/collector model has yet to be developed. This is particularly important in the context of the study of complex hunter-gatherers (e.g., Price and Brown 1985; Price and Feinman 1995), in which long-term changes in subsistence and settlement may play a critical role in explaining evolutionary changes in hunter-gatherer cultural complexity, including the development of social inequality (e.g., Fitzhugh 1996, 2002).

Binford's original formulation of the forager/collector model was subsequently critiqued and expanded by Polly Wiessner (1982), who argued that people regularly construct social relationships to mediate spatiotemporal resource variation, and that these social relationships are as significant in hunter-gatherer settlement strategies as the environmental parameters emphasized by Binford. Subsequent development of this line of reasoning in ecological anthropology has focused on the contexts in which exchange, mobility, and storage are differentially pursued (e.g., Blurton Jones 1987; Bettinger 1999; Goland 1991; Gould 1982; Hawkes 1992; Hegmon 1991; O'Shea 1981; Rowley-Conwy and Zvelebil 1989; Smith 1988; Speth 1990; Winterhalder 1986).

Binford himself has presented revisions to his original model, arguing, for example, that increased costs of pursuing terrestrial game should affect residential patterns in the absence of population pressure (Binford 1990). In such cases, investment in productive and predictable aquatic resources and the development of technologically intensive methods for improving the foraging efficiency of these prey items should lead to more residential stability. The addition of technological intensification to these models provides a mechanism for significant systemic change in the relative benefits of residential mobility that is generated, at least proximately, by internal developments in the technoeconomic system. Because the original version of the forager/collector model was framed in strictly environmental terms, any extension of the model to address long-term/evolutionary change would necessarily invoke environmental change as the primary cause of changes in residential and logistical strategies. By adding technological change in combination with environmental change, the forager/collector model leaves more room for the strategic input of individual decision makers and becomes more appropriate to the theme of evolutionary change (see Fisher, this volume; Fitzhugh, this volume).

Paralleling the forager/collector distinction, a separate but overlapping set of models has explored the social implications of hunter-gatherer modes of production and consumption. Woodburn's (1980) distinction between immediate-return and delayed-return hunting and gathering has been nearly as influential as Binford's forager/collector model. Highlighting the social consequences of immediate consumption compared to storage systems, this model has further engaged hunter-gatherer theory to consider the embedded contexts of environmental and social domains. It is significant that immediate-return hunter-gatherers share many basic elements with Binford's concept of "foragers," whereas delayed-return foragers are very similar to Binford's "collectors," and the two models are often combined in application (for an exception, see Kelly 1995). Unlike the forager/collector model, the immediate/delayed-return distinction has more often been central in models of long-term systemic or cultural change (e.g., Testart 1982). Nevertheless, it can be argued that the model is insufficient because it lacks a mechanism to explain the economic change from immediate to delayed return and thus is little improvement over the original forager/collector model in the evolutionary dimension.

Bettinger's traveller/processor model (1999) draws together elements of the forager/collector model and the immediate/delayed-return model. Inspired in part by optimal foraging models, Bettinger proposes that a critical phase shift occurs when mobile hunter-gatherers find mobility increasingly costly relative to investment in processor-intensive subsistence pursuits. For him, a key shift occurs when people begin to invest their limited energy in resources that entail considerable processing costs to be useful. In his model, population growth and social circumscription are identified as proximate causes of increased mobility costs. In some ways, Bettinger's model comes closest to the goals of this volume in theorizing and indeed demonstrating that systemic (evolutionary) change is an expected consequence of long-term hunter-gatherer sequences (see Fitzhugh, this volume for similar argumentation).

Given these contexts, this edited volume pushes the range of hunter-gatherer theory and brings together a diverse set of authors and perspectives toward their goal of expanding our understanding of hunter-gatherer settlement dynamics and change. Within this context, this book seeks to contribute to (1) the development of new models that can explain variability in hunter-gatherer settlement and land use and (2) theoretical discussions of the mechanisms of long-term changes in hunter-gatherer settlement systems.

REEVALUATION OF THE FORAGER/COLLECTOR MODEL

The first dimension of this book concerns the reevaluation of Binford's forager/collector model (Binford 1980). The authors in this book take the pulse of the forager/collector model twenty years after its introduction. In particular, we assess the strengths and weaknesses of the model as it has evolved during this period. The authors are unified in the conviction that Binford's model has been, and continues to be, one of the best tools for understanding a major source of variation in hunter-gatherer subsistence-settlement dynamics. Nevertheless, several authors see a need to modify the model to make it applicable to cases outside of the rather restrictive set on which the model was developed (e.g., Ames 1991), as well as to make it applicable to evolutionary scale changes in settlement system (e.g., Aldenderfer, Cannon, Fisher). In addition, this volume also provides an opportunity to subject the forager/collector model to rigorous archaeological evaluation.

Several authors in this volume point out the complexity of human–environment interactions and suggest that, in addition to the distribution pattern of critical resources as suggested by Binford (1980), other ecological, economic, technological, social, and ideological factors may have played an important role in determining subsistence-settlement systems. For example, for several authors, evolutionary ecology and its strict economic logic and formal modeling machinery is an excellent framework for formalizing the forager/collector model into a more testable set of hypotheses. David Zeanah, using optimal foraging models as his point of departure, suggests that the presence of unanticipated variability among Great Basin subsistence-settlement systems is a result of local trade-offs between diet breadth, transport costs, and central place location. Ben Fitzhugh draws on the patch choice model to suggest that maritime hunter-gatherers of the North Pacific might not always have been residentially stable "collectors," as is often assumed. Using a modified diet breadth model (Schmidt 1998), Lynn Fisher's chapter on the Paleolithic–Mesolithic transition in southern Germany suggests that hunter-gatherers may alter search modes (e.g., between focal pursuit of big and small game) in response to threshold conditions related to the costs and benefits of subsistence-based mobility. Merging environmental and social considerations with the help of evolutionary ecological risk theory, Renato Kipnis argues that late Pleistocene and early Holocene Brazilian rock art sites reflect changes in the context of intergroup information sharing and territoriality.

Ken Ames critiques the applicability of the forager/collector model to boat-using hunter-gatherers, suggesting that regular access to boats revolutionizes mobility strategies, residential patterns, and processing patterns of procured food, resulting in both longer foraging radii and longer logistical forays. He suggests that none of these changes can be accommodated by the classic forager/collector model. Ames' treatment reaffirms the value of comparative ethnography for refining archaeological models, and his conclusions are generally compatible with the archaeological applications of Cannon and Fitzhugh, who also consider boat-based hunter-gathering around the greater Pacific Northwest of North America.

One aspect of the forager/collector model that is not given sufficient discussion in any single chapter, but which emerges in the comparison between the chapters, pertains to the analytical meaning of the central concepts of the forager/collector model: foraging, collecting, residential mobility, and logistical mobility. According to its original formulation, foragers are supposed to be residentially mobile, "mapping onto" resources, whereas collectors are supposed to be logistically mobile, extracting resources at a distance from stable residential camps. Ames' analysis suggests that this dichotomy may be invalid in certain contexts. Boat use leads to greater camp sedentism but fewer logistical forays (because the daily foraging radius is enlarged, thus reducing the need for overnight forays) and fewer specialized extraction facilities (because bulk transport enables processing tasks to occur at the base camp). More complex subsistence organization then can yield archaeological patterns structurally similar to classic foragers (see also Fitzhugh).

Case studies presented by other authors clearly reflect recent shifts in theoretical interests away from the study of subsistence-settlement systems themselves. For example, Aubrey Cannon's chapter on the Central Northwest Coast of North America suggests that, although the forager/collector model provides a useful first step in organizing the complex array of settlement systems, it does not take into account important factors such as social constraints and opportunities or the role of ideology. Similarly, Mark Aldenderfer suggests structural modifications to the model in terms of history, agency, contingency, and cultural logic. In addition, Jim Savelle, who in one sense uses the forager/collector model without revision, shows how social changes can limit the flexibility of a cultural system to respond to critical changes in the availability of resources. He suggests that amongst the Thule culture, collecting systems with more highly structured social organization may have been less responsive to environmental change than less socially structured foraging systems, leading to their collapse (and presumed extinction of the local population) during the Little Ice Age.

Finally, many of the chapters in this volume indicate that hunter-gatherers in the past may have shifted along the forager/collector continuum quite frequently (e.g., chapter by Habu) and that the settlement patterns may have varied even at a local level (e.g., chapters by Bar-Yosef, Habu, and Zeanah). Although the presence of such diversity is not contradictory to Binford's original model, these case studies nevertheless warn of the danger of oversimplification in interpreting hunter-gatherer subsistence and settlement.

LONG-TERM CHANGES IN SETTLEMENT SYSTEMS

Second, the reexamination of the forager/collector model leads to theoretical considerations of long-term changes in hunter-gatherer subsistence-settlement systems. As discussed above, this is a topic intentionally neglected in Binford's original treatment. We are interested in the mechanisms of changes in the geographical distribution of hunter-gatherer settlement and activity locations and also in the theoretical implications of these changes for evolutionary processes, such as the development of hunter-gatherer cultural complexity, the transition from hunting-gathering to food-producing societies, and/or changes in the role of exchange and dependencies between hunter-gatherers and farmers. The forager/collector model is essentially an ecological model, in which hunter-gatherer settlement systems are related to the seasonal and spatial distribution of critical resources. By expanding the discussion to include the evolutionary implications of changes from foragers to collectors or vice versa, contributions in this volume enhance our understanding of hunter-gatherer subsistence-settlement systems and their causal conditions.

The use of the term "evolution" in the title of this volume and throughout raises important semantic questions. Current archaeological theory has tended to restrict the use of the word "evolution" to strictly Darwinian forms of heritable change. This is partly due to the growing dissatisfaction with the theory of "social evolution" in the tradition of Spencer, Morgan, Marx, White, and others.

The sense in which we are using the term here, following the tradition that Binford and others have been instrumental in developing, is more systemic than particularistic, and therefore not directly related to Darwinian mechanisms of evolutionary change. We feel that the term remains useful in characterizing systemic change in relationships within and between social/environmental parameters and individual strategic behaviors, where this change results in unprecedented organizations (see Fitzhugh, this volume). Traditional views of hunter-gatherer variation have failed to consider the role of cumulative evolutionary process. In studies of social evolution, hunter-gatherers are assumed merely to reflect the ecological contexts in which they find themselves. They are at the baseline of evolutionary models of food production, village sedentism, complexity, and the additional trappings that have dominated archaeological interest during the past half-century. Several of the authors in this volume would probably agree that a goal of anthropological archaeology is identifying the mechanisms (including Darwinian ones) that contribute to systemic evolutionary change. Others probably would not.

A focus on evolution does not need to demand a unilinear or progressive view, as is commonly assumed. The paper by Junko Habu makes this point effectively by showing that systemic change is reversible. Savelle's paper shows that increasing complexity can be hazardous in the face of environmental change, dispelling any lingering myths that evolved complexity is in any way "adaptive" in a progressive sense.

"Evolution" in the sense used here is sometimes posed in opposition to "history," where systemic evolutionary theory is concerned with explicating predictable processes of change in general terms, whereas history comes to represent idiosyncratic change, pernicious, unpredictable, and unique (see Schrire 1995). In some cases, history is also used to place individual creativity, intellect, and agency into the picture of change. The authors in this volume vary in the degree of emphasis placed on the evolutionary process versus historical idiosyncratic effects.

Several authors in this volume identify historical factors as important in long-term hunter-gatherer change. Aldenderfer argues that archaeological models of social evolution should include explicit recognition of the importance of history, contingency, agency, and cultural logic. Cannon feels that the pattern of stasis and rapid change on the Northwest Coast is best explained as a result of ideological conservatism.

GLOBAL COMPARISON OF LONG-TERM CHANGES IN HUNTER-GATHERER SETTLEMENT AND LAND USE

In addition to the two objectives discussed above, this volume also seeks to broaden our understanding of regional diversity in hunter-gatherer subsistence and settlement at the global level. By assembling case studies that document long-term changes in hunting and gathering settlement and land use around the world and in varying social and ecological contexts, the theoretical points made in this collection have significant empirical grounding. The contributions also serve as a source of comparative data from which readers may build a more generalized understanding of evolutionary dynamics of hunter-gatherer settlement systems in the future. Toward this goal, an international group of authors has been recruited to discuss relevant aspects of their research on five continents (plus insular Japan) that temporally ranges from the late Pleistocene to the recent past.

Comparison of long-term changes among case studies presented in this volume indicates that evolutionary trajectories of hunter-gatherer settlement systems may vary significantly in different parts of the world. For example, although the cases from the Northwest Coast (chapter by Cannon) and from the Subarctic (chapter by Fitzhugh) represent the development of logistical strategies and storage through time, the case from Jomon Japan (chapter by Habu) indicates that such directional changes may not necessarily have been the norm in other parts of the world. Laura Junker's chapter on hunter-gatherers in Southeast Asia provides a valuable archaeological example in which long-term changes in hunter-gatherer strategies may have been closely related to forager-farmer trade.

SUMMARY

In sum, this volume seeks (1) to assess the successes and limitations of the forager/collector model after twenty years of currency, (2) to push the explanation of hunter-gatherer settlement pattern organization and evolution beyond the confines of Binford's original treatment, and (3) to present globally diverse samples of archaeological case studies of the evolution of hunter-gatherer settlement systems. Although the case studies presented here cover only part of the global diversity in hunter-gatherer settlement systems and the quality and quantity of currently available archaeological data limit our ability to understand past hunter-gatherer settlement systems more fully, we nevertheless hope that this volume will help stimulate active discussions on long-term changes in hunter-gatherer subsistence, settlement, and land use.

In his Afterword to this volume, Doug Price notes the recent resurgence in interest in archaeological studies of hunter-gatherers. This volume suggests that hunter-gatherers remain a significant and vital topic of archaeological inquiry. And how could it be otherwise? Hunter-gatherers have been leaving archaeological remains for more than a million years and have expanded into most habitable landscapes around the globe. Far from the stereotypes of a century ago, we now recognize that variability is the dominant characteristic of hunter-gatherer economic, social, and political organization, and we are better positioned than ever to develop comprehensive explanations of much of this variability. Binford's forager/collector model remains a vital tool for understanding a portion of this variability across space, and it provides a useful starting point for modeling evolutionary change through time.

In the temporal dimension, the papers in this volume further disabuse us of the notion that evolutionary trajectories are unilinear, progressive, or inevitable. Instead, they suggest greater fluidity along the forager/collector continuum, greater importance of high local scale variation in environmental and social parameters, and a greater role for chance (i.e., history and/or contingency) in local trajectories than has previously been considered. These papers also clearly show that although hunter-gatherers can vary the way they organize themselves residentially and logistically according to an array of local variables, nevertheless, common tendencies emerge from the mix. This suggests that continued effort in documenting and explaining the evolutionary dimension of hunter-gatherer settlement and land use will bring us closer to an understanding of the causes underlying systemic change in hunter-gatherer societies.

REFERENCES

Ames, K. M., 1991, The Archaeology of the *Longue Durée*: Temporal and Spatial Scale in the Evolution of Social Complexity on the Southern Northwest Coast. *Antiquity* 65: 935–945.

Bang Anderson, S., 1996, Coast/Inland Relations in the Mesolithic of Southern Norway. *World Archaeology* 27(3): 427–443.

Bettinger, R. L., 1999, From Traveler to Processor: Regional Trajectories of Hunter-Gatherer Sedentism in the Inyo-Mono Region, California. In *Settlement Pattern Studies in the Americas: Fifty Years Since Viru*, edited by B. R. Billman and G. M. Feinman, pp. 39–55. Smithsonian Institution Press, Washington DC.

Binford, L. R., 1978, *Nunamiut Ethnoarchaeology*. Academic Press, New York.

Binford, L. R., 1980, Willow Smoke and Dogs' Tails: Hunter-Gatherer Settlement Systems and Archaeological Site Formation. *American Antiquity* 45(1): 4–20.

Binford, L. R., 1982, The Archaeology of Place. *Journal of Anthropological Archaeology* 1(1): 5–31.

Binford, L. R., 1983 Long-Term Land-Use Patterning: Some Implications for Archaeology. In *Working at Archaeology*, pp. 379–386. Academic Press, New York.

Binford, L. R., 1990, Mobility, Housing, and Environment: A Comparative Study. *Journal of Anthropological Research* 46(2): 119–152.

Blurton Jones, N. G., 1987, Tolerated Theft, Suggestions about the Ecology and Evolution of Sharing, Hoarding and Scrounging. *Social Science Information* 26:31–54.

Cowan, F. L., 1999, Making Sense of Flake Scatters: Lithic Technological Strategies and Mobility. *American Antiquity*, 64(4): 593–607.

Fitzhugh, (J.) B., 1996, The Evolution of Complex Hunter-Gatherers in the North Pacific: An Archaeological Case Study from Kodiak Island, Alaska. Unpublished Ph.D Dissertation, University of Michigan, Ann Arbor.

Fitzhugh, B., 2002, The Evolution of Complex Hunter-Gatherers on the Kodiak Archipelago. In *Hunter-Gatherers of the North Pacific Rim*, edited by J. Habu, S. Koyama, J. Savelle, and H. Hongo. Senri Ethnological Studies, Osaka, Japan (in press).

Goland, C., 1991, The Ecological Context of Hunter-Gatherer Storage: Environmental Predictability and Environmental Risk. In *Foragers in Context: Long Term, Regional, and Historical Perspectives in Hunter-Gatherer Studies*, edited by P. T. Miracle, L. E. Fisher, and J. Brown, pp. 107–125. Michigan Discussions in Anthropology, Vol. 10. Ann Arbor.

Gould, R., 1982, To Have and Have Not: The Ecology of Sharing among Hunter-Gatherers. In *Resource Managers: North American and Australian Hunter-Gatherers*, edited by N. M. Williams and E. S. Hunn, pp. 69–91. Westview Press, Boulder.

Hawkes, K., 1992, Sharing and Collective Action. In *Evolutionary Ecology and Human Behavior*, edited by E. Smith and B. Winterhalder, pp. 269–300. Aldine de Gruyter, New York.

Hegmon, M., 1991, Risks of Sharing and Sharing as Risk Reduction: Interhousehold Food Sharing in Egalitarian Societies. In *Between Bands and States*, edited by S. A. Gregg, pp. 309–329. Occasional Paper No. 9, Center for Archaeological Investigations, Southern Illinois University, Carbondale.

Kelly, R. L., 1983, Hunter-Gatherer Mobility Strategies. *Journal of Anthropological Research* 39(3): 277–306.

Kelly, R. L., 1995, *The Foraging Spectrum*. Smithsonian Institution Press, Washington DC.

O'Shea, J., 1981, Coping with Scarcity: Exchange and Social Storage. In *Economic Archaeology: Towards an Integration of Ecological and Social Approaches*, edited by A. Sheridan and G. Bailey, pp. 167–186. British Archaeological Reports, International Series 96, Oxford.

Price, T. D. and Brown, J. A. (eds.), 1985, *Prehistoric Hunter-Gatherers*. Academic Press, Orlando.

Price, T. D. and Feinman, G. M. (eds.), 1995, *Foundations of Social Inequality*. Plenum, New York.

Rowley-Conwy, P. and Zvelebil, M., 1989, Saving It for Later: Storage by Prehistoric Hunter-Gatherers in Europe. In *Bad Year Economics*, edited by P. Halstead and J. O'Shea, pp. 40–56. Cambridge University Press, Cambridge.

Savelle, J. M., 1987, *Collectors and Foragers: Subsistence-Settlement System Change in the Central Canadian Arctic, A.D. 1000–1960.* BAR International Series 358. British Archaeological Reports, Oxford.

Savelle, J. M. and McCartney, A. P., 1988, Geographical and Temporal Variation in Thule Eskimo Subsistence Economies: A Model. In *Research in Economic Anthropology*, Vol. 10, edited by B. L. Issac, pp. 21–72. JAI Press, Greenwich.

Schalk, R. F., 1981, Land Use and Organizational Complexity among Foragers of Northwestern North America. In *Affluent Foragers: Pacific Coasts East and West*, edited by S. Koyama and D. H. Thomas, pp. 53–75. Senri Ethnological Studies 9, National Museum of Ethnology, Osaka.

Schmidt, K. A., 1998, The Consequences of Partially Directed Search Effort. *Evolutionary Ecology* 12: 263–277.

Schrire C., 1995, *Digging Through Darkness: Chronicles of an Archaeologist.* University Press of Virginia, Charlottesville.

Silberbauer, G. S., 1972, The G/wi Bushmen. In *Hunters and Gatherers Today*, edited by M. G. Bicchieri, pp. 271–325. Holt, Rinehart and Winston, New York.

Smith, E. A., 1988, Risk and Uncertainty in the "Original Affluent Society": Evolutionary Ecology of Resource Sharing and Land Tenure. In *Hunters and Gatherers 1: History, Evolution, and Social Change*, edited by T. Ingold, D. Riches, and J. Woodburn, pp. 222–252. Berg, Oxford.

Speth, John D., 1990, Seasonality, Resource Stress, and Food Sharing in so-called "Egalitarian" Foraging Societies. *Journal of Anthropological Archaeology* 9: 148–188.

Testart, A., 1982, The Significance of Food Storage among Hunter-Gatherers: Residence Patterns, Population Densities, and Social Inequalities. *Current Anthropology* 23: 523–537.

Thomas, D. H., 1981, Complexity among Great Basin Shoshoneans: The World's Least Affluent Hunter-Gatherers? In *Affluent Foragers: Pacific Coasts East and West*, edited by S. Koyama and D. H. Thomas, pp. 19–52. Senri Ethnological Studies 9, National Museum of Ethnology, Osaka.

Thomas, D. H., 1983, The Archaeology of Monitor Valley 1, Epistemology. *Anthropological Papers of the American Museum of Natural History* 58(1): 1–194.

Wiessner, P., 1982, Beyond Willow Smoke and Dogs' tails: A Comment on Binford's Analysis of Hunter-Gatherer Settlement Systems. *American Antiquity* 47(1): 171–178.

Winterhalder, B. P., 1986, Diet Choice, Risk, and Food Sharing in a Stochastic Environment. *Journal of Anthropological Archaeology* 5: 369–392.

Woodburn J., 1980, Hunters and Gatherers Today and Reconstruction of the Past. In *Soviet and Western Anthropology*, edited by E. Gellner, pp. 95–117. Duckworth, London.

Part I

Regional Scale Processes of Settlement Pattern Change

Introduction to Part I

Regional Scale Processes of Settlement Pattern Change

The four chapters in Part I discuss regional scale processes of settlement pattern change using archaeological examples of classic "collectors" in four areas of the world: the North Pacific Coast of North America, the Canadian Arctic, the Jomon of the Japanese Islands, and the Natufians in the Eastern Mediterranean. We have assembled these chapters in this first part, because they share the theoretical orientation and operating assumptions of the original forager/collector model more closely than many other chapters, some of which radically depart from the model's founding principles. Looked at from another angle, these four chapters represent a middle ground between more formal approaches associated with evolutionary ecology (chapters in Part II) and approaches of social archaeology in which individual actions and historical contexts are emphasized (chapters in Part III).

The theoretical and methodological implications of these four chapters cover a wide range. First, they demonstrate that, despite recent theoretical shifts toward an emphasis on history and ideology on the microscale (such as individuals and households), the analysis of subsistence and settlement on the macroscale will continue to enrich our understanding of past hunter-gatherer societies. Using the forager/collector model as their point of departure, the authors of these chapters propose that additional factors be considered in our examination of hunter-gatherer cultural landscapes. Although each author does so in a different way, the results all indicate that, with these additional considerations, we can draw more realistic pictures of hunter-gatherer lifeways while still acknowledging the importance of subsistence activities and settlement organizations as the hub of the hunter-gatherer landscape.

Second, the case studies presented in these chapters indicate that, with modifications and extensions, the forager/collector model, which was originally developed to explain synchronic system dynamics, can be applied to investigating diachronic change. During the 1970s and early 1980s, when Binford's *Nunamiut Ethnoarchaeology* and the original versions of the forager/collector model were published, major research foci were on examining synchronic diversity under different natural environments, rather than on studying diachronic change from an evolutionary perspective. In fact, one of the major contributions of the model at that time was that it detached the concept of sedentism (low residential mobility) from the unilinear perspective of cultural evolution. More than 20 years after the publication of the original model, authors of these chapters highlight various ways in which the model can be operationalized to examine subsistence-settlement characteristics at a particular point in time and also to provide explanations for long-term system change. Though this was not the original intent of the model, the case studies presented here demonstrate the utility of the model in this new direction.

Third, the examinations of long-term settlement change presented in these four chapters reveal considerable regional diversity of evolutionary trajectories. The role of organizational complexity, including intensive subsistence activities, sedentism, and social inequality, in the course of cultural evolution has been a topic of debate among researchers (e.g., see volumes edited by Arnold 1996; Gregg 1991; Koyama and Thomas 1981; Price and Brown 1985; Price and Feinman 1995). Bar-Yosef's emphasis on cultural complexity among Early Natufian collectors, which suggests the presence of sedentary, nonegalitarian hunter-gatherers before the advent of agriculture in the region, fits nicely in the context of the traditional picture of "Neolithic Revolution." On the other hand, the chapters by Habu and Savelle indicate that the direction of long-term change may not necessarily be unilinear, nor does it always form the foundation for the transition to food-producing societies. Although Ames' chapter does not directly address the issue of long-term change, the large body of ethnographic evidence that he assembled can also be evaluated in the context of long-term system change. In this regard, comparisons among these case studies will allow us to reexamine the nature and role of hunter-gatherer cultural complexity in relation to the traditional view of evolutionary stages of culture.

Altogether, the four chapters in Part I explore one possible avenue through which we can expand our understanding of the dynamics of change in the hunter-gatherer cultural landscape. Needless to say, archaeological insights into the causes and consequences of dynamic system change presented in these case studies will be a rich source of information for area specialists regardless of their theoretical orientations. (Eds.)

ACKNOWLEDGMENT

The editors would like to thank Larkin Hood, Michael Konzak, and an anonymous reviewer for their assistance in the production of this volume.

REFERENCES

Arnold, J. E., (ed.), 1996, *Emergent Complexity: The Evolution of Intermediate Societies.* International Monographs in Prehistory, Archaeological Series 9, Ann Arbor.

Gregg, S. A., (ed.), 1991, *Between Bands and States.* Occasional Paper No. 9, Center for Archaeological Investigations, Southern Illinois University, Carbondale.

Koyama, S., and Thomas, D. H., (eds.), 1981, *Affluent Foragers: Pacific Coasts East and West.* Senri Ethnological Studies 9, National Museum of Ethnology, Osaka.

Price, T. D., and Brown, J. A., (eds.), 1985, *Prehistoric Hunter-Gatherers.* Academic Press, Orlando.

Price, T. D., and Feinman, G. M., (eds.), 1995, *Foundations of Social Inequality.* Plenum, New York.

Chapter **2**

Going by Boat

The Forager–Collector Continuum at Sea

KENNETH M. AMES

INTRODUCTION

In 1990, Binford argued that logistical mobility strategies...

> are the consequence of two major evolutionary changes that occurred long ago: (1) the "aquatic resource revolution" with its early occurrence primarily in higher latitudes, and (2) the perfection of transport technologies, particularly water transport vessels and the use of pack and draft animals. (Binford 1990: 138)

Hunter-gatherers that pursue aquatic resources will be strongly logistical in their mobility strategies. In fact, they are virtually obligatory collectors (Binford 1990). Following Binford, I will accept these statements as guides to further research. The purpose of the current paper is to explore two broad implications of these statements.

First, it has long been recognized that in contrast with terrestrial, pedestrian hunter-gatherers, aquatic hunter-gatherers tend to have higher population densities (e.g., Renouf 1988; Keeley 1988), to be residentially more stable, i.e., more sedentary, and to be perhaps more socially and economically complex than most terrestrial hunter-gatherers. Explanations for the causes of these apparent distinctions commonly focus on the relative or absolute productivity and dietary value of aquatic relative to terrestrial resources. In these discussions, lip service is paid to the importance of

KENNETH M. AMES • Department of Anthropology, Portland State University, Portland, Oregon 97207.

"transport technologies," but their importance is rarely investigated. However, access to waterborne transportation alone can have a significant positive impact on population size and stability (Batten 1998). This implies a more complex (and interesting) interplay between environment and technology in the evolution of mobility strategies than is generally appreciated.

Secondly, most aquatic hunter-gatherers are collectors, following Binford's basic definition that collectors move resources to people. However, archaeological expectations for recognizing collectors and foragers are based on terrestrial hunter-gatherers. There are important differences between aquatic and terrestrial hunter-gatherers, due, in part, but not exclusively, to differences in transport. These differences mean that the archaeological record of terrestrial and aquatic collectors may be quite different from each other, at least in degree. A question arising from these considerations is whether these differences in degree are, cumulatively, differences in kind. If this is so, a further question is whether comparative analyses, such as those of Binford (1980, 1990), Keelley (1988), and Kelly (1995) that are based on ethnographic samples with significant numbers of aquatic hunter-gathers, are therefore flawed (see Yesner 1980 for an early argument to this effect).

This paper is about boats, but it is about boats as transportation, about the integration of boats into production on a daily basis, about boats being used to haul material 100 meters across a lake or 1000 kilometers over difficult seas. It is not about the evolution of boats or about evidence for the earliest boats, or about boats as the only way to get to Australia, or as a means to people the Americas. It is about boats as instruments of production and whether the use of boats is theoretically important.

AQUATIC HUNTER-GATHERERS

What to call the people who are the focus of this paper? Binford (1990) uses the term "aquatically oriented hunter-gatherers." These are people who are "dependent" on aquatic resources, by which he means aquatic hunting and fishing, but not collecting aquatic or hydrophytic plants. When Kelly (1995) speaks of aquatic resources, he clearly means fishing and sea-mammal hunting. My meaning of the term is somewhat broader. Aquatic hunter-gatherers are those whose production activities rely on water for procuring food, other resources, and for transportation. I use this term because there is no alternative that clearly distinguishes these from terrestrially based economies. "Maritime hunter-gatherer" generally refers to people who exploit marine environments, be they close inshore littoral environments or distant pelagic ones. The phrase has always struck me as both too broad and too narrow. It is too broad because it can be applied, and often is, equally to people, like the Aleut, who hunt whales in open water with highly evolved tackle, including their vessels, and to those who are essentially strandloopers, collecting mollusks and exploiting near-shore environments without specialized gear or tackle. Lyman (1991) calls the latter a "littoral" adaptation and reserves the title "maritime" for the former, with its specialized gear and knowledge. This distinction is at the heart of Lyman's debate with Hildebrandt and Jones over whether people along the Oregon coast hunted seals from boats in open water or clubbed or speared them in their rookeries (Hildebrandt and Jones 1992; Lyman 1995; Jones and Hildebrandt 1995). Lyman's distinction, useful as it is, does not capture my meaning. Maritime is also too narrow because it does not apply to people who exploit wetlands, rivers, lakes etc.

The Chinookan peoples who lived in the Wapato Valley region of the lower Columbia River are a case in point. Although riverine fishing was central to their economy, they harvested a wide array of terrestrial, wetland, lacustrine, and riverine resources. It would, in fact, be difficult to categorize them as either terrestrial or aquatic hunter-gatherers based solely on the sources of their food resources. They, however, relied very heavily on canoes of a variety of shapes and sizes. They used these canoes to move resources and themselves across the landscape. It is as much this dependence on canoes that makes them aquatic hunter-gatherers as it is the salmon and sturgeon they harvested from their boats.

Binford's seminal paper on collectors and foragers (Binford 1980) was published in the same year as Yesner's on "Maritime Hunter-Gatherers: Ecology and Prehistory (Yesner 1980)." Yesner's article was something of a manifesto for the study of and theory building about maritime hunter-gatherers. He argued that modern hunter-gatherer studies may have little direct relevance for understanding ancient hunter-gatherers because, according to Yesner, most modern hunter-gatherers occupy marginal environments. He also criticized the methodology used by Binford and others to look for broad regularities or correlations between environment and economy, suggesting that such studies masked important dimensions of variability. He strongly implied that these studies are flawed by the use of the wrong scales of "cultural and ecological units for analysis (to avoid spurious correlations) (Yesner 1980: 728)." Yesner went on to argue that maritime hunter-gatherers are and were significantly different from modern, terrestrial hunter-gatherers in marginal environments. Like many others, Yesner also addressed the productivity of marine environments as a source of dietary calories, protein, and nutrients relative to terrestrial environments.

For Binford (1990: 134), the apparent increasing use of aquatic resources in the Holocene is "one of the major problems archaeologists have yet to address realistically in terms of the issues of complexity and human evolution." Debate over this question generally focuses on the ecological productivity of aquatic (usually understood as marine) and terrestrial environments (e.g., Erlandson 1988, 1994; Keeley 1988; Osborne 1997; Perlman 1980; Schalk 1981; Yesner 1980). Here, I do not directly address the relative merits of the arguments of these authors; rather I explore an alternative or supplementary position, which is that the availability of efficient (or effective) transportation can have a significant positive impact on the net productivity of aquatic environments.

AQUATIC TRANSPORT

Batten examined the population histories of 327 European cities in the period from 1500 to 1800 A.D. to determine the relationship of city growth (as a measure of urbanization) and transportation. His actual concern was the impact of transportation on urbanization in ancient Mesoamerica. However, the European data provided (1) good estimates of population size, (2) precise estimates of age, and (3) good control over systems of transportation. Cities were assigned to three "transportation" categories: landlocked, river, and ocean (see Batten 1998: 494–496, for the way these seemingly broad categories relate to transportation). As a result of his analyses, Batten concluded:

1. Ocean cities (those accessible from the ocean, port cities) exhibited faster and more sustained growth during the period of his study.
2. Median populations of cities on water (both ocean and river) far surpassed those of landlocked cities.
3. Population sizes of landlocked cities were more subject to fluctuations. They were the only cities to lose populations during this period.
4. "Population should be proportional to the size of the food producing hinterland ... if ... food supply is an important part of the relationship between transport and population size (Batten 1998: 510).

Batten's results cannot be translated directly to hunter-gatherers. He was not certain they could be directly applied to urbanization in Mesoamerica from Europe. However, the implications are suggestive and do find some support from recent comparative work on modern hunter-gatherers. For example, both Keeley's (1988) and Kelly's (1995) data sets indicate that population densities for coastal hunter-gatherers are generally higher than among terrestrial peoples, although Keeley indicates that the differences are not large. However, if one compares coastal groups with adjacent terrestrial groups, rather than comparing them globally, the differences are more marked.

I calculated median, mean, and standard deviations (Table 2.1) and constructed box and whisker plots (Fig. 2.1) of the population densities for

Table 2.1. Descriptive Statistics of Population Densities (individuals/100/km²) for Aquatic and Terrestrial Hunter-Gatherers in Western North America (density estimates from Kelly 1995, Table 6-4, 222–226)

Region	N	Mean Density	Std. Dev.	Median	Minimum	Maximum
California						
Aquatic	18	203.7	199.8	173	25	843
Terrestrial	25	125	91	103	12	103
Northwest Coast/Plateau						
NW Coast	16	68.3	49.2	61	10	195
Plateau	10	14.6	13.1	9.5	2	38
Arctic/Subarctic						
Coast	16	11	16.9	3.4	.5	65
Interior	23	1.5	1.9	.8	.2	7.6

aquatic and terrestrial hunter-gatherers in western North America, including California (aquatic and terrestrial[1]), the Northwest Coast and Plateau,[2] and the Arctic and Subarctic.[3] I used the density figures in Kelly's Table 6-4

[1]I assigned groups to the "aquatic" and "terrestrial" classes based on my reading of ethnographies, particularly the accounts in the Handbook (Heizer 1978). This assignment was not always straightforward. Some groups located on or near the Pacific coast made almost no use of coastal resources, for example. On the other hand, salmon was an important resource for many peoples living in the interior of central and northern California. In this latter case, I generally assigned these groups to the terrestrial category because fishing was usually their only "aquatic" harvest, the rivers did not provide usable transportation routes, and they made little use of water plants.

[2]This comparison is between coastal peoples and those of the continental interior, rather than a clear-cut comparison between aquatic and terrestrial economies. The peoples along the Northwest Coast are straightforward examples of aquatic hunter-gatherers, although many groups generally considered maritime by anthropologists may have relied more heavily on terrestrial plant foods than typically thought (Duer 1999). The Plateau is not so straightforward. The Plateau of North America is the topographically complex region between the Cascade/Coast Ranges of the Pacific coast and the Rocky Mountains from southern Oregon to central British Columbia. Many peoples (but not all) of the Plateau heavily relied on salmon and other fish; canoe travel was also important, although how important is not known. However, Plateau groups also were very dependent on roots (e.g., Thoms 1989; Peacock 1998) and terrestrial mammals. Rather than trying to class Plateau groups as more or less aquatic and terrestrial, I compare the two large regions.

[3]All of the groups in the "aquatic" class are coastal, and most are found in the North American Arctic. Kelly's sample includes only the Aleut and Chugach from the Pacific coast of Alaska, not the Koniag of Kodiak Island and adjacent mainland areas. Virtually all the "terrestrial" groups are subarctic hunters, and most are Athabaskans. Interior Inuit groups include Nunamuit and Copper Eskimo.

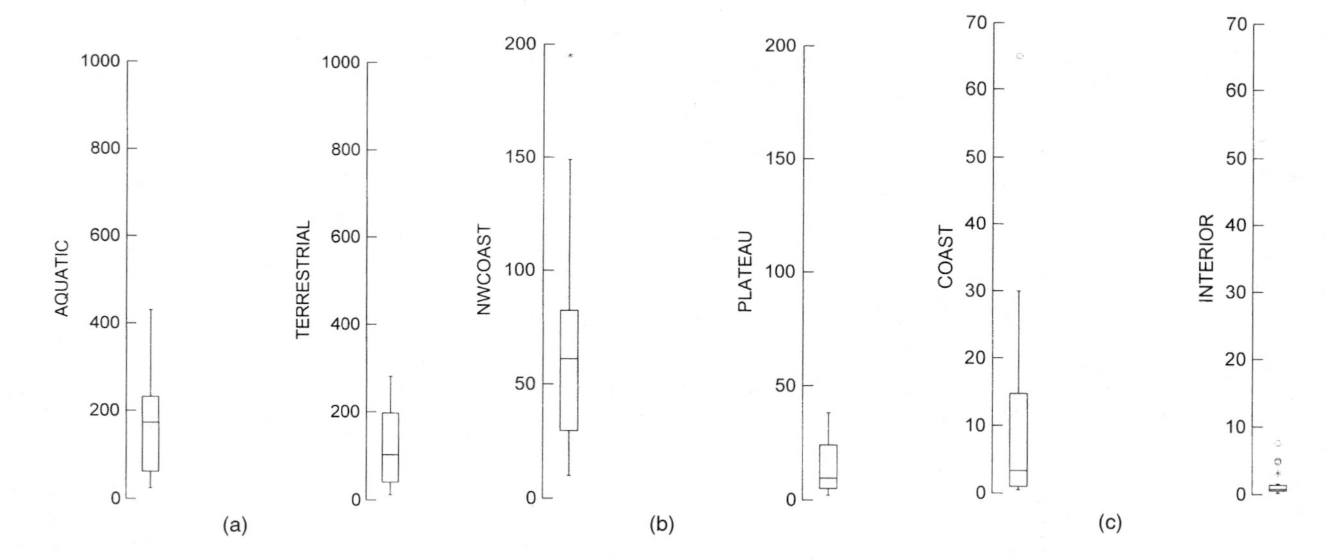

Figure 2.1. Box and whisker plot of the population densities of aquatic and terrestrial hunter-gatherers in western North America, including California (a), the Northwest Coast and Plateau (b), and the Arctic and Subarctic (c). Note differences in scale of the y axes.

(Kelly 1995) and the descriptions of territories, subsistence practices, and material culture in the relevant volumes of the Handbook of North American Indians (Damas 1984; Heizer 1978; Helm 1981; Suttles 1990b; Walker 1998). It is not possible to separate the effects of environmental productivity and those of transportation by using these data. However, they are quite clear on a number of points.

In North America, aquatic hunter-gathers consistently have higher mean and median population densities, as suggested by Batten. There is consistently greater variation in population density among aquatic hunter-gatherers, although the distribution of variation around the mean and median differs among the three aquatic groups. The patterns of variation among the terrestrial groups are more regular, and the median is consistently smaller than the mean. Population densities of both aquatic and terrestrial groups can be equivalently low, whereas the higher densities of aquatic groups are consistently much larger than among terrestrial groups.

These data do not address Batten's first conclusion: that populations (cities) on the ocean will grow faster than those in the continental interior. This would also be difficult, although not impossible, to test archaeologically. In any event, that test is beyond the scope of the current work.

The data do support his second conclusion: that peoples on coasts or using aquatic environments have higher median (and mean) population densities than those in continental interiors or those who are adjacent to coasts but who are primarily terrestrial in economy and transportation. Thus, the pattern he detects for preindustrial cities can be generalized more broadly.

Taken at face value, these data seem to contradict his third conclusion that there is greater variability in population sizes among interior groups or put in another way, that coastal groups have more stable populations. However, his conclusion is about variation through time; these data show variation in space. Further, his evidence indicates that interior cities are likely to lose population in times of stress. What these data do show is that the upper limits on population densities of terrestrial hunter-gatherers are consistently lower than those on aquatic population densities, regardless of local population size. Population densities of terrestrial groups in California are huge relative to those of the Subarctic, but the pattern is the same.

The rest of this paper addresses, at least indirectly, Batten's fourth point.

TRANSPORT TECHNOLOGY OF THE NORTH PACIFIC

In this section, I will briefly look at these aspects of boat use: freight capacities, universality, distances and speed, and their disadvantages. This discussion draws on two sets of examples: umiaks and kayaks in the North

American Arctic and Northwest Coast dugout canoes. These examples also illustrate two contrasting approaches to boat use: using a general-purpose boat for many tasks[4] versus using multiple specialized hulls. They also illustrate two quite different forms of hull construction. Umiaks are hide boats, whereas Northwest Coast vessels were dugouts made from logs.

Vessels

North Pacific Skin Boats (This Discussion is Based Primarily on Durham (1960), Chappelle (1994), and Rousselot (1994) Unless Otherwise Cited.)

Skin boats were widely used along the most northerly shores of the North Pacific and in the North American Arctic. They have a variety of names, but they are most widely known as umiaks and kayaks. Umiaks are open, generally flat-bottomed boats with an internal frame covered by sea-mammal hides. Walrus hide was preferred because of its toughness. The vessels ranged in size from 4.6 m (15 ft) to more than 12 m (40 ft), although early European sailors off Greenland observed umiaks 18 m (60 ft.) long (Chappelle 1994). Chappelle suggests that there is no practical reason why even longer skin vessels would not be possible. The most common size apparently was about 9 m (30 ft) long by 2 m (6 ft) wide, and about 15 cm (6 in) deep. Although umiaks could carry as many as 40 people, 10 to 20 people, with their gear, was the usual number. Umiaks are seaworthy, fast, and easily maneuvered, particularly when empty. They are also light, and, because of their construction, easily taken apart. In the recent past, hulls apparently varied greatly in detail, depending upon the vessel's function and local water and wind conditions. Kayaks are decked boats, usually with single cockpits for their one-person crews. Two-person kayaks were also used in some areas. Kayaks were used principally for hunting, sometimes in seawaters, but also in rivers and lakes. Kayak hulls varied markedly through their distribution. Rousselot (1994) documents 22 different hulls in North America and adjacent Siberia.

Northwest Coast Dugouts [This is Based on Durham 1960; Waterman and Coffin 1920; Drucker 1951; Swan 1967; Sproat 1987; Holm 1994; and Olson (1927) Unless Otherwise Cited.]

Northwest Coast canoes, with exceedingly rare exceptions (e.g., de Laguna 1972), were "dugouts" made from logs of western red cedar (*Thuja*

[4]Of course, many groups using umiaks also used kayaks, a highly specialized hunting vessel, although some did not. However, the point remains that for many purposes, one craft was used in contrast with the Northwest Coast.

plicata). The techniques used to make these vessels are described in many sources (e.g., Boas 1909; Drucker 1951) and need not be reviewed here.

There was great diversity in hull form and size along the coast, although this variety probably declined during the nineteenth century. Suttles (1990b) illustrates 10 nineteenth century hull forms but does not exhaust the variability. This diversity has a number of dimensions, including geography, water conditions, function, and hull size. Regional specialization in boat building overlay and partially obscured local variation in vessel form. The coast had two major boat-building groups: the Haida people of the Queen Charlotte Islands on the northern Coast and the NuuChahNulth peoples along the west coast of Vancouver Island. Both made canoes for export, and their craft were widely distributed (Fig. 2.2). Haida canoes were widespread on the northern coast, although both Coast Tsimshian and Tlingit also made their own. The Tlingit appear to have been particularly dependent on this trade for their large, seagoing vessels. NuuChahNulth hulls were traded broadly on the central and southern coasts. As with the Haida, they were probably their region's primary source for large freight and seagoing canoes. Distinctive vessels were found among the Coast Salish and the Chinookans of the lower Columbia River. In both cases, the vessels were adapted to the particular conditions of the local waters.

Despite this variability, there were three or perhaps four basic hulls. The largest canoes were what Durham (1960) calls great canoes, a term I will use here. Great canoes included so-called "war" and freight canoes. Freight canoes were 10 or more meters long, broad-beamed, and high-sided. War canoes were narrower, but usually as long, if not longer. A nineteenth century Haida canoe is described as almost 25 m (80 ft) long although Durham (1960) suggests that a length of 18 m (60 ft) was their practical limit, and 12 m (40 ft) the most common. War canoes had beams of about 2 m (6–7 ft), whereas those of freight canoes were somewhat wider. War canoes were used for warfare and also as the coast's seagoing vessels, both far from shore and on long voyages. It is not altogether clear that "war" and freight canoes were separate classes of hulls everywhere, although hulls were designed specifically for warfare in some places.

There was a range of medium-sized vessels [length: 5.5–11 m (18–35 ft); beam: 1–2 m (3–6 ft)] that some authors (e.g., Durham 1960; Shackleton and Roberts 1983) term "family canoes." The hulls of these canoes were apparently very similar to those of the larger styles. Additionally, there were even smaller vessels that carried one, two, or three people. The smaller vessels were evidently quite numerous, and though the capacities of the largest vessels were spectacular, it may be that these small vessels were more important. The smallest vessel on the lower Columbia River, for example, was only about 4 m long (10–14 ft) with a beam of 0.5 m (21 in.) and a depth of 0.2 m (9 in.) and was used by one person.

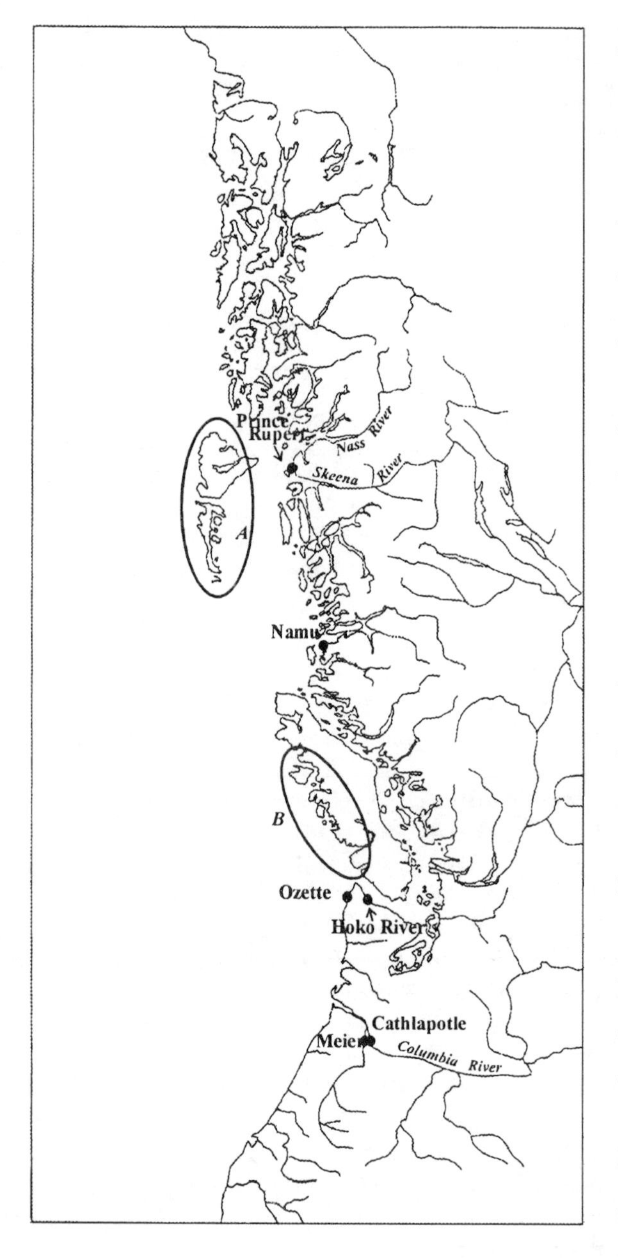

Figure 2.2. Archaeological sites and locations on the Northwest Coast mentioned in the text. The ovals enclose groups specializing in making and exporting canoes: A: Haida; B: NuuChahNulth.

Specialized hulls included hunting and whaling canoes and canoes designed for use in rivers and in relatively placid waters. These specialized hulls varied in size; some were as large as great canoes, although most were usually closer to the family canoes in size.

Capacities

Umiaks were quite commodious. Larger ones could hold as many as 60 passengers or 5 tons of cargo (Durham 1960). Ten to 20 people were a more typical crew. Small, one to two-person umiaks were used in some places. Jochelson left this description of the capacity of the Koryak version of the umiak:

> The Koryak skin boat can carry fairly heavy loads ... [I]n two boats ... we carried about two thousand pounds of cargo, and our party consisted of twenty-five members ...In addition, each boat carried eight dogs in harness, which lay in the stern. Notwithstanding this heavy load, the boats were not more than half in the water. (Jochelson 1908: 538; as quoted in Rousselot 1994: 248)

The largest Northwest Coast vessels carried equally enormous loads. Meriwether Lewis (Moulton 1990: 267–272) estimated that freight canoes on the lower Columbia River could carry as much as 4 to 6 tons. Ethnohistoric sources more often give an indication of capacities in terms of the numbers of people carried rather than the loads. Great canoes usually carried 20 to 30 people, and the "household canoes" carried 10 to 15 with their loads. For example, the Reverend Myron Eells traveled up Puget Sound in January 1878 with a party of 65 in seven canoes, or about nine per vessel (Castile 1985). The group included women, children, and men on their way to a potlatch. In 1813, a group arrived at Fort Astoria, the fur-trading post at the mouth of the Columbia, in six canoes and two boats. The party included 53 people and "116 packs Beaver, Baggage, etc. etc. (Jones 1999: 190)."

On the coast, freight canoes ferried house planks from one village to another. Sometimes two freight canoes were lashed together about 4 m (12 ft) apart, and the house plank placed across them. Household goods were then stacked on the planks. According to Drucker (1951: 88), these rafts were "slow and cumbersome" but in good weather "transported heavy and very bulky cargoes in good style."

In discussing capacities, it is easy to focus on canoes for hauling freight. However, they were employed in a variety of productive ways, including whaling and sea-mammal hunting. They were also central to many less spectacular, but probably more important tasks. Meriwether Lewis observed Chinookan women harvesting the corms of *Sagittaria latifolia* by wading out into a lake where the plant grew, loosening their bulbs with their feet, and, when the bulb floated to the lake's surface, throwing them into the smallest style of Chinookan canoe (Moulton 1991: 30). In this instance, the women had to portage the full canoes a short distance to a river to take the corms away, but many lakes and wetlands in this area had outlets.

Two additional points need to be made here. First, both the skin boats and dugouts are described as very seaworthy, even when fully loaded. Their crews were also quite skilled. Secondly, heavily loaded traveling groups included dogs and children as well as women and men. Both men and women handled boats, including paddling and navigating.

Distances and Speed

Although the peoples of the North Pacific undertook very long voyages, I am interested here in daily distances under normal travel conditions. I could find little information on distances traveled in umiaks. Durham (1960) comments that people sailed only in good weather, and "progress over long distances was usually very slow (p. 23)." In contrast, there is better evidence for distances traveled on the Northwest Coast. Gilbert Sproat, for example, commented in his 1866 memoir of his time on Vancouver Island that two people could easily "paddle a medium-sized canoe 40 miles on a summer's day (Sproat 1987: 83)."

Myron Eells provides what may be the best data on distances traveled and speed for the Northwest Coast. He made two trips along his route of January 1878, the January trip and an earlier one in the summer of 1876 (Castile 1985). The route was from Skokomish, Washington, in Puget Sound to Dungeness at the eastern end of the Straits of Juan de Fuca. He gives the mileage as 90 miles. The January trip, with the large party, took 21 "traveling" hours from Skokomish to Dungeness and 33 hours back. The quicker trip was with favorable winds. The 1876 summer trip was done in one boat with two women and two men. The women paddled, one of the men steered, and the other "rowed." The outbound leg took 31 hours and the home trip 23 hours, again with favorable winds. The faster speeds for both trips, with favorable winds, averaged 6.5 km/h (3.5 knots/h), and the slower average speed was 4.4 km/h (2.4 knots/h). At these speeds, Sproat's two men in a medium boat would take 10 to 13 hours to travel their 40 miles. Croes and Hackenberger (1988), using somewhat different sources, estimated canoe speeds of 4.5 km/h in good weather and 2.7 km/h in bad. These translate to somewhat slower estimated speeds of 2.5 and 1.5 knots/h.[5] These estimates may be a better reflection of speeds in

[5]The lowest speed for which I can find a record is a passage across the 15-mile wide Strait of Juan de Fuca that took place in a severe storm. The trip took 11 hours (Durham 1960), for a

"normal" short distance travel than mine, which may be traveling speeds. It does not appear from Eells' figures that crew or boat size had much affect on speed, at least for smaller vessels. The very largest canoes may have been underpowered for their size and slower than smaller boats (Durham 1960).

Of course, people do not always just travel, and the habits or patterns of movement can affect speed and distance. Hunter-gatherers particularly embed information gathering in their movements. The experiences of Swan, Sproat, and Eells may tell us something about the way canoe trips were conducted. Both Swan (1857) and Sproat (1987) observed a distinctive rhythm to canoe travel. Crews would paddle hard for some distance, and then stop to discuss things (in water or on land) that they had observed or to recall a story. It is likely that these pauses were crucial to collecting information about water conditions and for navigation. Once the discussion was ended, paddling resumed. In Eells' summer trip, the party traveled straight through, arose before dawn, and paddled for 12 or more hours with few, brief pauses. In the winter trip, the group traveled for 6 to 8 hours and then made camp. On both trips, they were forced to remain in camp to allow unfavorable winds to die down.

Universality

By this I mean two related things: (1) the number of boats available and access people had to them and (2) their centrality to the conduct of everyday business and life. On the North Pacific, boats were numerous, ubiquitous, and central to life. It is difficult to assess the numbers of umiaks. However, in 1881, Edward Nelson (1899) observed a camp of about 150 conical lodges with more than 60 umiaks and 200 kayaks. Early traveler's accounts suggest swarms of kayaks.

The numbers of boats are better documented for the Northwest Coast. Durham (1960) asserts, based on his readings of early accounts, that villages usually had enough to be completely waterborne. He cites, for example, George Simpson's account in the 1840s of encountering 800 Comox in 40 to 50 canoes in the Gulf of Georgia. In November 1805, William Clark counted 52 canoes at a Chinookan village in what is now Portland, Oregon. He states that the village contained 200 men. In the spring of 1806, the Lewis and Clark expedition passed another village in the same general area and recorded 100 small canoes of the kind used by women to collect *Sagittaria latifolia* and other aquatic resources. The village, it was estimated, had 100 people. In short, boats were both very numerous and ubiquitous.

speed of 1.3 knots/hour. This suggests to me that Croes and Hackenburger's bad weather estimate is probably too slow for normal travel.

It also seems clear that literally everyone, at least on the Northwest Coast, had one. One class of canoe made by the Nuuchahnuulth was called the "children's canoe." Sproat (1987) describes small children managing boats in swift streams. People also had access to good boats. In many cases, the hulls of "family canoes" were simply somewhat smaller versions of the largest seagoing canoes. It also appears that access to umiaks was unrestricted.

The largest Northwest Coast vessels were made by specialists, who worked on commission. Thus, only chiefs could command the resources to pay for a great canoe. Durham (1960) suggests that chiefs might have one or two such vessels that were used for a variety of purposes. Among the Tlingit, at least, canoes belonged to the household (Oberg 1973). It is not clear that specialists made all of the canoes on the coast, but as Swan put it,

> "The manufacture of a canoe is a matter of great moment ... It is not every man among them that can make a canoe, but some are, like our white mechanics, more expert than their neighbors." (Swan 1967; 80)

It may be that many of these expert boat builders were what I have termed "embedded specialists (Ames 1996)."

I could find no indications whether specialists made umiaks and kayaks. My sense is that they did not. However, I would suspect that, as with canoes, some makers were more expert than others, and were probably sought after, particularly given the potential costs of an ill-made vessel.

The general accessibility and widespread use of well-made, seaworthy craft on the North Pacific is in marked contrast to the availability of seagoing boats among the Chumash of southern California. The Chumash made a seagoing plank canoe, a canoe type unique to them, at least in western North America. These canoes were owned and controlled by the individuals who could afford to pay for them. Durham estimates that a village of 500 might have only 10 of these canoes. Arnold (1995) has argued that control of these canoes and related production were central to the development of social inequality among the Chumash. That kind of control over sea-worthy craft had no obvious parallels on the Northwest Coast and Arctic. It was always possible on the Northwest Coast to vote with your feet, or your paddle, when chiefs became too autocratic, as they did. There is some evidence, however, that the Chumash also made dugout and oceangoing balsa canoes (Grant 1978). If so, then there may have been wider access to more humble vessels.

Disadvantages of Boats

Boats are trouble, as anyone who has owned one knows. They are used in environments that can be quite hazardous and unpredictable. Even seemingly placid waters can drown both the unwary and the highly skilled. A storm that is mildly inconvenient ashore can be exceedingly dangerous in an open boat.

Boats are also subject to wear and tear and design problems. They require constant attention and maintenance. They can represent a major initial investment and a high ongoing one. For skin boats, the major maintenance problem is the skin hull. The skins are waterproofed by oiling them. They require drying and reoiling after about 4 days of use (Chappelle 1994). This may explain, in part, the slow progress of long journeys in these vessels noted above. Groups on long trips sometimes took a spare umiak and alternated them, with the wet boat turned hull up to dry over the one in use. Durham (1960) estimates that an umiak crew blown out to sea had 10 days to get back to land before the vessel failed completely. Differences in skin quality also affected the seaworthiness of boats. Walrus hides were better than fur-seal hides, for example.

The hulls of Northwest Coast canoes were made extremely thin so that the vessels would be light, and easy to paddle. Western red cedar splits readily, and these thin hulls required constant care and repair. Boats were wet down on sunny days, even in use, and kept moist and covered when beached to prevent splitting and checking. The maximum life of a Northwest Coast hull may have been about 12 years (Durham 1960). The thin hulls did not have internal frames and could flex. As a result, under certain sea conditions, the canoes might split open lengthwise (Drucker 1951; Durham 1960) and sink. The rafts made by lashing house planks across freight canoes broke up easily in choppy seas (Drucker 1951). Such an event could spill an entire village into the sea. Both wooden and skin vessels had to be launched and beached carefully to protect their hulls.

Umiak frames were wooden, and in some parts of the Arctic, finding the right wood for the vessel might have been problematic. On the Northwest Coast, western red cedar did not grow everywhere. The Tlingit homeland of southeastern Alaska is north of the red cedar's range, and they had to trade for boats, particularly for larger vessels. The Makah, famous whalers, had no boat-quality red cedar in their territories and had to trade for whaling boats from their Nuchahnuulth kin. The Coast Salish, Chinook, and others also traded for hulls from Vancouver Island. In short, boats are costly to construct and maintain.

Summary

These major points emerge from the preceding: seaworthy watercraft were common and ubiquitous on the North Pacific; they could carry enormous loads; a wide range of vessels was employed, whose variability reflects differences in the purpose of the boat, wind and water conditions, and available material; everyone, men, women, and children, had access to and used boats. This use and dependence on boats came with costs. These costs included the hazards of boat use, and the demands of boat-building and of ongoing attention and repair.

The preceding discussion has been a lengthy but necessary background to discussing aquatic collectors, the subject of the next section.

COLLECTORS AT SEA

Introduction

Foragers and Collectors

The distinctions between foragers and collectors are well known: foragers move people to resources; collectors move resources to people. Foragers exploit resources within a foraging radius of the residential base, and processing of resources occurs within that radius (6 km) or at the residential base. Collectors also exploit resources within the foraging radius, but, more importantly, they acquire more distant resources by sending out task groups that harvest or collect the resource (whatever it is), process it in the field, and return to the home base with the processed product. Mass processing, therefore, usually will occur away from home.

The predicted archaeological record for collectors and foragers is equally well known. Foragers are expected to produce only two kinds of sites, residential bases, and locations. Locations are places where "extractive tasks are exclusively carried out" (Binford 1980: 9). Among foragers, these have low archaeological visibility. Collectors generate residential bases, field camps, stations, and caches. These are likely to have higher archaeological visibility because bulk processing occurs in the field, often repeatedly at the same place. Stations are information-gathering places, and caches are places where high-bulk resources are stored or cured in the field using temporary storage facilities.

A crucial difference between foragers and collectors, as described by Binford (1980), is that collectors, who are delayed-return foragers in Woodburn's terms (1980), do bulk extraction and processing of resources, which foragers do not do. Because of this, Binford regards that differential transportation cost is the central consideration in whether groups are collectors or foragers (Binford 1990: 139), except under conditions of demographic packing, under which logistical strategies will evolve regardless of cost. Issues of cost in time and energy are central to the discussion in the next section, and bulk processing to that in the following section.

Implications of Boats for Hunter-Gatherer Mobility

Kelly (1995: Chap. 4) provides a useful basis for examining the impact of boats on hunter-gatherer mobility strategies, and this section follows that discussion. Kelly identifies five dimensions that structure the variation in hunter-gatherer mobility strategies: the number of residential moves/year, the average distance moved/residential move, the total distance moved cumulatively/year, the total area used/year, and the average length of logistical forays, as he terms them (Kelly 1995: 120–121). I will discuss these and two additional topics he raises: foraging radius and sedentism.

Foraging Radius, Size of Areas Exploited, and Length of Logistical Forays

The concept of foraging radius is central to considerations of mobility patterns. The point of diminishing returns sets this distance. After a camp is established, foragers explore and harvest resources within an easy walk of camp. As nearby resources are consumed, people must go farther afield. At some point, the distance is judged too great, and the camp is moved. Kelly concludes,

> The effective foraging radius ... is largely a product of the return rates of the available resources and the degree of dependence on them (which, in turn, is a function of how many people are foraging for each family, family size, and per capita caloric needs). (Kelly 1995: 135)

Collectors, of course, deal with these problems by sending task groups beyond the foraging radius. Kelly's data indicate that this radius is generally about 4–6 km. He also suggests that 20–30 kilometers is the maximum distance hunter-gatherers can comfortably walk in a day across varied terrain. This is a walking speed of around 3 km/h, which is about the same as the slower canoe speeds calculated above. The 20–30 km figure can be taken as the maximum distances task groups traveled on foot in a day.

The foraging radii of boat-using groups could theoretically be considerably greater than those of pedestrian hunter-gatherers. The potential area that waterborne groups can exploit is large, particularly at radii greater than 10 km (Fig. 2.3). Four to six kilometers are 1–2 hours by boat, and 20–30 kilometers are a 4–10 hour trip one-way. Therefore, the foraging radius of a settlement, depending on water conditions, could be as great as 30 kilometers (Fig. 2.3).

The potential distance and scope of multiday logistical forays also significantly increase with boats, again depending on conditions. A fast-moving party, in good conditions, might cover 50 kilometers or more in a single day. Such parties could include men, women, children, and dogs, a composition that is somewhat different from our expectations for a "fast-moving" task group.

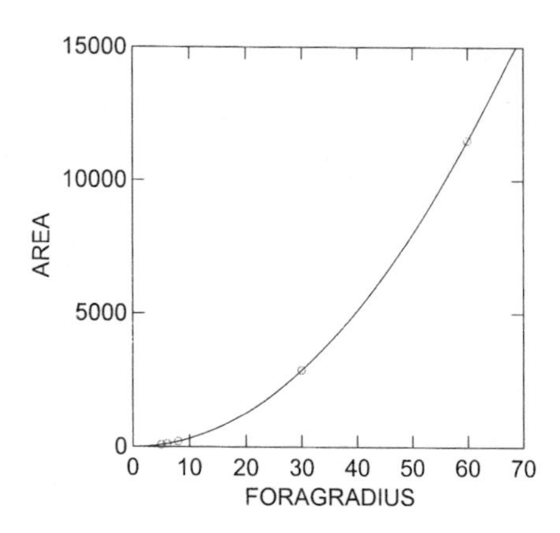

Figure 2.3. Total areas in square kilometers within foraging radii of 5, 6, 8, 30, and 60 km.

Despite this potential, foraging areas on the Northwest Coast, at least, are not much larger than those expected for pedestrian hunter-gatherers. I calculated foraging radii based on the size of village territories and group territories on the Northwest Coast (Table 2.2), using figures developed by Schalk (1978). Village foraging radii are usually less than 10 kilometers. They commonly range from 3.5 to 8 km. Only in three instances are they greater. Generally, the foraging radii based on tribal territories represent a single day's one-way trip, with the clear exception of the Southern Tlingit. The radii for three other groups, the Gulf Salish, Kwakwaka'wakw, and Northern Tlingit probably represent 2-day trips, except under good weather and sea conditions. In comparing these distances with the foraging radii for pedestrian foragers (4–6 km) and the distance they can cover in a day (up to 20–30 km), we see that though they are not much larger overall (and there is considerable overlap), they are somewhat larger. In addition, the potential for long logistical forays is clear, even if such forays are not the rule.

Figure 2.4 shows the areas within three levels of trips from the Coast Tsimshian principal towns in Prince Rupert Harbor, British Columbia. The smallest circle encloses a 15-kilometer radius, or a round trip of 30 kilometers, a one-way trip of 3–5 hours. The next circle encloses a 30-kilometer radius, or a 6 to 10 hour trip. This probably represents the maximal daily foraging area. The outer circle has a 60-km radius, a one-way trip of 10–20 hours. This is likely to be the extreme maximum for a 1-day one-way foraging trip (i.e., out to an island where a base camp is located). Interestingly enough, the boundaries of almost all of Coast Tsimshian territory fall between the middle and outer circles (Halpin and Sequin 1990: 268), with the exception of the Skeena River, where the boundary is farther upstream.

Table 2.2. Estimated Tribal Territories, Numbers of Winter Villages, Village Areas, and Approximate Foraging Radii for Some Northwest Coast Groups (from Schalk 1978, Table 4-5)

Group	Area	Foraging Radius (km)	Winter Villages	Villages Area (km²)	Foraging Radius (km)
Wiyot	1300	11.5	41	31.7	1.8
Yurok	1900	13.9	54	35.2	1.9
Karok	3200	18	108	29.6	1.7
Tolowa	2100	14.6	23	91.3	3.0
Chinook	3190	18	27	118.2	3.5
Quineault	2190	14.0	20	109.5	3.3
Puyallup-Nisqually	6480	25.6	34	190.7	4.4
Makah	950	9.8	5	190.1	4.4
Quileute	1190	11	6	185	4.3
Twana	2960	17.3	14	211.7	4.6
Upper Skagit	7300	27.2	36	202.9	4.5
Nooksack	3200	18	9	355.8	6.0
Straits Salish	3190	18	55	58	2.4
Gulf Salish	34080	58.8	54	631.1	8.0
Kwakwaka'wakw[a]	21100	46.3	29	727.6	8.6
Oowekenoo	4470	21.3	7	639	8.1
NuuChahNulth	8550	29.4	24	356.4	6.0
Nuxalk	15000	39	24	625	8.0
Haisla	8000	28.5	2	4000	20.1
Haida	10300	32.3	17	926.9	9.7
S. Tlingit	74200	86.8	38	1952.6	14.1
N. Tlingit	25000	50.4	10	2500	15.9
Mean		27.8			6.6
Median		21.3			4.6

[a]A number of the tribal names used by Schalk and Kelly have been changed or are presently spelled differently. Kwakwaka'wakw has replaced Southern Kwakiutl, Oowekeeno is preferred to Owkino, Nuxalk for Bella Coola, and NuuChahNulth for Nootka.

Some groups did make extremely long residential moves. Mitchell (1971) calculates the total distance traveled annually by households of three Coast Salish groups: the Cowichan, Nanaimo, and West Saanich. The Cowichan made five household moves/annum at a total distance of 450 km (280 miles). The Nanaimo moved four to five times and 483 km (280 miles) annually, and the West Saanich moved three to five times and up to 328 km (200 miles) annually. All three groups lived on the east side of Vancouver Island and moved across the Gulf of Georgia to the Fraser River to exploit the salmon runs, a round trip that would exceed 100 km for each group.

What all these data show is that using boats can increase foraging areas and allow long logistical forays, but it does not inevitably do so. Foraging areas and forays may be the same size among both aquatic and

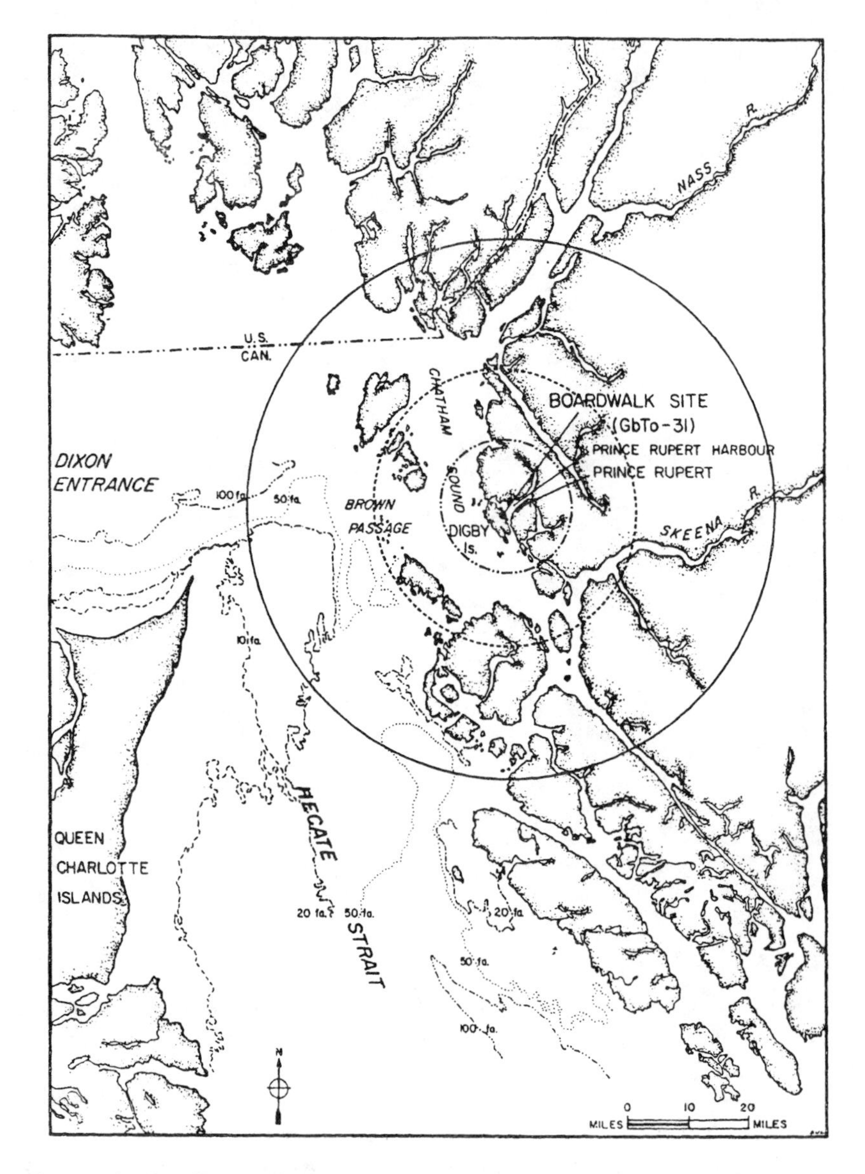

Figure 2.4. Map of the northern British Columbia coast. The circles are the areas of the 15, 30, and 60 km foraging radii from Prince Rupert Harbor.

terrestrial hunter-gatherers. Of course, aquatic hunter-gatherers who exploit interior wetlands and lakes may not be able to make long logistical trips. An obvious and not very startling implication of this is that the importance of boats rests with the much greater bulk and weight they can carry than can humans on foot. Another, perhaps less obvious, implication is that boats can facilitate efficient use of small foraging areas by permitting multiple trips on a single day in small areas and the deployment of larger task groups.

Transport Costs and Bulk Processing

One of the traits that define collectors is large-scale field or bulk processing by task groups. It is this aspect of collectors that creates a distinctive land use pattern and archaeological record. Decisions to field process are based on a variety of factors, including transport costs.

Transport costs are determined by, among other things, the weight and bulkiness of the material to be carried, the available transportation technology, and the transport distance. Pedestrian hunter-gatherers can reduce weight and bulk by field processing before transport. Field processing also increases the net nutritional return from transporting food resources. O'Connell et al. (1990) suggest that three factors affect transport decisions (for meat and bones) among terrestrial foragers: carcass size, size of carrying party, and the distance from the kill site to the residential base. In their work with the Hadza, they found that hunters begin field processing if (1) the piece of meat weighs more than about 15 kg and (2) the residential site is more than about a 2-hour walk away. Transport decisions are also affected by whether the material can be carried in a container and the bulk (weight and volume) of the container (Jones and Madsen 1989; Rhodes 1990). Jones and Madsen, for example, show that the nutritional returns of various Great Basin seeds and insects varied according to the distance they were carried in a burden basket. Some resources, such as grasshoppers, may have very high returns only if the transport distance is very short. The nutritional returns of some resources may increase as transport distance increases. Field processing may also be necessary to permit carrying a resource in a burden basket or pack.

Rhodes (1990) does criticize their modeling on the basis that foragers do not carry foods some of the extreme distances produced in the Jones–Madsen model (e.g., 829 km). However, this does not diminish the point that transport costs affect the net nutritional gains from harvesting foods for people on foot. For aquatic hunter-gatherers, these decisions are based on the load-bearing capacities of boats. Weights that are daunting on foot are trivial in many boats: what is 15 kg in a boat that can easily carry 2000 kg? Even when aquatic hunter-gatherers harvest terrestrial resources, the calculations about whether to transport overland or to process in the field may not be based on the distance to the residential site, but the distance to the boat.

Much of the research on field processing focuses on mammal remains, reflecting in part Binford's focus in his ethnoarchaeological work and his original utility curves (Binford 1978). As noted above, animals taken far from home are butchered in the field, and only the most useful cuts are carried home; the greater the distance, the more rigorous the setting of priorities. In contrast, for aquatic hunter-gatherers, it appears that a great deal of animal processing occurs at the residential site. Not all of it does, but a significant amount will. To a degree, this reflects the realities of fishing and hunting sea-mammals. If a party kills a sea lion in open water, they cannot settle down at the kill site and butcher it, as they might an elk or caribou. The carcass is attached to floats and brought home. It is not surprising then to find archaeological evidence for the return of entire sea-mammal carcasses to residential sites and butchering them there, often indoors (e.g., Huelsbeck 1994). Interestingly, the record on the Northwest Coast is more mixed for terrestrial mammals. At some sites [e.g., Hoko River (Croes 1995) and Ozette (Huelsbeck 1994)], land mammals comprise a trivial part of the faunal assemblage. At Ozette (Fig. 2.2), deer were field butchered, and selected parts were brought back. However, the terrestrial portion of Ozette's catchment was not accessible by boat. At two sites with terrestrial hinterlands accessible by boat—Boardwalk (Stewart 1976; Stewart and Stewart 1996; Ames n.d.) and Namu (Cannon 1991)—deer carcasses came in complete. Where weight or bulk are not significant limiting factors on transport, nutritional returns may not be the prime factor that determine bulk processing and carcass transport. Terrestrial mammals seem to have been hunted as much or more as a source of tool bone as meat in some places. There is evidence of the transport of tool bone to insular logistical sites on the Northwest Coast and Mesolithic Scotland (Ames n.d.; Smith 1992), and to mainland (Huelsbeck 1994) and insular (Ames n.d.) residential sites on the Northwest Coast.

There is little readily available evidence on the spatial distribution of plant processing by aquatic hunter-gatherers. Data from my projects in the Wapato Valley of the lower Columbia River suggests quite strongly that bulk processing occurred primarily at residential sites, at least in the Wapato Valley. At present, the primary evidence for this comes from two Chinookan residential sites located in wetlands close to the Columbia River, near Portland, Oregon. One of these is the Meier site, which contains evidence of a single, large plank structure (Ames et al. 1992), and the other is Cathlapotle, a town of 900 to 1400 people in 1806 that contained at least six very large plank structures (Ames et al. 1999). Both sites were occupied from about 1400 to 1830 A.D. or so.

As observed above, roots contributed significantly to the Chinookan diet. These roots include *Sagitaria latifolia* (wapato) and *Camassia quamash* (camas). Both plants produce large, nutritious corms or bulbs. Wapato grows along lake margins (Darby 1996), whereas camas is a meadow plant (Thoms 1989). There is extensive wapato habitat close to these sites (Darby 1996), but there is no camas habitat close by. In addition to these plants, the Chinook also exploited nut mast, primarily acorns (*Quercus* sp.) and hazelnuts (*Corylus cornuta*). Oak habitat occurs close to both sites, but hazelnut habitat is farther away.

The lake where Lewis and Clark observed Chinookan women filling their canoes with wapato is about 5 kilometers south of Cathlapotle, the closest known Chinookan town to this lake. Cathlapotle is accessible to the lake by foot, either across gentle terrain, or by boat, along a usually placid river. As Lewis and Clark observed, the canoes had to be portaged perhaps 300 meters across a flat meadow to the river.

Ethnographic sources suggest that the standard Chinook storage basket was about 0.8 m in diameter, about the same depth, and had a capacity of 115 liters. Burden baskets were certainly smaller. I have no estimates for Chinookan carrying baskets, but a large Great Basin basket has a capacity of 64 liters (Jones and Madsen 1989). The capacity of the smallest Chinookan canoe—that used for collecting wapato and other wetland resources—was about 400 liters, or almost three storage baskets and seven to eight burden baskets. In addition to its size, the canoe had other advantages. Women collected the corms directly in the canoe, which functioned as both container and transporter. The corms did not have to be collected in a waterborne container, transferred to carrying baskets, and carried to the village, or transferred to canoes. In this instance, the canoes did have to be portaged, an easy task with one person at either end of the full vessel. It is also likely that many more trips were possible in a single day using canoes than going by foot. Further, the number and range of people who could be involved in moving the harvested roots by boat was probably larger. A child who might find a fully laden basket a difficult, awkward load across 5 km could handle a loaded canoe. In this instance, as we will see below, processing was done in town.

The handling of camas (*Camassia quamash*) in other parts in western North America is an illustrative contrast. Both camas and wapato roots were processed by roasting them in earth ovens (see Peacock 1998 for a description of cooking camas; Darby 1996 for wapato). Camas was a significant resource for the peoples of the Columbia Plateau, east of the region discussed here, and in the Willamette Valley to the south. In these areas, camas ovens are associated with camas meadows, sometimes in large numbers (e.g., Thoms and Burtchard 1986). Such features are very rarely associated with residential sites on the Plateau. Camas bulbs were roasted and than made into flour that was transported in baskets to the residential base. Raw bulbs were also transported, at least after the introduction of the horse. Residential bases were often a few days away from the camas grounds. Thus, although a basket of camas flour probably weighed more than the same basket full of unprocessed roots, the nutritional return was higher, despite the additional work. In contrast, earth ovens are primarily (although not exclusively) associated with village sites in the Portland area. There are not concentrations of earth ovens around wapato and camas habitat there as there are around camas meadows on the Columbia Plateau and in the Willamette Valley.

At Cathlapotle, earth ovens and earth oven debris are densely distributed immediately outside the houses. The ovens contain charred camas roots, acorns, and hazelnuts. Wapato tissue has yet to be identified, but finding it will be analogous to finding a charred, peeled potato that is 400 years old. However, preliminary analyses indicate that the densities of camas bulbs in ovens sampled at Cathlapotle are almost as high as those at major interior root grounds (Stenholm 1995). Using canoes, large volumes of raw roots could easily be transported to the residential site. Ovens were then used to process multiple resources, including roots and nuts, rather than single resources.

Coupled with the evidence for bulk processing at these sites is evidence of stockpiling raw materials. Unmodified stone nodules and cobbles were transported to both Meier and Cathlapotle, where they were tested for suitability as tool stone (Hamilton 1994) and either stored or discarded as unsuitable. Raw material and raw material tests form a significant minority of the lithic artifacts at the Meier site, for example. Antler and bone for tools were also curated.

I do not know at present whether these local patterns can be extended to aquatic hunter-gatherers more generally. However, at least on the Northwest Coast, there does seem to be much more bulk processing at residential sites than would be anticipated based on the collector–forager model.

It is possible from the foregoing to propose a model for the aquatic forager landscape. I suggest that bulk processing activities are likely to be strongly focused at the residential site. Harvesting occurs elsewhere; the raw resources (food, tool stone, etc.) are transported to the residential base and processed (or discarded) there, rather than in the field. Thus, there may be more evidence for processing at the residential sites of aquatic hunter-gatherers, even if overall levels of processing are equivalent to those of terrestrial groups. Further, using boats, more resources and bulkier unprocessed resources can be transported to the residential site for processing. This pattern is predictable for maritime resources, but it may extend to any resources that are accessible by boat. This does not preclude the existence of specialized task camps. Such sites do exist. However, the implication here is that specialized task camps are embedded in a different archaeological landscape than that predicted by the classic collector models.

Residential Moves and Residential Bases

Aquatic hunter-gatherers generally make relatively few residential moves per year. This must reflect their ability to position themselves centrally and make multiple, short- and long-distance logistical forays. They can also field larger task groups with broad mixes of people, which means that a wide array of tasks can be undertaken from camps. However, they can and do make major residential moves, sometimes for only short periods. On the Northwest Coast, the distances for these moves can be short or rather long (Table 2.3). These moves were usually done seasonally to reposition people near available resources. The 33-km Coast Tsimshian move, for example, is the one-way distance between Prince Rupert Harbor and summer fishing villages on the Skeena River. In this move, entire households and their possessions were freighted to the summer village by the rafts described before. In the fall, they returned with the winter's store of salmon. Such moves were not unusual on the coast. The early nineteenth century NuuChahNulth great chief Maquinna annually moved his entire village lock, stock, and barrel to a fishing locality for a month in midwinter (Jewitt 1967). I described before the lengthy annual moves made by the Cowichan and others. In a system like this, it becomes hard to distinguish a logistical foray from a residential shift involving an entire settlement.

The capacity of boats to transport a range of personnel, men, women, children, and dogs, also suggests that the makeup of even small task-oriented

Table 2.3. Numbers and Distances of Residential Moves by Some Northwest Coast Groups (from Schalk 1978, Table 4-11)

Group	Annual # of Residential Moves	Mean and Range	Mean Distance (km)	Mean Annual Distance (km)
Makah	2	2	4–6	9.2
Gulf Salish	2.2	0–3	22	48
Straits Salish	3	2–4		
Kwakwaka'wakw	2.6	2–4	8.5	22
NuuChahNulth	2.8	2–4	8.5	22
Tsimshian	3	3–4	33	100

camps may differ from, and the range of activities there may be much broader than those of pedestrian collectors. On the Northwest Coast, for example, it is notoriously difficult to distinguish village from nonvillage sites in the absence of architectural evidence. The potential availability of large parties will also affect decisions about overland transport and processing. The implications of boats, however, go beyond the nature of residential and task group sites.

One of the factors sometimes thought of as leading to increased residential stability (sedentism) is tethering. Hunter-gatherers in arid climates, for example, may be tethered, tied to, water sources. In a similar way, people can be thought of as tethered by winter stores, both by their bulk and by the facilities required to store them. The availability of large boats, however, would either ease or eliminate this limitation.

The ability to move bulk goods also makes available some very unlikely places for villages. For example, in Prince Rupert Harbor, British Columbia, villages were located on three small islands. One of these, Garden Island, has been extensively excavated (Ames 1976, n.d.). The island is located in shallow salt water that probably supported a kelp forest and is easily defended, but it has no fresh water. Garden Island was regularly occupied after about 1400 B.C., and was a village after 1000 A.D., if not much earlier. Rainwater might have been captured in watertight boxes, but it seems more likely that water was ferried in. The site is not far from a small stream, but fresh-water would have to have come in by boat. This must be very unusual, but it does indicate what is possible with boats.

TRADE, INTERACTION, AND REGIONAL ECONOMIES

Seaworthy boats permit the movement of people and goods, including very bulky items, over long distances. I will briefly explore two implications of this. The first of these is biological. A central problem for any human group, particularly one that is thinly scattered, is finding appropriate mates (e.g., Wobst 1974). Boats can facilitate the formation of potentially far-flung mating networks [and access to distant resources (e.g., Kelly 1995)], ensuring demographic stability and even growth under conditions that would otherwise lead to population declines. This may be in part what underlies Batten's observation that cities on rivers and the ocean were demographically more stable than those in the interior. The formation of far-flung networks would also be part and parcel of the formation of wide social and exchange networks.

On the Northwest Coast, regional systems of exchange seem to have begun evolving as early as 10,000 years ago, and recognizable interaction spheres may have been present as early as 6000 B.C. (Ames and Maschner

1999). In contrast, on the Columbia Plateau in the interior, interaction spheres may not have formed until after 1000 B.C. (Erickson 1990; Galm 1994). Items exchanged on the Plateau were all small and readily transported by foot (Hayden and Schulting 1997). Although such things were traded along the coast, large, bulky, and awkward items were also traded. Boats were not only central in the evolution of regional systems; they were no doubt equally central to the development of specialized production of goods for trade. I have already noted regional specialization and trade of canoe hulls by the Haida and NuuChahNulth. The trade in Haida canoes was part of what is the best example of the role of canoes in this context on the Northwest Coast—the eulachon trade at the mouth of the Nass River.

Eulachon (*Thaleichthys pacificus*) are a particularly oily smelt. They are anadramous, like salmon, but spawn near the mouths of rivers. The major eulachon run on the northern Northwest Coast is the late February, early March run at the mouth of the Nass River. The eulachon arrive in vast numbers at a time when winter stores are low, and what is left is old, dried fish. The smelt can be the first fresh food of the season in this area. Their oil is highly prized. It was a key part of the regional diet and was traded widely across the northern Northwest Coast. It was also carried by porters well into the interior along grease trails defended by small fortresses (MacDonald 1984). The oil was a prestige good (e.g., Oberg 1973) that was served at feasts and burned at potlatches. The Coast Tsimshian controlled the eulachon trade because they owned the spawning grounds. During the run, groups came from all over the northern coast and traded for the oil. They traded slaves, copper, furs, hides, ocher, dentalium, and halibut for it. As they traded for the oil, they also traded among themselves. The Haida brought their freight and war canoes to exchange. All groups returned home with canoes loaded to the gunnels with oil stored in large watertight boxes. To return home, the Haida crossed the stormy waters of Dixon entrance in their largest freight canoes.

SUMMARY AND CONCLUSIONS

In 1980, Binford defined collectors and foragers entirely in terms of terrestrial mobility. A decade later, he amplified his discussion of collectors and specifically argued for the importance of transport technology in the evolution of collector strategies (Binford 1990). He was particularly concerned in that paper with looking at the relationship between house form and transportation. In this paper, I have examined several related questions, including the role of boats in shaping mobility and bulk processing decisions and the potential relationship between waterborne transport, higher population densities, and greater demographic stability among aquatic hunter-gatherers. I was also interested in the degree to which the archaeological record of aquatic hunter-gatherers can be expected to be similar to or differ from that of terrestrial collectors, and ultimately whether the collector–forager continuum is even relevant in this case.

The evidence developed here indicates that aquatic hunter-gatherers do have higher median and mean population densities than those of their terrestrial numbers. I was unable to explore whether population densities among aquatic hunter-gatherers are more stable through time than those of terrestrial hunter-gatherers, although it seems likely, given the large mating networks that boats potentially facilitate.

It appears that the hinterlands, or foraging areas, of aquatic and terrestrial hunter-gatherers can be the same size, even when the former includes extensive bodies of water. I confess that this result was something of a surprise. I was expecting consistently larger foraging areas for aquatic hunter-gatherers. For day-to-day production decisions, this means that the capacity of boats to allow more time spent foraging, multiple daily trips, and to haul large and bulky loads is crucial. One consequence of this capacity is that bulk processing is concentrated at residential sites, rather than in the field.

A further implication is that aquatic hunter-gatherers may create few or no task specific sites and that residential sites will contain evidence for what may appear to be intensive processing, but which may simply be localized processing. There may also be evidence for an even greater array of activities than is commonly associated with collector settlements.

For example, Aikens et al. (1986) argued that Jomon settlement patterns should reflect increasing levels of logistical organization, as population growth during the Jomon led to increased settlement packing. Bleed (1992) tested this expectation against the artifact contents of the Yagi site, an Early Jomon settlement in southwest Hokkaido, the northernmost island of Japan. He concluded that the diversity of the stone tool assemblage reflected a diverse range of activities and subsistence pursuits, including fishing and marine mammal hunting (Crawford and Bleed 1998). He also suggested that if these people were collectors, they were an odd sort of collector, because there was no evidence of task-oriented sites in the area. He argued that the site's inhabitants were able to acquire all of their food within the standard foraging radius. Bleed is probably correct in this. His characterization of the Yagi occupation nicely fits the expectations for aquatic hunter-gatherers developed in this chapter; most unprocessed foods were brought into the residential site and processed there, and most activities were based there. However, it may be that task specific sites do exist and that they are some distance from Yagi, or that they are indistinguishable from small residential sites, or that they are other residential sites.

There is another methodological issue beyond the obvious one of distinguishing logistical moves from residential moves. Thomas (1989) showed that in the archaeological literature of the Great Basin of western North America, the distinction between a residential site and a task site is often a consequence of how much of a site is sampled, rather than any intrinsic qualities of the sites themselves. A big sample yields a residential site, a small sample a task-specific site. Lyman (1991) shows that for the Oregon coast, volumetric samples in shell midden excavations of $100\,m^3$ are necessary to ensure recovery of architectural features such as postholes—or to ensure that the absence of these features is not sampling error. Thus sampling issues may compound what appears to be the inherent ambiguity of the sites that aquatic hunter-gatherers may produce.

The final questions to be addressed are whether the collector–forager continuum is even relevant to boat-using peoples and whether aquatic hunter-gatherers are different in kind from terrestrial hunter-gatherers, at least in ways that are theoretically important. In one sense, they are collectors, because they move resources to people. However, they may not generate archaeological records easily accommodated to expectations based on the distinction. Further, they tend to have population densities higher than we might otherwise expect. Using the distinction, then, may obscure variability in the record.

Boats are theoretically important. This importance rests, I think, on the capacity to move and process large amounts of resources even across small distances, thus easing potential problems in intensification of production and simultaneously opening possibilities for intensification which would otherwise not be economical. The eulachon described before is a very small-bodied fish that travels (or traveled) in vast schools. Herring is another such fish. To be effectively exploited, they must be taken in vast numbers with small gauge nets or some other device. Such nets are costly in time and labor (Lindstrom 1996). Moseley (Moseley 1975; Moseley and Feldman 1988) argues that the initial basis for the rise of Andean civilization was fishing for small-bodied schooling fish off the coast of Peru. In any case, these are resources that are really useful only if harvested in very large numbers; large numbers require the capacity to move them once caught or once processed. Presumably, the peoples on the coast of Peru transported the dried fish inland by porter. But boats are more effective.

Does the forager–collector continuum even apply? A yes or no answer does not really matter at this point. What matters is that the distinction has given archaeologists an exceptionally productive means for exploring the variation in hunter-gatherer mobility strategies. By learning that it does not fit and why, we learn a great deal.

ACKNOWLEDGMENTS

I thank Ben Fitzhugh and Junko Habu for the invitation and opportunity to write this chapter, which has been percolating in the back of my brain for several years. I also thank T. Douglas Price and Peter Rowley-Conwy who provided assistance to the first and far more ambitious version of the chapter and encouragement to pursue the topic. The final version of this chapter benefited from the comments of Virginia Butler, who read it with her usual thoughtfulness and rigor. Finally, I thank the interlibrary loan department of Portland State University's Branford Price Millar Library, without whose timely assistance, the chapter could never have been written. All errors, of course, are mine.

REFERENCES

Aikens, C. M., Ames, K. M., and Sanger, D., 1986, Affluent Collectors at the Edges of Eurasia and North America: Some Comparisons and Observations on the Evolution of Society among North-Temperate Coastal Hunter-Gatherers. In *Prehistoric Hunter-Gatherers in Japan; New Research Methods*, edited by T. Akazawa and C. M. Alkens, pp. 3–26. Bulletin 27, The University Museum, University of Tokyo, Tokyo.

Ames, K. M., 1976, *The Bone Tool Assemblage from the Garden Island Site, Prince Rupert Harbor, British Columbia: An Analysis of Assemblage Variation through Time*. Ph.D. Dissertation, Washington State University.

Ames, K. M., 1996, Chiefly Power and Household Production on the Northwest Coast. In *Foundations of Inequality*, edited by T. D. Price and G. M. Feinman, pp. 155–187. Plenum, New York.

Ames, K. M., n.d., *The North Coast Prehistory Project Excavations in Prince Rupert Harbour, British Columbia: The Artifacts.*

Ames, K. M., and Maschner, H. G. D., 1999, *Peoples of the Northwest Coast: Their Archaeology and Prehistory*. Thames and Hudson, London.

Ames, K. M., Raetz, D. F., Hamilton S. C., and McAfee, C., 1992, Household Archaeology of a Southern Northwest Coast Plank House. *Journal of Field Archaeology* 19:275–290.

Ames, K. M., Smith, C. M., Cornett, W. L., Sobel, E. A., Hamilton, S. C., Wolf J., and Raetz, D., 1999, *Archaeological Investigations at 45CL1 Cathlapotle (1991–1996): Ridgefield National Wildlife Refuge, Clark County, Washington; A Preliminary Report*. Cultural Resources Series Number 13. U.S. Department of Interior, Fish and Wildlife Service, Region 1, Portland.

Arnold, J. E., 1995 Transportation Innovation and Social Complexity among Maritime Hunter-Gatherer Societies. *American Anthropologist* 97(4):733–747.

Batten, D. C., 1998, Transport and Urban Growth in Preindustrial Europe: Implications for Archaeology. *Human Ecology* 26(3):489–516.

Binford, L. R., 1978, *Nunamuit Ethnoarchaeology*. Academic Press, New York.

Binford, L. R., 1980, Willow Smoke and Dogs' Tails: Hunter-Gatherer Settlement Systems and Archaeological Site Formation. *American Antiquity* 45(1):4–20.

Binford, L. R., 1990, Mobility, Housing, and Environment: A Comparative Study. *Journal of Anthropological Research* 46(2):119–152.

Bleed, P., 1992, Ready for Anything: Technological Adaptations to Ecological Diversity at Yagi, an Early Jomon Community in Southwestern Hokkaido, Japan. In *Pacific Northeast Asia in Prehistory: Hunter-Fisher-Gatherers, Farmers and Sociopolitical Elites*, edited by C. M. Aikens and S. N. Rhee, pp. 47–52. Washington State University Press, Pullman.

Boas, F., 1909, *Ethnology of the Kwakiutl*. Memoirs of the American Museum of Natural History, Vol. 8, pt. 2, pp 111–126.

Cannon, A. 1991, *The Economic Prehistory of the Namu*. Archaeology Press, Simon Fraser University, Burnaby.

Castile, G. P. (ed.), 1985, *The Indians of Puget Sound: The Notebooks of Myron Eells*. University of Washington Press, Seattle.

Chapelle, H. I., 1994, Chapter Seven: Arctic Skin Boats. In *The Bark Canoes and Skin Boats of North America*, edited by E. T. Adney and H. I. Chapelle. Smithsonian Institution, Washington DC.

Crawford, G. W., and Bleed, P., 1998, Scheduling and Sedentism in the Prehistory of Northern Japan. In *Seasonality and Sedentism: Archaeological Perspectives from Old and New World Sites*, edited by T. R. Rocek and O. Bar-Yosef, pp. 109–128. Peabody Museum Bulletin 6, Peabody Museum of Archaeology and Anthropology, Harvard University, Cambridge.

Croes, D. R., 1995, *The Hoko River Archaeological Complex: the Wet/Dry Site (45CA213), 3,000–1,700 B.P.* Washington State University Press, Pullman.

Croes, D. R., and Hackenberger, S., 1988, Hoko River Archaeological Complex: Modeling Prehistoric Northwest Coast Economic Evolution. *Research in Economic Anthropology* supplement 3:19–86.

Damas, D. (ed.), 1984, *Handbook of North American Indians, Vol. 5, The Arctic*. Smithsonian Institution, Washington DC.

Darby, M. C., 1996, *Wapato for the People: An Ecological Approach to Understanding the Native American Use of Sagittaria latifolia on the Lower Columbia*, M.A. Portland State University.

Drucker, P., 1951, *The Northern and Central Nootkan Tribes*. Bureau of American Ethnology Bulletin 144, Smithsonian Institution, Washington DC.

Duer, D., 1999, Salmon, Sedentism and Cultivation: Toward an Environmental Prehistory of the Northwest Coast. In *Northwest Lands, Northwest Peoples: Readings in Environmental History*, edited by D. D. Goble and P. W. Hirt, pp. 129–158. University of Washington Press, Seattle.

Durham, B., 1960, *Canoes and Kayaks of Western America*. Copper Canoe Press, Seattle.

Emmons, G. T., 1991, *The Tlingit Indians (edited with additions by F. de Laguna)*. University of Washington Press, Seattle.

Erickson, K., 1990, Marine Shell in the Plateau Culture Area. *Northwest Research Notes* 24(1): 91–144.

Erlandson, J. M., 1988, The Role of Shellfish in Coastal Economies. *American Antiquity* 53(1):102–109.

Erlandson, J. M., 1994, *Early Hunter-Gatherers of the California Coast*. Plenum, New York.

Galm, J. R., 1994, Prehistoric Trade and Exchange in the Interior Plateau of Northwestern North America. In *Prehistoric Exchange Systems in North America*, edited by T. G. Baugh and J. E. Ericson, pp. 275–305. Plenum, New York.

Grant, C., 1978, Island Chumash. In *Handbook of North American Indians, Vol. 8, California*, edited by R. F. Heizer, pp. 524–529. Smithsonian Institution, Washington DC.

Halpin, M., and Sequin, M., 1990, Tsimshian Peoples: Southern Tsimshian, Coast Tsimshian, Nisga, and Giksan. In *Handbook of North American Indians, Vol. 7: The Northwest Coast*, edited by W. Suttles, pp. 267–284. The Smithsonian Institution, Washington DC.

Hamilton, S. C., 1994, *Technological Organization and Sedentism: Expedient Core Reduction, Stockpiling and Tool Curation at the Meier Site (35CO5)*, M.A. Portland State University.

Hayden, B., and Schulting, R., 1997, The Plateau Interaction Sphere and Late Prehistoric Cultural Complexity. *American Antiquity* 62(1):51–85.

Heizer, R. F. (ed.), 1978, *Handbook of North American Indians, Vol. 8, California*. Smithsonian Institution, Washington DC.

Helm, J. (ed.), 1981, *Handbook of North American Indians, Vol. 6, Subarctic*. Smithsonian Institution, Washington DC.

Hildebrandt, W. L., and Jones, T. L., 1992, Evolution of Marine Mammal Hunting: A View from the California and Oregon Coasts. *Journal of Anthropological Archaeology* 11(4): 360–401.

Holm, B., 1994, Canoes of the Northwest Coast. In *Anthropology of the North Pacific Rim*, edited by W. W. Fitzhugh and V. Chaussonnet, pp. 259–264. Smithsonian Institution Press, Washington DC.

Huelsbeck, D. R., 1994, Mammals and Fish in the Subsistence Economy of Ozette. In *Ozette Archaeological Project Research Reports, Vol. II, Fauna*, edited by S. R. Samuels, pp. 17–92. Reports of Investigations 66. Department of Anthropology, Washington State University, Pullman.

Jewitt, J. R., 1967, *Narrative of the Adventures and Sufferings of John R. Jewitt (Reprint)*. The Galleon Press, Fairfield.

Jochelson, W., 1908, *The Koryak*. Memoir of the American Museum of Natural History 10, New York.

Jones, K. T., and Madsen, D. B., 1989, Calculating the Cost of Resource Transportation: A Great Basin Example. *Current Anthropology* 30(4):529–534.

Jones, R. F. (ed.), 1999, *Annals of Astoria: The Headquarters Log of the Pacific Fur Company on the Columbia River, 1811–1813*. Fordham University Press, New York.

Jones, T. L., and Hildebrandt, W. R., 1995, Reasserting a Prehistoric Tragedy of the Commons: Reply to Lyman. *Journal of Anthropological Archaeology* 14(2):78–98.

Keeley, L. H., 1988, Hunter-Gatherer Economic Complexity and "Population Pressure": A Cross-Cultural Analysis. *Journal of Anthropological Archaeology* 7(4):373–411.

Kelly, R. L., 1995, *The Foraging Spectrum: Diversity in Hunter-Gatherer Lifeways*. Smithsonian Institution Press, Washington DC.

Kelly, R. L., 1996, Foraging and Fishing. In *Great Basin Fisherfolk: Optimal Diet Breadth Modeling the Truckee River Aboriginal Subsistence Fishery*, edited by M. G. Plew, pp. 208–214. Boise State University, Boise.

Kelly, R. L., 1998, Foraging and Sedentism. In *Seasonality and Sedentism: Archaeological Perspectives from Old and New World Sites*, edited by T. R. Rocek and O. Bar-Yosef, pp. 9–24. Peabody Museum Bulletin 6, Peabody Museum of Archaeology and Ethnology, Cambridge.

Laguna, F. de, 1992, *Under Mount Saint Elias: The History and Culture of the Yakutat Tlingit*. Smithsonian Contributions to Anthropology Vol. 7, Washington DC.

Lindstrom, S., 1996, Great Basin Fisherfolk: Optimal Diet Breadth Modeling the Truckee River Aboriginal Subsistence Fishery. In *Prehistoric Hunter-Gatherer Fishing Strategies*, edited by M. G. Plew, pp. 114–179. Boise State University, Boise.

Lyman, R. L., 1991, *Prehistory of the Oregon Coast*. Academic Press, New York.

Lyman, R. L., 1994, *Mammalian Zooarchaeology of the Meier Site (35CO5)*. Portland State University. Report on file, Portland State University.

Lyman, R. L., 1995, On the Evolution of Marine Mammal Hunting on the West Coast of North America. *Journal of Anthropological Archaeology* 14(1):45–77.

MacDonald, G. F., 1984, The Epic of the Nekt: The Archaeology of Metaphor. In *The Tsimshian: Images of the Past Views of the Present*, edited by M. Seguin, pp. 65–81. University of British Columbia Press, Vancouver.

Metcalf, D., and Jones, K. T., 1988, A Reconsideration of Animal Body-Part Utility Indices. *American Antiquity* 53(3):486–503.

Mitchell, D.H., 1971, Archaeology of the Gulf of Georgia area, A Natural Region and Its Cultural Type. *Syesis* 4: Supplement 1.

Moseley, M. E., 1975, *The Maritime Foundations of Andean Civilization*. Cummings, Menlo Park.

Moseley, M. E., and Feldman, R. A., 1988, Fishing, Farming and the Foundations of Andean Civilization. In *The Archaeology of Prehistoric Coastlines*, edited by G. Bailey and J. Parkington, pp. 125–134. University of Cambridge Press, Cambridge.

Moulton, G. E. (ed.), 1990, *The Journals of the Lewis & Clark Expedition, Vol. 6, November 2, 1805–March 22, 1806*. University of Nebraska Press, Lincoln.

Moulton, G. F. (ed.), 1991, *The Journals of the Lewis & Clark Expedition, Vol. 7, March 23–June 9, 1806*. University of Nebraska Press, Lincoln.

Nelson, E. W., 1899, *The Eskimo about Bering Strait*. Eighteenth Annual Report of the Bureau of American Ethnology, 1896–1897. Washington DC.

Oberg, K., 1973, *The Social Economy of the Tlingit Indians*. University of Washington Press, Seattle.

O'Connell, J. F., Hawkes, K., and Jones, N. B., 1990, Reanalysis of Large Mammal Body Part Transport among the Hadza. *Journal of Archaeological Science* 17(3):301–316.

Olson, R. L., 1927, *Adze, Canoe, and House Types of the Northwest Coast (1967 reprint)*. University of Washington Press, Seattle.

Osborne, A. J., 1977, Strandloopers, Mermaids and Other Fairy Tails: Ecological Determinants of Marine Resource Utilization. In *For Theory Building in Archaeology*, edited by L. R. Binford, Academic Press, New York.

Peacock, S. L., 1998, *Putting Down Roots: The Emergence of Wild Plant Food Production on the Canadian Plateau*. University of Victoria.

Perlman, S., 1980, An Optimum Diet Model, Coastal Variability, and Hunter-Gatherer Behavior. *Advances in Archaeological Method and Theory* 3:256–310.

Raymond, S., 1981, The Maritime Foundation of Andean Civilization: A Reconsideration of the Evidence. *American Antiquity* 46(4).

Renouf, M. A. P., 1988, Sedentary Coastal Hunter-Gatherers: An Example from the Younger Stone Age of Northern Norway. In *The Archaeology of Prehistoric Coastlines*, edited by G. Bailey and J. Parkington, pp. 102–116. Cambridge University Press, Cambridge.

Rhodes, D., 1990, On Transportation Costs of Great Basin Resources: An Assessment of the Jones–Madsen Model. *Current Anthropology* 31(4):413–418.

Rousselot, J.-L., 1994, Watercraft in the North Pacific: A Comparative View. In *Anthropology of the North Pacific Rim*, edited by W. W. Fitzhugh and V. Chaussonnett, pp. 243–258. Smithsonian Institution Press, Washington DC.

Schalk, R. F., 1978, *Foragers of the Northwest Coast of North America: The Ecology of Aboriginal Land Use Systems*. Ph.D. Thesis, University of New Mexico.

Schalk, R. F., 1981, Land Use and Organizational Complexity among Foragers of Northwestern North America. In *Affluent Foragers: Pacific Coasts East and West*, edited by S. Koyama and D. H. Thomas, pp. 53–76. Senri Ethnological Studies, No. 9. National Museum of Ethnology, Osaka.

Shackleton, P., and Roberts, K. G., 1983, *The Canoe: A History of the Craft from Panama to the Arctic*. Macmillan of Canada, Toronto.

Smith, C., 1992, *Late Stone Age Hunters of the British Isles*. Routlege, London.

Sproat, G. M., 1987, *The Nootka, Scenes and Studies of Savage Life (Reprint of Scenes and Studies of Savage Life, 1868, published by Smith, Elder: London)*, Sono Nis Press, Victoria.

Stenholm, N. A., 1995, *Botanical Analysis of Cathlapotle Floral Samples*. Report on file, Portland State University.

Stewart, F. L., 1977, *Vertebrate Faunal Remains from the Boardwalk Site (GbTo 31) of Northern British Columbia*. Archaeological Survey of Canada, Manuscript No. 1263, ASC Archives.

Stewart, F. L., and Stewart, K., 1996, The Boardwalk and Grassy Bay Sites: Patterns of Seasonality and Subsistence on the Northern Northwest Coast, B.C. *Canadian Journal of Archaeology* 20:39–60.

Suttles, W. (ed.), 1990a, *Handbook of North American Indians, Vol. 7: The Northwest Coast*. The Smithsonian Institution, Washington DC.

Suttles, W., 1990b, Introduction. In *Handbook of North American Indians, Vol. 7: The Northwest Coast*, edited by W. Suttles, pp. 1–15. The Smithsonian Institution, Washington DC.

Swan, J. G., 1857, *The Northwest Coast (1967 reprint, University of Washington Press, Seattle)*. Harper & Brothers, New York.

Thomas, D. H., 1989, Diversity in Hunter-Gatherer Cultural Geography. In *Quantifying Diversity in Archaeology*, edited by R. T. Leonard and G. T. Jones, pp. 85–91. Cambridge University Press, Cambridge.

Thoms, A. V., 1989, *The Northern Roots of Hunter-Gatherer Intensification: Camas and the Pacific Northwest*, Ph.D. Dissertation, Washington State University.

Thoms, A. V., and Burtchard, G. C., 1986, *Calispell Valley Archaeological Project: Interin Report for 1984 and 1985 Field Seasons*. Contributions in Cultural Resource Management 10, Washington State University, Pullman.

Walker, D. E. (ed.), 1998, *Handbook of North American Indians, Vol. 12, Plateau*. Smithsonian Institution, Washington DC.

Waterman, T. T., and Coffin, G., 1920, *Types of Canoes on Puget Sound*. Indian Notes and Monographs, Museum of the American Indian, Heye Foundation, New York.

Wobst, M., 1974, Boundary Conditions for Paleolithic Social Systems: A Simulation Approach. *American Antiquity* 39:147–179.

Woodburn, J., 1980, Hunter-Gatherers Today and Reconstructing the Past. In *Soviet and Western Anthropology*, edited by E. Gellner, pp. 94–118. Columbia University Press, New York.

Yesner, D. R., 1980, Maritime Hunter-Gatherers: Ecology and Prehistory. *Current Anthropology* 21(6):727–750.

Chapter 3

Jomon Collectors and Foragers

Regional Interactions and Long-term Changes in Settlement Systems among Prehistoric Hunter-Gatherers in Japan

JUNKO HABU

INTRODUCTION

The purpose of this chapter is to expand the utility of the forager/collector model (Binford 1980, 1982) by examining the dynamics of long-term system change on an interregional scale. Among numerous models of hunter-gatherer behavior, Binford's (1980, 1982) forager–collector continuum has been one of the most frequently cited models of subsistence and settlement organization during the past two decades. As with most formal models of subsistence and settlement (such as optimal foraging models), the forager/collector model assumes that economic rationality is the basic principle that determines hunter-gatherer subsistence strategies and residential mobility. However, unlike optimization models, which are deductive and formal in their structure, Binford's model was inspired by ethnographic examples. Because of its informal and inductive origins, the

JUNKO HABU • Department of Anthropology, University of California, Berkeley, California 94720

forager/collector model is flexible enough to account for various anomalies. The concept of serial foragers (Binford 1980: 16–17), which refers to cold-environment hunter-gatherers who adopt "mapping-on" strategies to position themselves so that they can exploit seasonally fluctuating resources, is a good example of this. Similarly, various nonenvironmental factors that can influence hunter-gatherer subsistence-settlement practice, such as population pressure, trade/exchange, and group alliance, are not necessarily ignored by the model. We may need to modify the model to incorporate these factors as part of the system, but the core of the model that outlines the basic principles of labor investment versus return with regard to resource distribution, subsistence strategies, and residential mobility can still be operational. Thus, despite some scholars' critical views (e.g., Wiessner 1982), the forager/collector model remains useful.

This does not imply, however, that the model does not need expansion or elaboration. Issues that have not been fully addressed since the original appearance of the model include the causes, mechanisms, and consequences of long-term system change in the context of regional interactions between groups. Though Binford (1983) was apparently interested in long-term hunter-gatherer behavior, what he was able to infer from Nunamiut ethnographic data and oral history was temporally and spatially limited. As a result, we have very little knowledge about the way changes in one system might affect neighboring systems, and what the long-term effects of these changes might be during the course of several hundred years.

In this regard, archaeological studies of prehistoric Jomon hunter-gatherers on the Japanese Archipelago provide us with an excellent opportunity to examine long-term settlement pattern change at the interregional level. In Japan, a large number of large-scale salvage excavations took place during and after the 1960s following the implementation of the national government's land development policy (Habu 1989). Tens of thousands of Jomon sites have been excavated with systematic financial support from national, prefectural, and municipal governments, as well as private developers. In many cases, the results of these excavations are available as published reports. As a result, we have hundreds, and sometimes even thousands, of excavated sites from each of the Jomon subperiods, which can be used to examine the course of long-term change in subsistence and settlement practice.

This chapter examines long-term change in regional settlement patterns from the Early to the Middle Jomon periods (ca. 6100–4000 uncalibrated b.p.; for the rest of this paper, lower-case b.p. will be used for uncalibrated ^{14}C dates) in the context of the forager/collector model (Binford 1980, 1982). Through this case study, possible factors that triggered changes between collecting and foraging systems are inferred, as are their mechanisms. The implications of these changes are discussed in relation to the development of hunter-gatherer cultural complexity.

BACKGROUND TO THE STUDY

Jomon is the name of a prehistoric culture in the Japanese Archipelago that followed the Palaeolithic period and preceded the agricultural Yayoi period. Unlike many other prehistoric hunter-gatherer cultures, the Jomon culture is characterized by the production and use of pottery. The Jomon period is conventionally divided into six subperiods: Incipient, Initial, Early, Middle, Late, and Final. The appearance of pottery (ca. 13,000 b.p.) marks the beginning of the Jomon period (Nakamura and Tsuji 1999; Taniguchi 1999), but not all of the characteristics that researchers commonly associate with the Jomon culture were present during the Incipient and Initial Jomon periods. By the Early Jomon, however, a distinctive set of cultural traits that characterize the rest of the Jomon period began to emerge. These include the presence of large settlements, various kinds of ceremonial features and artifacts, food storage, and long-distance trade. In this regard, the Early to Final Jomon cultures share a number of characteristics with so-called "complex" hunter-gatherers in various parts of the world (Price and Brown 1985). For this reason, researchers in the broader field of hunter-gatherer archaeology have been interested in the study of the Jomon culture (Aikens 1981; Aikens and Dumond 1986; Aiken et al. 1986; Cohen 1981; Hayden 1990; Pearson 1977; Price 1981; Price and Brown 1985; Soffer 1989).

Recent developments in Jomon studies have revealed that regional and temporal variability within the Jomon culture was far greater than scholars once assumed (Ikawa-Smith 1998). For example, at the Initial Jomon Uenohara site in Kagoshima Prefecture in southern Kyushu, sophisticated pottery, such as jars with long necks, and ornaments, such as clay earrings, were recovered (Okamura 1995). Neither of these two types of artifacts had been reported from Initial Jomon sites in other parts of Japan. Other lines of evidence, such as feature types and lithic assemblage characteristics, also indicate that the Incipient and Initial Jomon cultures in this region were quite different from those in the rest of the Japanese Archipelago (Amemiya 1999; Shinto 1995, 1999). At the Early and Middle Jomon Sannai Maruyama site in Aomori Prefecture, northern Japan, an extraordinarily large settlement associated with more than 700 pit-dwellings has been recorded (Habu et al. 2001; Kidder 1998; Okada 1995a,b; Okada and Habu 1995). Radiocarbon dates indicate that the site was occupied from approximately 5050 to 3900 b.p., or 5900 to 4300 calibrated B.P. (Tsuji 1999; see also M. Imamura 1999). Because the site was occupied for more than 1500 years,

Figure 3.1. Map of Japan showing the location of the Kanto and Chubu Mountain regions.

it is unlikely that all of the 700 pit-dwellings were inhabited contemporaneously. Nevertheless, the number of pit-dwellings and other features from the site is unusually large compared to other Jomon sites.

Among these regional varieties, the Middle Jomon culture in the Kanto region (eastern side of central Honshu Island including Tokyo) and the adjacent Chubu Mountain region (inland part of central Honshu; Fig. 3.1) have attracted considerable attention because of the extremely high site density, large site size, and the complex pottery decoration in these sites. According to K. Imamura (1996: 93), 70% of all excavated Jomon pit-dwellings in the Kanto and Chubu Mountain regions belong to the Middle Jomon period, and 50% of all excavated pit-dwellings belong to the latter half of this period. Many researchers have also pointed out that the Middle

Jomon sites in these regions are characterized by an abundance of so-called "chipped stone axes." Although these stone tools are called "axes," most researchers believe that they were used as hoes for digging, either to collect wild plant roots (Watanabe 1976) or possibly for incipient plant cultivation (e.g., Fujimori 1950; Oyama 1927, 1934).

Archaeologists have long recognized the prosperity of the Middle Jomon culture in the Kanto and Chubu Mountain regions, but very few studies have systematically examined the processes of its development. Because of the presence of large settlements during and after the Middle Jomon period in these regions, scholars have assumed that the development of sedentary life in these regions began in the Early Jomon period and that the degree of Jomon cultural complexity increased gradually and smoothly from the Early to the Middle Jomon (e.g., Wajima 1948, 1958). However, detailed examination of changes in regional settlement patterns at the end of the Early Jomon period indicates that the long-term development from the Early to the Middle Jomon may have been more complex than previously assumed.

SETTLEMENT PATTERNS OF THE EARLY JOMON MOROISO PHASE

In a previous study, I examined regional settlement patterns of the Early Jomon Moroiso phase (ca. 5000 b.p.) of the Kanto and Chubu Mountain regions (Habu 1996, 2001) from the perspective of the forager/collector model (Binford 1980, 1982). The Moroiso phase is the second to the last phase of the Early Jomon period and is divided into three subphases: Moroiso-a, -b and -c from the earliest to the latest. The duration of the Moroiso phase is estimated to have been approximately 200–300 years (Habu 2001).

According to the forager/collector model, subsistence-settlement systems of hunter-gatherers can be classified into two basic systems: (1) forager systems that are characterized by high residential mobility and (2) collector systems that are characterized by low residential mobility. Foragers tend to acquire food on a day-to-day basis near their residential base, whereas collectors tend to organize their subsistence activities logistically (i.e., they send specialized task groups to acquire food resources located far away from their residential base, who then bring these resources back and store them). Forager systems are commonly found in environments in which the distribution of critical resources is seasonally and spatially homogeneous, whereas collector systems are adapted to environments in which the distribution of critical resources is seasonally or spatially uneven (for a description of the forager/collector model, see also Kelly 1995).

Table 3.1. Expected Patterns of Residentially Used Sites

Type	Intersite Variability in Lithic Assemblages	Intersite Variability in Site Size	Site Distribution Pattern
Fully sedentary collectors	Small	Small	Clustered
Collectors with seasonal moves	Large	Large	Clustered
Foragers	Small	Small	Dispersed

In my previous study (Habu 2001), I pointed out that collector systems would include fully sedentary hunter-gatherers who occupied a single residential base throughout the year, as well as relatively sedentary hunter-gatherers who moved their residential bases seasonally. Because the presence of large residential sites is characteristic of collector systems, the Jomon people in general were probably closer to the collector end of the forager–collector spectrum. However, the issue of whether they moved their residential bases seasonally has yet to be determined.

To address this issue, I suggested that examining intersite variability in lithic assemblages and site size could prove useful (Habu 1996, 2001; Table 3.1). Relatively sedentary collectors, who move their residential bases seasonally, would have used each of their residential bases for seasonally different subsistence activities. Accordingly, the lithic assemblages from each residential base should reflect these seasonal differences. Therefore, we can expect large variability in the lithic assemblages among residential bases. Furthermore, the seasonal movement of relatively sedentary collectors is often associated with the dispersion and amalgamation of residential groups. Therefore, we can also expect large variability in the size of residential bases. Accordingly, if the Jomon people were collectors who moved their residential bases seasonally, we can expect to find large intersite variability in both (1) lithic assemblages and (2) site size. In contrast, in the case of fully sedentary collectors, we can expect that lithic assemblage and site size variability among residential bases will be relatively small because residential bases of fully sedentary collectors were occupied year-round, thus reflecting generally similar activities. For both kinds of collectors, we would expect to find clusters of residential bases because collectors' residential bases are usually located near primary resource concentrations. Finally, in a forager system, lithic assemblage and site size variability among residential bases should be small because functions of residential bases are relatively similar to each other. This means that, in variability among residential bases, foragers and fully sedentary collectors may show similar characteristics. However, it is expected that foragers' residential bases would be much

Table 3.2. Summary of Subsistence-Settlement Systems in Each Region[a]

Region	Moroiso-a subphase (Oldest)	Moroiso-b subphase	Moroiso-c subphase (Youngest)
Northwestern Kanto region	Relatively sedentary collectors	Relatively sedentary collectors	Relatively sedentary collectors
Southwestern Kanto region	Relatively sedentary collectors	Relatively sedentary collectors	Foragers
Chubu Mountain region	Unknown (sample size was too small)	Relatively sedentary collectors	Relatively sedentary collectors

[a]From Habu 2001.

smaller than those of collectors, possibly with no permanent dwelling structures. Furthermore, foragers' residential bases would be dispersed throughout the research area, because the distribution of resources, which is closely related to the location of residential bases, is dispersed evenly.

In my previous work (Habu 2001), I applied this model to settlement pattern data from the Early Jomon Moroiso phase in the Kanto and Chubu Mountain regions of Japan. The results of this analysis (Table 3.2) suggested that the settlement patterns of the Moroiso Phase generally correspond well to the model of relatively sedentary collectors. The only apparent exception was that of the Moroiso-c subphase in the southwestern Kanto region (Fig. 3.2), which was characterized by a scarcity of dwelling sites (i.e., sites with pit-dwellings). Of the 278 sites identified from this subphase, only five sites (1.8%) are dwelling sites. However, the presence of a number of nondwelling sites (i.e., sites that have no pit-dwellings) indicates that the region was still actively used by Jomon hunter-gatherers. These nondwelling sites were interpreted as residential bases in a forager system. The fact that these nondwelling sites are dispersed throughout the southwestern Kanto region also supports the hypothesis that this was a forager system. None of the settlement patterns examined corresponded to the model of fully sedentary collectors. Based on these results, I suggested that the people of the Moroiso phase did not stay at the same site throughout the year and that the degree of sedentism in southwestern Kanto decreased through time as the subsistence-settlement systems in the region changed from relatively sedentary collector systems to forager systems (Habu 2001).

LONG-TERM CHANGES IN THE SOUTHWESTERN KANTO REGION

The primary purpose of the analysis described above was to examine the degree of sedentism of the Moroiso phase people, but the result of the analysis also provides insights in terms of long-term change in

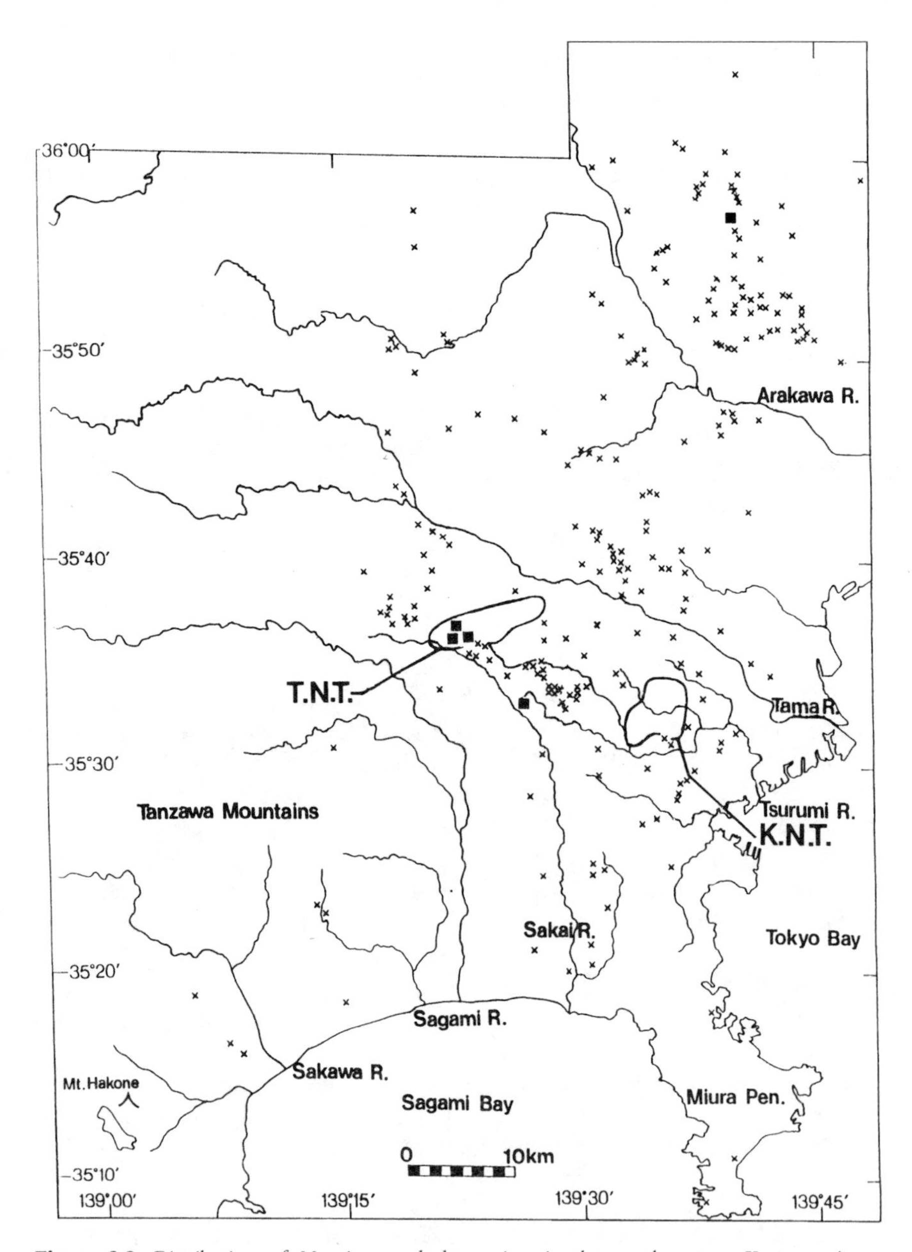

Figure 3.2. Distribution of Moroiso-c subphase sites in the southwestern Kanto region. ■: dwelling sites. x: nondwelling sites. In addition to the sites represented in the diagram, there are 37 nondwelling sites in the Tama New Town area (indicated as T.N.T.) and five nondwelling sites in the Kohoku New Town area (indicated as K.N.T.), respectively (modified from Habu 2001).

subsistence-settlement systems. First, the shift in southwestern Kanto from relatively sedentary collectors to foragers indicate that the degree of sedentism of hunter-gatherers in this region decreased through time. This shift contradicts the general expectation that increasing sedentism through time is an important indicator of cultural development.

One possible explanation for this shift is the change in the natural environment. The Moroiso phase roughly coincides with the "Climatic Optimum," the time of maximum sea level transgression (Matsushima 1979; Matsushima and Koike 1979). The sea level reached its maximum height during or just before the Moroiso-a subphase and then gradually retreated through the rest of the Moroiso phase (Fuji 1984; Sakamoto and Nakamura 1991; but see Horiguchi 1983). Some researchers suggest that the retreat of the sea level resulted in a significant decrease in the habitat of littoral shellfish species (e.g., Matsushima and Koike 1979). As shown in Table 3.3, the frequency of shell-midden sites in southwestern Kanto decreased dramatically from the Moroiso-b to the Moroiso-c subphases. Of all the Moroiso-a sites in southwestern Kanto, 8.4% have shell-middens, whereas only 2.2% of the Moroiso-b sites have shell-middens. No shell-middens have been reported from the Moroiso-c subphase (Habu 2001).

Note here that the difference in the total number of sites among subphases may reflect the difference in the duration of each subphase. In particular, the fact that the total number of sites in the Moroiso-b subphase is twice that for the other two subphases may indicate that the Moroiso-b subphase lasted longer than the other two subphases. Nonetheless, the decrease in the relative proportion of shell-midden sites through time is evident.

It seems that the decrease in the frequency of shell-middens is a direct response to a decline in shellfish resources. If so, the decline may have caused significant changes in the overall resource distribution patterns in southwestern Kanto. More specifically, the decline in the availability of shellfish would have resulted in less densely clumped distributions of available resources because the distribution of shellfish is spatially limited to the coastal area and they are available primarily during the spring to early summer.

Table 3.3. Frequencies of Shell-Midden Sites in Southwestern Kanto

	Moroiso-a	Moroiso-b	Moroiso-c
Number of all sites	273	631	278
Number of shell-midden sites	23	14	0
Percentage of shell-midden sites	8.4	2.2	0.0

By suggesting this, I do not necessarily imply that the people of the Moroiso-a subphase relied primarily on shellfish. As indicated by Suzuki (1979), the caloric value of shellfish is relatively low compared to other kinds of food available within the Japanese Archipelago. Furthermore, many Jomon researchers agree that plant food must have been the staple food source for most of the Jomon people (e.g., Watanabe 1976), with the exception of Hokkaido Jomon (e.g., Minagawa and Akazawa 1992). However, the primary importance of shellfish lies in its seasonality; it is one of the few food items in Japan that are available from the late winter to the spring, when other food resources are scarce (see Erlandson 1988). In this regard, it is possible that changes in the availability of shellfish may have had a significant effect on overall subsistence-settlement systems at that time. More specifically, lack of shellfish resources at the end of the winter may have prevented these people from forming large aggregation settlements for the winter. If this was the case, even though the caloric contribution of shellfish was small in the overall Jomon people's diet, the decrease in shellfish may have had a catastrophic effect on the overall subsistence-settlement systems in the region. Further analysis will be necessary to investigate the interrelationships among environmental change, the disappearance of shell-middens, and the collapse of logistically organized subsistence-settlement systems in this region.

DEVELOPMENT OF A NEW SYSTEM ASSOCIATED WITH AN ABUNDANCE OF CHIPPED STONE AXES

As discussed in the previous section, currently available data indicate that environmental adaptation may account for the shift from collecting strategies to foraging strategies in southwestern Kanto. However, it does not mean that the change had no evolutionary implications for long-term changes in the Jomon culture. On the contrary, several lines of evidence suggest that the change played a critical role in the development of the "Middle Jomon type" of subsistence-settlement systems that were characterized by an abundance of chipped stone axes.

First, the shift from collecting to foraging systems in southwestern Kanto seems to have been associated with a significant population decrease in this region. According to K. Imamura (1992) and others (e.g., Kobayashi 1973; Shibue and Kuro'o 1987), the scarcity of dwelling sites in southwestern Kanto is characteristic of the Moroiso-c subphase and also of the following Jusanbodai phase. In particular, Imamura (1992) points out that settlement patterns of the Moroiso-c subphase and the Jusanbodai phase in the southwestern Kanto region share a number of similarities, including

(1) scarcity of dwelling sites and (2) small site size in terms of both the number of associated dwellings and the amount of artifacts. Based on these observations, Imamura (1992) concludes that the decrease in the number of dwelling sites and site size during the Moroiso-c subphase and the Jusanbodai phase in this region represents a significant population decrease.

Assuming that this was the case, a question then arises whether this occurred through a gradual or catastrophic increase in the mortality rate or through emigration to other areas. Changes in the number of sites in each region seem to indicate the possibility of population migration from the Kanto to the Chubu Mountain regions. As shown in Table 3.4, the total numbers of sites from the Moroiso-b to Moroiso-c subphases in the northwestern and southwestern Kanto regions decreased, whereas the number in the Chubu Mountain region increased.

Detailed analysis of settlement pattern changes in the southwestern Kanto region also indicates that the shift from the collector system to the forager system occurred gradually throughout the Moroiso phase along with population movement from coastal to inland areas. As shown in Table 3.5, the settlement pattern of the Moroiso-a subphase in the southwestern Kanto region is characterized by a relative abundance of large and medium dwelling sites compared to the other two phases. The presence of large settlements is characteristic of relatively sedentary collectors who form large residential groups during a certain time of the year (often fall to early spring) and disperse in smaller settlements for the rest of the year. However, by the Moroiso-b subphase, the percentages of large and medium sites

Table 3.4. Changes in the Total Number of Sites in Each Region

Region	Moroiso-b	Moroiso-c
Northwestern Kanto region	128	69
Southwestern Kanto region	631	278
Chubu Mountain region	76	97

Table 3.5. Site Size Variability among Dwelling Sites in Southwestern Kanto

Site Size	Number of associated pit-dwellings	Moroiso-a	Moroiso-b	Moroiso-c
Small	1–4	41 (83.7%)	84 (92.3%)	5 (100.0%)
Medium	5–9	5 (10.2%)	6 (6.6%)	0
Large	10≦	3 (6.1%)	1 (1.1%)	0
Total		49 (100.0%)	91 (100.0%)	5 (100.0%)

decreased. This indicates that the settlement system of this subphase was less closely associated with the annual amalgamation of residential groups. Furthermore, many of the dwelling sites of the Moroiso-b subphase are located in the inland area of the southwestern Kanto region (i.e., closer to the Chubu Mountain region). This is in sharp contrast to the settlement pattern of the Moroiso-a subphase, in which dwelling sites are primarily concentrated in the coastal area of southwestern Kanto. Thus, by the time of the Moroiso-b subphase, there was a major population shift from the coastal to the inland areas within southwestern Kanto. Because the inland area is the gateway from the coastal area to the Chubu Mountain region, such a population shift is consistent with the assumption of emigration from southwestern Kanto to the Chubu Mountain region.

Second, examination of lithic assemblage changes through time in the Chubu Mountain region indicates that, in this region, the shift to the "Middle Jomon type" of subsistence-settlement systems (i.e., those characterized by an abundance of chipped stone axes) occurred as early as the Moroiso-c subphase (the second to the last phase of the Early Jomon period). None of the dwelling sites from the Moroiso-a and Moroiso-b phases is characterized by an abundance of chipped stone axes, but they are the most abundant type of stone tools in four out of the thirteen dwelling sites from the Moroiso-c subphase (Fig. 3.3). This is the first phase in which the dominance of chipped stone axes in site assemblages can be observed in this region. As noted above, an abundance of chipped stone axes is often considered a characteristic of Middle Jomon settlements in this region (e.g., Fujimori 1950). Many archaeologists suggest that this reflects the importance of either plant food collecting or possibly plant cultivation in the subsistence systems of the Middle Jomon people in this region (Fujimori 1950, 1970; Oyama 1927, 1934; Tsuboi 1962).

Unlike the southwestern Kanto region, where environmental deterioration in the coastal area may explain the change from collecting to foraging systems, no clear evidence of environmental change in the Chubu Mountain region has been found. Instead, the previously described lines of evidence indicate the possibility of population migration from the southwestern Kanto to the Chubu Mountain region. If this were the case, the migration would have resulted in a significant increase in population pressure within the Chubu Mountain region. It is quite possible that the increased population pressure caused a major system change, which resulted in the beginning of the "Middle Jomon type" of subsistence-settlement systems characterized by an abundance of chipped stone axes. Because both the previous and the new systems were collector systems, the shift may have initially seemed insignificant. Nevertheless, from the perspective of long-term development within the Jomon culture, the shift was critical. Though this shift first

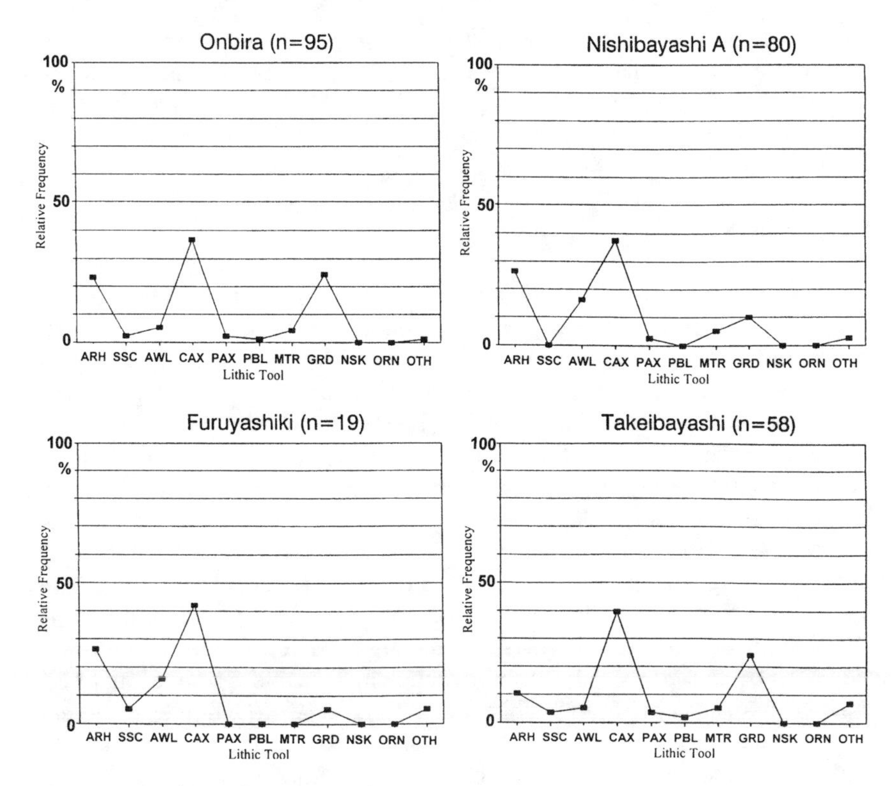

Figure 3.3. Relative frequencies of lithic tools per category in assemblages from four sites (Onbira, Nishibayashi, Furuyashiki, and Takeibayashi) in the Chubu Mountain region. ARH: arrowheads; SSC: stemmed scrapers; AWL: awls; CAX: chipped stone axes; PAX: polished stone axes; PBL: pebble tools; MTR: stone mortars; GRD: grinding stones; NSK: net sinkers; ORN: ornaments; OTH: others.

occurred in the Chubu Mountain region at the end of the Early Jomon period, by the Middle Jomon period, a similar system was adopted by the people of the southwestern Kanto region. As a result, the subsistence-settlement systems in southwestern Kanto shifted back from foraging to collecting systems.

K. Imamura's (1992) analysis is quite suggestive of the nature of the subsistence-settlement systems characterized by an abundance of chipped stone axes. According to his analysis, Middle Jomon sites associated with a large number of chipped stone axes are distributed only in the southwestern Kanto and Chubu Mountain regions. In contrast, Middle Jomon sites in the northeastern Kanto region are characterized by an abundance of large storage pits. Imamura points out that the distribution of these two types of sites is mutually exclusive and suggests that the chipped stone axes were used to collect wild root crops such as yam (*Discorea japonica*), whereas storage pits were more closely related to the heavy reliance on nuts such as acorns and chestnuts. Further study is necessary to identify the type(s) of staple food associated with these two types of subsistence-settlement systems.

DISCUSSION

The analysis presented above has several implications for studying long-term changes in hunter-gatherer settlement patterns. First, the results indicate that the shift between collecting strategies and foraging strategies may have occurred fairly frequently. In particular, the archaeological data from southwestern Kanto indicates that the shift from collecting strategies to foraging strategies through the Moroiso phase was gradual and continuous. This corresponds well with Binford's (1980) suggestion that forager and collector systems are not polar opposites of settlement systems but rather are a graded series from simple (forager) to complex (collector). Also, if the shift from collecting strategies to foraging strategies was triggered by a decrease in the availability of shellfish, which probably was not the staple food of the Moroiso phase people, it implies that the shift between collectors and foragers can be caused by only a minor change in available resources, not necessarily by major environmental changes. In other words, hunter-gatherers' decision-making processes for their subsistence strategies may be extremely intricate; a minor change in the quality and/or quantity of available resources may result in fundamental system changes.

Previous studies have emphasized such Jomon cultural characteristics as the presence of large settlements and food storage, both of which are typical of collecting systems. However, given the evidence of system fluidity discussed above, it is probable that the Jomon subsistence-settlement systems were quite flexible and included varying combinations of collecting and foraging strategies. Similar variability in subsistence-settlement systems may also have been present among many other prehistoric hunter-gatherers, who had to cope with constantly changing environments.

Second, the case study presented here suggests that regional interactions may have played a critical role in long-term system changes and that archaeological analyses of settlement patterns at the interregional level can reveal the mechanisms of these changes within the context of a dynamic cultural landscape. Because the core of the Binford's model is ecological, one tends to think that the model assumes a direct relationship between resource distribution and hunter-gatherer subsistence and settlement organization. However, through the case study presented here, it is clear that hunter-gatherer systems do not exist in a cultural vacuum and that interactions between regional systems can be an important factor in determining the course of long-term system changes. This implies that we may need to reevaluate the scale of regional settlement pattern analysis in archaeological studies. Since Willey's (1953) pioneering Viru Valley project, numerous studies of regional settlement patterns have been published. In many cases, however, regional settlement pattern analyses are based on the results of surveys conducted at the "local" level, for example, within a single valley or river drainage basin (e.g., Adams 1965; Sanders et al. 1979; Willey et al. 1965). The case study presented here clearly demonstrates the research potential of settlement pattern analysis at the interregional level.

Finally, the previous explanation for the development of the "Middle Jomon type" systems provides insights in the light of previous discussions on the evolution of hunter-gatherer sedentism and cultural complexity. For example, several researchers have suggested population pressure (i.e., necessity) as a major condition for, or cause of, the development of hunter-gatherer sedentism and cultural complexity (e.g., Cohen 1981), whereas others have suggested that resource abundance (i.e., opportunity) was a more critical factor (e.g., Hayden 1995) (for discussions on opportunity vs. necessity, see also Cannon 1998; Yesner 1987). Assuming that the "Middle Jomon type" systems, which are characterized by an abundance of chipped stone axes and large settlements, are highly logistically organized systems, the present study can be used as an example to support the population pressure theory. However, supporters of this theory typically assume that the population pressure increases naturally (i.e., with no specific causes or conditions), and they tend to interpret the process of the development in terms of a general model. The analysis presented here, on the other hand, describes a case in which population increase and resulting system changes in one region was a consequence of population migration caused by system changes in an adjacent region. In this regard, the present study suggests that the development of the "Middle Jomon type" systems was a historically unique event.

CONCLUDING REMARKS

This chapter presented a case study of long-term settlement pattern changes in which a shift from collectors to foragers in one region (southwestern Kanto) triggered a system change in another region (Chubu Mountain). As a result of this system change, the "Middle Jomon type" of subsistence-settlement systems developed first in the Chubu Mountain region and were then adopted in the southwestern Kanto region. These new systems were characterized by an abundance of chipped stone axes, which suggests some kind of subsistence intensification, possibly intensive collecting of wild root crops, and also by the presence of extremely large settlements and high site density. These characteristics indicate that the systems were highly logistically organized collector systems, even though detailed settlement pattern analyses from this period have yet to be conducted. The combination of environmental factors and population pressure was suggested as the cause for the development of this new system.

It should be noted that, unlike prehistoric hunter-gatherer cultures in California (e.g., Arnold 1992, 1995) and the Northwest Coast of North America (e.g., Ames 1995; Ames and Maschner 1999; Matson and Coupland 1995), currently available data from the Middle Jomon period in the Kanto and Chubu regions lack clear evidence of social stratification. Accordingly, even though the Middle Jomon culture shares a number of characteristics with so-called "complex" hunter-gatherers, it may not be "complex" in a strict sense if we take the position that hunter-gatherer cultural complexity should be defined on the basis of the presence of hereditary social inequality (e.g., Arnold 1996; see also Hayden 1995). Nevertheless, the Middle Jomon culture is characterized by logistically organized collector systems, and also by sophisticated material culture, large ceremonial features, and long-distance trade. In this regard, systematic examination of the causes and consequences of the development of this unique culture will help answer a number of general questions regarding the long-term development of hunter-gatherer cultural complexity in a dynamic landscape.

ACKNOWLEDGMENTS

I thank Kenneth M. Ames, Ben Fitzhugh, and James M. Savelle, all of whom read earlier drafts of this paper and gave me invaluable comments and suggestions. The paper also benefited greatly from the comments of T. Douglas Price and other participants of the "Beyond Foraging and Collecting" symposium at the 2000 SAA meeting. Raw data were obtained principally through library research at the Nara National Cultural Properties Research Institute, the Archaeology Museum of Meiji University, and the Department of Archaeology of the University of Tokyo. Funding for the data collection was provided through the Social Sciences Research Committee of McGill University. I extend my gratitude to these organizations.

REFERENCES

Adams, R. M., 1965, *Land behind Baghdad: A History of Settlement on the Diyala Plains*. University of Chicago Press, Chicago.

Aikens, C. M., 1981, The Last 10,000 Years in Japan and Eastern North America: Parallels in Environment, Economic Adaptation, Growth of Social Complexity, and the Adoption of Agriculture. In *Affluent Foragers: Pacific Coasts East and West*, edited by S. Koyama and D. H. Thomas, pp. 261–273. Senri Ethnological Studies, No. 9, National Museum of Ethnology, Osaka.

Aikens, C. M., Ames, K. M., and Sanger, D., 1986, Affluent Collectors at the Edges of Eurasia and North America: Some Comparisons and Observations on the Evolution of Society among North-Temperate Coastal Hunter-Gatherers. In *Prehistoric Hunter-Gatherers in Japan*, edited by T. Akazawa and C. M. Aikens, pp. 261–273. University of Tokyo Press, Tokyo.

Aikens, C. M., and Dumond, D. E., 1986, Convergence and Heritage: Some Parallels in the Archaeology of Japan and Western North America. In *Windows on the Japanese Past: Studies in Archaeology and Prehistory*, edited by R. J. Pearson, G. L. Barnes, and K. L. Hutterer, pp. 261–273. Center for Japanese Studies, University of Michigan, Ann Arbor.

Amemiya, M., 1999, Minami Kyushu ni Miru Jomon Teiju Shuryo-Saishu-min Sekai [Jomon Sedentary Hunter-Gatherers in Southern Kyushu]. *Kikan Kokogaku* [Archaeology Quarterly] 64:19–24 (in Japanese).

Ames, K. M., 1995, Chiefly Power and Household Production on the Northwest Coast. In *Foundations of Social Inequality*, edited by T. D. Price and G. M. Feinman, pp. 155–188. Plenum, New York.

Ames, K. M., and Maschner, H., 1999, *Peoples of the Northwest Coast: Their Archaeology and Prehistory*. Thames and Hudson, London.

Arnold, J. E., 1992, Complex Hunter-Gatherer-Fishers of Prehistoric California: Chiefs, Specialists, and Maritime Adaptations of the Channel Islands. *American Antiquity* 57(1):60–84.

Arnold, J. E., 1995, Social Inequality, Marginalization, and Economic Process. In *Foundations of Social Inequality*, edited by T. D. Price and G. M. Feinman, pp. 155–188. Plenum, New York.

Arnold, J. E., 1996, The Archaeology of Complex Hunter-Gatherers. *Journal of Archaeological Method and Theory* 3(2):77–126.

Binford, L. R., 1980, Willow Smoke and Dogs' Tails: Hunter-Gatherer Settlement Systems and Archaeological Site Formation. *American Antiquity* 45(1):4–20.

Binford, L. R., 1982, The Archaeology of Place. *Journal of Anthropological Archaeology* 1(1):5–31.

Binford, L. R., 1983, Long-Term Land-Use Patterning: Some Implications for Archaeology. In *Working at Archaeology*, edited by L. R. Binford, pp. 155–188. Academic Press, New York.

Cannon, A., 1998, Contingency and Agency in the Growth of Northwest Coast Maritime Economies. *Arctic Anthropology* 35(1):57–67.

Cohen, M. N., 1981, Pacific Coast Forager: Affluent or Overcrowded? In *Affluent Foragers: Pacific Coasts East and West*, edited by S. Koyama and D. H. Thomas, pp. 155–188. Senri Ethnological Studies No. 9. National Museum of Ethnology, Osaka.

Erlandson, J. M., 1988, The Role of Shellfish in Prehistoric Economies: A Protein Perspective. *American Antiquity* 53(1):102–109.

Fuji, N., 1984, *Koko Kafungaku* [Archaeo-Palynology]. Yuzankaku, Tokyo (in Japanese).

Fujimori, E., 1950, Nihon Genshi Rikuko no Sho-mondai [On Primitive Dry Field Cultivation in Japan]. *Rekishi Hyoron* [Critiques in History] 4(4):41–46 (in Japanese).

Fujimori, E., 1970, *Jomon Noko* [Jomon Plant Cultivation]. Gakusei-sha, Tokyo (in Japanese).

Habu, J., 1989, Contemporary Japanese Archaeology and Society. *Archaeological Review from Cambridge* 8(1):36–45.

Habu, J., 1996, Jomon Sedentism and Intersite Variability: Collectors of the Early Jomon Moroiso Phase in Japan. *Arctic Anthropology* 33(2):38–49.

Habu, J., 2001, *Subsistence-Settlement Systems and Intersite Variability in the Moroiso Phase of the Early Jomon Period of Japan*. International Monographs in Prehistory, Archaeological Series 14, Ann Arbor, MI.

Habu, J., Kim, M., Katayama, M., and Komiya, H., 2001, Jomon Subsistence-Settlement Systems at the Sannai Maruyama Site. *Bulletin of the Indo-Pacific Prehistory Association* 21:9–21.

Hayden, B., 1990, Nimrods, Piscators, Pluckers, and Planters: The Emergence of Food Production. *Journal of Anthropological Archaeology* 9(1):31–69.

Hayden, B., 1995, Pathways to Power. In *Foundations of Social Inequality*, edited by T. D. Price and G. M. Feinman, pp. 15–86. Plenum, New York.

Horiguchi, M., 1983, Saitama-ken Juno Deitanso Iseki no Gaikyo to Shizen Kankyo ni Kansuru 2, 3 no Mondai [Summary of Juno Peaty Site and Some Problems of Its Natural Environment]. *Daiyonki Kenkyu* [Quaternary Research] 22(3):231–244 (in Japanese with English title and summary).

Ikawa-Smith, F., 1998, Gender in Japanese Prehistory. Paper Presented at the Worldwide Perspective on Women and Gender Conference, Bellagio, Italy.

Imamura, K., 1992, Jomon Zenki Matsu no Kanto ni okeru Jinko Gensho to Sore ni Kanren-suru Sho-gensho [Population Decrease and Associated Phenomena at the End of the Early Jomon Period in the Kanto Region]. In *Musashino no Kokogaku: Yoshida Itaru Sensei Koki Kinen Ronbunshu* [Essays in Honour of Dr. I. Yoshida's 70th Birthday], edited and published by Yoshida Itaru Sensei Koki Kinen Ronbunshu Kanko kai [Editorial Committee of Essays in Honour of Dr. I. Yoshida's 70th Birthday], pp. 85–115. Tokyo (in Japanese).

Imamura, K., 1996, *Prehistoric Japan*. University of Hawaii Press, Honolulu.

Imamura, M., 1999, Koseido ^{14}C nendai Sokutei to Kokogaku [High Precision ^{14}C Dating and Archaeology]. *Chikyu* [The Earth], Special Issue No.26:23–31 (in Japanese).

Kelly, R. L., 1995, Foraging and Mobility. In *The Foraging Spectrum*, by R. L. Kelly, pp. 111–160. Smithsonian Institution Press, Washington DC.

Kidder, J. E., 1998, The Sannai Maruyama Site: New Views on the Jomon Period. *Southeast Review of Asian Studies* 20:29–52.

Kobayashi, T., 1973, Tama New Town no Senjusha: Shu to shite Jomon Jidai no Settlement System ni tsuite [Previous Inhabitants of the Tama New Town Area: Settlement Systems during the Jomon Period]. *Gekkan Bunkazai* [Monthly Journal of Cultural Property] 112:20–26 (in Japanese).

Matson, R. G., and Coupland, G., 1995, *The Prehistory of the Northwest Coast*. Academic Press, San Diego.

Matsushima, Y., 1979, Minami Kanto ni okeru Jomon Kaishin ni Tomonau Kairui Gunshu no Hensen [Littoral Molluscan Assemblages during the Post-Glacial Jomon Transgression in the Southern Kanto, Japan]. *Daiyonki Kenkyu* [Quaternary Research] 17(4):243–265 (in Japanese with English title and summary).

Matsushima, Y., and Koike, H., 1979, Shizen Kaiso ni yoru Naiwan no Kankyo Fukugen to Jomon Jidai no Iseki [Reconstruction of the Littoral Environment Using the Result of Natural Shell Deposit Analysis, with Special Reference to the Distribution of Jomon Sites]. *Kaizuka* [Shell-Midden] 22:1–9 (in Japanese).

Minagawa, M., and Akazawa, T., 1992, Dietary Patterns of Japanese Jomon Hunter-Gatherers: Stable Nitrogen and Carbon Isotope Analyses of Human Bones. In *Pacific Northeast Asia in Prehistory*, edited by C. M. Aikens and S. N. Rhee, pp. 53–57. Washington State University Press, Washington.

Nakamura, T., and Tsuji, S., 1999, Aomori-ken Higashi-Tsugaru-run Kanita-machi Odai Yamamoto I Iseki shutsudo no Doki Hahen Hyomen ni Fuchaku shita Biryo Tanka-butsu no Kasokuki ^{14}C Nendai [AMS ^{14}C Dates of Carbon Samples Collected from the Surface of Potsherds Recovered at the Odai Yamamoto I Site at Kanita Town, Higashi-Tsugaru County, Aomori Prefecture]. In *Odai Yamamoto Iseki no Kokogaku Chosa: Kyusekki Bunka no Shumatsu to Jomon Bunka no Kigen ni kansuru Mondai no Tankyu* [Archaeological Research at the Odai Yamamoto I Site: Inquiry into the Question of the End of the Palaeolithic Culture and the Beginning of the Jomon Culture], edited by Odai Yamamoto I Iseki Hakkutsu Chosa-dan [The Odai Yamamoto I Site Excavation Team], pp. 107–111. Odai Yamamoto I Iseki Hakkutsu Chosa-dan, Tokyo. (in Japanese).

Okada, Y., 1995a, Ento Doki Bunka no Kyodai Shuraku [A Large Settlement from the Ento Pottery Culture: The Sannai Maruyama Site in Aomori Prefecture]. *Kikan Kokogaku* [Archaeology Quarterly] 50:25–30 (in Japanese).

Okada, Y., 1995b, Nihon Saidai no Jomon Shuraku "Sannai Maruyama Iseki" [The Largest Jomon Site in Japan, The "Sannai Maruyama Site"]. In *Jomon Bunka no Hakken: Kyoi no Sannai Maruyama Iseki* [Discovery of the Jomon Civilization: Amazing Discoveries at the Sannai Maruyama Site], edited by T. Umehara and Y. Yasuda, pp. 107–111. PHP Kenkyujo, Tokyo (in Japanese).

Okada, Y., and Habu, J., 1995, Public Presentation and Archaeological Research: A Case Study from the Jomon Period Sannai Maruyama Site. Paper Presented at the 1995 Chacmool Conference, Calgary.

Okamura, M., 1995, Uenohara Iseki [The Uenohara Site]. In *Hakkutsu Sareta Nihon Retto: >'95 Shin-hakken Koko Sokuho* [Excavations in the Japanese Archipelago: New Archaeological Discoveries in the 1995 Fiscal Year], edited by Bunka-cho [Agency of Cultural Affairs], pp. 14–15. Asahi Shinbun-sha, Tokyo (in Japanese).

Oyama, K., 1927, *Shizen Kenkyukai Shuho 1: Kanagawa-kenka Araiso-mura Aza Katsusaka Ibutsu Hoganchi Chosa Hokoku* [Report of the Association of Prehistory, No. 1: Excavation Report of the Open Site at Katsusaka, Araiso Village in Kanagawa Prefecture]. Shizen Kenkyukai [Association of Prehistory], Tokyo (in Japanese).

Oyama, K., 1934, Nihon Sekki Jidai no Seigyo Seikatsu [Subsistence Activities during the Japanese Stone Age]. *Kaizo* 16(1): 69–83 (in Japanese).

Pearson, R. J., 1977, Paleoenvironment and Human Settlement in Japan and Korea. *Science* 197 (No. 4310):1239–1246.

Price, T. D., 1981, Complexity in 'Non-Complex' Societies. In *Archaeological Approaches to the Study of Complexity*, edited by S. E. van der Leeuw, pp. 14–15. Albert Egges van Giffen Instituut voor Prae-en Protohistorie, Universiteit van Amsterdam, Amsterdam.

Price, T. D., and Brown, J. A., 1985, Aspects of Hunter-Gatherer Complexity. In *Prehistoric Hunter-Gatherers*, edited by T. D. Price and J. A. Brown, pp. 3–20. Academic Press, Orlando.

Sanders, W. T., Parsons, J. R., and Santley, R. S., 1979, *The Basin of Mexico: Ecological Processes in the Evolution of a Civilization*. Academic Press, New York.

Sakamoto, A., and Nakamura, W., 1991, Jomon Kaishin-ki no Jukyoshi Fukudo-nai Kaiso: Yokohama-shi Nishinoyato Kaizuka J30-go Jukyoshi to Sono Kaiso ni tsuite [A Shell-Midden Associated with a Pit-Dwelling from the Jomon Marine Transgression Period: Dwelling J30 of the Nishinoyato Site, Yokohama City and Its Associated Shell-Midden], *Chosa Kenkyu Shuroku* [Excavation and Research Results], Vol. 8, pp. 61–130. Yokohama-shi Maizo Bunkazai Center [Buried Cultural Property Center of Yokohama City], Yokohama (in Japanese).

Shibue, Y., and Kuro'o, K., 1987, Jomon Jidai Zenki Matsuyo no Kyoju Keitai: Yosatsu [Settlement Patterns at the End of the Early Jomon Period: A Preliminary Analysis]. *Kaizuka* [Shell-Midden] 39:1–9 (in Japanese).

Shinto, K., 1995, Minami Kyushu no Shoki Jomon Bunka [Incipient and Initial Jomon Cultures in Southern Kyushu]. *Kikan Kokogaku* [Archaeology Quarterly] 50:56–61 (in Japanese).

Shinto, K., 1999, Minami Kyushu no Tokushu-sei: Sosoki o Chushin ni [The Uniqueness of the Jomon Culture in Southern Kyushu, with Special Reference to the Incipient Jomon]. *Kikan Kokogaku* [Archaeology Quarterly] 69:18–22 (in Japanese).

Soffer, O., 1989, Storage, Sedentism and the Eurasian Palaeolithic Record. *Antiquity* 63:719–732.

Suzuki, K., 1979, Jomon jidai ron [The Jomon Period]. In *Nihon Kokogaku o Manabu* 3: *Genshi Kodai no Shakai* [Studying Japanese Archaeology Vol. 3: Ancient, Prehistoric Societies], edited by H. Otsuka, M. Tozawa, and M. Sahara, pp. 178–202. Yuhikaku, Tokyo (in Japanese).

Taniguchi, Y., 1999, Chojakubo Bunka-ki no Sho-mondai [Problems of the Chojakubo Culture Period]. In *Odai Yamamoto Iseki no Kokogaku Chosa: Kyusekki Bunka no Shumatsu to Jomon Bunka no Kigen ni kansuru Mondai no Tankyu* [Archaeological Research at the Odai Yamamoto I Site: Inquiry into the Question of the End of the Palaeolithic Culture and the Beginning of the Jomon Culture], edited by Odai Yamamoto I Iseki Hakkutsu Chosa-dan [The Odai Yamamoto I Site Excavation Team], pp. 178–202. Odai Yamamoto I Iseki Hakkutsu Chosa-dan, Tokyo (in Japanese).

Tsuboi, K., 1962, Jomon Bunka-ron [Controversy over the Jomon Culture]. In *Iwanami Koza Nihon no Rekishi* 1: *Genshi oyobi Kodai* [Iwanami Lectures in Japanese History, Vol.1, pp. 109–138. Prehistoric and Protohistoric Periods]. Iwanami Shoten, Tokyo (in Japanese).

Tsuji, S., 1999, Koseido ^{14}C Nendai Sokutei ni Yoru Sannai Maruyama Iseki no Hennen [Chronology at the Sannai Maruyama Site Using a High Precision ^{14}C Dating Method]. *Chikyu* [The Earth], Special Issue No.26:32–38 (in Japanese).

Wajima, S., 1948, Genshi shuraku no kosei [The Organization and Composition of Prehistoric Settlements]. In *Nihon Rekishigaku Koza* [Lectures in Japanese History], edited by Tokyo Daigaku Rekishigaku Kenkyukai [Historical Association of the University of Tokyo], pp. 1–32. Tokyo Daigaku Rekishigaku Kenkyukai, Tokyo (in Japanese).

Wajima, S., 1958, Nanbori Kaizuka to Genshi Shuraku [The Nanbori Shell-Midden and Settlements of the Prehistoric Period]. In *Yokohama Shi-shi* [The History of Yokohama City], Vol. 1, pp. 29–46. Yokohama City, Yokohama (in Japanese).

Watanabe, M., 1976, *Jomon Jidai no Shokubutsu-shoku* [Plant Food during the Jomon Period]. Yuzankaku, Tokyo (in Japanese).

Wiessner, P., 1982, Beyond Willow Smoke and Dogs' Tails: A Comment on Binford's Analysis of Hunter-Gatherer Settlement Systems. *American Antiquity* 47(1):171–178.

Willey, G. R., 1953, *Prehistoric Settlement Patterns in the Viru Valley, Peru.* Bureau of American Ethnology, Bulletin 155, Smithsonian Institution, Washington DC.

Willey, G. R., Bullard, W. R., Glass, J. B., and Gifford, J. C., 1965, Prehistoric *Maya Settlements in the Belize Valley.* Papers of the Peabody Museum of Archaeology and Ethnology, Harvard University, 54, Cambridge, MA.

Yesner, D. R., 1987, Life in the "Garden of Eden." In *Food and Evolution*, edited by M. Harris and E. B. Ross, pp. 285–310. Temple University Press, Philadelphia.

Chapter 4

Logistical Organization, Social Complexity, and the Collapse of Prehistoric Thule Whaling Societies in the Central Canadian Arctic Archipelago

JAMES M. SAVELLE

INTRODUCTION

The investigation of prehistoric Thule Inuit logistical and social organization in the Canadian Arctic is a relatively recent topic of interest in Canadian Arctic archaeology, and represents what McCartney (1980) has termed a "second phase" research objective in this region. This follows from and depends on the prior development of local and regional chronological frameworks. With few exceptions, earlier "first phase" interpretations of Thule social characteristics tended to be generic characterizations based very loosely on ethnographic analogy. In contrast, and as noted by Grier and Savelle (1994: 95–96), the "investigation of Thule social organization as a specific goal has been a relatively recent phenomenon, and as

JAMES M. SAVELLE • Department of Anthropology, McGill University, Montreal, Quebec H3A 2T7 CANADA.

a consequence tends to be rather intuitive, and its attendant methodologies exploratory." In this paper, Binford's (1980) collector–forager model forms the basis for an essentially exploratory investigation into the social and logistical context within which Thule culture underwent extensive modification in some regions of the Canadian Arctic and disappeared completely in others. These events have been referred to as the "Classic to Modified Thule transition" by McCartney (1977) and as the "Postclassic Modification" by Maxwell (1985).

OVERVIEW OF THULE CULTURE

Following is a brief overview of Thule culture to establish the context of the present study. It is based primarily on the summaries provided in McGhee (1984), Maxwell (1985), and the more recent detailed overview in Whitridge (2000). Prehistoric Thule Eskimo culture originated in northern Alaska shortly before 1000 A.D., and within one or two hundred years, by about 1000–1100 A.D., a "pioneering" Thule culture had apparently spread into, at least, the western Canadian Arctic. The exact relationship of slightly later (beginning 1200 A.D.) Thule occupations further east in the Canadian Arctic to the pioneering Thule groups is to some extent still unclear, but certainly by 1200–1450 A.D., much of the central Canadian Arctic Archipelago (CAA), adjacent coastal mainland (Fig. 4.1), and northwestern Greenland was occupied by Thule societies. These societies constitute what McCartney (1977) referred to as "Classic" Thule (although note that the Ruin Island Thule phase of northern Ellesmere and northwestern Greenland probably represents an independent expansion at about the same time; see, e.g., McCullough 1989; Morrison 1989). Throughout much of the central CAA region during this period, Thule subsistence was centered to varying degrees on bowhead whales (Fig. 4.2), the bones of which were typically incorporated as construction materials in dwellings and other features (Figs. 4.3 and 4.4). It should be stressed here that some researchers (e.g., Freeman 1979; Yorga 1979) have suggested that whale bones at Thule sites may reflect whale bone use (scavenging) as opposed to active whaling. Some opportunistic scavenging of bowhead bones may have taken place, but the available data accumulated and discussed by McCartney (1980), McCartney and Savelle (1985, 1993), Savelle and McCartney (1994, 1999), Whitridge (1999), and Savelle (2000) leave little doubt that the Thule were active bowhead whale hunters in those areas where extensive bowhead whale bone is found incorporated within structures.

Elsewhere in the central CAA during the "Classic" Thule period—primarily the Arctic mainland coastal regions, but also in the King William

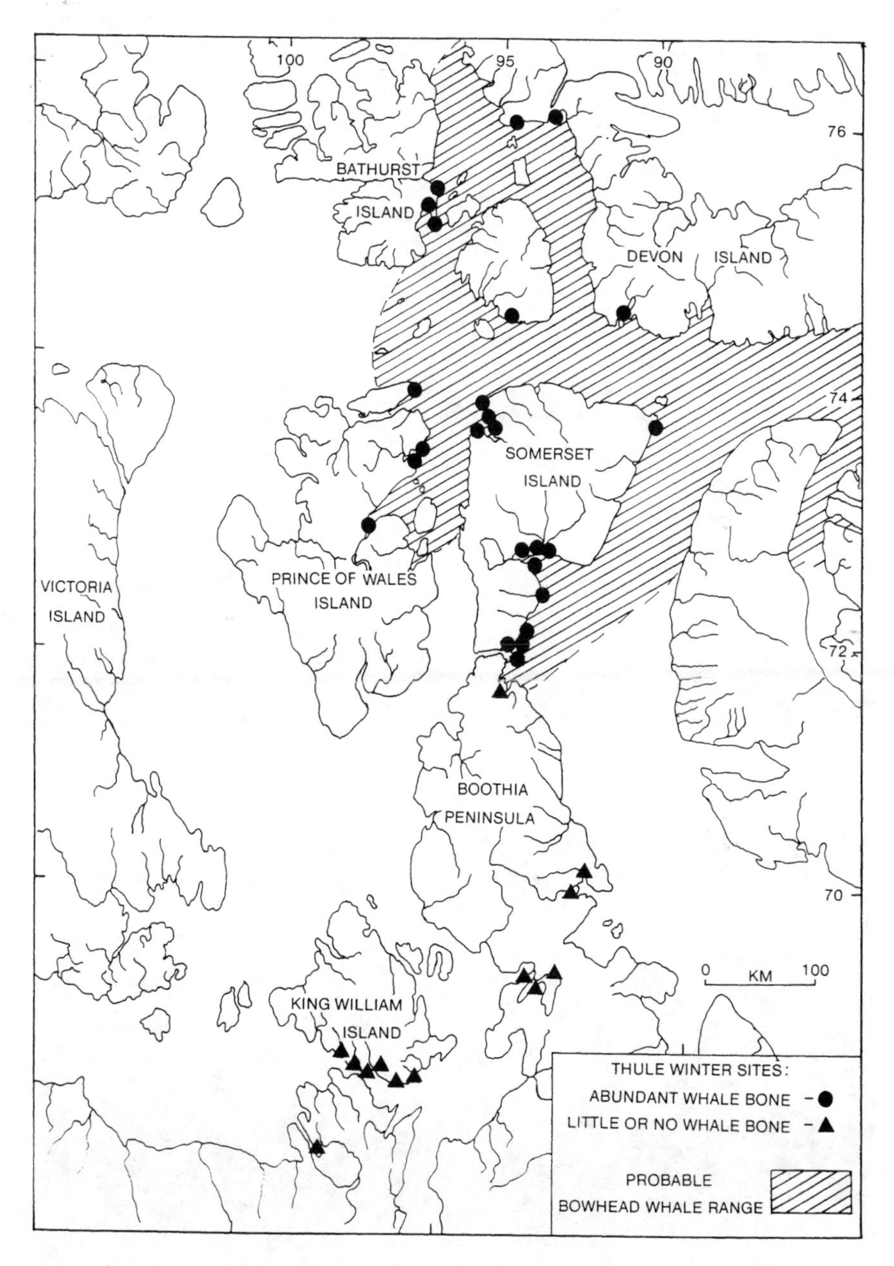

Figure 4.1. Map of central Canadian Arctic showing whaling and nonwhaling sites (after McCartney and Savelle 1985).

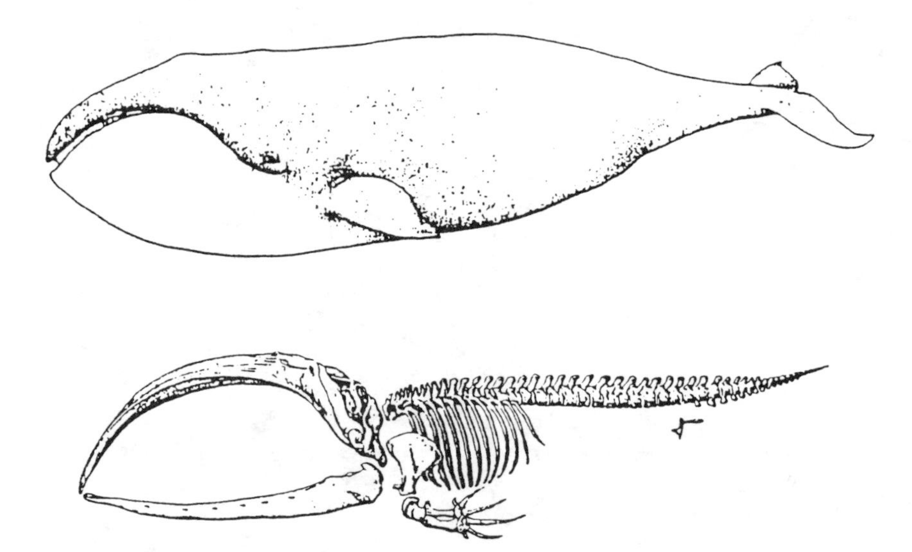

Figure 4.2. Diagram of bowhead whale and skeleton (from McCartney 1980).

Figure 4.3. Photograph of whale bone house before excavation, Deblicquy site, Bathurst Island (photograph by J.M. Savelle).

Figure 4.4. Photograph of excavated Thule dwelling at Hazard Inlet, Somerset Island (photograph by J.M. Savelle).

Island and Victoria Island regions—with few exceptions, bowheads represented either relatively minor components of Thule diets or were completely absent (Fig. 4.1). Instead, walrus, various seals, caribou, musk ox, smaller mammals, and fish and birds, in varying proportions, represented the primary subsistence species (see, e.g., the various tables and charts in Savelle and McCartney 1988; Savelle 1994).

CLASSIC THULE AS COLLECTORS

Regardless of subsistence, most village sites dating to the "Classic" Thule period in the CAA consist of the remains of substantial dwellings, which in most instances have been interpreted as representing seasonal, primarily winter, maximum band aggregates.

Within the whaling region (i.e., within the region in Fig. 4.1 characterized by sites or site clusters containing abundant whale bone) are typically found the largest villages; 15–25 dwellings are not unusual, and one site, PaJs-2 on southeastern Somerset Island, consists of at least 60 dwellings (Whitridge 1999). Several of these sites in the "core" whaling area of southeastern Somerset Island (sensu Savelle and McCartney 1994) contain the remains of more than 100 individual bowhead whales (McCartney 1978; Savelle 2000), and in PaJs-2, 261 individual bowheads (Whitridge 1999). Given that these figures are based on surface bone counts and that typically from only 15–50% of structural whale bone is exposed (McCartney 1979), it is apparent that bowhead whaling was conducted on a major scale.

Furthermore, a number of Thule winter village sites within the whaling zones contain communal features that served as workshops and sites of whaling ceremonialism and are interpreted as equivalent to the historically documented *kariyit* (sing. *karigi*) of northern Alaska (see, e.g., Savelle 1987; Habu and Savelle 1994; Whitridge 1999).

Associated tent ring sites (Fig. 4.5) are generally representative of dispersed, more mobile summer microband units, and the smallest such sites (one to two tent rings) are tentatively interpreted as field camps sensu Binford (1980). Other site types within the whaling region include specialized fall whaling camps (Fig. 4.6), and bowhead whale processing and caching sites (Fig. 4.7). Note that *kariyit* have also been recorded in association with these fall whaling camps; at least six occur in the Hazard Inlet region of southeastern Somerset Island (see, e.g., Savelle and Wenzel n.d. and Fig. 4.8).

Figure 4.5. Photograph of tent ring site, Aston Bay, Somerset Island (photograph by J.M. Savelle).

Figure 4.6. Photograph of excavated qarmat at fall whaling camp, Hazard Inlet, Somerset Island (photograph by J.M. Savelle).

Figure 4.7. Photograph of whale bones and cache, Hazard Inlet, Somerset Island. Arrows indicate bowhead crania (photograph by J.M. Savelle).

Figure 4.8. Photograph of excavated *karigi*, Hazard Inlet, Somerset Island (photograph by J.M. Savelle).

Figure 4.9. Aerial photograph of caribou drive system, Sherman Inlet, Adelaide Peninsula, Arctic mainland coast (photograph by J.M. Savelle).

Outside the whaling regions, associated sites include major caribou drive systems and related processing and caching sites (Fig. 4.9) in addition to tent ring residential and field camp sites.

As discussed in considerable detail elsewhere (e.g., Savelle 1987; Savelle and McCartney 1988; Savelle and Habu 1998), in the context of Binford's collector–forager model, these "Classic" Thule societies can be considered representative of fully developed collecting societies. The permanent (winter) residential bases, extensive field processing and caching sites, and specialized late summer/fall whaling camps, for example, are all characteristic of collecting systems. The spatial distribution of these various site types on a regional level also strongly suggest collecting systems: for example, winter village sites or sites clusters typically center what can be interpreted as inner foraging and outer logistical zones, and attendant site types within these zones form patterns consistent with collecting systems (Fig. 4.10).

THE POSTCLASSIC THULE MODIFICATION AND ABANDONMENT OF THE NORTHERN CAA

Beginning approximately 1400–1450 A.D. and generally coinciding with the cooling trend that led to the Little Ice Age, many areas of the previously

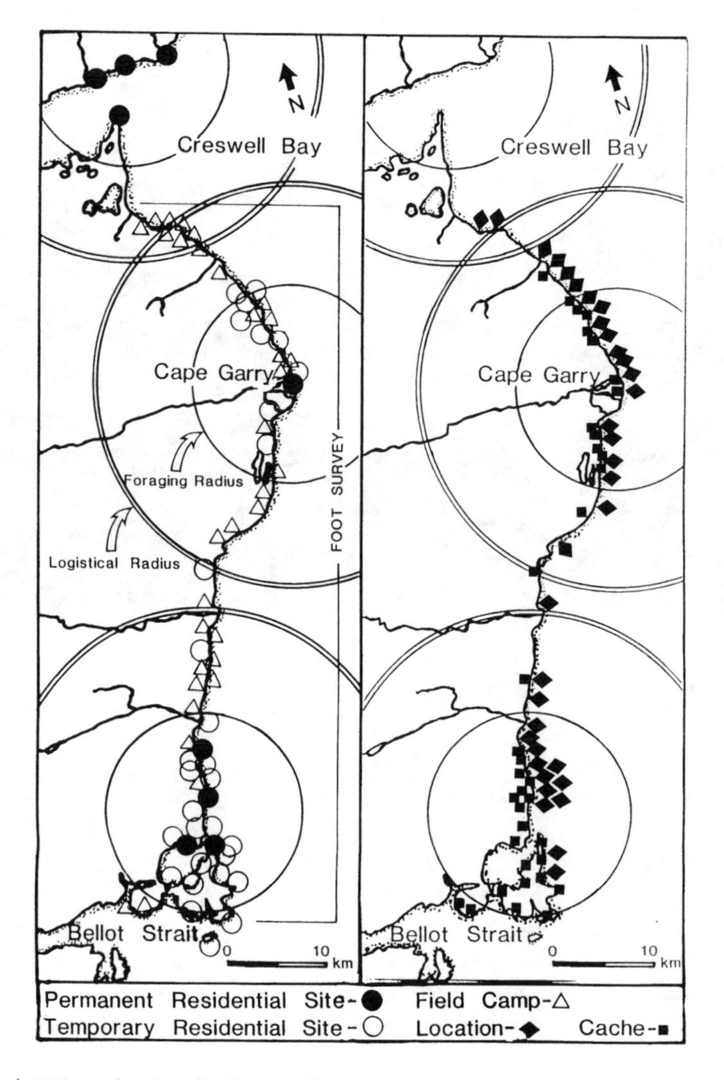

Figure 4.10. Map showing distribution of site types on Somerset Island according to foraging and collecting zones (from Savelle and McCartney 1988).

inhabited central CAA were completely abandoned. The available radiocarbon dates (see, e.g., Morrison 1989) and artifact assemblage characteristics (see, e.g., Park 1994) make it difficult to put precise terminal dates on the abandonment, and presumably the terminal dates varied within the overall area. Certainly by the early nineteenth century, however, when the

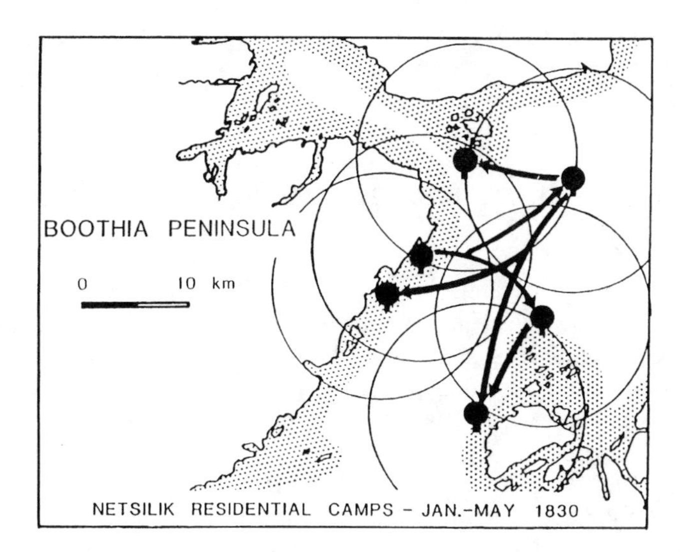

Figure 4.11. Locations and movements of early historic Netsilik Inuit residential units during the winter and spring of 1829–30 at Lord Mayor Bay, Boothia Peninsula (from Savelle and McCartney 1988).

first explorers and whalers entered the area, the entire northern part of the central CAA was uninhabited.

In the areas further south for which we have evidence of continued occupation, the substantial dwellings of the winter villages were abandoned in favor of less substantial and less permanent dwellings. These in turn were eventually replaced, for the most part, by snow igloos characteristic of the highly mobile historical Central Eskimo (see, e.g., Boas 1888). Finally, evidence of highly organized activities such as sophisticated caribou drives, and also of significant food storage, decreases substantially.

A typical example of the winter movements associated with an early nineteenth century Inuit society in the southern part of the study area (southern Boothia Peninsula) is illustrated in Figure 4.11. Note that these represent a fusion–fission sequence of igloo-based settlements on the sea ice.

The reason commonly cited for the shift from permanent winter villages to much more temporary villages and the attendant increase in seasonal residential movements is the depletion of many marine resources during the Little Ice Age. Specifically, an increase in the extent and duration of summer sea ice coverage resulted in fewer migratory sea mammals, in particular the bowhead whale (see, e.g., McGhee 1969/70; McCartney 1977). Conversely, this same increase in the summer sea ice may have resulted in an increase in nonmigratory ringed seals (see, e.g., McGhee 1978: 105), which are highly dependent on sea ice. In addition, these same cooling conditions may have been detrimental, especially, to the large migratory caribou herds in the south part of the region (see, e.g., Kelsall 1968; Parker 1972).

POTENTIAL PROBLEMS WITH THE TRADITIONAL ABANDONMENT SCENARIO

Essentially then, Thule societies responded to a decrease in large, abundant resources and an increase in smaller, dispersed resources by shifting from a collecting system to a foraging system (see Savelle 1987 for a detailed treatment of this hypothesis).

This is a simple, tidy explanation for these changes. However, consider again the distribution and characteristics of Thule winter village sites in the CAA region (Fig. 4.1). What is of interest here is that the entire area delineated by the whaling sites—i.e., all of those within the bowhead range—was completely abandoned by or shortly after 1450–1500 A.D. The southern nonwhaling area, on the other hand,—i.e., Boothia Peninsula, King William Island, and neighbouring coastlines—was inhabited continuously through to the historic period.

The immediate question is, What happened to the Thule whaling societies? Certainly the most parsimonious answer would be that they simply moved south and amalgamated with nonwhaling Thule societies and/or moved into other areas where bowheads remained abundant (see, e.g., McGhee 1969/70, Whitridge 2000). There are several potential problems with the amalgamation scenario, one of which is that the nonwhaling Thule societies would themselves have been going through some fairly substantial adjustments of their own. Specifically, the transition from collecting to foraging strategies in the southern region would in itself likely have resulted in the reduction in the ability to maintain previous population levels, let alone absorb a substantial population influx from neighboring areas.

In terms of a migration out of the area, there is certainly clear evidence for a Thule expansion south onto southern Baffin and northern Hudson Bay coast at about this time (see, e.g., Schledermann 1975; McCartney 1977; Sabo 1991; Stenton 1989; McGhee 1996). However, even if there was movement of the Thule out of the northern area following the decline of whaling, why didn't at least some of these societies simply readjust their subsistence systems to ringed seal-based (and/or other resource-based) foraging strategies within the same area, as did the Thule to the south? Certainly there is evidence that the pre-Thule inhabitants of this and adjacent regions, i.e., various Paleoeskimo groups, on occasion survived through episodes of similar climatic deterioration (see, e.g., Bielawski 1988; Helmer 1991). Further, these

Paleoeskimo groups possessed a less sophisticated marine mammal hunting technology than the Thule.

LOGISTICAL ORGANIZATION, SOCIAL COMPLEXITY, AND INTERNAL CONSTRAINTS

One possible answer to the problem of the lack of subsistence-settlement readjustment by the northern whaling societies may lie in the nature of the organization of Thule whaling societies, at least in the central CAA region. As noted by Ames (1985), collecting societies tend to have relatively larger population levels, and typically greater organizational complexity, than foragers. This organizational complexity in turn results in a concomitant need for increased information processing efficiency, of which the formation of vertical hierarchies is one such response (Ames 1985; see also Johnson 1982).

There is substantial evidence in the northern whaling areas to suggest that such hierarchies existed. Specifically, the larger settlements, typically consisting of 15–25 dwelling, are often organized into groups of several dwelling clusters. One or more of these dwelling clusters are often centered around distinct dwellings that are most certainly the equivalent of the North Alaskan Eskimo *kariyit*, or communal/ceremonial dwellings, as noted above (Fig. 4.12; see Spencer 1959; Burch 1980, 1981; Sheehan 1985, 1995, 1997; Whitridge 1999 for detailed discussions of the structure and function of *karigi*-related social groups). These same patterns are evident at least at some of the fall whaling camps (Fig. 4.13). Each *karigi* in turn would have

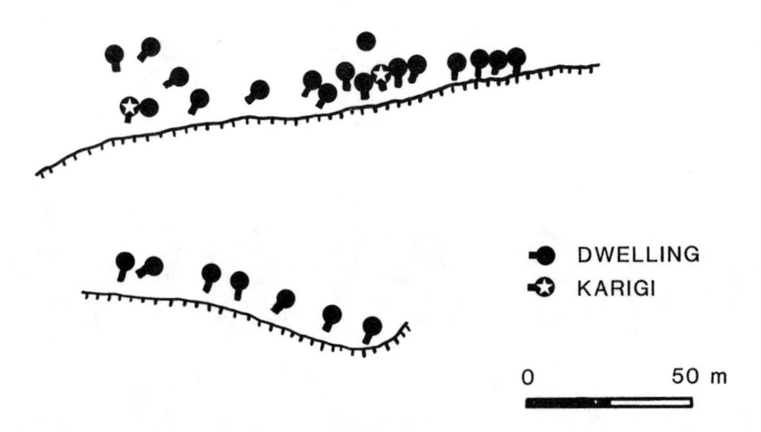

Figure 4.12. Site map of winter whale bone house site, Cape Garry, Somerset Island.

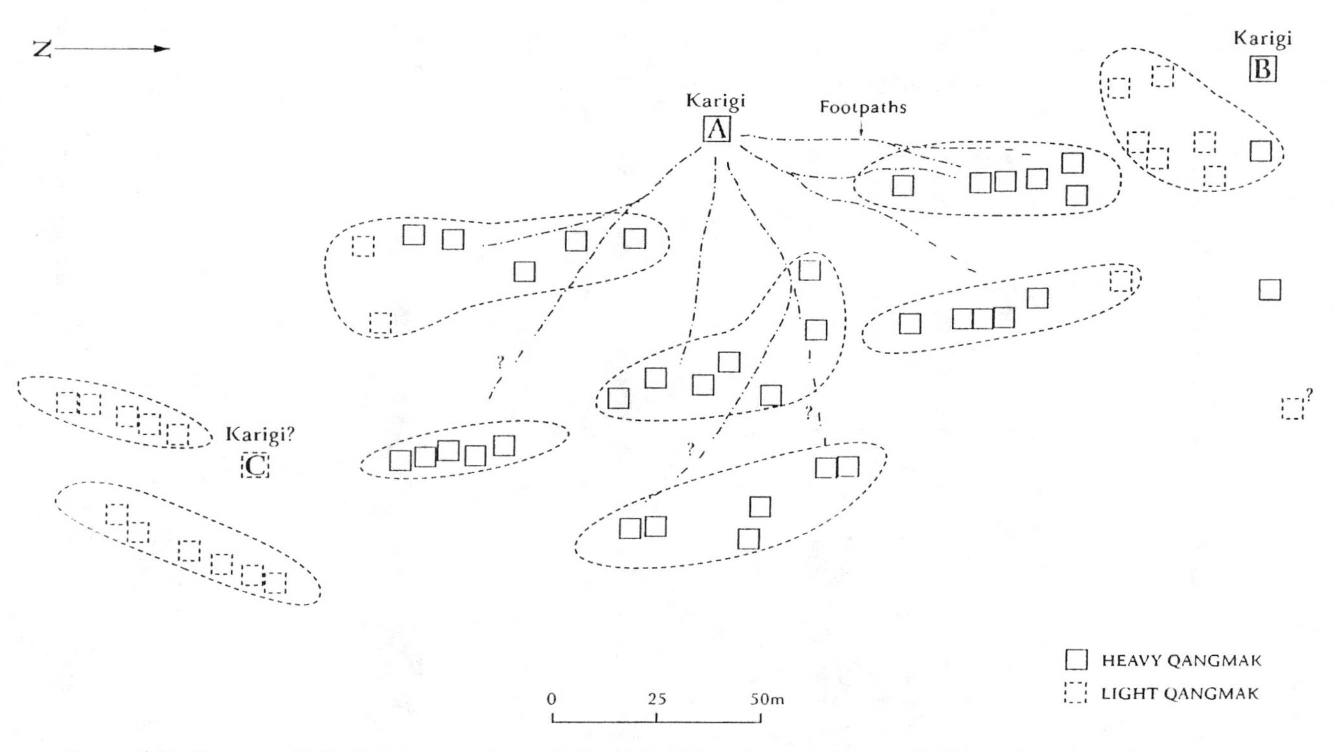

Figure 4.13. Site map of fall whaling camp, Somerset Island. Dwelling clusters indicate probable whaling crews (from Savelle and Wenzel n.d.).

been associated with several whaling crews. As noted by Burch (1981), in such circumstances amongst the North Alaskan Eskimo, not all whaling crew leaders would have been considered equal. Thus, we can anticipate a hierarchy of whaling crew leaders within *karigi* associations and also within individual villages, and very likely, within several villages in locally circumscribed territories. Furthermore, the competitiveness and building and maintaining alliances that these types of hierarchies entail result in considerable internal social constraints (Sheehan 1985, 1997; Whitridge 1999). These constraints may eventually have become so restrictive that, in the end, they did not allow sufficient flexibility for rapid or effective response to changing external—that is, environmental—constraints. The result was essentially either (1) a collapse of the system and/or (2) forced movement out of the area to maintain the system, rather than a relatively smooth transition from a collecting system to a foraging system. Thule societies in the southern, non-whaling region, though still initially engaged in collecting strategies, would nevertheless have been logistically much simpler. As a result, village or intervillage level hierarchies likely never developed to anywhere near the extent of those amongst the whaling societies. Accordingly, internal social constraints did not impose an impediment to flexibility in adapting to external constraints.

DISCUSSION

As stated at the beginning of this chapter, this approach to the study of Thule subsistence-settlement system change and the collapse of Thule whaling in the central CAA is very exploratory. In particular, the exact nature of the possible constraints and the mechanisms by which they prevented a smooth collector–forager transition have not been examined in any detail. These constraints may possibly be viewed in the general context of socially driven "subsistence conservatism", such as that which apparently led to the disappearance of the Greenland Norse colonies (see, e.g., Sheehan 1997: 36–39 for a discussion of similarities between Norse and North Alaskan subsistence conservatism). Other avenues of inquiry that might prove instructive are the examination of potential constraints among hierarchically organized hunter-gatherers in the context of, for example, costly signalling theory (see, e.g., Bliege Bird et al. 2001; Smith and Bliege Bird 2000) or Hawkes' (1991) "show off" model.

Overall, however, the chapter presents a very general working hypothesis of Thule subsistence-settlement system change from other than a rigidly environmental approach.

ACKNOWLEDGMENTS

I would like to thank Ben Fitzhugh and Junko Habu for the invitation to contribute to the volume, and to both those individuals and a third, anonymous, referee for their many valuable comments on an earlier draft of this paper.

REFERENCES

Ames, K. M., 1985, Hierarchies, Stress, and Logistical Strategies Among Hunter-Gatherers in Northwestern North America. In *Prehistoric Hunter-Gatherers: The Emergence of Cultural Complexity*, edited by T. D. Price and J. A. Brown, pp. 155–180. Academic Press, Orlando.

Bielawski, E., 1988, PaleoEskimo Variability: The Early Arctic Small Tool Tradition in the Central Canadian Arctic. *American Antiquity* 53:52–74.

Binford, L. R., 1980, Willow Smoke and Dog's Tails: Hunter-Gatherer Settlement Systems and Archaeological Site Formation. *American Antiquity* 45:4–20.

Bliege Bird, R., Smith, E. A., and Bird, D. W., 2001, The hunting handicap: Costly Signaling in Human Foraging Strategies. *Behavioral Ecology and Sociobiology* 50(1):9–19.

Boas, F., 1888, *The Central Eskimo.* Bureau of American Ethnology, Sixth Annual Report, Smithsonian Institution, Washington, DC, pp. 409–669.

Burch, E. S., Jr., 1980, Traditional Eskimo Societies in Northwest Alaska. In *Alaska Native Culture and History*, edited by Y. Kotani and W. B., Workman, pp. 253–304. Senri Ethnological Series, No. 4. National Museum of Ethnology, Osaka, Japan.

Burch, E. S., Jr., 1981, *The Traditional Hunters of Point Hope, Alaska: 1800–1875.* North Slope Borough, Alaska.

Freeman, M. M. R., 1979, A Critical View of Thule Culture and Ecological Adaptation. In *Thule Eskimo Culture: an Anthropological Retrospective*, edited by A. P. McCartney, pp. 278–285. National Museum of Man Mercury Series, Archaeological Survey of Canada, Paper 88.

Grier, C., and Savelle, J. M., 1994, Intersite Spatial Patterning and Thule Social Organization. *Arctic Anthropology* 31(2):95–107.

Habu, J., and Savelle, J. M., 1994, Construction, Use and Abandonment of a Thule Whale Bone House, Somerset Island, Arctic Canada. *The Quaternary Research* 33(1):1–18 (Japanese Association for Quaternary Research).

Hawkes, K., 1991, Show Off: Tests of a Hypothesis About Men's Foraging Goals. *Ethological Sociobiology* 12:29–54.

Helmer, J. W., 1991, The Paleo-Eskimo Prehistory of the North Devon Lowlands. *Arctic* 44:301–327.

Johnson, G. A., 1982, Organizational Structure and Scalar Stress. In *Theory and Explanation in Archaeology*, edited by C. Renfrew, M. J. Rowlands, and B. A. Seagraves, pp. 389–421. Academic Press, New York.

Kelsall, J. P., 1968, *The Migratory Barren-Ground Caribou of Canada.* Queen's Printers, Ottawa.

Maxwell, M. S., 1985, *Prehistory of the Eastern Arctic.* Academic Press, Orlando.

McCartney, A. P., 1977, *Thule Eskimo Prehistory Along Northwestern Hudson Bay.* National Museum of Man Mercury Series, Archaeological Survey of Canada, Paper 70.

McCartney, A. P., 1979, 1976 Excavations on Somerset Island. In *Archaeological Whale Bone: A Northern Resource*, edited by A. P. McCartney, pp. 285–314. University of Arkansas Anthropological Papers No.1. University of Arkansas, Fayetteville.

McCartney, A. P., 1980, The Nature of Thule Eskimo Whale Use. *Arctic* 33:517–541.

McCartney, A. P., and Savelle, J. M., 1985, Thule Eskimo Whaling in the Central Canadian Arctic. *Arctic Anthropology* 22(2):37–58.

McCartney, A. P., and Savelle, J. M., 1993, Bowhead Whale Bones and Thule Eskimo Subsistence-Settlement Patterns in the Central Canadian Arctic. *Polar Record* 29(168):1–12.

McCullough, K. M., 1989, *The Ruin Islanders: Early Thule Culture Pioneers in the Eastern High Arctic*. Canadian Museum of Civilization, Ottawa.

McGhee, R., 1969/70, Speculations on Climatic Change and Thule Culture Development. *Folk* 11–12:173–184.

McGhee, R., 1978, *Canadian Arctic Prehistory*. Van Nostrand Reinhold, Toronto.

McGhee, R., 1984, Thule Prehistory of Canada. In *Handbook of North American Indians, Vol. 5. Arctic*, edited by D. Damas, pp. 369–376. Smithsonian Institution Press, Washington, DC.

McGhee, R., 1996, *Ancient People of the Arctic*. UBC Press, Vancouver.

Morrison, D. A., 1989, Radiocarbon Dating Thule Culture. *Arctic Anthropology* 26:48–77.

Park, R. W., 1994, Approaches to Dating the Thule Culture in the Eastern Arctic. *Canadian Journal of Archaeology* 18:29–48.

Parker, G. R., 1972, Biology of the Kaminuriak Population of Barren-Ground Caribou, Part 1: Total Numbers, Mortality, Recruitment and Seasonal Distribution. Canadian Wildlife Service, Report Series 40.

Sabo, G. III, 1991, Long Term Adaptations Among Arctic Hunter-Gatherers: A Case Study from Southern Baffin Island. Garland, New York.

Savelle, J. M., 1987, Collectors and Foragers: Subsistence-Settlement System Change in the Central Canadian Arctic, A.D. 1000–1960. BAR International Series No. 358, British Archaeological Reports, Oxford.

Savelle, J. M., 1994, Prehistoric Exploitation of White Whales (*Delphinapterus leucas*) and Narwhals (*Monodon monoceros*) in the Eastern Canadian Arctic, Meddelelser om Gronland. *Bioscience* 39:101–117.

Savelle, J. M., 2000, Information Systems and Thule Eskimo Bowhead Whaling. In *Animal Bones: Human Societies*, edited by P. Rowley-Conwy, pp. 74–86. Oxbow Books, Oxford.

Savelle, J. M., and Habu, J., 1998, Thule Inuit Settlement Systems and Artifact Assemblage Variability. Paper presented at the 63rd Annual Meeting of the Society for American Archaeology, Seattle.

Savelle, J. M., and McCartney, A.P., 1988, Geographical and Temporal Variation in Thule Eskimo Subsistence Economies: A Model. *Research In Economic Anthropology* 10:21–72.

Savelle, J. M., and McCartney, A. P., 1994, Thule Inuit Bowhead Whaling: A Biometrical Analysis. In *Threads of Arctic Prehistory: Papers in Honour of William E. Taylor*, edited by D. Morrison and J. -L. Pilon, pp. 281–310. Canadian Museum of Civilization, Mercury Series, Archaeological Survey of Canada, Paper 149.

Savelle, J. M., and McCartney, A. P., 1999, Thule Eskimo Bowhead Whale Interception Strategies. *World Archaeology* 30(3):437–451.

Savelle, J. M., and Wenzel, G. W., n.d., Out of Alaska: Reconstructing Prehistoric Canadian Thule Culture Social Structure. Submitted for inclusion in *Hunter-Gatherers of the North Pacific Rim*, edited by J. Habu, S. Koyama, J. M. Savelle, and H. Honto. Senri Ethnological Series. National Museum of Ethnology, Osaka, Japan.

Schledermann, P., 1975, *Thule Eskimo Prehistory of Cumberland Sound, Baffin Island, Canada*. National Museum of Man, Mercury Series, Archaeological Survey of Canada, Paper 38.

Sheehan, G. W., 1985, Whaling as an Organizing Focus in Northwestern Alaska Eskimo Society. In *Prehistoric Hunter-Gatherers: The Emergence of Cultural Complexity*, edited by T. D. Price and J. A. Brown, pp. 123–154. Academic Press, Orlando.

Sheehan, G. W., 1995, Whaling Surplus, Trade, War and the Integration of Prehistoric Northern and Northwestern Alaskan Economies, A.D. 1200–1826. In *Hunting the Largest Animals: Native Whaling in the Western Arctic and Subarctic*, edited by A. P. McCartney, pp. 185–206. Studies in Whaling No.3. Canadian Circumpolar Institute, Edmonton.

Sheehan, G. W., 1997, *In the Belly of the Whale*. Aurora, VI. Alaska Anthropological Association, Anchorage.

Smith, E. A., and Bliege Bird, R. L., 2000, Turtle Hunting and Tombstone Opening: Public Generosity as Costly Signaling. *Evolution and Human Behavior* 21:245–261.

Spencer, R. F., 1959, *The North Alaskan Eskimo: A Study in Ecology and Society*. Bureau of American Ethnology Bulletin 171, Smithsonian Institution, Washington, DC.

Stenton, D. R., 1989, Recent Archaeological Investigations in Frobisher Bay, Baffin Island, N.W.T. *Canadian Journal of Archaeology* 11:13–48.

Whitridge, P., 1999, The Construction of Social Difference in a Prehistoric Inuit Whaling Community, Ph.D. Dissertation, Arizona State University.

Whitridge, P., 2000, The Prehistory of Yupik and Inuit Whale Use. *Journal of American Archaeology* 16:99–154.

Yorga, B., 1979, Migration and Adaptation: a Thule Culture Perspective. In *Thule Eskimo Culture: an Anthropological Retrospective*, edited by A. P. McCartney, pp. 286–300. National Museum of Man, Mercury Series, Archaeological Survey of Canada, Paper 88.

Chapter 5

Natufian

A Complex Society of Foragers

OFER BAR-YOSEF

OPENING REMARKS

The interest in the significance of the Natufian culture to the study of social evolution was ignited some 70 years ago when the contents of the first excavated sites, Shukbah Cave and El-Wad Cave and Terrace (Garrod 1932; Garrod and Bate 1937, 1942) indicated that they may represent the socioeconomic phase that preceded the Neolithic Revolution. According to the terminology of the day, the proliferation of microliths and in particular the lunates led Garrod and others to define the Natufian assemblages as Mesolithic. This label, in Europe, meant early post-Glacial hunters. However, Garrod also uncovered numerous sickle blades (identifiable by their special sheen) and mortars and pestles. Her intuitive conclusion was that "it may seem surprising that we get the evidence of agriculture at such an early date among people who possess no pottery and do not appear to have domesticated animals" (Garrod 1932: 268). This interpretation that Natufian sites were farming communities was later challenged, but the importance of these discoveries as evidence in the search for the origins of agriculture in western Asia did not escape the eyes of most scholars.

OFER BAR-YOSEF • Peabody Museum, Harvard University, Cambridge, Massachusetts 02138.

For example, G. Childe noted that "only one culture is known today that can lay claim plausibly to temporal priority and transitional status. The chronological position of the Natufian before the local Neolithic in the Palestinian culture sequence has been established stratigraphically at Jericho... The Jerichoan Neolithic could be derived from the Natufian Mesolithic." (Childe 1953: 28–29, 226). Hence, the relative archaeological sequence was already established in the excavations of the 1930s. Recently, we were able to dendrocalibrate the chronology of the Natufian culture (Fig. 5.1), which demonstrates its great antiquity and long duration.

The evolutionary position of the Natufian culture as a hunting and gathering society, the first to disappear from the global Pleistocene milieu and give way to farming communities, attracted the attention of anthropologists. The inquiry concerning its emergence in a world of foragers remains controversial as does its transformation to a Neolithic population subsisting on cultivation, hunting, and fishing. Proposed explanations are derived

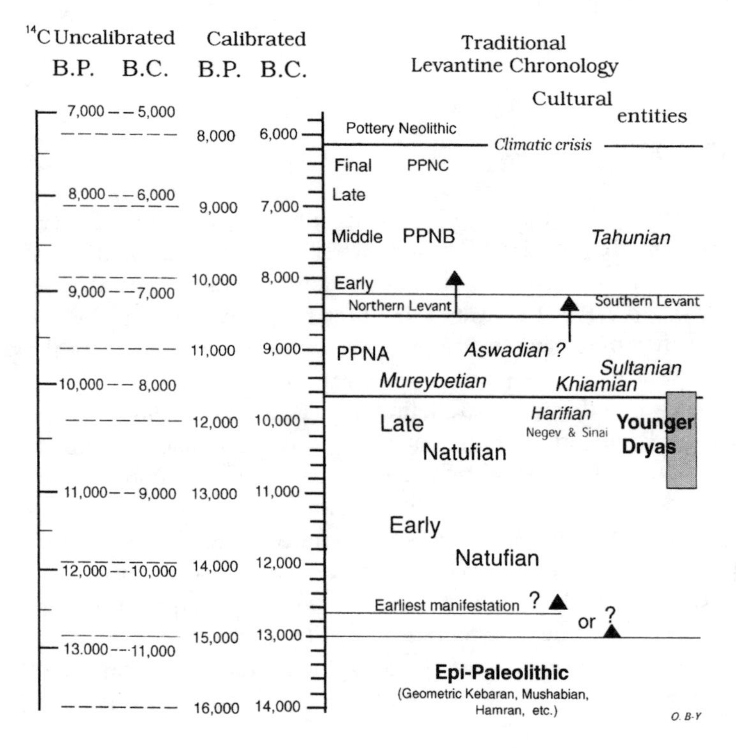

Figure 5.1. Uncalibrated and calibrated radiocarbon chronologies (INTCAL98.4.01) of the Levant. Please note the uncertainity of the earliest dates of the Natufian and the duration of the Younger Dryas.

from the realm of ecological adaptations and/or culture-history currently enhanced by stressing the social factors in societal changes such as the role of human agency and symbols (Bar-Yosef and Belfer-Cohen 1989, Belfer-Cohen and Bar-Yosef 2000; Cauvin 2000a; Goring-Morris and Belfer-Cohen 1997; Henry 1989). Consequently, the need to identify the social structure and institutions within the Natufian culture lately became a target for intensified research (Bar-Yosef, in press; Belfer-Cohen 1995; Byrd and Monahan 1995; Henry 1989; Valla 1999).

Tracing the various types of social organizations and institutions in the archaeological records is not an easy task. We make inferences from known historical and ethnographic records. In particular it is more arduous for us to "read" the material evidence of prehistoric populations of the Late Paleolithic than that of the Neolithic in southwest Asia. Preservation varies in almost all the excavated sites, and often the bulk of perishables such as wooden tools, baskets, garments, cords, edible and nonedible plants are not available to the archaeologist. Thus, we are ignorant of the role played by women and children and certain activities that were performed by men. Indeed, we must be satisfied with the remains that are preserved such as stone tools, animal bones, dwelling structures, burials, and in rare cases, carbonized plant remains, pollen spectra, and phytoliths that provide insights into the use and consumption of vegetal substances. However, to test our evolutionary models, we must design testable hypotheses. For this purpose, we conduct actualistic or replicative studies that hopefully provide insights for interpreting the available remains and the missing data. For example, searching for ashes was proven possible through mineralogical analyses when the observable hearths are gone (Schiegl et al. 1994). In sum, if we are to identify social institutions, group organization, degree of equality among members of a given population, and trends of change, we need to seek a range of information aggressively to obtain plausible inferences.

The current evidence points to the Natufians as those who initiated intentional cultivation, which culminated in the establishment of Neolithic villages where plants and animals were domesticated (Bar-Yosef 1998; Cauvin 2000b; Harris 1998). In a simplified version of the evolutionary trajectory, it seems that the new social structures of Early Natufian sedentary hamlets replaced the former so-called egalitarian mobile foragers and enjoyed rapid population growth, as well as an increasing degree of inequality. Assuming that this scenario is correct (and further testing is provided in the following pages), an interesting task for prehistorians would be to find "how" and "why" these unpredictable consecutive changes were triggered and sustained. The question can be alternately phrased as, What happened within the Natufian society during the three millennia of its duration that resulted in the emergence of farming communities?

MODELING THE EMERGENCE OF SMALL-SCALE SOCIETIES

Before delving into the "when," "where" and "why" the previously described process took place, we need to review briefly the theoretical aspects involved in the study of "small-scale," "intermediate," or "middle-range" societies, the category into which the Natufian falls. In the course of classifying the socioeconomic state of a given social entity and its place along a continuum from presumably an egalitarian to a nonegalitarian society, most scientists develop their models from the ethnographies of various regions. Prominent among these are the North American northwest coast, the New Guinean and the Pacific Islands field studies, as well as a few sub-Saharan African examples (Arnold 1996a,b; Hayden 1990, 1995b; Johnson and Earle 2000 and papers therein; Price and Feinman 1995 and papers therein; Upham 1990).

Not surprisingly, while analyzing the intricate social structure of nonegalitarian societies, whether related to power and labor (Arnold 1996; Earle 1997) or household sizes and social stability (Ames 1995; Coupland 1996), the authors rely on the recorded regional continuity between the ethnohistorical information and the immediately preceding archaeological remains. However, in most, if not all of these cases, the transition from simple hunting and gathering to nonegalitarian foragers is hardly explained, perhaps because the historical records, including most oral traditions, do not provide ample evidence for the primordial time of the local population discussed. Therefore, we are left with a somewhat static summary of the social components and organization of small-scale societies, be it of foragers or farmers, and need to hypothesize the processes that led to increasing social complexity (Ames 1995; Hayden 1995b; Johnson and Earle 2000; Price 1995).

In an interesting effort to determine how egalitarian societies evolved to be controlled by a chief, Hayden (1995b) proposed a series of new descriptive terms for the various phases of social organization. Hayden built a double model based, on the one hand, on the North American northwest coast and, on the other hand, on cases from New Guinea. Hayden's model, like others, suffers from two unresolved problems. The first is that we are rarely instructed how to bridge the theoretical and practical gap between the ethnographic samples and the archaeological observations to validate or refute the proposed interpretations of the social organization. An obvious example is the issue of feasting, which Hayden correctly views as an important social activity. How can one securely derive the necessary evidence from the flimsy residues of a hunting and gathering group (often just a lithic assemblage and a pile of bones)? The information one hopes to retrieve from the exposures in Neolithic villages (such as the presence of public open areas, particular dumping zones, uncovered sanctuaries, and the like) is far richer from late Paleolithic or Natufian contexts in the Near East.

The second difficulty emerges from the ethnographic sources themselves. For example, we have almost no archaeological or historical data to show how the New Guinean populations became a suite of variable nonegalitarian societies, as recorded by western ethnographies (Roscoe 2000). In part, this observation is also correct for the North American northwest coast, although archaeologically this region is much better known (Ames 1995, 1999; Ames and Maschner 1999; Hayden 1997; Matson 1985). Hence, to examine the process that led to the emergence of the Natufian culture as an entity that would fall under the category of a complex society of hunter-gatherers (Henry 1985, 1989), we must turn to the basic models available in the literature (Bettinger 1991; Binford 1980, 1996; Keeley 1988; Kelly 1992, 1995).

The determinant factors in the lifeways of "hunter-gatherers" (I follow Kelly's approach and use this term as interchangeable with "foragers") include the needs of every biological entity for long-term survival, namely, subsistence and reproduction. From this follows that existence in a certain area requires knowledge of the spatial distribution of resources, the technology to exploit some or most of them, and a social organization that will be sufficiently flexible to face seasonal and annual shortages. Each of these categories is briefly detailed.

To survive on the available resources in a given environment, annual monitoring of the degree of reliability, accessibility, and predictability of every exploitable resource, whether vegetal or animal, is required. A certain technical level of food acquisition and preparation is needed for populations who cannot practice the simple strategy of "feed as you go" (Isaac 1984; Oswalt 1972). The technical information as well as the spatial, annual, and decadal fluctuations of resources are stored in the collective long-term memory of the group, particularly in environments prone to major climatic vagaries (Minc and Smith 1989). Finally, maintaining the mating system even across large territories is crucial for the biological viability of the population. Thus, consideration of these variable components led Binford to suggest the by now famous "forager–collector" model (Bettinger 1991; Binford 1980; Kelly 1995).

In classifying the activities and movements of bands of hunter-gatherers, it was proposed that "foragers" are those who survive in their environment by moving their camp from one resource location to another, known as Binford's "residential mobility" or in the Near East as Mortensen's "circulating settlement pattern" (Binford 1980; Mortensen 1972). The "collectors" differ by locating their camps in strategic places from where they make forays by employing task groups or individuals. Through these forays, they provide consumers with food and materials in a strategy known as "logistical mobility." The radius of daily trips was the basis for the "site catchment analysis" proposed by Vita-Finzi and Higgs (1970) and further elaborated by Bailey and Davidson (1983) in the Mediterranean basin.

Unfortunately, the ethnographic evidence has its limits. We need to keep in mind that except for the Mediterranean-type environments in Australia (Lourandos 1997) and certain parts of California (Kelly 1995), most examples of hunting-gathering groups cannot serve as simple analog to the Eastern Mediterranean situations. From the recorded ethnographic data, we know that the range from high to low residential mobility depends on the level of primary biomass of a given region and the type of available resources such as game animals, fish, fruits, seeds, and tubers. These resources vary from the arctic to the tropical forest (Kelly 1995, Table 3-1). A measure for resource productivity was suggested to incorporate the calculated "primary production" (PP) and the "effective temperature" (ET; Bailey 1960; Bettinger 1991; Binford 1980; Kelly 1995). Scholars agree that there is generally a good fit between the ET and the relative frequencies of dietary sources (hunting, gathering, and fishing) among the various groups of foragers. However, the calculations of both ET and PP do not allow for the effects of CO_2 fluctuations during the Terminal Pleistocene. In principle, higher CO_2 favors plants with C_3 pathways, such as cereals and numerous tree species, whereas lower ratios would support the C_4 pathways (Sage 1995).

In describing the archaeological data from the Levant and its social interpretation, I recognize, like other modelers, that the common denominator for all societies is human nature. In the so-called simple forager societies, this means inter- and intragroup competition, the presence of alpha males, the acceptance of temporary or permanent coercion (*sensu* Durkheim 1938), and occasional violent conflicts (Keeley 1996; Knauft 1987). These activities that dominate daily life are classified as social interactions interwoven with technology, subsistence, and ritual (Service 1962).

Beyond the basic societal traits, the rudimentary types of social organizations employed in this paper will be "egalitarian" and "nonegalitarian" societies. The supposedly simple organizational state is well known (Johnson and Earle 2000; Sahlins 1968). I will use the second as defined through a series of attributes by Kelly (1995). Originally, a nonegalitarian society was considered a stage in social evolution, but to keep the terminology separate from unfounded assumptions, it was named "complex society" (Price and Brown 1985), and later as "small-scale" or "intermediate" society with the understanding that this class demonstrates a considerable inherent variability (Arnold 1996a; Price and Feinman 1995).

Among pristine nonegalitarian societies in a region where resources allowed a semi- to almost fully sedentary annual cycle, we may expect a headman to emerge, although not necessarily as a hereditary position. Hierarchies are better recorded for hamlets and villages, where local headmen probably wielded power.

Applying the various social definitions to the archaeological remains—even when preservation is excellent (i.e., littered with organic remains) and well recorded by today's standards—is open to diverse interpretations. Therefore, I feel obliged to explicitly enumerate the archaeological attributes that signify the social state (for the society) and/or the presumed status of the individual. For this purpose, similarly to others, I rely primarily on information from graveyards, size and contents of houses, intrahamlet organization, and the nature and degree of long-distance exchange (or trade). A few examples are in place.

The presence or absence of decorated skeletons (with the distinctions of male/female and age at death) among sedentary or semisedentary foragers is often interpreted as a status symbol. In village societies, a seriation from simple to elaborate burials with rich body decorations and grave offerings reflects individual (and/or household) status within the community. The lack of decorated skeletal relics in the village community is interpreted as motivated by the need to negotiate equality among members (Kuijt 1996). Variability in the size of domestic buildings is interpreted as an expression of the wealth of a family or the household (Byrd 2000).

Driven by the need to interpret phenomenon larger than a particular site, we review the data gathered from several sites. In such an overview, we expect to detect differences between a "core area" and its "periphery." This kind of socioeconomic distinction, which could but did not always overlap a sociopolitical classification, relates to world systems and industrial advances, and also to socially construed concepts, ideologies, and material elements. The notion that "core areas" and "peripheries" existed in the prehistoric past was not uniformly accepted. Three decades ago, archaeologists considered that all past hunter-gatherer groups across the continents formed a social and biological continuum (Wobst 1976). Every proposed subdivision on the basis of comparative lithic analysis was considered an artificial construct of past societies by researchers who wanted to identify "archaeological cultures" (Binford and Sabloff 1982; Chilton 1999, and papers therein; Dobres 2000). Those who study knapping techniques and stone-tool typologies were often blamed for "creating prehistoric ethnicities" to build a cultural-historical sequence exclusively reflected in material elements. The role of the human agency in configuring the material production through social interactions and/or the role of gender were presumably neglected or ignored by most lithic analysts. However, in recent years the approach that advocates the importance of the symbolic role embedded in the various forms of material culture that reflect social boundaries, while avoiding the artificial dichotomy between style and function, is resurfacing (Chilton 1999; Wobst 1999). Moreover, the realization that genes, languages, and material culture do not necessarily covary facilitated the return to the basic aim of investigating archaeological cultures and their subsequent changes. Such studies are aided by advances in the anthropology of technology and improved understanding of the nature of human actors in the past (Clark 1999; Dobres and Hoffman 1994; Lemonnier 1992). None of these approaches to "archaeology as history" pretends to identify "ethnicity," language, or the evidence for gene drift versus gene flow.

The investigation of material culture may seem tedious and often an unrewarding exercise, but it enables us to recognize various important aspects of human behaviors. These include the study of subsistence systems which incorporates the seasonal exploitation of different resources gleaned through patterns of hunting, trapping, fishing, fowling, and butchering. Lithic studies record stone toolmaking and the variable usage of selected items, raw material procurement from local or remote sources, core reduction techniques, curation of selected artifacts, and tool functions deducted from microwear and edge damage investigations. Building techniques and the variable use of natural shelters are complemented by studies of home ranges and territories. As the following pages will demonstrate, these research approaches that were practiced in the variable landscape of the Levant document aspects of the Natufian culture. Therefore, I begin by attempting to convey briefly the information on the natural features and sources of this region.

THE LATE PLEISTOCENE–EARLY HOLOCENE LEVANTINE LANDSCAPE AND ITS RESOURCES

The Mediterranean Levant is about 1100 km long and about 250–350 km wide and it incorporates a variety of landscapes from the southern hilly flanks of the Taurus–Zagros arc in Turkey to the Sinai Peninsula (Bar-Yosef and Meadow 1995). The variable topography encompasses a narrow coastal plain and two parallel and almost continuous mountain ranges with a Rift Valley in between. The eastward sloping plateau is dissected by the river valleys such as the Euphrates, a series of wadis, and a few additional shorter mountain ranges. In the northern Levant, the area between the Euphrates and the Tigris, where two river valleys (the Balikh and the Khabur) descend as the north–south tributaries of the Euphrates, is also known as Upper Mesopotamia.

The region is characterized by marked seasonality: winters are cold and rainy, and summers are hot and dry. Mediterranean forests and woodlands (dominated by various species of oaks) grade into open parkland vegetation (the oak–terebinth belt). The latter changes into terebinth–almond woodland–steppe that becomes shrubland, steppic vegetation. Annual precipitation varies from 400 to 1200 mm per year and decreases to under 200 mm per year where arid plant associations (Saharo-Arabian) proliferate (Hillman 1996; Zohary 1973).

Today, the annual patterns of winter storm tracks alternate infrequently. One carries humidity from the Mediterranean Sea to the southern Levant, and the second crosses from northern Europe and turns to the northern Levant, leaving the southern portion dry. Chemical studies of the beds of the Upper Pleistocene lake Lisan and the early Holocene distribution of C_3 and C_4 plants in the Negev demonstrated that during the Late Pleistocene and the early Holocene, the geographic pattern of annual rainfall was similar to today's although the amount varied.

Detailed paleoclimatic information is derived from the records of oxygen isotope fluctuations registered in ice cores, deep sea cores, and the oxygen $^{16/18}O$, ^{13}C wiggles in which stalagmites reflect rainfall fluctuations (Bar-Mathews et al. 1997, 1999; Frumkin et al. 1999). The terrestrial vegetational reconstructions based on pollen cores from lakes were recently reanalyzed by Rossignol-Strick (1995, 1997) who demonstrated that the radiocarbon dates in most of the inland pollen cores (Baruch and Bottema 1991; van Zeist and Bottema 1991) are aberrant due to the effects of hard water in the lakes. Summarizing the various sources, the following climatic fluctuations can be discerned:

1. During the last glacial maximum (abbreviated as LGM), a series of cold events occurred from about 22,000 cal. B.P. to 19,000 cal. B.P. The entire region was cold and dry, but the hilly coastal areas enjoyed winter precipitation and were covered by forests.
2. Precipitation increased from about 19,000 cal. B.P. onward, and more rapidly from 15/14,500 cal. B.P., peaking around 13,500 cal. B.P. This period corresponds to the warmer Bölling/Alleröd in Europe.
3. Rainfall decreased considerably during the Younger Dryas (ca. 13/12,800–11,600/500 cal. B.P.), and CO_2 values fell to earlier levels.
4. Pluvial conditions returned around 11,500/300 cal. B.P. but did not reach the previous peak in the central and southern Levant. A wetter Early Holocene (11,300–8400/8200 cal. B.P.) allowed the establishment of systematic cultivation during the Early Neolithic and later allowed the PPNB civilization to flourish.

Among the physical effects was the gradual rise in sea level after the LGM until the mid-Holocene which reduced the flat sandy coastal plain of the Levant by a strip 5–20 km wide and 500 km long. Given that this corner of the Mediterranean Sea is poor in aquatic resources, the rising water affected mainly the size of the foraging territories and the collection of marine shells commonly used for decoration.

The dominant factor in determining the biomass in the Levant is the amount and distribution of winter rains (Shmida et al. 1986) more than the temperature. Hence, local resources are seasonal, seeds and leaves are the most abundant from late February through July, and fruits from September to November. Edible tubers are rare. Among the three vegetational zones, the Mediterranean oak forest is the richest, and has more than a hundred species of edible fruits, seeds, leaves, and tubers (Zohary 1973).

Procurement techniques depend on the availability, reliability and predictability of resources. Although the *r*-resources (mostly vegetal) could be obtained almost everywhere within the main vegetational belts, the *K*-resources, the game animals, had variable home ranges judged historically collected information (Harrison and Bates 1991).

In the northern Levant, sheep (*Ovis orientalis*) and goats (*Capra aegagrus*), were the common species, as evidenced by the faunal assemblages from the Taurus and Zagros ranges, the foothills and parts of Upper Mesopotamia. Other taxa there and further south were wild cattle (*Bos primigenius*), wild boar (*Sus scrofa*), fallow deer (*Dama mesopotamica*), red deer (*Cervus elaphus*), and roe deer (*Capreolus capreolus*). The mountain gazelle (*Gazella gazella*) was common in the parkland. This gazelle is a stationary antelope with a small home range that varies from a few to as many as 25 square kilometers. A similar pattern, perhaps with a larger seasonally determined home range, can be inferred for goitered gazelle (*Gazella subgutturosa*), the dominant species in the Syro-Arabian desert. Other ungulate forms commonly exploited by Late Pleistocene/Holocene peoples were wild equids (*Equus hemionus* and *Equus africanus*). Ibex (*Capra nubiana*) occupied the craggy areas (Uerpmann 1987). Of these, only the equids—inhabitants of the desert-steppe such as the goitered gazelle—are likely to have migrated over significant distances. The other species—similar to the mountain gazelle—had more restricted ranges within the forest, woodland, and parkland belts and thus formed reliable, accessible, and even predictable and abundant resources for human exploitation.

If societal structure is determined by its past ecology, then the geography and the spatial environmental pattern exploited by a given population could be homogeneous and/or heterogeneous. Estimating, especially in the latter case, the carrying capacity is hardly a matter of a mathematical formula. Resources could be available, but exploitation depends on the practiced technology, group size, scheduling, and social interactions.

Hence, features of physical geography, the size of the region measured in square kilometers, as well as the nature of the resources, should be taken into account. The Levant is a subregion of western Asia, where within a short transect (e.g., 80–150 kilometers), one finds an almost globally uncoupled topographic and vegetational heterogeneity. This observation is illustrated, for example, by the number of plant species per square kilometer (Danin 1988, Table 1). Compared to other Mediterranean countries, the Levant, has 0.0855 species per square kilometer, a little more than twice the number of those in coastal California, Greece, or countries in temperate Europe. Hence, the Levant is a region that differs in size, resources, and seasonality from other regions where hunter-gatherers were studied, although is probably similar to a few small territories in California. However, analogies for societal evolution are often taken from groups that inhabit the larger territories, where foragers were not displaced by farmers. As a cautionary remark, we note that the Levant is about one-third the size of the North American northwest coast and about one-fifth that of New Guinea. In sum, comparisons among these three regions should take into account the number, kind, and distribution of potential food resources, whether vegetal and/or animal, and their seasonal availability, before we use the latter two as a basis for analogies.

APPLYING THE MODEL TO LEVANTINE CONDITIONS

The optimal settlement pattern of Late Pleistocene hunter-gatherers in the Eastern Mediterranean, as will be demonstrated below, combines both residential and logistical movements, and was probably the most efficient strategy (Bar-Yosef and Belfer-Cohen 1989, 1992; Bar-Yosef and Meadow 1995). Topography made anticipated moves of social units or task forces along east–west transects (and north–south in the Taurus mountains region) easier, because this transit took advantage of the wadis or small rivers descending from the mountain and hilly ranges. In addition, the vegetational belts stretched north–south in the Levant and east–west in the Taurus made the crossing of at least two phytogeographical zones most profitable.

Elsewhere, we estimated the optimum territory for the long-term survival of a band of hunter-gatherers within the Mediterranean woodland/parkland belt at about 300–500 square kilometers, whereas foragers in the steppic and/or desertic region were required to monitor a larger area of 500–2000 square kilometers as a buffer against annual fluctuations (Bar-Yosef and Belfer-Cohen 1989).

In this settlement pattern, decreasing annual precipitation and shifts in the regional distribution of rains, caused diminishing yields of wild fruits, seeds, and game animals, and generated stress first on groups that survived in the steppe and semiarid belts. In contrast, resources in the woodland/parkland zone would have been stable if the climatic regime did not change and, in times of severe droughts or cold and harsh winters, required additional larger territories for survival. However, this vegetational belt was much less affected than the steppic-arid region. For example, a time of major climatic impact was the dry and cold Younger Dryas (Bar-Yosef 2000; Bar-Yosef and Belfer-Cohen in press).

To alleviate short- and long-term stress, Levantine foragers would have had a number of options: (1) to aggregate population into the woodland/parkland belt and to form sedentary communities, (2) to reorganize the social and technoeconomic structure within the same territories which would have an impact on those occupying neighboring areas, (3) to migrate to adjacent regions northward or southward along the coastal ranges, or (4) to take over other territories by force especially where bands did not belong to the same alliance (Bar-Yosef and Belfer-Cohen 1991).

Each of these strategies or a combination of several would have led to the formation of new social organizations and eventually to adjusted ideologies. Hence, the conditions of the particular environment would result in a spatial realignment, growth, or decline of the settlement pattern. As the new social regime succeeded, one may expect to trace the social realignments or structures reflected in the ideological realm.

In the following pages, I will try to demonstrate how modes of production, fluid and/or conservative societal structures, and environmental fluctuations intertwined to form what could be a reasonably coherent social history of the Natufian culture. But before we delve into the archaeological documentation, the issue of sedentism requires clarifications.

THE CYCLICAL NATURE OF SEDENTISM

Sedentism is regarded as the essential first step for the evolution of complex societies. The term "sedentism" implies that the presence of humans in a given location is permanent and shows signs of ownership by the particular community. Such places could be caves, rock shelters and, in particular, open-air sites. It is often assumed that the archaeological evidence is sufficiently explicit to ensure identifying a site as sedentary. The most common attribute is the built-up environment of human communities. As such, and identified as either hamlets or villages, it has been shown that they embody the inhabitants' symbols of social structure and land ownership. Basic architecture, kinship, economic relationships and cosmology within "House Societies," (Lévi-Strauss 1983) were generally studied in contemporary villages or in historical contexts (Carsten and Hugh-Jones 1995; Oliver 1971; Rapoport 1969). Therefore, it is not surprising that a cluster of houses, recognized as a hamlet or village, was taken to represent a sedentary community (Rafferty 1985). However, given the ambiguities in the interpretation of the Pleistocene prehistoric remains, one can assume that in general villages and towns of Chalcolithic or Bronze Age periods can be taken as evidence of sedentism.

Ethnoarchaeological research, as well as historical documents, caution us against employing the presence of pit-houses or above ground dwellings as the sole determinant evidence of sedentism. There are known cases of mobile groups of herders who invest in constructing shelters within the confines of their anticipated year-round routes (Kent 1989). Recognizing this difficulty means that labeling prehistoric sites sedentary requires a careful approach.

The pioneering works of Tchernov (1991a,b, 1993a, 1997), demonstrated that the degree of sedentism should rely on the presence of high frequencies of commensals ("self-domesticated species") among the bones of microvertebrates and birds in the prehistoric context. The best candidates among the commensals include the house mouse, rats, and the house sparrow. In semiarid areas such as the Sinai Peninsula, the spiny mouse was found as a commensal (Haim and Tchernov 1974).

Scholars agree that sedentism has both cultural and biological affects (Belfer-Cohen and Bar-Yosef 2000; Bentley et al. 1993; Rosenberg 1998 and references therein). However, there are divergent interpretations for the reasons human groups became sedentary. Interestingly, authors who propose one model or another rarely discuss the particular population in isolation, regardless who were the neighbors.

One explanation holds that sedentism is caused by the attraction of humans to spatially rich, stable, and restricted resources, where the "law of least effort" would enhance the attraction to stay for many months in the same area. This process is also called the "pull" model (Stark 1986; Rosenberg 1998). The second interpretation suggests that economic and social circumstances enforce sedentism, and therefore it is labeled the "push" model (Bar-Yosef and Belfer-Cohen 1989, 1991; Henry 1989; Keeley 1988; Kelly 1991; Rosenberg 1998 and references therein). This scenario is applicable when abrupt climatic change may force foragers to map onto recourses in another geographic area. The decision to become sedentary seems to occur when a seasonal camp within a mobility pattern under favorable conditions becomes a permanent camp that allows for optimal exploitation of resources (both *K*- and *r*-selected) with or without developing storage techniques. There is plenty of room for the role of human agency in both strategies for optimal socioeconomic solutions in times of stress. For example, pregnant females joined by the older members of the group may decide to stay a foot and thus force the active males to drop their plans to relocate the camp to another location.

In each discussion of "sedentism," the nature and degree of "population pressure" is raised. Archaeologists often interpret this to mean population growth and acknowledge that foragers, in spite of their high fertility potentials tend to regulate their population growth to survive in a given environment. The role of disease is not considered in most of these calculations. Evidently, the Pleistocene world of foragers did not fill the entire world, especially before the last glacial maximum. The improved ecological conditions in many parts of the world after the LGM (Straus et al. 1996 and papers therein) resulted in population sprawl. Populations were larger relative to the previous period when cold and dryer conditions prevailed in most regions. Although the data from the Terminal Pleistocene period is plagued with uncertainties that emanate from the degree of site visibility and recovery techniques, various archaeologists observed a population growth during this time in the Levant (Bar-Yosef and Belfer-Cohen 1989; Henry 1989; Moore 1989). However, none of the observable increases resemble the rapid population growth recorded since the expansion of the agricultural systems. Therefore, the evidence sought in archaeological data for "population pressure," for example, before the Natufian sedentism, is simply not there. In spite of numerous publications of surveys and site sizes in the Levant, population densities are not sufficient to stipulate an objective measure of "population pressure" leading to sedentism. Therefore, we need to incorporate in our considerations the social concept of what "too many people around" means within each society. It is this kind of social perception that could influence the decisions of groups and/or leaders.

"Population pressure" can be viewed as a culturally constructed concept that arises in a territorially bounded population which must decide, when their mobility options are reduced, either to settle down or expand and face physical conflicts with their neighbors. Today, a similar a social concern is reflected by the road signs in the US that read "thickly settled" in areas that would be considered "sparsely settled" by Near Eastern or Chinese standards (Bar-Yosef 1997).

Sedentism has sociopolitical, economic, and health ramifications. Daily life in a village larger than a foragers' band heralds the restructuring of the social organization (Flannery 1972) because it imposes more limits on the individual and on entire households. To ensure the long-term predictability of livable conditions in a village, members accept certain rules of conduct that include, among others, leaders or headmen (possibly the richest in the community), shamans (practicing in closed shrines), ceremonies (conducted publicly in an open space), and the like (Wolf 1966). The archaeological correlates for most of these aspects are commonly uncovered in Neolithic sites, as demonstrated by site reports and syntheses

(Aurenche and Kozlowski 1999; Cauvin 1997, 2000b; Hayden 1995a; Kuijt 2000, Kuijt (ed.) 2000; Voigt 1990, 2000).

The organizational resiliency embedded in contemporary human societies, at least since the Upper Paleolithic, demonstrates that foragers have no sense of dichotomy between year-round sedentism and mobility. When conditions were right the range of options within the system of residential and/or logistical mobility (Binford 1980) allowed establishing sedentary camps. In such cases, the group stayed in the same general location, and winter and summer dwellings were built close to each other such as in the Russian plains in the Dnepr-Desna area (Serguin 1999; Soffer 1989). Thus, the eastern Gravettian seem to have been considerably less mobile than the groups in Western Europe. Similar proposals were suggested for the post-LGM sites along the Dnestr river (Borziyak 1993).

One may assume that when additional seasonal information from Upper Paleolithic sites and later ones is gathered or reanalyzed, we will recognize that the state of human occupation was cyclical and that sedentism occurred in the past in more than one region or one time. Therefore, the decisive transition from sedentary communities of foragers to those of cultivators in the Levant becomes an important change, a "point of no return," in the history of humankind. However, this does not mean that farmers are always year-round inhabitants. The ethnological evidence indicates that under certain circumstances farmers continue to be partially mobile (as well as hunters), and we should expect the same when early Neolithic communities are examined in detail.

In sum, when sedentism is positively identified in the archaeological record (with priority given to the biological indicators), based on evidence for well-built aboveground dwellings or pit-houses, as well as storage facilities, it could have been a temporary or permanent socioeconomic solution. In certain regions, where climate and topographic variability play an important role in determining the carrying capacity, there could have been a close relationship between the climatic fluctuations and the settlement pattern. However, in every case of demonstrated paleoclimatic shift, we need to examine the combination of the local social structure, social alliances, and technology in evaluating the causal relationship between environmental and cultural changes.

FROM MOBILE HUNTER-GATHERERS TO SEDENTARY FORAGERS IN THE LEVANT

The archaeology of the Late Paleolithic (Epi-Paleolithic) foragers in the Levant is relatively well known (Bar-Yosef and Belfer-Cohen 1989, 1992; Byrd 1998; Goring-Morris 1987, 1995; Goring-Morris and Belfer-Cohen 1997).

Social units have been identified on the basis of selective analysis of stone artifacts combined with other attributes such as site size, site structure, distribution of settlements, and the reconstructed pattern of seasonal mobility. For instance, the Kebaran sites (ca. 19,000–15,500 cal. B.P.) were limited geographically to the coastal Levant and isolated oases due to the prevailing cold and dry climate. Geometric Kebaran foragers took advantage of the climatic amelioration around 16,500–14,500 cal. B.P. and expanded into the formerly desertic belt that had become a lusher steppe. Besides a rich microlithic industry, groundstone mortars, bowls and cup-holes that first appeared in the earlier Upper Paleolithic contexts are considered to indicate vegetal food processing. Their invention marks a revolutionary departure from Middle Paleolithic plant food preparations. It heralds the "broad spectrum exploitation" that was conceived as a prerequisite for the agricultural revolution (Flannery 1969) and is supported by a recent discovery of carbonized plant remains in a waterlogged site, Ohallo II, dated to 21,000 cal. B.P. (Kislev et al. 1992). The assemblage contains a rich suite of seeds and fruits, already known from the basal layers of Abu Huriera, dated to ca. 12,000 cal. B.P. (Hillman et al. 1989). Both collections that were preserved due to specific conditions reflect intensified gathering of *r*-resources from a variety of habitats and plant associations.

Fallow deer, gazelle, and wild boar were hunted in the central Levant, and gazelle, ibex, and hare were the common game in the steppic belt. Unfortunately, the sandy sediments of desert sites did not preserve animal bones, yet these sites are well dated due to the presence of wood charcoal. Plant remains were also not preserved in most sites, leading to a lack of data on the vegetal diet of these groups.

The climatic improvement of the Bölling–Alleröd after ca. 15,000/14,500 cal. B.P. facilitated the expansion of human occupation into the steppic/desertic belt. Groups ventured into areas that were previously uninhabited from the Mediterranean steppe into the Syro-Arabian desert. Others probably came from the Nile valley (Bar-Yosef and Phillips 1977). However, the documented lithic variability from the Azraq basin and southern Jordan (Byrd 1998; Garrard et al. 1994; Henry 1997) indicates that we are still far from recognizing all of the various Late Paleolithic groups that inhabited the Levant at that time.

THE NATUFIAN CULTURE- SETTLEMENT PATTERN AND SITE SIZE

The emergence of the Natufian culture around ca. 15,000/14,500 cal. B.P. was a major turning point in the history of the Near East. D. Garrod and R. Neuville were among the pioneers of prehistoric research who identified (on the basis of the lithic, bone, and ground stone industries as well as the presence of burials in Mount Carmel and the Judean desert sites) the Natufian culture (Garrod 1932; Neuville 1934). This culture has continued to attract the attention of archaeologists since the 1930s. Excavations in the 1950s and 1960s in Eynan (Mallaha), Nahal Oren, Hayonim Cave, Rosh Zin, and later in Wadi Hammeh 27, exposed a series of pit-houses, whose walls were aligned in stones. The cluster of such dwellings at Eynan led J. Perrot (1966) to interpret the site as the remains of a village. Additional excavations of the lower layers at Beidha, Rosh Horesha, Wadi Judayid, Tabaqa, Azaraq 18, and Gebel Saide provided new data that have led to the recognition that the original Natufian "homeland" was probably in the central Levant (Bar-Yosef and Valla 1991; Belfer-Cohen 1991b; Henry 1989; Stordeur 1981, 1992; Valla 1995; and see Fig. 5.2).

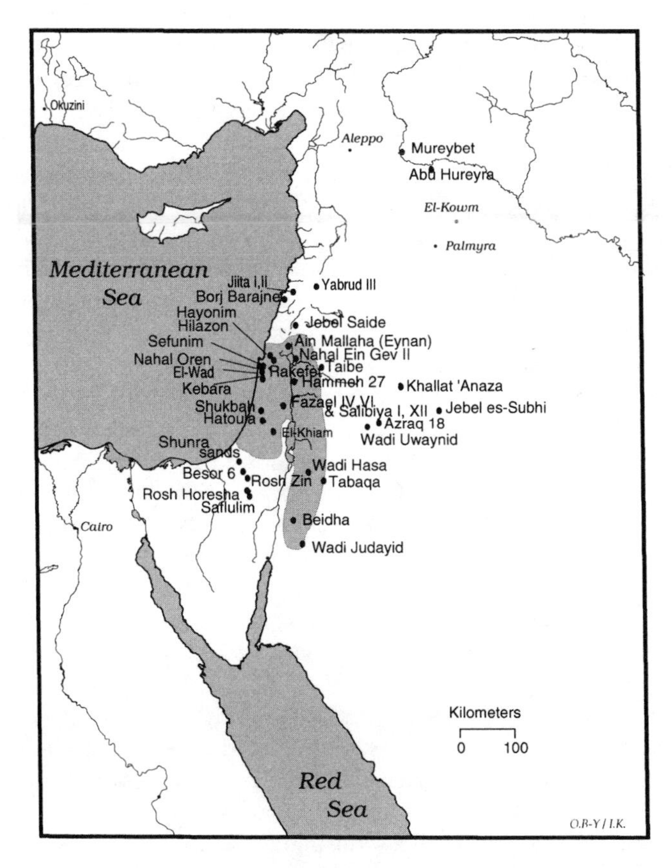

Figure 5.2. The location of most known Natufian sites. Note the gray area indicating the location of the Natufian homeland.

In classifying the Natufian sites, those that had dwelling structures, burials, proliferations of grinding tools, rich lithic industries, animal bone assemblages, and identifiable remains spread over a relatively large area were interpreted as base camps. We also refer to these sites as hamlets to differentiate between small clusters of habitations and larger ones common in the early Neolithic. Thus, Natufian sites fall into three size categories: small (15–100 square meters), medium (400–500 square meters), and large (greater than 1000 square meters). In areas where the information is somewhat more complete, such as Mt. Carmel, one can more easily identify the settlement pattern, although historical agricultural activities and the rapid development since the late nineteenth century erased the evidence of many of the small, ephemeral sites.

All Natufian base camps in the "homeland" area were located in the woodland–parkland belt, where both oak and pistachio were the dominant species. The herbaceous undergrowth of this open forest was at sometime dominated by high frequencies of cereals. The high mountains of Lebanon and the Anti-Lebanon, the steppic areas of the Negev and Sinai, and the Syro-Arabian desert in the east accommodated only small Natufian sites. It seems that in these the Natufians faced other groups of foragers. Only during the Late Natufian several large sites were established within the steppic belt.

Natufian base camps are characterized by pit-houses, whose foundations are stone retaining walls that supported the circumference of the dwelling. The upper structure was probably just wood and brush with rare wattle, and by employing these materials, the Natufian structures differ considerably from those that used mud bricks and became common practice during the ensuing Pre-Pottery Neolithic A (PPNA) period. The circular form of the pit-houses was a common feature, but the investment in the walls and ceiling was entirely different, indicating that the Natufians did not build solid roofs as did their descendants, possibly due to differences in the amount of time spent at the base camp. The best examples of Natufian houses were uncovered in Eynan (Mallaha), Wadi Hammeh 27, Hayonim Terrace, and small rooms inside Hayonim Cave probably served for more than just domestic activities. Note that every base camp produced ample evidence that houses were rebuilt, indicating periods of abandonment of unknown duration.

Domestic dwellings, recognized either by their rounded or squarish hearths and often mundane sets of tools and other finds, were pit-houses about 3–6 meters in diameter. The fills of all dwellings contained rich assemblages, but identifying specific floors was not easy.

A special building is the semicircular house 131 in Eynan (Mallaha) (Fig. 5.3), 9 meters in diameter, where a series of postholes was preserved (Valla 1988). Clusters of artifacts were noticed in certain areas of the floor.

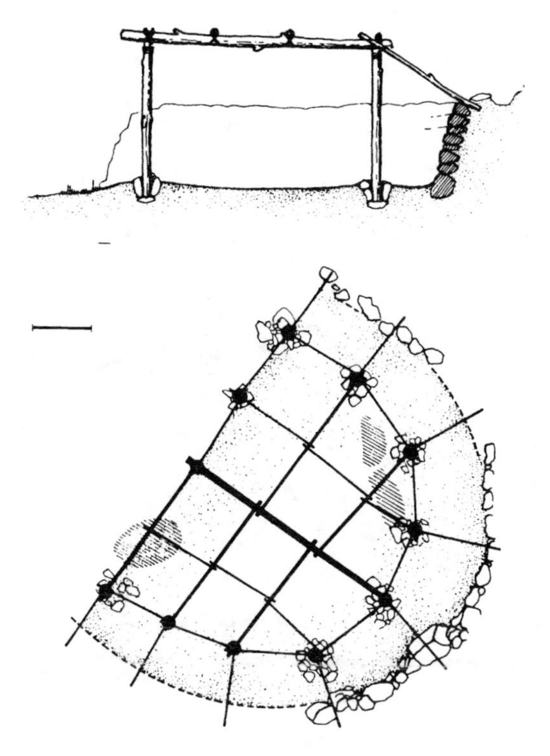

Figure 5.3. The plan and reconstruction of the large public building at Eyan (Mallaha) redrawn after Valla 2001.

These clusters, which included a group of pestles with a human cranium, a group of 24 pebbles with three shaft straighteners, and large amounts of lithics and faunal remains near the hearths, seem to have remained in their original positions. Worth noting is another small building in Eynan (Mallaha), where a rounded bench covered with lime plaster was preserved.

A series of small adjoining oval rooms inside Hayonim Cave (2.5–3.5 meters in diameter) as built of undressed stones (Fig. 5.4). There was a hearth or two in each room except for one. Finds from the lower fill of every room indicated its domestic use, although this function seems subsequently to have changed: one room was a kiln for burning limestone and later was the site of bone tool production (Bar-Yosef 1991b).

At Nahal Oren Terrace, a late Natufian site, elongated enclosure walls were uncovered, somewhat similar in their outline to a wall exposed by D. Garrod on the terrace of El-Wad Cave. In a lower level of this site, a series of postholes surrounded a large fireplace were found amid a cemetery area. The postholes resemble the series found inside the large building in

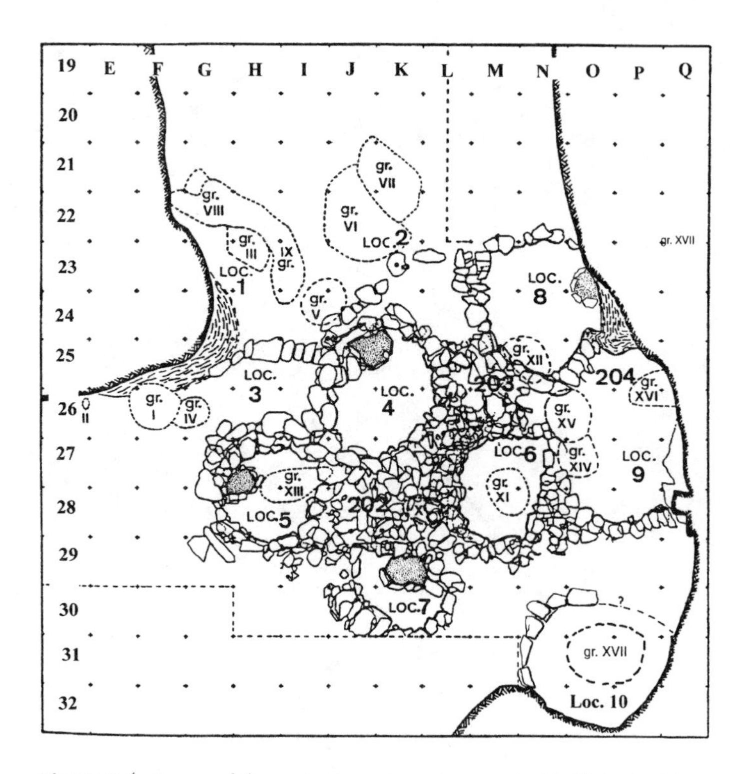

Figure 5.4. A map of the various rooms and graves inside Hayonim Cave.

Eynan (Mallaha), where the posts supported the roof. In the Nahal Oren case, the relationship between the various features is poorly known, and the wooden posts could have been part of a structure related to rituals carried out in the cemetery. Recently, Late Natufian structures were uncovered by F. Valla; some of them served for purposes other than domestic habitation (Fig. 5.5; Valla et al. 2001).

At Jebel Saaîdé, a late Natufian site in the Bekaa Valley (Lebanon), the remains of collapsed walls were identified, despite much destruction caused by modern terracing (Schroeder 1991). Circular structures (pit-houses?) were also exposed in Rosh Zin (Henry 1976). One room had a slab pavement and a limestone monolith 1 meter tall erected at its edge. The unusual use of the monolith possibly indicates that it was erected for specific ritual purposes. Unfortunately, due to the small number of Late Natufian excavations, only a limited amount of information about site structures is available.

Despite the expectations for a society that is said to be sedentary, storage installations are rare in Natufian sites. The few examples include a

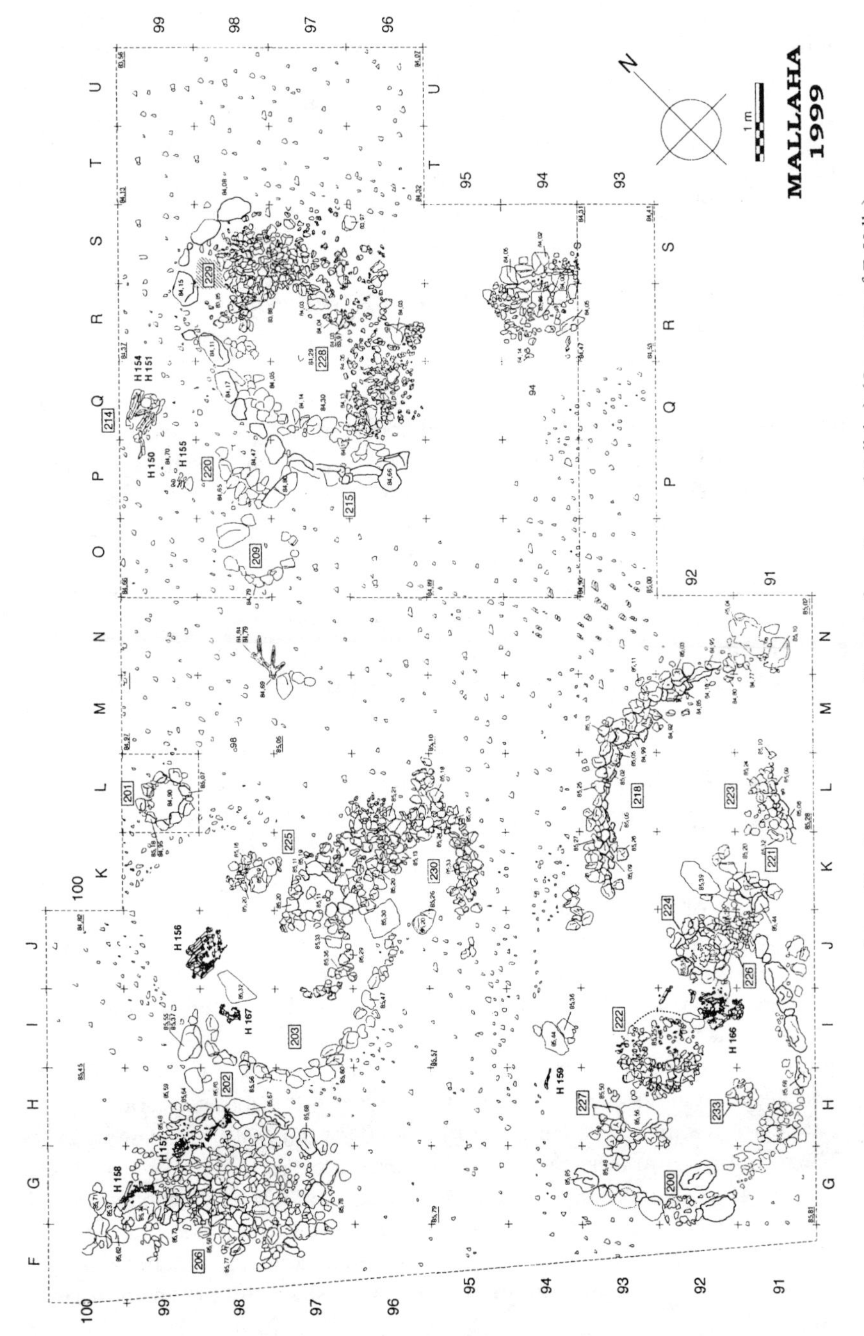

Figure 5.5. A map of Late Natufian structures and burials from Eynan (Mallaha) (Courtesy of F. Valla).

paved bin in Hayonim Terrace (Valla et al. 1991) and several plastered pits at Eynan (Mallaha) (Perrot 1966) that could have served as underground facilities. It is possible that baskets could also have been used for aboveground storage. It is reasonable to speculate that the indirect evidence provided by special bone tools may indicate the manufacturing of baskets (Campana 1989). Because the charred remains of twining were found in PPNA sites, this is a plausible hypothesis.

MORTUARY PRACTICES AND THE EVIDENCE FOR A RANKED SOCIETY

Mortuary practices are considered a major source of societal information. Since the introduction of systematic analyses of graves, burials and grave goods, (Saxe 1970; Binford 1971), various scholars have examined the archaeological reports to recover information concerning social structure, the presence or absence of an elite, and the like. The Natufian burials received the same treatment.

Natufian graves were uncovered in all of the hamlets. In most places, the graves were dug in deserted pit-houses outside the domestic structures or in specific areas of the sites (such as inside Hayonim and El-Wad Caves). Burials under the floors are definitely an exception (Belfer-Cohen 1995; Byrd and Monohan 1995). The pits are either shallow or deep and rarely paved with stones or lime-coated, as they seem to have been in a few cases in Eynan (Mallaha). The location of the graves in Hayonim Cave was marked by small cupholes on one of the encircling stones.

The positions of the corpses reflect the nature of changing mortuary practices. The skeletons are found supine, semiflexed, or flexed with various orientations of the heads and the upper limb bones. The number of inhumations varies from single to collective, although the latter are more common in the Early Natufian. Several cases of skull removals were observed in the Late Natufian contexts that heralded common treatment of adults in the early Neolithic (Belfer-Cohen 1988). Secondary burials were either in special graves or mixed with primary burials. In sites where reasonably large areas were excavated, there are indications of some sort of spatial arrangement of burials. One large burial was uncovered inside El-Wad Cave, and numerous individual graves were located on the terrace (Garrod and Bate 1937). Two concentrations of burials were identified in Eynan (Mallaha) (Perrot and Ladiray 1988). It is worth mentioning that scattered human bones occur within occupational deposits which means that disturbing older burials was not viewed as a shameful act in the eyes of Natufian people.

The presence of decorated skeletons raises the question of social status. About 8% of Early Natufian skeletons had body decorations composed of marine shells (mainly *Dentalia* sp.), bone, and animal tooth pendants. These could have been the residues of garments, belts, headgear and the like, given their position in relation to skulls, upper and lower limb bones, and the waist. In a few graves, additional objects were recognized as grave offerings. These include mundane bone tools, decorated bone objects, and figurines (see below).

This aspect of social hierarchy within Natufian communities was first investigated by Wright (1978). In interpreting the El-Wad cemetery, he proposed that the differentiation in decoration and grave goods within a large group of Natufians was needed to maintain the social order. Such a structure emerged within a population that was composed of more than one extended family or clan and relied on storable surpluses. Coordination and distribution of foodstuffs seem to have necessitated the social hierarchy. Henry (1989), employing an observation from the North American northwest coast, stressed the nonegalitarian nature of the Natufian hamlets as an intergroup phenomenon, which he viewed as generating intensified hunting that led in due course to depletion of local resources. Olszewski (1991) adopted a somewhat similar attitude.

Recent analyses (Belfer-Cohen 1995; Byrd and Monahan 1995) suggest that there is no evidence to support the contention that the Natufian was a ranked society. The decorated skeletons of the Early Natufian are of all ages but from only a quarter of the total uncovered population. They are buried in each hamlet either in abandoned pit-houses or in open areas (Valla 1996, 1999; Tchernov and Valla 1997). Concentrations of graves within the site were recognized in Eynan (Mallaha) (Perrot and Ladiray 1988). There are several differences among the sites in the common elements of body decoration, especially that of the bone beads and pendants, which seem to mark the presence of particular local groups (Belfer-Cohen 1995). If the Early Natufian was a society of egalitarian foragers, it would be difficult to explain the presence of individuals rich and poor in body decorations, although one may claim that these attributes only marked individual choices. However, taking into account the overall richness of the sites, their size, location within optimally exploited environments, it seems that the Natufian society as an entity was ranked.

Late Natufian burials, such as in the large cemetery at Nahal Oren (Mt. Carmel), had no decorated burials and only a few grave goods (Noy 1991). In addition, there are more secondary burials from Late Natufian contexts. Most authorities agree that the Late Natufian society was more mobile compared to its ancestors, probably due to the harsher environmental conditions created by the Younger Dryas. Therefore, the shift in mortuary practices probably reflects a change from what was a nonegalitarian society, back to a more egalitarian society. In sum, it seems that the contents of the burials and the difference between the Early and the Late Natufian burials point to the emergence of a nonegalitarian society during the Early Natufian, formerly called "complex hunter-gatherers" (Kelly 1995). Another difference between the two periods is the joint inhumations of humans and domestic dogs which occur only in the early times in Mallaha (Eynan) and Hayonim Terrace (Davis and Valla 1978; Tchernov and Valla 1997; Valla 1990).

LITHIC ASSEMBLAGES: STYLE AND FUNCTION

The production of stone tools is one of the most conservative human activities. Research into late Paleolithic assemblages in the Levant has demonstrated that it is almost impossible to relate changes in core reduction strategies and the morphology of the particular artifacts to environmental conditioning. Good quality flint is available in every area within this region at a distance that never exceeds 5–10 km (O. Bar-Yosef 1991a; Henry 1989). Therefore, the characteristics of blank production from nodules, the particular technique of snapping bladelets (known as the "microburin technique"), types of retouch (Helwan, backed, fine), as well as the forms of the retouched pieces, were employed in the search for identifiable social entities (Figs. 5.6 and 5.7). In spite of insufficient radiocarbon dates, Natufian stratigraphies indicate a basic subdivision into Early and Late Natufian. Further subdivisions into three or four phases were offered (Neuville 1951; Valla 1984). In addition, it was proposed that territorial groups be recognized on the basis of the presence or absence of products of the "microburin technique," as well as the dominant type of retouch that shaped the lunates (Bar-Yosef and Valla 1979; Belfer-Cohen 1991b; Henry 1974). This pattern was refined to include the regional-ecological location of the sites (Olszewski 1986).

Extensively used cores and the production of small, short and wide bladelets and flakes characterized the Natufian lithic industry. Among the retouched pieces, the frequencies of end scrapers and burins fluctuate considerably. Backed blades grade into the retouched, and backed bladelets defined as microliths. Microliths and geometrics comprise 40% or more of every assemblage. In the Early Natufian, geometrics include high frequencies of Helwan lunates and fewer of the backed types, trapeze-rectangles, and triangles. In the Late Natufian, backed lunates replace the Helwan type among the geometrics.

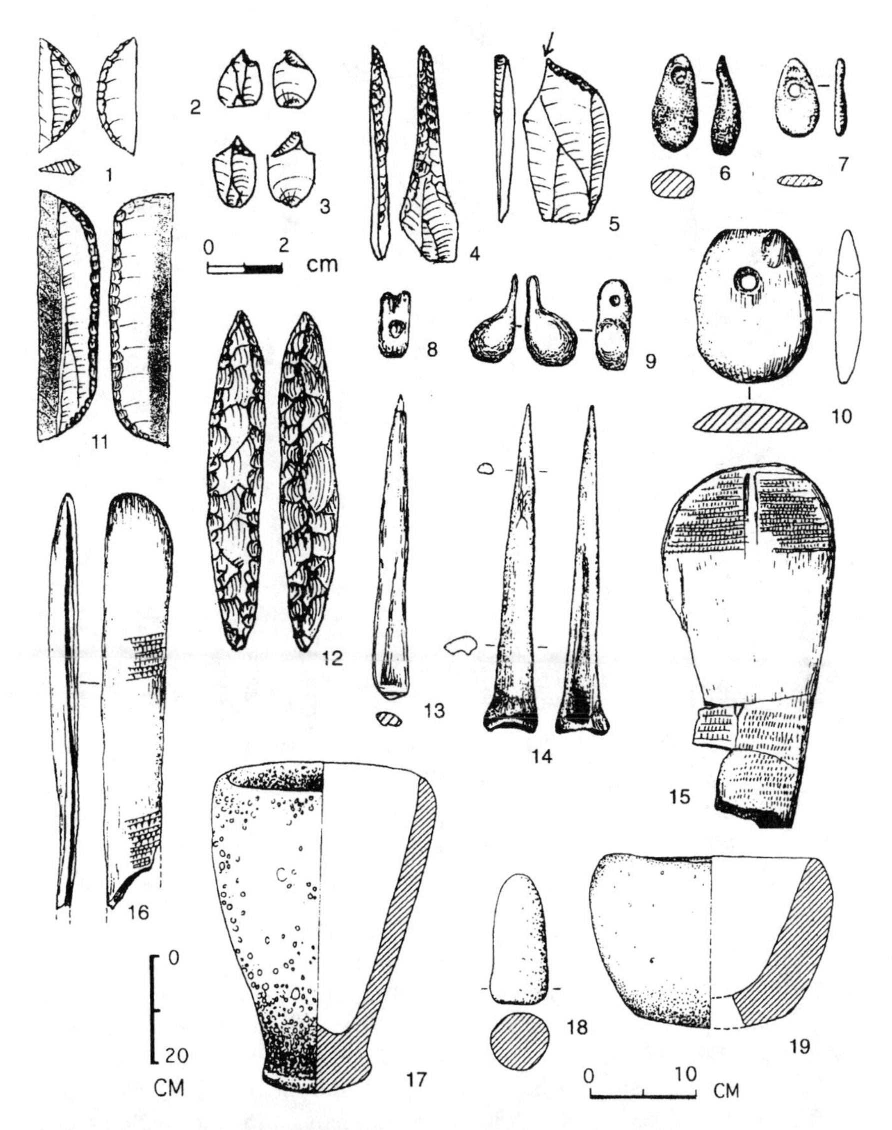

Figure 5.6. Early Natufian assemblage from Hayonim cave and El-Wad: 1. Helwan lunate; 2, 3. Microburian; 4. Perforator; 5. Burin on truncation; 6, 7. Bone pendant; 8. Bone bead; 9. Bone pendant from El-Wad; 10. Bone pendant; 11. Helwan sickle blade; 12. Pick; 13, 14. Bone points; 15. Bone spatula; 16. Broken bone sickle haft; 17. Goblet-shaped basalt mortar (note the different scale); 18. Basalt pestle; 19. Limestone mortar.

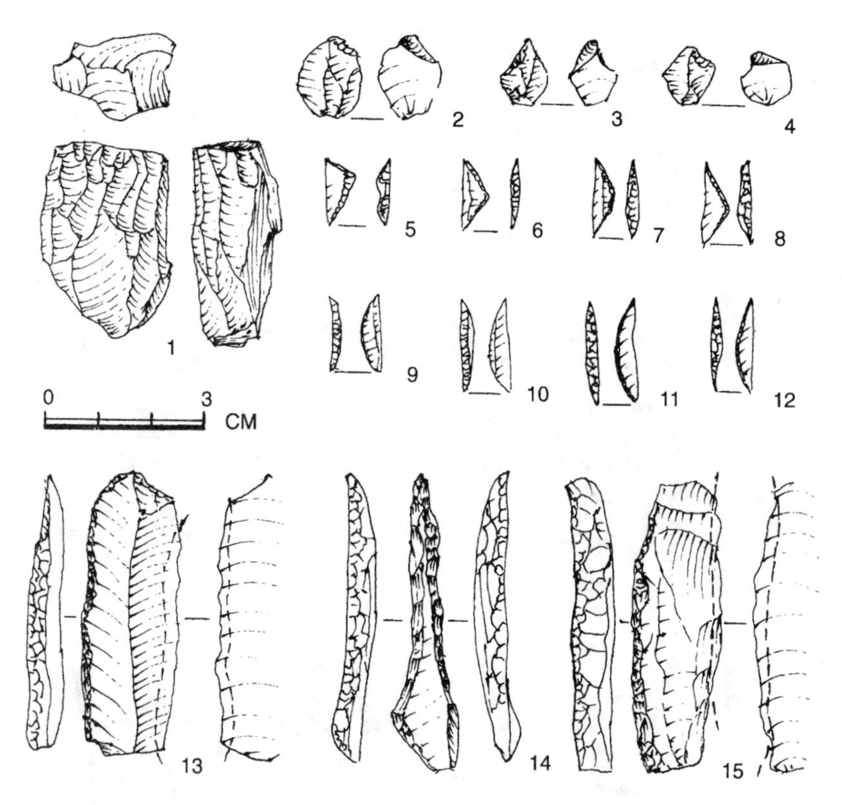

Figure 5.7. Late Natufian assemblage from Fazael IV (redrawn after Grosman et al. 1999): 1. Bladelet core; 2–4. Microburins; 5–8. Triangles; 9–12. Backed lunates; 13, 15. Backed sickle blades; 14. Perforator.

Picks and sickle blades are special tools that occur for the first time in the Natufian. The first, considered the forerunner of the axe-adzes group of the Neolithic period, are elongated (8–10 cm) and bifacially or trifacially flaked (Belfer-Cohen and Goring-Morris 1996). These could have been used in shaping limestone-ground stone tools or in heavy work with wood.

The second type, often considered a Natufian trait is the sickle blades. These items, known also as the "glossy pieces," are abundant in sites within the Natufian homeland. As demonstrated through experimental and microscopic studies, the gloss that is visible on a relatively wide area on both faces of these blades appears to be the result of harvesting cereals (Anderson 1994; Anderson and Vallas 1996; Unger-Hamilton 1991; Yamada 2000). The sickle blades were hafted in bone or, more often, in wooden handles. Employing sickles instead of beaters and baskets indicates that the

Natufians were interested in maximizing the yields of seeds from a limited area (Hillman and Davies 1990). Such actions can be interpreted as reflecting control over defined fields and is a notion that was suggested as an artistic expression in the form of an incised "map" (Bar-Yosef and Belfer-Cohen 1999). If such an interpretation is supported by additional data, it would also mean that the establishment of sedentary Early Natufian hamlets resulted in a territorial division of the Natufian homeland. This conclusion would imply that a network of social alliances was maintained as a buffer against annual shortages that could have occurred on either a decadal scale or from an abrupt major climatic change (Bar-Yosef and Belfer-Cohen, in press).

GROUND STONE TOOLS: THE UNKNOWN NATURE OF FOOD PREPARATION TECHNIQUES

"Ground stone tools" refers to all of the objects that are not made of flint (the commonest raw material in the Levant). The most distinctive types are bedrock mortars, "goblet" type mortars, portable daily size mortars, bowls of various types, cupholes, mullers, and pestles (Fig. 5.6; Wright 1992). They occur in large numbers in hamlets and in sites that are interpreted as seasonal camps with repeated occupations (noted by Olszewski (1991) as "base camps" without dwellings) and ephemeral stations.

Among these, the boulder mortars, which weigh up to 100 kg, are 70–80 cm deep, and are sometimes called "stone pipes," stand out. There is no direct evidence of the uses of the boulder mortars as domestic utensils in daily life. Their specific role is better known from their final depository. When breached at their bottoms, they were found standing at the edges of graves (e.g., Nahal Oren), resembling tombstones. In Jericho, in what seems from the report as a domestic context, Kenyon (1957) suggested that they were sockets for wooden totem posts. However, postholes in the large building in Eynan described above indicate that using pecked or deep drilled conical holes in a limestone boulder for this purpose would be a wasteful act.

The technique of making the mortars, especially the deeper ones, has not been fully explored, but research is currently being performed on this time- and labor-intensive activity (L. Dubrieul, personal communication). Hence, the number of uncovered mortars in a site could be used as a measure of the relative wealth of the local inhabitants. In addition, we should not forget that wooden mortars and bowls may have been in use, but as the unique 8000-year-old example from the Kfar Samir waterlogged site indicates, they were not preserved (Galili and Schick 1990).

There are at least three large "goblet-shaped" basalt mortars (one ca. 60 cm high and two ca. 40 wide, two of which appear in Eynan) that seem to have served purposes other than the domestic types. Figure 5.6:17 is an illustration of the third goblet-shaped mortar discovered in a grave in Hayonim Cave; the fragments were used (similarly to other undressed stones) to cover the dead. Although the contextual evidence is lacking, it does seem that the large mortars could be attributed to food preparation in the course of a public activity, perhaps on the occasion of annual feasting (Hayden 1995a).

The large mortars, a few small ones, and most of the pestles were made from basalt. Archaeometric study indicates that those objects found in the Mt. Carmel and western Galilee sites were transported from the lava plateau of the Golan heights, east of the Jordan Rift Valley, a distance of about 100 km (Weinstein-Evron et al. 1995). Eynan in the Hula Valley was much closer to the sources (some 10 km away). This could mean either that members of each community went seeking the proper source rock and shaped the object in a quarry site or that these items were exchanged within kinship-based hamlets across the central Levant.

In sum, microscopic observations have demonstrated that ground stone utensils were employed for food processing and also for crushing burned limestone and red ochre (Perrot 1966; Weinstein-Evron and Ilani 1994). Unfortunately, without the wooden objects, our picture of Natufian kitchen equipment is incomplete.

The last category included in this section is the grooved stones. Those made of sandstone were used as whetstones for shaping bone objects (Campana 1989). The other type, classified as "shaft straighteners," has a deep, parallel-sided groove which sometimes bears clear signs of burning. From ethnographic comparisons, it is identified as the tool employed for straightening wooden or reed shafts. The presence of this tool supports the contention that bows were used by the Natufians (Valla 1987). What remains unknown is whether the shafts had flint, bone, or wooden projectile points mounted at their tips. Lunates were traditionally considered the best candidates for this category.

BONE AND HORN-CORE INDUSTRY: A CULTURAL TRAIT OF THE NATUFIAN

The Natufian assemblages are marked by a preponderance of bone and horn-core industries richer in quantity and types than any earlier or later Levantine archaeological entity. Objects were made of bone shafts of large mammals, teeth, and horn-cores of gazelles, wolves, fallow deer, roe deer, and birds (Campana 1989; Stordeur 1988, 1991, 1992). The number of items and the typological variability seem to be richer in the Mt. Carmel-Galilee (including the Hula Valley) and less in all other directions (Stordeur 1992). Use-wear analysis indicates that bone tools were used for hide-working and basketry (Campana 1989, 1991). Barbed items are reconstructed as parts of hunting devices such as spears or arrows (Stordeur 1988), hooks and gorgets for fishing, and hafts for sickle blades.

Bone beads and pendants were made from shafts, gazelle finger bones, as well as tibiotarsus bones of partridge (Pichon 1983), and were shaped by grinding and drilling. Their social interpretation is discussed below.

Sickle hafts (Fig. 5.8:2, 4) and spatulas were made of ribs or long limb bones and sometimes bear specific decorations—sometimes a series of incisions or an image of an animal (El-Wad, Kebara, and Hayonim Caves). The famous Natufian examples are the animal figures carved on the edge of the hafts from El-Wad and Kebara, mostly identified as young ruminants (Bar-Yosef 1997; Garrod 1957). The special net patterns on objects typologically classified as "spatulas" from Hayonim, Kebara and Iraq ed-Dubb Caves are intriguing (Bar-Yosef and Belfer-Cohen 1999; Kuijt et al. 1992), although a suggested interpretation is that these items are musical instruments (I. Yalçinkaya, personal communication).

ORNAMENTATIONS AND GROUP IDENTITY

Body decorations and ornaments appear for the first time among early Upper Paleolithic foragers in the Near East. Their presence across Natufian sites is uneven, and it demonstrates both inter- and intrasite variability as well as a change over time. A variety of marine mollusks, bone beads, and pendants, perforated teeth, greenstone, and limestone beads and pendants were used by the Natufians for decorating headgear, belts, and possibly parts of garments, as well as for elements of necklaces, bracelets, and earrings. The original form of some of these decorations has been reconstructed from those found adhered to skeletons in graves at El-Wad, Eynan, and Hayonim Caves, all of Early Natufian age (Fig. 5.6:6–10).

Marine shells for Natufian jewelry were collected from the shores of the Mediterranean Sea or, more rarely, were brought from the Red Sea (D. E. Bar-Yosef 1991). The Natufians sought mainly the *Dentalium* sp. which today is scarcely found on the beaches of the Eastern Mediterranean. Although the shells could have been collected after winter storms, we need to consider an alternative means of acquisition. Native Americans of the Northwest Coast used boats to collect these shells from depths as low as 40 meters below sea level. This raises the possibility that a similar method

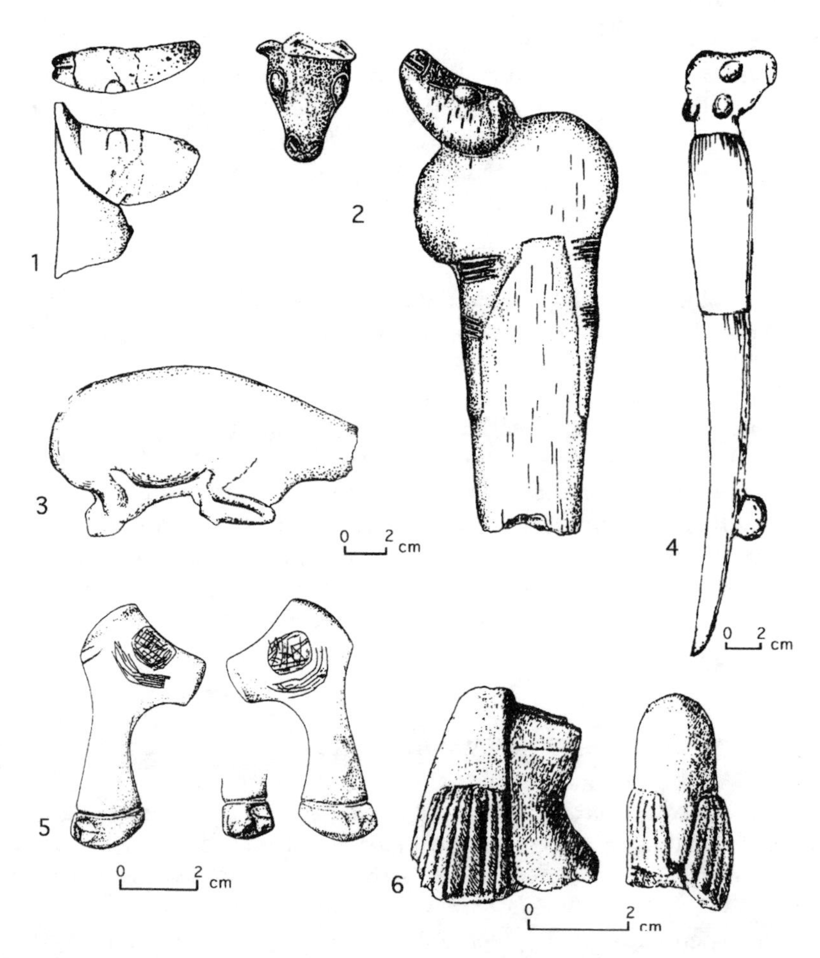

Figure 5.8. Animal figures: 1. An Ungulate head From Nahal Oren (after Stekelis and Yizraeli 1963); 2. Ungulate on a broken sickle haft from El-Wad (after Garrod and Bate 1937); 3. Kneeling ungulate made of lime stone from Um ez-Zuweitina (after Neuville 1951); 4. Ungulate on a sickle haft from Kebara Cave (after Turville-Petr 1932); 5. A double end figurine made of limestone from Nahal Oren (after Stekelis and Yizraeli 1963); A "baboon" made of limestone from Nahal Oren (after Noy 1991). Redrawn from Bar-Yosef and Belfer-Cohen 1998.

was used by the Early Natufians. We know this is possible because sea crossing was documented from the Aegean when obsidian from the island of Melos was obtained by 12,000 cal. B.P. (Perlès 1979) and seafaring enabled the *en masse* colonization of Cyprus by 9600 cal. B.P. with evidence for Epi-Paleolithic landing in Akrotiri-Aetokremnos at least a millennium earlier (Simmons and Wigand 1994).

Other marine shells such as *Columbella* sp. and *Nassarius* sp., are found in very low frequencies. In spite of the overwhelming dominance of the *Dentalium* sp. shells, the straight shapes of these tusk shells in contrast to the rounded forms of all the rest of the assemblages are interpreted as gender symbols (Valla 1999). This argument is derived from the known natural symbols where straight lines or objects are identified with males while the rounded forms are considered as representing females (see also Cauvin 2000a,b).

Several sites are distinct for rare shells from distant localities. In Eynan, there is a *Dentalium* shell from the Atlantic Ocean and a fresh water bivalve (*Aspatharia* sp.) from the Nile River (Mienis 1987). In Salibiya I, there is an Indo-Pacific shell that does not occur today in the Red Sea but could have been there in the past (Crabtree et al. 1991). Other rare items are small pieces of Anatolian obsidian uncovered solely at Eynan in the uppermost layers. These and other lines of evidence may reflect long range exchanges among Natufian communities. To the south, this is not surprising, as Helwan lunates were found in Egypt. In addition, the bones of fruit bats (*Rousettus aegyptiacus*, which prey on fruit trees and are also closely associated with sycamore trees, *Ficus sycamorus*) were found only in Natufian contexts and later times. Sycamores, natives of East Africa, grow in the Nile Valley as their secondary habitat (Galil et al. 1976). Although the latter connection cannot be confidently established, one may speculate that the climatic conditions during the Early Natufian—contemporary with the Bölling/Alleröd ameliorated conditions—facilitated human movements through the Sinai Peninsula.

Greenstone and malachite beads were found in various sites but their exact sources are not yet determined. Altogether, the differences in the suites of jewelry among the sites are considered as identity markers of distinct groups.

NATUFIAN IMAGERY

Natufian imagery is expressed in various ways. Often the domestic items such as spatulas, sickle hafts, bowls, and shaft straighteners bear incised decorations (Bar-Yosef and Belfer-Cohen 1998; Noy 1991). The now famous animal bone carvings on sickle hafts are generally rare (Belfer-Cohen 1991a,b). There are also small figurines of various animals such as gazelle, "baboon," tortoise, a double end with a dog and an owl, and one with a human and an ungulate (Figs. 5.8, 5.9; Bar-Yosef and Belfer-Cohen 1998, 1999; Belfer-Cohen 1991a,b; Goring-Morris 1998; Noy 1991).

The attention given to young ruminants and their appearance as decoration on sickles is rather curious but perhaps can be interpreted as representing a totemic group idol.

The human figure is rare, although a schematic limestone figurine was found in El-Wad, along with a few more recent schematic ones (Fig. 5.9; Garrod 1957; Weinstein-Evron and Belfer-Cohen 1993). Figurines with a schematic face or full body were uncovered in Eynan and Hayonim Terrace (Garrod and Bate 1937; Perrot 1966; Valla 1999; Weinstein-Evron and Belfer-Cohen 1993). The Ain Sakhri limestone figurine which is

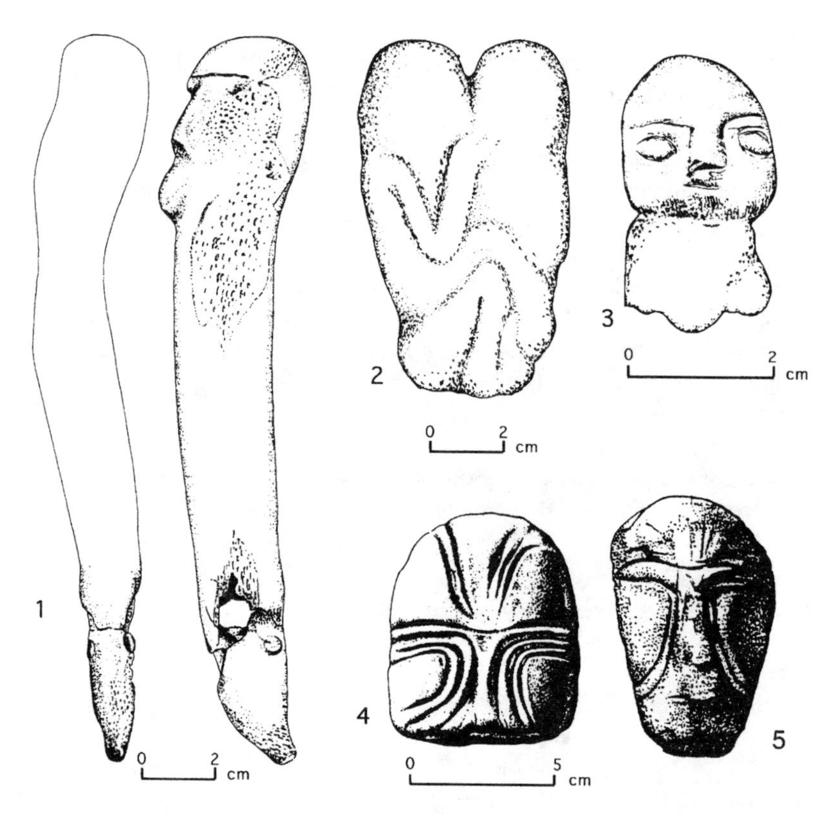

Figure 5.9. Human figurines: 1. A double end figurine of a human head and Ungulate on a horn core from Nahal Oren (after Noy 1991); 2. Copulating couple made of limestone from Ain Sakhri (after Neuville 1951); 3. A schematic human figure chipped from a flint nodule from El-Wad (After Weinstein-Evron and Belfer-Cohen 1993); 4. Two schematic human heads made of Limestone from Eynan (After Perrot 1966); 5. Genderless human head made of Limestone from El-Wad (after Garrod and Bate 1937). Redrawn from Bar-Yosef and Belfer-Cohen 1998.

interpreted as representing a mating couple (today in the British Museum) was found in a collection in a monastery near Bethlehem and is attributed to the small Natufian station in Ain Sakhri in the Judean Desert (Boyd and Cook 1993).

The interpretation of these items is not simple, as shown by the various approaches. It seems that small imagery objects are more common in Late Natufian sites, which by other criteria, it is thought, generally represent a return to a more mobile way of life. In the chronologically subsequent Neolithic contexts, the range of figurines includes numerous representations of humans (mostly females), ungulates, and birds of prey. These images in farming contexts are interpreted as emblems of various natural powers (or their representation) which inhabit the overall cosmology. In an evolutionary approach, Cauvin (Cauvin 1972, 2000a) views the Natufian figurative expressions as heralding what he calls the "revolution of symbols," created within the society of the last Levantine hunter-gatherers.

Several limestone slabs, recovered from the rounded structures inside Hayonim Cave and Terrace, are incised mostly with the "ladder" pattern motif. They are interpreted as the accumulated effects of notational marks (Fig. 5.10; Bar-Yosef and Belfer-Cohen 1999; Belfer-Cohen 1991a). Marshack (1997) examined several in detail and concluded that the markings were undoubtedly notational. A new slab uncovered recently bears the pattern of a series of "fields" (Bar-Yosef and Belfer-Cohen 1999), an interpretation that would accord the Natufian territoriality, as demonstrated above, and had similar patterns from the Upper Paleolithic European world (Züchner 1996).

In addition, worth noting is a large slab on which the rough form of a fish was deeply incised. This incision refers to the minimal amount (as we know today) of marine fishing by the Natufians, or rather it represents the exploitation of fish in the Hula lake, as recorded in Eynan (Desse 1987).

The meander pattern, which represents water (and snakes) in almost every culture, was carved on large limestone slabs uncovered in Wadi Hammeh 27 (Edwards 1991), as well as on basalt bowls from Eynan and Shukbah Cave (Noy 1991).

In sum, the particular incised patterns found on both bone and stone objects include the net, chevron (or zigzag), and meander patterns (Fig. 5.11). Most appear on spatulas, stone bowls, shaft-straighteners, and the rare ostrich-egg shell containers (found as broken pieces) uncovered in Rash Zin in the Negev highlands (Henry 1976). Because they differ from site to site, one may use them to further our identification of different Natufian groups. For the time being, we know that their frequencies are higher within the Natufian "homeland" in the central Levant.

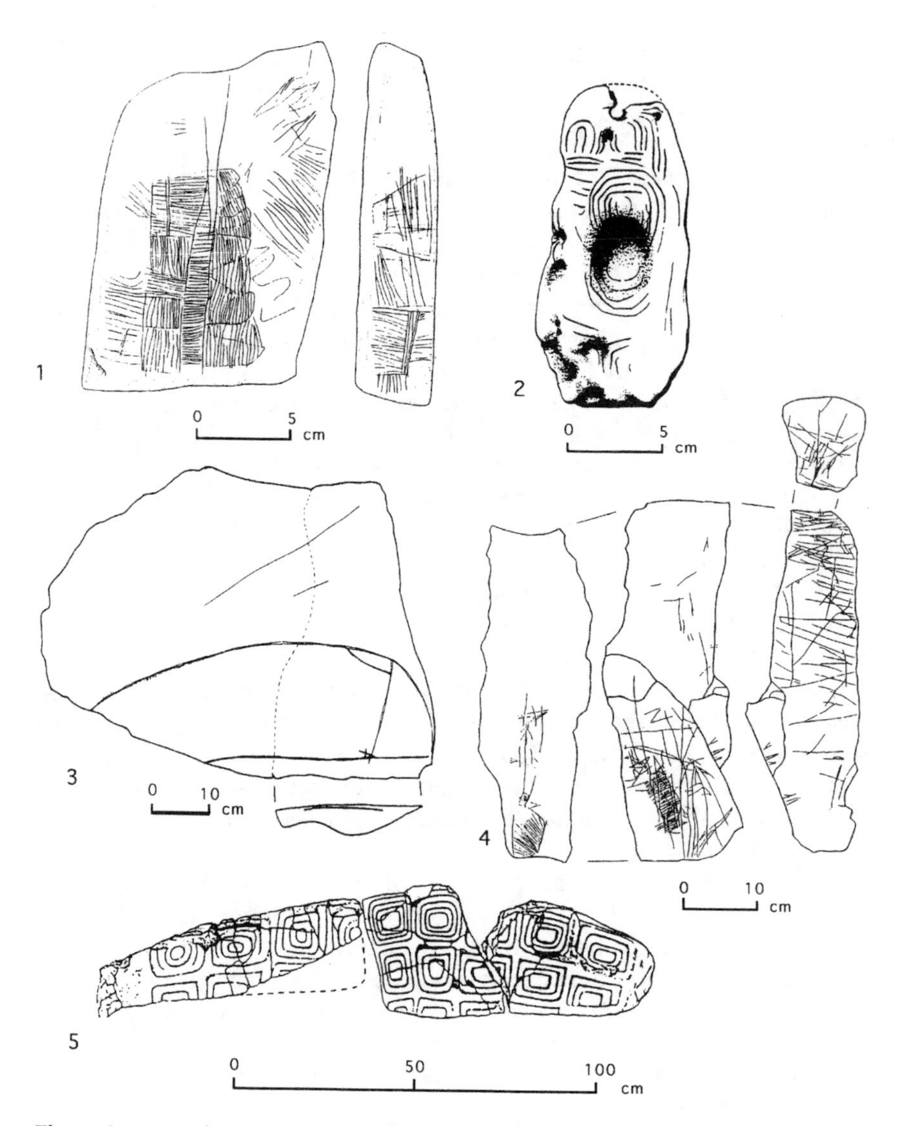

Figure 5.10. 1, 3, 4. Incised limestone slabs from Hayonim Cave; 2. Eynan; 5. Wadi Hammeh 27 (after Bar-Yosef and Belfer-Cohen 1999; Belfer-Cohen 1991; Perrot 1966; Edwards et al. 1988). Redrawn from Bar-Yosef and Belfer-Cohen 1998.

SUBSISTENCE AND SEASONAL ACTIVITIES

Most Natufian sites were excavated before recovery techniques such as systematic dry sieving and flotation were introduced in the late 1960s. However, even in recent excavations at Nahal Oren, Hayonim Cave and

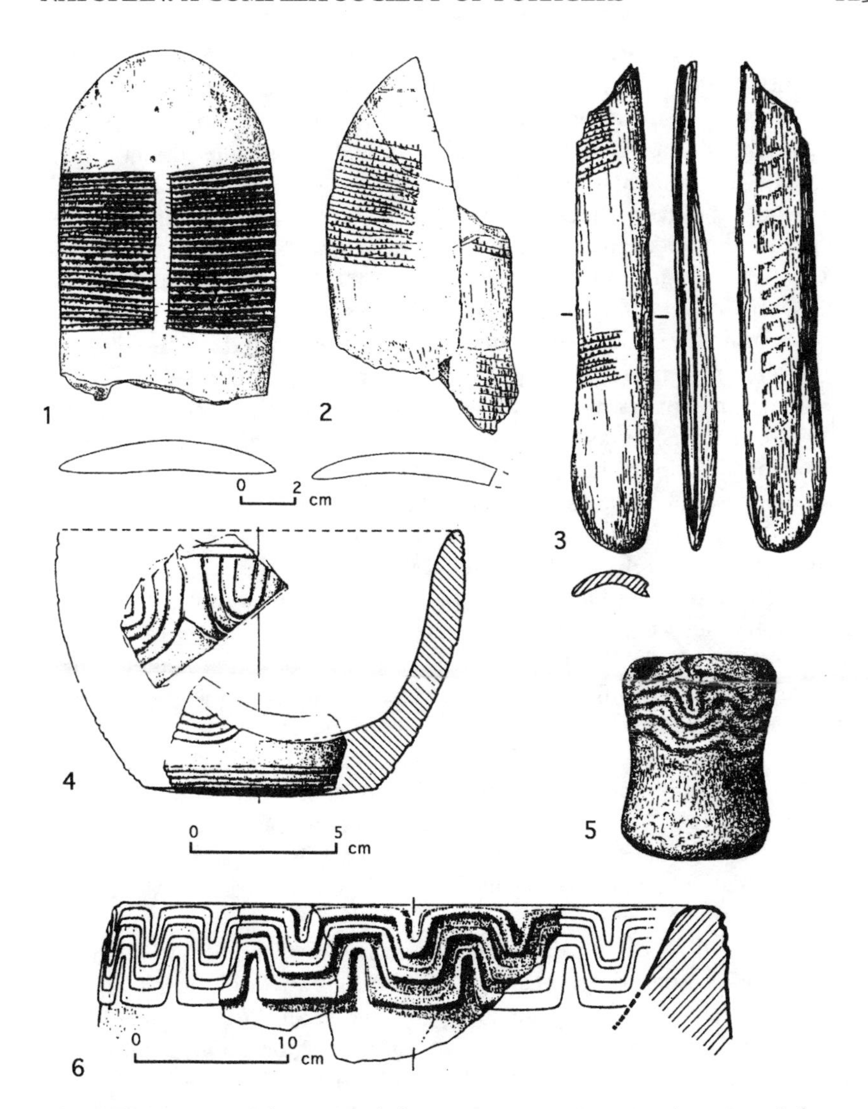

Figure 5.11. Decorated domestic tools from 1, 3. Hayonim Cave; 2. Kebara Cave; 4, 6. Eynan; 5. Nahal Oren (after Bar-Yosef and Belfer-Cohen 1999; Bar-Yosef and Tchernov 1970; Perrot 1966). 1, 2. Bone spatulas; 3. Sickle haft; 4, 6. Broken basalt bowls; 5. Shaft-straightener. Redrawn from Bar-Yosef and Belfer-Cohen 1998.

Terrace, Eynan, Rosh Zin, and Wadi Hammeh 27, water flotation failed to retrieve sufficient quantities of floral remains, and in some cases the few grains found were later dated by AMS to recent times (Colledge in Edwards et al. 1988; Hopf and Bar-Yosef 1987; Legge and Rowley-Conway 1986).

The poor preservation of vegetal remains in Natufian sites located within the Mediterranean woodland belt seems to have been the result of two processes: First, the prevailing sediment type in most of the open-air sites, namely terra rossa, becomes wet every winter and subsequently dries and cracks in summer; thus, plant remains are destroyed, and charcoal, small bones, and even lithics are subjected to both downward and upward movements. Better charcoal preservation was noted in the desertic loess deposits of the Negev (Goring-Morris et al. 1999; Marks 1977) and in drier, deeply buried layers at sites in the lower Jordan Valley. The second reason for the destruction of the plant remains is probably related to the nature of the building materials used by the Natufians. As mentioned above, it appears that the superstructure above the stone wall of the pit-houses was built of wood, brush, grass, and perhaps even wattle and daub. Following a period of abandonment, the walls collapsed, and as their volume was essentially made of organics, the material rotted, deteriorated, and leached, resulting in volume reduction. For example, white ashes in Natufian hearths are preserved but rarely are the charcoal flecks. The scant charred remains sometimes found under flat stones explains the current scarcity of charcoal radiocarbon dates of Natufian sites. Therefore, one may expect to discover a well-preserved Early Natufian site only in the lower Jordan Valley or the Damascus Plain. Under these circumstances, we resort to the rarely found charred seeds to reflect the more complete suite of vegetal sources exploited by the Late Natufians in Abu Hureyra and Mureybet (Colledge 1998; Hillman in Moore et al. 2000; Hillman et al. 1989; van Zeist and Bakker-Herres 1986). Interestingly, the list of plant species collected from these sites corroborates in general what is known from the earlier, water-logged late Paleolithic site of Ohallo II, which dates to ca. 21,000 cal. B.P. (Kislev et al. 1992).

The overall body of data supports the contention that legumes, cereals, various nuts, different vegetables, fruits, and only rarely acorns were extensively gathered (Hillman in Moore et al. 2000; Miller 1997). There is insufficient evidence to demonstrate the wide range exploitation of acorns as staple food, as occurred in California where acorn consumption most likely was the result of the depletion of other resources (McCorriston 1994). Although this conclusion seems probable, in a detailed examination S. Mason describes the difficulties in formulating analogies between the Levant and California (Mason 1995). In addition, there is no archaeological evidence in Natufian sites that reflects the long and tedious process necessary to transform raw acorns into an edible food.

Tools for food acquisition, such as sickles, food processing, and cooking/parching installations, such as mortars, bowls, pestles, and hearths, are generally interpreted as evidence for harvesting and processing of wild cereals and legumes. The idea that the Natufians were the earliest agriculturalists was suggested, as mentioned above, by D. Garrod in 1932 and despite later criticism, was revived by others (Moore 1982). Indirect supportive evidence for the systematic harvesting of cereals in either wild stands or intentionally planted fields is derived from the experimental studies of sickle blades (Anderson 1991, 1998; Korobkova 1992; Unger-Hamilton 1989, 1991; Yamada 2000).

Recently, a new analysis of Late Natufian plant remains from Mureybet and Abu Hureyra indicates the intentional cultivation of the wild species (Colledge 1998). Most authorities agree that systematic cultivation of the wild progenitors would have caused the domestication of wheat, barley, and other species (Zohary 1996). An interesting hypothesis which suggests that the process of domestication was due to unconscious selection by the Late Natufians or their descendants, the early PPNA farmers, was recently proposed (Zohary et al. 1998). However, this proposal ignores the ample evidence that hunter-gatherers interfere in their environment in various ways. For example, foragers set fire to grasslands (Lewis 1972), dig irrigation trenches (Lourandos 1997), and tend fruit trees (Laden 1992). Assuming that humans did not intentionally plant wild cereals (einkorn, emmer, barley) requires better proof than an unfounded model.

We witnessed a shift in recent years in identifying the state of prehistoric crops. The traditional quantitative morphological attributes for recognizing domesticated cereals, such as the frequencies of the broken internodes, were put in doubt (Kislev 1989, 1997). So was the length of time that it took most of the fields to produce the domesticated varieties. Instead of several decades or a few centuries, it seems that the process took almost a thousand years (Kislev 1992) and was accomplished by the end of the PPNA period (ca. 10,500 cal. B.P.). Current DNA analyses may now verify the conclusions drawn from the ancient morphological evidence that has already provided interesting clues. The first study showed that the oldest variety of einkorn was present in the area of the northern Levant and the later varieties spread to the Zagros foothills (Heun et al. 1997). Barley appears to have been domesticated in the southern Levant (Badr et al. 2000). Missing yet is the evidence concerning emmer wheat which seems, by the charred remains, to have been of greater importance in the Damascus Basin (van Zeist and Bakker-Heeres 1985). Hence, the presence of the cereals in the earliest Neolithic contexts tells us indirectly about processes that began at least during the Late Natufian.

Good bone preservation in most sites has made the faunal evidence a subject of numerous studies (Bar-El and Tchernov 2000; Campana and Crabtree 1990; Churcher 1994; Cope 1991; Crabtree et al. 1991; Davis 1982, 1983; Davis et al. 1994; Dayan and Simberloff 1995; Ducos 1978; Ducos and

Kolska Horwitz 1998; Garrard et al. 1996; Lieberman 1991; Tchernov 1991a, 1993b,c). Natufians hunted gazelles along with other game, depending on the geographical location of each site (Fig. 5.12). Among the larger species were deer, cattle, and wild boar in the coastal ranges; in the steppic belt, equids and ibex were more often the common prey. Smaller mammals such as hare were also important, as well as various birds (mainly partridges and ducks), and fish (mostly fresh water from Lake Hula).

The attempt to explain the Natufian faunal assemblages as the result of net hunting (Campana and Crabtree 1990) presents an interesting idea, but this proposal does not fully conform to the ethnographic evidence which indicates that such a technique is best suited for forested areas where the degree of visibility is rather low (Roscoe 1990).

Water fowl undoubtedly formed part of the Natufian diet, especially in sites along the Jordan Valley, where migratory, seasonally nesting ducks of various species were trapped mainly during the winter stress season (Pichon 1991; Tchernov 1994). Species of freshwater fish were caught seasonally at Eynan, where thousands of fish vertebrae were documented (Desse 1987). Fishing seems to have been less important along the Mediterranean coast, as witnessed by scarce saltwater fish bones from Hayonim Cave and the new excavations at El-Wad and Hatoula. However, this scant evidence may solely reflect old excavation techniques which often rendered an incomplete record. Hooks have a distinct form, but the gorgets could have been used as projectiles or barbs. The very few bone hooks and the gorgets are generally interpreted as fishing gear.

Faunal and vegetal remains are used to determine seasonal occupation of particular sites and/or the seasonal mobility practiced by a prehistoric

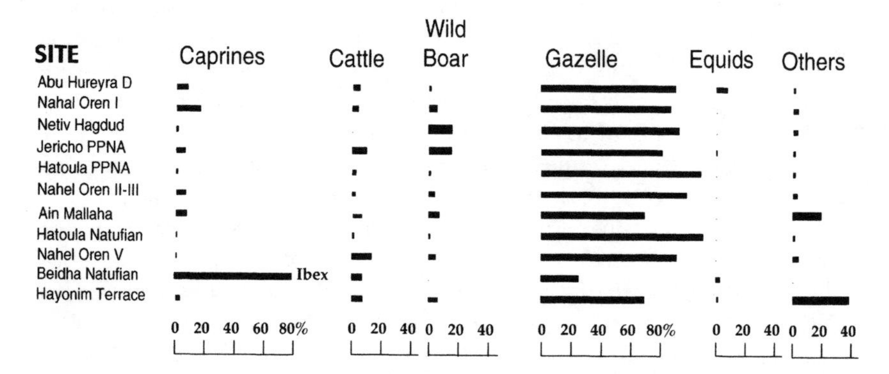

Figure 5.12. Faunal remains from selected Natufian and PPNA sites demonstrating the continuous exploitation of gazelle. Note that the small game such as hare, partridge, and tortoise are not represented (Stiner et al. 1999).

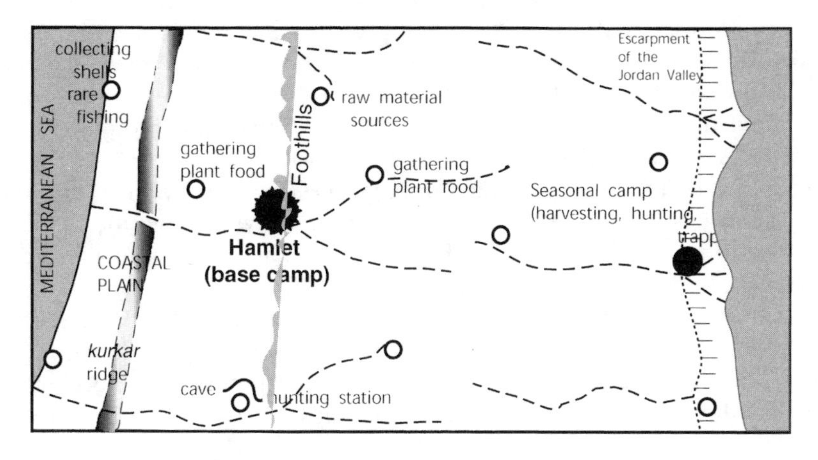

Figure 5.13. A reconstructed settlement pattern within the Natufian homeland representing the different types of sites in a cross section between the Mediterranean and the Jordan Valley.

entity (Rocek and Bar-Yosef 1998, and papers therein). In a thought provoking paper, F. Valla (Valla 1998) suggested that the main settlement pattern in the central Levant incorporated an aggregation site ("base camp") and several ephemerally occupied seasonal camps. He made use of the cementum data produced by Lieberman (Lieberman 1993) as well as the evidence from the seasonal pattern of fishing in Lake Hula (Desse 1987). Valla stresses the observation that the Natufian settlement pattern was not uniform. In sum, one can envision the seasonal movements from western base camps in Mt. Carmel or the Judean Hills to the Jordan Valley where foraging stations (Fig. 5.13) were established. Further north in the Galilee, the area was subdivided into two territories, one in the western Galilee and the other in the area of the Hula Valley. A similar west–east mobility can be reconstructed for the sites along the Jordanian plateau. Both the Negev in the south and the northern mountainous areas in Lebanon and Syria accommodated other trajectories.

THE YOUNGER DRYAS AND THE COLLAPSE OF THE EARLY NATUFIAN

The climatic crisis of the Younger Dryas, mentioned above, was first recorded in the pollen cores of Scandinavia and later in deep-sea cores and ice cores. The length of this period is ca. 1300±70 years, or from 12,900 to 11,600 or 12,800 to 11,500 cal. B.P. (Alley et al. 1993; Landmann et al. 1996; Mayewski and Bender 1995; Rossignol-Strick 1995). However, there are certain discrepancies between European varve chronology and the ice cores (Goslar et al. 1995). For example, several core studies report the end of the Younger Dryas at 11,300 to 10,700 cal. B.P. Closer to the Levant, a similar problem was noted in Lake Van, where the varve chronology suggests that the Younger Dryas lasted from 12,600 to 10,000 or 10,510 cal. B.P. (Lemcke and Sturm 1997). Certain discrepancies can be attributed to missing varves in the lake records. However, from our viewpoint, the reaction of humans to the abrupt climatic change could have taken a few decades of human life before being noticed by foragers.

Complex hunter-gatherer societies such as the Early Natufian are considered unstable social entities (Arnold 1996a and references therein; Bar-Yosef and Belfer-Cohen in press; Byrd 1998; Henry 1991). The climatic crisis of the Younger Dryas differed in its impact between the Levant and other regions (Straus et al. 1996). However, major changes in settlement patterns were recorded everywhere. In the Near East, for example, the sites of Hallan Çemi Tepesi (12,900–10,900 cal. B.P.), and further east, Zawi Chemi Shanidar in Iraqi-Kurdistan, indicate that groups shifted from more mobile hunting and gathering to semisedentary settlements sometime between ca. 12,000 and 10,200 cal. B.P. Their rounded pit-houses do not differ from those of the Natufian (Rosenberg 1999; Solecki 1981), and their exchange routes covered large regions (Aurenche and Kozlowski 1999).

The impact of the Younger Dryas on the Natufian culture probably caused the inhabitants of some of the hamlets to disperse and become more mobile. The increasing degree in mobility is possibly reflected in the lithic industry, by the size diminution of lunates, exhaustion of the cores, and the like, which became the markers of the Late Natufian (Belfer-Cohen and Grossman 1997; Goring-Morris 1987; Grosman et al. 1999; Valla 1984). Greater mobility meant the dismantling of the nonegalitarian structure, as indicated by the almost total disappearance of decorated skeletons, the increase in multi-individual graves, and secondary burials in the Late Natufian. Though the change may reflect an attempt to erase social differentiation between kin groups and to emphasize the unity of the population by giving similar treatment to all community members (Kuijt 1996), stress is recorded in the skeletal data (Belfer-Cohen et al. 1991). It seems that the Early Natufian hamlets ceased to serve the needed mode of production to sustain population growth under contemporary environmental deterioration. This was especially felt in the belts of the terebinth–almond woodland and the steppe. Growing competition in the marginal belt (where other entities subsisted during the prior millennia) favored large groups. However, when returns diminish, the costs of maintaining a large group are high; thus a smaller group becomes economically advantageous (Kosse 1994).

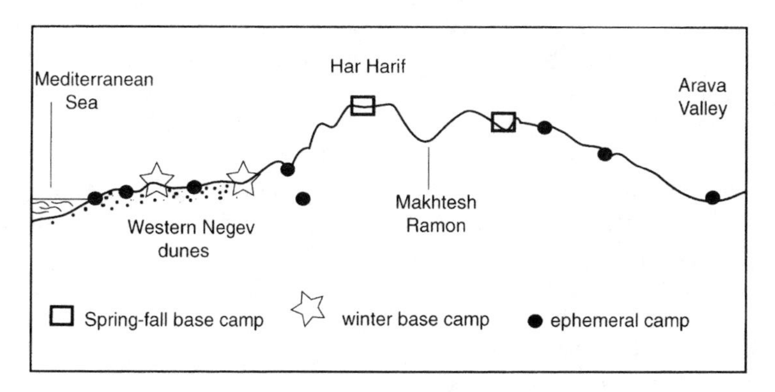

Figure 5.14. A reconstruction settlement pattern in the arid steppic zone more arid of the Negev between the Mediterranean Sea and the Arava Valley.

The archaeological manifestations of the Harifian entity in the Negev and northern Sinai exemplify a particular local adaptation. The local Late Natufian tool kit changes by incorporating a new type of projectile point that characterizes the Harifian (Goring-Morris 1991). The overall mobility of the Negev-northern Sinai groups centers on aggregation camps often in the higher elevations during spring, summer, and even fall, followed by dispersals onto the sandy lowlands and coastal zones in the wintertime (Fig. 5.14). The Harifians practiced elaborate exchange networks with their neighbors, probably to enhance alliances as a buffer for risky years. This aspect is reflected in the marine shell jewelry collected from both the Mediterranean and the Red Sea shores; a larger number of species was from the latter (D. E. Bar-Yosef 1991). In spite of efforts by the bearers of the Harifian lithic industry, this entity did not last for more than 500 years, perhaps even less, and disappeared from its homelands. In spite of the current ambiguities involved in calibrated dates of that period, it appears that the latest Harifian occurrences are partially contemporary with the earliest Pre-Pottery Neolithic A communities in the Jordan Valley (Goring-Morris and Belfer-Cohen 1997).

THE NATUFIAN PEOPLE: WHO ARE THEY?

Providing the readers with generalizations concerning the Natufian culture and its material attributes, while mentioning particular sites and even unique finds, requires getting to know the people themselves. These were the individuals and the groups who were responsible for producing what we find and the perishables that we are not able to discover. Unfortunately, it is only through the skeletal remains that we obtain a glimpse of this past population (Ferembach 1961; Grognier 1974; Arensburg et al. 1975; Solivères 1976; Belfer-Cohen et al. 1991; Hershkovitz 1990; Karasik et al. 2000; Smith 1991; Perrot and Ladiray et al. 1988; Bocquentin et al. under review).

Similar to the study of every archaeologically discovered population, the questions asked concerning the Natufian people are not different. We would like to know the demographic structure, life expectancies, signs of trauma and stress as evidence for past injuries (either through physical activities or warfare), as well as the dietary aspects. Based on the evidence accumulated to date by various researchers, a picture of the Natufians as real human beings begins to emerge, in spite of the expected distortion caused by their mortuary practices and postdepositional effects.

Morphologically, the Natufians were a proto-Mediterranean populace, and other studies substantiated the evidence for continuity with later Neoilithic groups. A certain degree of morphological variability was discerned among the Late Natufians, an additional indication for the suggested claim of increasing mobility which would also have been expressed in geographically wider mating networks.

The Natufian adults, as recorded in the four largest skeletal collections (Eynan, Hayonim Cave, El-Wad and Nahal Oren), were 160–175 cm high, and males were often slightly taller than females. Robustness (measured in shaft circumference of limb bones) decreases from the Early to the Late Natufian (Belfer-Cohen et al. 1991). There is some evidence of frequencies of biological markers, especially those that can be interpreted as indicating a degree of kinship, such as the congenital absence of the third molar (Smith 1973, 1991) and the retention of the metopic suture (Belfer-Cohen et al. 1991). The information about the former and incomplete data for the latter support the interpretation that the Natufian site cemeteries generally represent close kin.

There are low frequencies of pathologies among the Natufian remains, but they show signs for suffering from malaria, tuberculosis, or syphilis (Smith et al. 1984; Smith 1991), as well as infrequent signs of violence. On the other hand, stress is expressed in various ways, either by the stature of the adults or the higher mortality of children ages 5–7 (Belfer-Cohen et al. 1991). In conclusion, most authorities agree that the Natufian population represent a biologically "transitional" group between foragers and farmers.

CONCLUDING REMARKS

A prerequisite for investigating the origin of the Neolithic Revolution and its first phase (sometime around 11,500–11,000 cal. B.P.) is to review the archaeological evidence from the Natufian culture and its contemporary entities that date to the previous two or three millennia (Fig. 5.15). The Natufian culture was a special entity among other prehistoric entities of foragers in Western Asia and Northeast Africa. Now, it seems that the emergence of intentional cultivation, either during the last phase of the Younger Dryas or immediately after, triggered the establishment of farming villages. Although beyond the scope of this paper, this activity is currently seen as the direct result of mixed social and economic responses to the effects of the Younger Dryas. The dating of the last phase of this global event around 11,700–11,600 cal. B.P. means that we should expect that the bearers of the Late Natufian within the Levantine Corridor, or their immediate northern neighbors, were involved in the process. It is not impossible that the poorly defined Khiamian entity, the duration of which is not well known but often estimated as a few centuries, was the first archaeological expression of Neolithic farming hamlets (Bar-Yosef and Meadow 1995; Cauvin 2000a; Goring-Morris and Belfer-Cohen 1997). The low degree of archaeological resolution of defining criteria for the Khiamian may reflect the difficulties

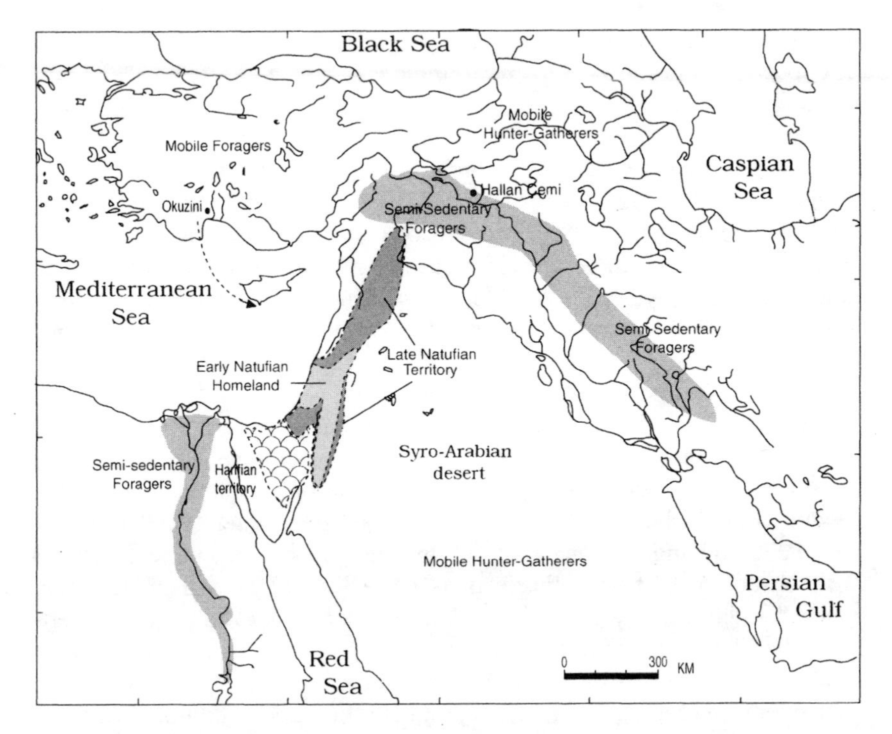

Figure 5.15. A general map of the Near East with the reconstructed distribution of Early and Late Natufian and Harifian as well as the territories of semisedentary and mobile foragers.

that faced the various groups in accepting a new socioeconomic order—and perhaps not incidentally led to the invention of the aerodynamically shaped arrowhead known as the Khiam point.

However, our story begins with the emergence of the nonegalitarian Early Natufian society. The paucity of radiocarbon dates for the first two millennia of its existence allows only general statements. We know that the first hamlets reflect a major energy investment compared to the earlier entities such as the geometric Kebaran and its contemporaries. The sites were occupied for a relatively long time but were eventually abandoned and then reoccupied. This observation raises the unresolved issues of overexploitation of particular territories and the role of warfare among rival communities. Numerous on-site graves were found, which in some examples hint to spatial concentrations (cemeteries?), but because areas outside the supposed limits of the hamlets were not excavated, there is no certainty that additional burials did not exist.

The complex burials of dead of all ages, the proliferation of body decorations, and their variability among the Early Natufians is interpreted as unambiguous evidence of social differentiation, group identification, and individual expression. In sum, the Early Natufian seem to have had a complex social organization compared to their ancestors and their descendants, the Late Natufians.

Large numbers of mortars and pestles reflect all sorts of grinding activities (food, burnt limestone, red ochre), and a portion of these utensils were brought from distances of up to 100 meters. Special "goblet type" mortars are imagined as serving in the course of communal feasting. Though no baskets were found, there are indirect indications that hint at storage and transportation in twined receptacles. Harvesting and gathering away from home is hypothesized based, for example, on the settlement pattern which incorporates the coastal areas and the Jordan Valley or the Jordanian plateau foothills and the eastern steppic areas.

In spite of the few available radiocarbon dates, the latter phase in the life history of this society took place during the Younger Dryas, the cold and relatively dry period, recognized also in the proxy climatic data in the Levant. In the lusher areas such as Mt. Carmel, the western Galilee, or the Hula Valley, changes in material culture and the nature of the dwellings were observed. However, these areas were less affected than the more open woodland and steppic belts. It is not surprising that most experts agree that the Late Natufian settlement pattern reflects greater mobility, or an almost full return to the late Epi-Paleolithic mode, which ranged from semisedentism to a highly mobile strategy. In the Negev and northern Sinai, the Harifian entity is seen as an expression of further adaptation of the Late Natufian to increasing aridity.

In the context of the Neolithic Revolution, which occurred later, it would be adequate to mention that the Natufian groups in the hilly western coastal area were probably harvesting wild cereals in natural stands in their immediate environments and also along the Jordan Valley. If yields of natural stands that provided the staple food decreased considerably during the Younger Dryas and relocation to other areas was hampered by potential physical conflicts with neighbors and the absence of positive social alliances, and the like, the motivation for intentional cultivation could have increased. In addition, it was advantageous to establish sedentary hamlets in the richest habitats along the Jordan Valley, the Damascus plain, and the Euphrates Valley, creating the so-called Levantine Corridor. Hence, it was this shift in settlement patterns that formed the largest Neolithic communities known as Pre-Pottery Neolithic A or the Sultanian and Mureybetian entities (Bar-Yosef 1998; Cauvin 2000a). The first cultivators exploited the alluvial lands as fields, resembling their Late Natufian ancestors, as well as lakeside flats. This choice of location suggests that the primary consideration was related to cereal cultivation and permanent water sources and not necessarily to the optimal foraging of other vegetal and animal resources. However, within the Natufian communities, adequate knowledge of the land, plant life cycles, and their natural habitats accumulated and served their decisions which resulted in an unexpected evolutionary, "point of no return," thus opening a new age in the history of humankind.

ACKNOWLEDGMENTS

This chapter is based on several previously published papers and in particular on Bar-Yosef 1998 and Bar-Yosef in press. In this version as in the others, I have benefited from the partnership of Anna Belfer-Cohen and the advice of Francois Valla and Nigel Goring-Morris. Wren Fournier skillfully edited this manuscript. Needless to mention that all shortcomings are mine.

REFERENCES

Alley, R. B., Meese, D. A., Shuman, C. A., Gow, A. J., Taylor, K. C., Grootes, P. M., White, J. W. C., Ram, M., Waddington, E. D., Mayewski, P. A., and Zielinski, G. A., 1993, Abrupt Increase in Greenland Snow Accumulation at the End of the Younger Dryas Event. *Nature* 362:527–529.

Ames, K. M., 1995, Chiefly Power and Household Production on the Northwest Coast. In *Foundations of Social Inequality*, edited by T. D. Price and G. M. Feinman, pp. 155–188. Plenum, New York.

Ames, K. M., 1999, Myth of the Hunter-Gatherer. *Archaeology* (September/October):45–49.

Ames, K. M., and Maschner, H. D. G., 1999, *Peoples of the Northwest Coast: Their Archaeology and Prehistory*. Thames and Hudson, New York.

Anderson, P., 1991, Harvesting of Wild Cereals During the Natufian as Seen from the Experimental Cultivation and Harvest of Wild Einkorn Wheat and Microwear Analysis of Stone Tools. In *The Natufian Culture in the Levant*, edited by O. Bar-Yosef and F. R. Valla, pp. 521–556. International Monographs in Prehistory, Ann Arbor.

Anderson, P., and Vallas, F., 1996, "Glossed Tools" from Hayonim Terrace: Blank Choice and Functional Tenancies. In *Neolithic Chipped Stone Industries of the Fertile Crescent and Their Contemporaries in Adjacent Regions: Proceedings of the Second Workshop on PPN Chipped Lithic Industries*, Warsaw 1995, Studies in Early Near Eastern Production, Subsistence, and Environment 3, Ex Oriente, Berlin, pp. 341–362.

Anderson, P. C., 1994, Reflections on the Significance of Two PPN Typological Classes in the Light of Experimentation and Microwear Analysis: Flint "Sickles" and Obsidian "Çayönü Tools". In *Neolithic Chipped Stone Industries of the Fertile Crescent: Proceedings of the First Workshop on PPN Chipped Lithic Industries*, Ex Oriente, Berlin, pp. 61–82.

Anderson, P. C., 1998, History of Harvesting and Threshing Techniques for Cereals in the Prehistoric Near East. In *The Origins of Agriculture and Crop Domestication*, edited by A. B. Damania, J. Valkoun, G. Willcox, and C. O. Qualset, pp. 145–159. ICARDA, Aleppo, Syria.

Arensburg, B., Goldstein, M., and Nathan, H., 1975, The Epipaleolithic (Natufian) Population in Israel. *Dos Arquivos de Anatomia e Antropologia* I:205–217.

Arnold, J. E., (ed.), 1996a, *Emergent Complexity: The Evolution of Intermediate Societies*. International Monographs in Prehistory, University of Michigan, Ann Arbor.

Arnold, J. E., 1996b, Understanding the Evolution of Intermediate Societies. In *Emergent Complexity: The Evolution of Intermediate Societies*, edited by J. E. Arnold, pp. 1–12. International Monographs in Prehistory, University of Michigan, Ann Arbor.

Aurenche, O., and Kozlowski, S. K., 1999, *La Naissance du Néolithique au Proche Orient ou Le Paradis Perdu*, Editions Errance, Paris.

Badr, A., Müller, K., Schäfer-Pregl, R., El Rabey, H., Effgen, S., Ibrahim, H. H., Pozzi, C., Rohde, W., and Salamini, F., 2000, On the Origin and Domestication History of Barley (*Hordeum vulgare*). *Molecular Biology and Evolution* 17(4):499–510.

Bailey, G. N., and Davidson, I., 1983, Site Exploitation of Territories and Topography: Two Case Studies from Paleolithic Spain. *Journal of Archaeological Science* 10(2):87–116.

Bailey, H. P., 1960, A Method of Determining the Warmth and Temperateness of Climate. *Geografiska Annaler* 43(1):1–16.

Bar-El, T., and Tchernov, E., 2000, Lagomorph Remains at Prehistoric Sites in Israel and Southern Sinai. *Paléorient* 26(1):93–110.

Bar-Mathews, M., Ayalon, A., and Kaufman, A., 1997, Late Quaternary Paleoclimate in the Eastern Mediterranean Region from Stable Isotope Analysis of Speleothems at Soreq Cave, Israel. *Quaternary Research* 47:155–168.

Bar-Mathews, M., Ayalon, A., Kaufman, A., and Wasserburg, G. J., 1999, The Eastern Mediterranean Paleoclimate as a Reflection of Regional Events: Soreq Cave, Israel. *Earth and Planetary Science Letters* 166:85–95.

Bar-Yosef, D. E., 1991, Changes in the Selection of Marine Shells from the Natufian to the Neolithic. In *The Natufian Culture in the Levant*, edited by O. Bar-Yosef and F. R. Valla, pp. 629–636. International Monographs in Prehistory, Ann Arbor.

Bar-Yosef, O., 1991a, Raw Material Exploitation in the Levantine Epi-Palaeolithic. In *Raw Material Economies Among Prehistoric Hunter-Gatherers*, edited by A. Montet-White and S. Holen, pp. 235–250. Publications in Anthropology, 19, University of Kansas, Lawrence.

Bar-Yosef, O., 1991b, The Archaeology of the Natufian Layer at Hayonim Cave. In *The Natufian Culture in the Levant*, edited by O. Bar-Yosef and F. R. Valla, pp. 81–93. International Monographs in Prehistory, Ann Arbor.

Bar-Yosef, O., 1997, Symbolic Expressions in Later Prehistory of the Levant: Why Are They So Few? In *Beyond Art: Pleistocene Image and Symbol*, edited by M. W. Conkey, O. Soffer, D. Stratmann, and N. G. Jablonski, Vol. 23, pp. 161–187. Memoirs of the California Academy of Science, San Francisco.

Bar-Yosef, O., 1998, On the Nature of Transitions: The Middle to Upper Palaeolithic and the Neolithic Revolution. *Cambridge Archaeological Journal* 8(2):141–163.

Bar-Yosef, O., 2000, The Impact of Radiocarbon Dating on Old World Archaeology: Past Achievements and Future Expectation. *Radiocarbon* 42(1):1–17.

Bar-Yosef, O., in press, From sedentary Foragers to Village Hierarchies: The emergence of Social Institutions. In *The Origin of Human Social Institutions*, edited by G. Runciman, Royal Society and British Academy, London.

Bar-Yosef, O., and Belfer-Cohen, A., 1989, The Origins of Sedentism and Farming Communities in the Levant. *Journal of World Prehistory* 3(4):447–498.

Bar-Yosef, O., and Belfer-Cohen, A., 1991, From Sedentary Hunter-Gatherers to Territorial Farmers in the Levant. In *Between Bands and States*, edited by S. A. Gregg, pp. 181–202. Occasional Paper No. 9, Center for Archaeological Investigations, Carbondale.

Bar-Yosef, O., and Belfer-Cohen, A., 1992, From Foraging to Farming in the Mediterranean Levant. In *Transitions to Agriculture in Prehistory*, edited by A. B. Gebauer and T. D. Price, pp. 21–48. Prehistory Press, Madison.

Bar-Yosef, O., and Belfer-Cohen, A., 1998, Natufian Imagery in Perspective. *Rivista di Scienze Prehistoriche* 49:247–263.

Bar-Yosef, O., and Belfer-Cohen, A., 1999, Encoding Information: Unique Natufian Objects from Hayonim Cave, Western Galilee, Israel. *Antiquity* 73(280):402–410.

Bar-Yosef, O., and Belfer-Cohen, A., in press, Facing Environmental Crisis—Societal and Cultural Changes at the Transition from the Younger Dryas to the Holocene in the Levant. In *The Transition from Foraging to Farming in Southwestern Asia*, edited by R. Cappers, S. Bottema, and U. Baruch, Ex Oriente, Berlin.

Bar-Yosef, O., and Meadow, R. H., 1995, The Origins of Agriculture in the Near East. In *Last Hunters, First Farmers: New Perspectives on the Prehistoric Transition to Agriculture*, edited by T. D. Price and A. B. Gebauer, pp. 39–94. School of American Research Advanced Seminar Series, School of American Research Press, Santa Fe.

Bar-Yosef, O., and Phillips, J. L., (eds.), 1977, *Prehistoric Investigations in Jebel Meghara, Northern Sinai*. Hebrew University, Jerusalem.

Bar-Yosef, O., and Valla, F. R., 1979, L'évolution du Natufien, Nouvelles Suggestions. *Paléorient* 5:145–152.

Bar-Yosef, O., and Valla, F. R., 1991, The Natufian Culture—An Introduction. In *The Natufian Culture in the Levant*, edited by O. Bar-Yosef and F. R. Valla, pp. 1–10. International Monographs in Prehistory, Ann Arbor.

Baruch, U., and Bottema, S., 1991, Palynological Evidence for Climatic Changes in the Levant ca. 17,000–9,000 B.P. In *The Natufian Culture in the Levant*, edited by O. Bar-Yosef and F. R. Valla, pp. 11–20. International Monographs in Prehistory, Ann Arbor.

Belfer-Cohen, A., 1988, The Natufian Graveyard in Hayonim Cave. *Paléorient* 14(2):297–308.

Belfer-Cohen, A., 1991a, Art Items from Layer B, Hayonim Cave: A Case Study of Art in a Natufian Context. In *The Natufian Culture in the Levant*, edited by O. Bar-Yosef and F. R. Valla, pp. 569–588. International Monographs in Prehistory, Ann Arbor.

Belfer-Cohen, A., 1991b, The Natufian in the Levant. *Annual Review of Anthropology* 20:167–186.

Belfer-Cohen, A., 1995, Rethinking Social Stratification in the Natufian Culture: The Evidence from Burials. In *The Archaeology of Death in the Ancient Near East*, edited by S. Campbell and A. Green, pp. 9–16. Oxbow Monograph 51, Oxford.

Belfer-Cohen, A., and Bar-Yosef, O., 2000, Early Sedentism in the Near East: A Bumpy Ride to Village Life. In *Life in Neolithic Farming Communities: Social Organization, Identity, and Differentiation*, edited by I. Kuijt, pp. 19–37. Plenum, New York.

Belfer-Cohen, A., and Goring-Morris, A. N., 1996, The Late Epipalaeolithic as the Precursor of the Neolithic: The Lithic Evidence. In *Neolithic Chipped Stone Industries of the Fertile Crescent and Their Contemporaries in Adjacent Regions*: Proceedings of the Second Workshop on PPN Chipped Lithic Industries, Warsaw 1995, Studies in Early Near Eastern Production, Subsistence, and Environment 3, Ex Oriente, Berlin, pp. 217–226.

Belfer-Cohen, A., and Grossman, L., 1997, The Lithic Assemblage of Salibiya I. *Journal of the Israel Prehistoric Society* 27:19–41.

Belfer-Cohen, A., Schepartz, L., and Arensburg, B., 1991, New Biological Data for the Natufian Populations in Israel. In *The Natufian Culture in the Levant*, edited by O. Bar-Yosef and F. R. Valla, pp. 411–424. International Monographs in Prehistory, Ann Arbor.

Bentley, G. R., Jasienska, G., and Goldberg, T., 1993, Is the Fertility of Agriculturalists Higher Than That of Nonagriculturalists? *Current Anthropology* 34:778–785.

Bettinger, R. L., 1991, Hunter-Gatherers: Archaeological and Evolutionary Theory. In *Interdisciplinary Contributions to Archaeology, The Language of Science*. Plenum, New York.

Binford, L. R., 1971, Mortuary Practices: Their Study and Potential. In *Approaches to the Social Dimensions of Mortuary Practices*, edited by J. A. Brown, Vol. Memoir 25, pp. 6–29. Society for American Archaeology, Washington, DC.

Binford, L. R., 1980, Willow Smoke and Dogs' Tails: Hunter-Gatherer Settlement Systems and Archaeological Site Formation. *American Antiquity* 45:4–20.

Binford, L. R., 1996, Hearth and Home: The Spatial Analysis of Ethnographically Documented Rock Shelter Occupations as a Template for Distinguishing Between Human and Hominid Use of Sheltered Space. In *Middle Palaeolithic and Middle Stone Age Settlement Systems*, edited by N. J. Conard and F. Wendorf, pp. 229–240. Proceedings of the XIII Congress of the International Union of Prehistoric and Protohistoric Sciences (UISPP), Vol. 6, Tome 1. A.B.A.C.O. Edizioni, Forlí.

Binford, L. R., and Sabloff, J. A., 1982, Paradigms, Systematics, and Archaeology. *Journal of Anthropological Research* 38(2):137–153.

Bocquentin, F., Sellier, P., and Murail, P., under review, La Population Natoufienne de Mallaha (Eynan, Israel): Denombrement, Age au Deces et Recrutement Funeraire.

Borziyak, I. A., 1993, Subsistence Practices of Late Paleolithic Groups along the Dnestr River and Its Tributaries. In *From Kostenki to Clovis: Upper Paleolithic – Paleo-Indian Adaptations*, edited by O. Soffer and N. D. Praslov, pp. 67–84. Plenum, New York.

Boyd, B., and Cook, J., 1993, A Reconsideration of the 'Ain Sakhri' Figurine. *Proceedings of the Prehistoric Society* 59:399–405.

Byrd, B. F., 1998, Spanning the Gap Between the Upper Paleolithic and the Natufian: The Early and Middle Epipaleolithic. In: *The Prehistoric Archaeology of Jordan*, edited by D. O. Henry, pp. 64–82. British Archaeological Reports (BAR) S705, Archaeopress, Oxford.

Byrd, B. F., 2000, Households in Transition: Neolithic Social Organization Within Southwest Asia. In *Life in Neolithic Farming Communities: Social Organization, Identity, and Differentiation*, edited by I. Kuijt, pp. 63–98. Plenum, New York.

Byrd, B. F., and Monahan, C. M., 1995, Death, Mortuary Ritual, and Natufian Social Structure. *Journal of Anthropological Archaeology* 14(3):251–287.

Campana, D., 1991, Bone Implements from Hayonim Cave: Some Relevant Issues. In *The Natufian Culture in the Levant*, edited by O. Bar-Yosef and F. R. Valla, pp. 459–466. International Monographs in Prehistory, Ann Arbor.

Campana, D. V., 1989, *Natufian and Protoneolithic Bone Tools*. British Archaeological Reports International Series (BAR) 494, Oxford.

Campana, D. V., and Crabtree, P. J., 1990, Communal Hunting in the Natufian of the Southern Levant: The Social and Economic Implications. *Journal of Mediterranean Archaeology* 3(2):223–243.

Carsten, J., and Hugh-Jones, S., 1995, *About the House: Lévi-Strauss and Beyond*. Cambridge University Press, Cambridge.

Cauvin, J., 1997, *Naissance des Divinités, Naissance de L'Agriculture*, 2nd ed. Centre National de Recherche Scientifique, Paris.

Cauvin, J., 2000a, *The Birth of the Gods and the Origins of Agriculture*. Translated by T. Watkins, Cambridge University Press, Cambridge.

Cauvin, J., 2000b, The Symbolic Foundations of the Neolithic Revolution in the Near East. In *Life in Neolithic Farming Communities: Social Organization, Identity, and Differentiation*, edited by I. Kuijt, pp. 235–251. Plenum, New York.

Cauvin, J. C., 1972, *Les Religions Néolithiques de Syro-Palestine*. Maisonneuve, Paris.

Childe, V. G., 1953, *New Light on the Most Ancient Near East*. Praeger, New York.

Chilton, E. S., 1999, *Material Meanings: Critical Approaches to the Interpretation of Material Culture*. Foundations of Archaeological Inquiry, University of Utah, Salt Lake City.

Churcher, C. S., 1994, The Vertebrate Fauna from the Natufian Level at Jebel es-Saaïdé (Saaïdé II), Lebanon. *Paléorient* 20(2):35–58.

Clark, J. E., 1999, Comments: On "Stone Tools" by G. Odell. *Lithic Technology* 24(2):126–135.

Colledge, S., 1998, Identifying Pre-Domestication Cultivation Using Multivariate Analysis. In *The Origins of Agriculture and Crop Domestication*, edited by A. B. Damania, J. Valkoun, G. Willcox, and C. O. Qualset, pp. 121–131. ICARDA, Aleppo, Syria.

Cope, C., 1991, Gazelle Hunting Strategies in the Southern Levant. In *The Natufian Culture in the Levant*, edited by O. Bar-Yosef and F. R. Valla, pp. 341–358. International Monographs in Prehistory, Ann Arbor.

Coupland, G., 1996, This Old House: Cultural Complexity and Household Stability on the Northern Northwest Coast of North America. In *Emergent Complexity: The Evolution of Intermediate Societies*, edited by J. E. Arnold, pp. 74–90. Vol. Archaeological Series 9, International Monographs in Prehistory, Ann Arbor.

Crabtree, P. J., Campana, D. V., Belfer-Cohen, A., and Bar-Yosef, D. E., 1991, First Results of the Excavations at Salibiya I, Lower Jordan Valley. In *The Natufian Culture in the Levant*, edited by O. Bar-Yosef and F. R. Valla, pp. 173–216. International Monographs in Prehistory, Ann Arbor.

Danin, A., 1988, Flora and Vegetation of Israel and Adjacent Areas. In *The Zoogeography of Israel*, edited by Y. Yom-Tov and E. Tchernov, pp. 129–158. Junk, Dordrecht.

Davis, S. J. M., 1982, Climatic Change and the Advent of Domestication of Ruminant Artiodactyls in the late Pleistocene-Holocene period in the Israel Region. *Paléorient* 8(2):5–16.

Davis, S. J. M., 1983, The Age Profile of Gazelles Predated by Ancient Man in Israel: Possible Evidence for a Shift from Seasonality to Sedentism in the Natufian. *Paléorient* 9:55–62.

Davis, S. J. M., Lernau, O., and Pichon, J., 1994, The Animal Remains: New Light on the Origin of Animal Husbandry. In *Le Gisement de Hatoula en Judée Occidentale, Israël*, edited by M. Lechevallier and A. Ronen, pp. 83–100. Memoires et Travaux du Centre de Recherche Français de Jerusalem, 8, Association Paléorient, Paris.

Davis, S. J. M., and Valla, F., 1978, Evidences for the Domestication of the Dog in the Natufian of Israel 12,000 Years Ago. *Nature* 276:608–610.

Dayan, T., and Simberloff, D., 1995, Natufian Gazelles: Proto-Domestication Reconsidered. *Journal of Archaeological Science* 22(5):671–676.

Desse, J., 1987, L'ichthyofaune. In *La Faune du Gisement Natoufien de Mallaha (Eynan) Israel*, edited by J. Bouchud, pp. 151–156. Mémoires et Travaux du Centre de Recherche Français de Jérusalem 4, Association Paléorient, Paris.

Dobres, M., 2000, *Technology and Social Agency*. Blackwell, Oxford, Malden.

Dobres, M., and Hoffman, C., 1994, Social Agency and the Dynamics of Prehistoric Technology. *Journal of Archaeological Method and Theory* 1(3):211–258.

Ducos, P., 1978, *Tell Mureybet, Étude Archéozoologique et Problèmes d'Écologie Humaine*. CNRS, Lyon.

Ducos, P., and Kolska Horwitz, R., 1998, The Influence of Climate on Artiodactyl Size During the Late Pleistocene-Early Holocene of the Southern Levant. *Paléorient* 23(2):229–247.

Durkheim, E., 1938, *The Rules of the Sociological Method*. Free Press, Glencoe.

Earle, T., 1997, *How Chiefs Come to Power: The Political Economy in Prehistory*. Stanford University Press, Stanford.

Edwards, P., 1991, Wadi Hammeh 27: An Early Natufian Site at Pella, Jordan. In *The Natufian Culture in the Levant*, edited by O. Bar-Yosef and F. R. Valla, pp. 123–148. International Monographs in Prehistory, Ann Arbor.

Edwards, P. C., Bourke, S. J., Colledge, S. M., Head, J., and Macumber, P. G., 1988, Late Pleistocene Prehistory in the Wadi al-Hammeh, Jordan Valley. In *The Prehistory of Jordan: The State of Research in 1986*, edited by A. N. Garrard and H. G. Gebel, pp. 525–565. British Archaeological Reports International Series 396, Oxford.

Ferembach, D., 1961, Squelettes du Natoufien d'Israel, Etude Anthropologique. *L'Anthropologie* 65(1–2):46–66.

Flannery, K. V., 1969, Origins and Ecological Effects of Early Domestication in Iran and the Near East. In *The Domestication and Exploitation of Plants and Animals*, edited by P. J. Ucko and G. W. Dimbleby, pp. 73–100. Duckworth, London.

Flannery, K. V., 1972, The Origins of the Village as a Settlement Type in Mesoamerica and the Near East: A Comparative Study. In *Man, Settlement and Urbanism*, edited by P. J. Ucka, R. Trigham, and G. W. Dimbleby, pp. 23–53. Duckworth, London.

Frumkin, A., Ford, D. C., and Schwarcz, H. P., 1999, Continental Oxygen Isotopic Record of the Last 170,000 Years in Jerusalem. *Quaternary Research* 51:317–327.

Galil, J., Stein, M., and Horovitz, A., 1976, On the Origin of the Sycomore fig (*Ficus sycomorus L.*) in the Middle East. *Gardens Bulletin* XXIX:191–205.

Galili, E., and Schick, T., 1990, Basketry and a Wooden Bowl from the Pottery Neolithic Submerged Site of Kefar Samir, Mitekufat Haeven. *Journal of the Israel Prehistoric Society* 23:142–151.

Garrard, A. N., Baird, D., and Byrd, B. F., 1994, The chronological basis and significance of the Late Paleolithic and Neolithic Sequence in the Azraq Basin, Jordan. In *Late Quaternary Chronology and Paleoclimates of the Eastern Mediterranean*, edited by O. Bar-Yosef and R. Kra, pp. 177–200. Radiocarbon and the Peabody Museum of Archaeology and Ethnology, Harvard University, Tucson and Cambridge.

Garrard, A. N., Colledge, S., and Martin, L., 1996, The Emergence of Crop Cultivation and Caprine Herding in the "Marginal Zone" of the Southern Levant. In *The Origins and Spread of Agriculture and Pastoralism in Eurasia*, edited by D. Harris, pp. 204–226. UCL Press, London.

Garrod, D. A. E., 1932, A New Mesolithic Industry: The Natufian of Palestine. *Journal of the Royal Anthropological Institute*, 62:257–270.

Garrod, D. A. E., 1957, The Natufian Culture: The Life and Economy of a Mesolithic People in the Near East. *Proceedings of the British Academy* 43:211–227.

Garrod, D. A. E., and Bate, D. M., 1937, *The Stone Age of Mount Carmel*. Clarendon Press, Oxford.

Garrod, D. A. E., and Bate, D. M., 1942, Excavations at the Cave of Shukbah, Palestine. *Proceedings of the Prehistoric Society* 8:1–20.

Goring-Morris, A. N., 1987, *At the Edge: Terminal Pleistocene Hunter-Gatherers in the Negev and Sinai*. British Archaeological Reports (BAR) 361, Oxford.

Goring-Morris, A. N., 1991, The Harifian of the Southern Levant. In *The Natufian Culture in the Levant*, edited by O. Bar-Yosef and F. R. Valla, pp. 173–216. International Monographs in Prehistory, Ann Arbor.

Goring-Morris, 1995, Complex Hunter-Gatherers at the End of the Palaeolithic (20,000–10,000 B.P.). In *The Archaeology of Society in the Holy Land*, edited by T. E. Levy, pp. 141–168. Leicester University Press, London.

Goring-Morris, 1998, Mobiliary Art from the Late Epipalaeolithic of the Negev, Israel. *Rock Art Research* 15(2):81–88.

Goring-Morris, A. N., and Belfer-Cohen, A., 1997, The Articulation of Cultural Processes and Late Quaternary Environmental Changes in Cisjordan. *Paléorient* 23(2):71–94.

Goring-Morris, A. N., Goldberg, P., Goren, Y., Baruch, U., and Bar-Yosef, D. E., 1999, Saflulin: A Late Natufian Base Camp in the Central Negev Highlands, Israel. *Palestine Exploration Quarterly* 131:36–64.

Goslar, T., Arnold, M., Bard, E., Kuc, T., Pazdur, M., Ralska-Jasiewiczowa, M., Rozanski, K., Tisnerat, N., Walanus, A., Wicik, B., and Wieckowski, K., 1995, High Concentration of atmospheric C14 during the Younger Dryas Cold Episode, *Nature* 377:414–417.

Grognier, E., and Dupouy-Madre, M., 1974, Les Natoufiens du Nahal Oren, Israel: Etude Anthropologique. *Paléorient* 2:103–121.

Grosman, L., Belfer-Cohen, A., and Bar-Yosef, O., 1999, A Final Natufian Site—Fazael IV, Mitekufat Haeven. *Journal of the Israel Prehistoric Society* 29:17–40.

Haim, A., and Tchernov, E., 1974, The Distribution of Myomorph Rodents in the Sinai Peninsula. *Mammalia* 38:201–223.

Harris, D. R., 1998, The Spread of Neolithic Agriculture from the Levant to Western Central Asia. In *The Origins of Agriculture and Crop Domestication*, edited by A. B. Damania, J. Valkoun, G. Willcox, and C. O. Qualset, pp. 65–82. Report No. 21 of the Genetic Resources Conservation Program, Division of Agriculture and Natural Resources, ICARDA, Aleppo, Syria.

Harrison, D. L., and Bates, P. J. J., 1991, *The Mammals of Arabia*. Harrison Zoological Museum, Sevenoaks, Kent.

Hayden, B., 1990, Nimrods, Piscators, Pluckers, and Planters: The Emergence of Food Production. *Journal of Anthropological Archaeology* 9:31–69.

Hayden, B., 1995a, A New Overview of Domestication. In *Last Hunters – First Farmers*, edited by T. D. Price and A. B. Gebauer, pp. 273–300. School of American Research Press, Santa Fe.

Hayden, B., 1995b, Pathways to Power: Principles for Creating Socioeconomic Inequalities. In *Foundations of Social Inequality*, edited by T. D. Price and G. M. Feinman, pp. 15–86. Fundamental Issues in Archaeology, Vol. Chapter 2. Plenum, New York.

Hayden, B., 1997, Observations on the Prehistoric Social and Economic Structure of the North American Plateau. *World Archaeology* 29:242–261.

Henry, D. O., 1974, The Utilization of the Microburin Technique in the Levant. *Paléorient* 2(2):389–398.

Henry, D. O., 1976, Rosh Zin: A Natufian Settlement near Ein Avdat. In *Prehistory and Paleoenvironents in the Central Negev*, edited by A. E. Marks, pp. 317–347. SMU Press, Dallas.

Henry, D. O., 1985, Preagricultural Sedentism: The Natufian Example. In *Prehistoric Hunter-Gatherers: The Emergence of Complex Societies*, edited by T. C. Price and J. A. Brown, pp. 365–384. Academic Press, New York.

Henry, D. O., 1989, *From Foraging to Agriculture: The Levant at the End of the Ice Age.* University of Pennsylvania Press, Philadelphia.

Henry, D. O., 1991, Foraging, Sedentism, and Adaptive Vigor in the Natufian: Rethinking the Linkages. In *Perspectives on the Past: Theoretical Biases in Mediterranean Hunter-Gatherer Research*, edited by G. A. Clark, pp. 353–370. University of Pennsylvania Press, Philadelphia.

Henry, D. O., 1997, Prehistoric Human Ecology in the Southern Levant East of the Rift from 20,000–6,000 B.P. *Paléorient* 23(2):107–120.

Hershkovitz, I., and Gopher, A., 1990, Paleodemography, Burial Customs, and Food-Producing Economy at the Beginning of the Holocene: A Perspective from the Southern Levant, Mitekufat Haeven. *Journal of The Israel Prehistoric Society* 23:9–47.

Heun, M., Schäfer-Pregl, R., Klawan, D., Castagna, R., Accerbi, M., Borghi, B., and Salamini, F., 1997, Site of Einkorn Wheat Domestication Identified by DNA fingerprinting. *Science* 278:1312–1314.

Hillman, G., 1996, Late Pleistocene Changes in Wild Plant-Foods Available to Hunter-Gatherers of the Northern Fertile Crescent: Possible Preludes to Cereal Cultivation. In *The Origins and Spread of Agriculture and Pastoralism in Eurasia*, edited by D. Harris, pp. 159–203. UCL Press, London.

Hillman, G., 2000, Abu Hureyra 1: The Epipalaeolithic. In *Village on the Euphrates: From Foraging to Farming at Abu Hureyra*, edited by A. M. T. Moore, G. C. Hillman, and A. J. Legge, pp. 327–399. Oxford University Press, Oxford.

Hillman, G. C., Colledge, S., and Harris, D. R., 1989, Plant Food Economy during the Epi-Palaeolithic Period at Tel Abu Hureyra, Syria: Dietary Diversity, Seasonality and Modes of Exploitation. In *Foraging and Farming: The Evolution of Plant Exploitation*, edited by D. R. Harris and G. C. Hillman, pp. 240–266. Hyman Unwin, London.

Hillman, G. C., and Davies, M. S., 1990, Domestication Rates in Wild-Type Wheats and Barley under Primitive Cultivation. *Biological Journal of the Linnaean Society* 39(1):39–78.

Hopf, M., and Bar-Yosef, O., 1987, Plant Remains from Hayonim Cave, Western Galilee. *Paléorient* 13(1):117–120.

Isaac, G. L., 1984, The Archaeology of Human Origins: Studies of the Lower Pleistocene in East Africa 1971–1981. In *Advances in World Archaeology*, edited by F. Wendorf and A. Close, Vol. 3, pp. 1–87. Academic Press, New York.

Johnson, A. W., and Earle, T., 2000, *The Evolution of Human Societies: From Foraging Group to Agrarian State*, 2nd ed. Stanford University Press, Stanford.

Karasik, D., Arensburg, B., and Pavlovsky, O. M., 2000, Age Assessment of Natufian Remains from the Land of Israel. *American Journal of Physical Anthropology* 113:263–274.

Keeley, L. H., 1988, Hunter-Gatherer Economic Complexity and "Population Pressure": A Cross-Cultural Analysis. *Journal of Anthropological Archaeology* 7:373–411.

Keeley, L. K., 1996, *War Before Civilization.* Oxford University Press, New York.

Kelly, R. L., 1991, Sedentism, Socio-Political Inequality, and Resource Fluctuations. In *Between Bands and States*, edited by S. A. Gregg, pp. 135–158. Center for Archaeological Investigations, Occasional Papers No.9, Southern Illinois University Press, Carbondale.

Kelly, R., 1992, Mobility/Sedentism: Concepts, Archaeological Measures, and Effects. *Annual Review of Anthropology* 21:43–66.

Kelly, R., 1995, *The Foraging Spectrum: Diversity in Hunter-Gatherer Lifeways*. Smithsonian Institution Press, Washington, DC.

Kent, S., (ed.), 1989, *Farmers as Hunters: The Implications of Sedentism*. Cambridge University Press, Cambridge.

Kenyon, K., 1957, *Digging Up Jericho*. Benn, London.

Kislev, M. E., 1989, Pre-Domesticated Cereals in the Pre-Pottery Neolithic A Period. In *People and Culture Change*, edited by I. Hershkovitz, pp. 147–152. British Archaeological Reports (BAR) 508i, Oxford.

Kislev, M. E., 1992, Agriculture in the Near East in the VIIth Millennium B.C. In *Préhistoire de L'Agriculture: Nouvelles Approches Experimentales et Ethnographiques*, edited by P. C. Anderson-Gerfaud, pp. 87–93. Monographie du CRA, No. 6, CNRS, Paris.

Kislev, M. E., 1997, Early Agriculture and Paleoecology of Netiv Hagdud. In *An Early Neolithic Village in the Jordan Valley Part I: The Archaeology of Netiv Hagdud*, edited by O. Bar-Yosef and A. Gopher, pp. 209–236. Peabody Museum of Archaeology and Ethnology, Harvard University, Cambridge.

Kislev, M. E., Nadel, D., and Carmi, I., 1992, Epi-Palaeolithic (19,000 B.P.) Cereal and Fruit Diet at Ohalo II, Sea of Galilee, Israel. *Review of Palaeobotany and Palynology* 71:161–166.

Knauft, B. M., 1987, Reconsidering Violence in Simple Human Societies: Homicide among the Gebusi of New Guinea. *Current Anthropology* 28(4):457–500.

Korobkova, G. F., 1992, The Study of Flint Tools and the Origin of Farming in the Near East. *Archaeological News* 3:166–180.

Kosse, K., 1994, The Evolution of Large, Complex Groups: A Hypothesis. *Journal of Anthropological Archaeology* 13(1):35–50.

Kuijt, I., 1996, Negotiating Equality Through Ritual: A Consideration of Late Natufian and Prepottery Neolithic A Period Mortuary Practices. *Journal of Anthropological Archaeology* 15:313–336.

Kuijt, I., 2000, Keeping the Peace: Ritual, Skull Caching, and Community Integration in the Levantine Neolithic. In *Life in Neolithic Farming Communities: Social Organization, Identity, and Differentiation*, edited by I. Kuijt, pp. 137–164. Plenum, New York.

Kuijt, I., (ed.), 2000, *Life in Neolithic Farming Communities: Social Organization, Identity, and Differentiation*. Plenum, New York.

Kuijt, I., Palumbo, G., and Mabry, J., 1992, Report on the 1990 Excavations of 'Iraq ed-Dubb, Jordan. *American Journal of Archaeology* 96:507–508.

Laden, G., 1992, Ethnoarchaeology and Land Use Ecology of the Efe (Pygmies) of the Ituri Rain Forest, Zaire: A Behavioral Ecological Study of Land Use Patterns and Foraging Behavior, Ph.D. Dissertation, Harvard University.

Landmann, G., Reimer, A., Lemcke, G., and Kempe, S., 1996, Dating Late Glacial Abrupt Climate Changes in the 14,570 Yr Long Continuous Varve Record of Lake Van, Turkey. *Palaeo* 122:107–118.

Legge, A. J., and Rowley-Conway, P. A., 1986, New Radiocarbon Dates for Early Sheep at Tell Abu Hureyra. In *Archaeological Results from Accelerator Dating*, edited by J. A. J. Gowlett and R. E. M. Hedges, pp. 23–35. Oxford University Committee for Archaeology, Oxford.

Lemcke, G., and Sturm, M., 1997, ∂ 18O and Trace Element Measurements as Proxy for the Reconstructions of Climate Changes at Lake Van (Turkey): Preliminary Results. In *Third Millennium B.C.* Climate Change and Old World Collapse, edited by H. N. Dalfes, G. Kukla, and H. Weiss, pp. 653–678. NATO ASI Series, Vol. I, 49, Springer-Verlag, Berlin.

Lemonnier, P., 1992, *Elements for an Anthropology of Technology*. Anthropological Papers, Museum of Anthropology, 88, University of Michigan, Ann Arbor.

Lévi-Strauss, C., 1983, *The Way of the Masks.* Translated by S. Modelski, Jonathan Cape, London.

Lewis, H. T., 1972, The Role of Fire in the Domestication of Plants and Animals in Southwest Asia: A Hypothesis. *Man* 7:195–222.

Lieberman, D. E., 1991, Seasonality and Gazelle hunting at Hayonim Cave: New Evidence for "Sedentism" During the Natufian. *Paléorient* 17(1):47–57.

Lieberman, D. E., 1993, The Rise and Fall of Seasonal Mobility among Hunter-Gatherers. *Current Anthropology* 34(5):599–631.

Lourandos, H., 1997, *Continent of Hunter-Gatherers: New Perspectives in Australian Prehistory.* Cambridge University Press, Cambridge.

Marks, A. E., (ed.), 1977, *Prehistory and Paleoenvironments in the Central Negev, Israel,* Vol. II. SMU Press, Dallas.

Marshack, A., 1997, Paleolithic Image Making and Symboling in Europe and the Middle East: A Comparative Review. In *Beyond Art: Pleistocene Image and Symbol,* edited by M. Conkey, O. Soffer, D. Stratmann, and N. G. Jablonski, Vol. 23, pp. 53–91. Memoirs of California Academy of Sciences, San Francisco.

Mason, S. L. R., 1995, Acorn-Eating and Ethnographic Analogies: A reply to McCorriston. *Antiquity* 69:1025–1030.

Matson, R. G., 1985, The Relationship Between Sedentism and Status Inequalities among Hunters and Gatherers. In *Status, Structure and Stratification: Current Archaeological Reconstructions,* edited by M. Thompson, M. T. Garcia, and F. J. Kense, pp. 245–252. University of Calgary Archaeological Association, Calgary.

Mayewski, P. A., and Bender, M., 1995, The GISP2 Ice Core Record—Paleoclimate Highlights. *Reviews of Geophysics* 33 (Supplement July):1287–1296.

McCorriston, J., 1994, Acorn Eating and Agricultural Origins: California Ethnographies and the Formal Use of analogy. *Antiquity* 69:1025–1030.

Mienis, H., 1987, Molluscs from the Excavation of Mallaha (Eynan). In *La Faune du Gisement Natoufien de Mallaha* (Eynan), edited by J. Bouchud, pp. 157–178. Mémoires et Travaux du Centre de Recherche Française de Jérusalem, No. 4, Association Paléorient, Paris.

Miller, N. F., 1997, The Macrobotanical Evidence for Vegetation in the Near East, c. 18 000/16 000 B.C. to 4 000 B.C. *Paléorient* 23(2):197–208.

Minc, L. D., and Smith, K. P., 1989, Spirit of Survival: Cultural Responses to Resource Variability in North Alaska. In *Bad Year Economics: Cultural Responses to Risk and Uncertainty,* edited by P. Halstead and J. O'Shea, pp. 8–39. Cambridge University Press, Cambridge.

Moore, A. M. T., 1982, Agricultural Origins in the Near East: A Model for the 1980s. *World Archaeology* 14:224–235.

Moore, A. M. T., 1989, The Transition from Foraging to Farming in Southwest Asia: Present Problems and Future Directions. In *Foraging and Farming: The Evolution of Plant Exploitation,* edited by D. R. Harris and G. C. Hillman, pp. 620–631. Unwin Hyman, London.

Mortensen, P., 1972, Seasonal Camps and Early Villages in the Zagros. In *Man, Settlement and Urbanism,* edited by P. Ucko, R. Tringham, and G. W. Dimbleby, pp. 293–297. Duckworth, London.

Neuville, R., 1934, Le Préhistoire de Palestine. *Revue Biblique* 43:237–259.

Neuville, R., 1951, *Le Paléolithique et le Mésolithique de Désert de Judée.* Archives de L'Institut de Paléontologie Humaine Mémoire 24, Masson, Editeurs, Paris.

Noy, T., 1991. Art and Decoration of the Natufian at Nahal Oren. In *The Natufian Culture in the Levant,* edited by O. Bar-Yosef and F. R. Valla, pp. 557–568. International Monographs in Prehistory, Ann Arbor.

Oliver, P., 1971, *Shelter in Africa.* Praeger, New York.

Olszewski, D. I., 1986, A Reassessment of Average Lunate Length as a Chronological Marker. *Paléorient* 12(1):39–44.

Olszewski, D. I., 1991, Social Complexity in the Natufian? Assessing the Relationship of Ideas and Data. In *Perspectives on the Past: Theoretical Biases in Mediterranean Hunter-Gatherer Research*, edited by G. A. Clark, pp. 322–340. University of Pennsylvania Press, Philadelphia.

Oswalt, W. H., 1972, *Habitat and Technology: The Evolution of Hunting*. Holt, Reinhart & Winston, New York.

Perlès, C., 1979, Des Navigateurs Méditerranéens Il Y A 10 000 Ans. *La Recherche* 10:82–83.

Perrot, J., 1966, Le Gisement Natoufien de Mallaha (Eynan), Israël. *L'Anthropologie* 70:437–484.

Perrot, J., and Ladiray, D., 1988, *Les Hommes de Mallaha (Eynan) Israel*. Mémoires et Travaux du Centre Recherche Française de Jérusalem No. 7, Association Paléorient, Paris.

Pichon, J., 1983, Parures Natoufiennes en Os de Perdrix. *Paléorient* 9(1):91–98.

Pichon, J., 1991, Les Oiseaux au Natoufien, Avifaune et Sédentarité. In *The Natufian Culture in the Levant*, edited by O. Bar-Yosef and F. R. Valla, pp. 371–380. International Monographs in Prehistory, Ann Arbor.

Price, T. D., and Brown, J. A., (eds.), 1985, *Prehistoric Hunter-Gatherers: The Emergence of Cultural Complexity*. Studies in Archaeology, Academic Press, Orlando.

Price, T. D., and Feinman, G. M., (eds.), 1995, *Pathways to Power: Principles for Creating Socioeconomic Inequalities*. Plenum, New York.

Rafferty, G. E., 1985, The Archaeological Record on Sedentariness: Recognition, Development and implications. In *Advances in Archaeological Method and Theory*, edited by M. B. Schiffer, Vol. 8, pp. 113–156. Academic Press, New York.

Rapoport, A., 1969, *House form and Culture*. Foundations of Cultural Geography Series, Prentice-Hall, Englewood Cliffs, NJ.

Rocek, T. R., and Bar-Yosef, O., (eds.), 1998, *Seasonality and Sedentism: Archaeological Perspectives from Old and New World Sites*. Peabody Museum of Archaeology and Ethnology, Cambridge.

Roscoe, P., 1990, The Bow and Spreadnet: Ecological Origins of Hunting Technology. *American Anthropologist* 92(3):691–701.

Roscoe, P., 2000, New Guinea Leadership as Ethnographic Analogy: A Critical Review. *Journal of Archaeological Method and Theory* 7(2):79–126.

Rosenberg, M., 1998, Cheating at Musical Chairs: Territoriality and Sedentism in an Evolutionary Context. *Current Anthropology* 39(5):653–681.

Rosenberg, M., 1999, Hallan çemi. In *Neolithic in Turkey: Cradle of Civilization*. New Discoveries, edited by M. Özdogan and N. Basgelen, pp. 25–34. Arkeoloji ve Sanat Yayinlari, Istanbul.

Rossignol-Strick, M., 1995, Sea-Land Correlation of Pollen Records in the Eastern Mediterranean for the Glacial-Interglacial Transition: Biostratigraphy Versus Radiometric Time-Scale. *Quaternary Science Reviews* 14:893–915.

Rossignol-Strick, M., 1997, Paléoclimat de la Méditerranée Orientale et de l'Asie du Sud-Ouest de 15,000 à 6,000 B.P. *Paléorient* 23(2):175–186.

Sage, R. F., 1995, Was Low Atmospheric CO_2 During the Pleistocene a Limiting Factor for the Origin of Agriculture? *Global Change Biology* 1:93–106.

Sahlins, M. D., 1968, *Tribesmen*. Foundations of Modern Anthropology, Prentice-Hall, Englewood Cliffs, NJ.

Saxe, A. A., 1970, *Social Dimensions of Mortuary Practices*. University of Michigan, Ann Arbor.

Schiegl, S., Lev-Yadun, S., Bar-Yosef, O., El Goresy, A., and Weiner, S., 1994, Siliceous Aggregates from Prehistoric Wood Ash: A Major Component of Sediments in Kebara and Hayonim Caves (Israel). *Israel Journal of Earth Sciences* 43:267–278.

Schroeder, B., 1991, Natufian in the Central Béqaa Valley, Lebanon. In *The Natufian Culture in the Levant*, edited by O. Bar-Yosef and F. R. Valla, pp. 42–80. International Monographs in Prehistory, Ann Arbor.

Serguin, V. I., 1999, Zhilishcha na Pamiatnikakh Vostochnogo Gravetta Russky Ravniny. In *Vostochny Gravett*, edited by A. A. Amirkhanov, pp. 151–176. Nauchni Mir, Moskva.

Shmida, A., Evenari, M., and Noy-Meir, I., 1986, Hot Desert Ecosystems: An Integrated View. In *Hot Deserts and Arid Shrublands*, edited by M. Evenari, pp. 379–387. Elsevier Science, Amsterdam.

Simmons, A. H., and Wigand, P. E., 1994, Assessing the Radiocarbon Determinations from Akrotiri Aetokremnos, Cyprus. In *Late Quaternary Chronology and Paleoclimates of the Eastern Mediterranean*, edited by O. Bar-Yosef and R. S. Kra, pp. 247–254. Radiocarbon and the Peabody Museum, Tucson.

Smith, P., 1973, Family Burials at Hayonim: A Brief Communication. *Paleorient* 1:69–71.

Smith, P., 1991, The Dental Evidence for Nutritional Status in the Natufians. In *The Natufian Culture in the Levant*, edited by O. Bar-Yosef and F. R. Valla, pp. 425–432. International Monographs in Prehistory, Ann Arbor.

Smith, P., Bar-Yosef, O., and Sillen, A., 1984, Archaeological and Skeletal Evidence for Dietary Change During the Late Pleistocene, Early Holocene in Levant. In *Paleopathology at the Origins of Agriculture*, edited by M. N. Cohen and G. Armelagos, pp. 101–136. Academic Press, New York.

Soffer, O., 1989, Storage, Sedentism and the Eurasian Palaeolithic Record. *Antiquity* 63:719–732.

Solecki, R. L., 1981, *An early village site at Zawi Chemi Shanidar*. Bibliotheca Mesopotamica 13, Undena, Malibu.

Solivères-Massei, O., 1988, Les Hommes de Mallaha: Etude Anthropologique. In *Les Hommes de Mallaha (Eynan) Israel*, edited by J. Perrot, Vol. 7, pp. 108–208. Memoires et Travaux du Centre de Recherche Francais de Jerusalem, Association Paleorient, Paris.

Stark, B., 1986, Origins of Food Production in the New World. In *American Archaeology Past and Future*, edited by D. J. Melzer, D. D. Fowler, and J. A. Sabloff, pp. 277–321. Smithsonian Institution Press, Washington, DC.

Stekelis, M., and Yizraeli, T., 1963, Excavations at Nahal Oren (Preliminary Report). *Israel Exploration Journal* 13(1):1–12.

Stiner, M. C., Munro, N. D., Surovell, T. A., Tchernov, E., and Bar-Yosef, O., 1999, Paleolithic Population Growth Pulses Evidenced by Small Animal Exploitation. *Science* 283:190–194.

Stordeur, D., 1981, La Contribution de L'Industrie de L'Os à la Délimination des Aires culturelles: L'Example du Natoufien. In *Préhistoire du Levant*, edited by J. Cauvin and P. Sanlaville, pp. 433–437. CNRS, Paris.

Stordeur, D., 1988, *Outils et Armes en Os du Gisement Natoufien de Mallaha (Eynan), Israël*. Mémoires et Travaux du Centre de Recherche Française de Jérusalem, No. 6, Association Paléorient, Paris.

Stordeur, D., 1991, Le Natoufien et Son Évolution à Travers les Artefacts en Os. In *The Natufian Culture in the Levant*, edited by O. Bar-Yosef and F. R. Valla, pp. 483–520. International Monographs in Prehistory, Ann Arbor.

Stordeur, D., 1992, Change and Cultural Inertia: From the Analysis of Data to the Creation of a Model. In *Representations in Archaeology*, edited by J. C. Gardin and C. S. Peebles, pp. 205–222. Indiana University Press, Bloomington.

Straus, L. G., Eriksen, B. V., Erlandson, J. M., and Yesner, D. R., (eds.), 1996, *Humans at the End of the Ice Age: The Archaeology of the Pleistocene-Holocene Transition*. Plenum, New York.

Tchernov, E., 1991a, Biological Evidence for Human Sedentism in Southwest Asia During the Natufian. In *The Natufian Culture in the Levant*, edited by O. Bar-Yosef and F. R. Valla, pp. 315–340. International Monographs in Prehistory, Ann Arbor.

Tchernov, E., 1991b. On Mice and Men: Biological Markers for Long-Term Sedentism: A Reply. *Paléorient* 17(1):153–160.

Tchernov, E., 1993a, The Effects of Sedentism on the Exploitation of the Environment in the Southern Levant. In *Exploitation des Animaux Sauvages à Travers le Temps*, edited by J. Desse and F. Audoin-Rouzeau, pp. 137–159. APDCA, Juan-les-Pins.

Tchernov, E., 1993b, Exploitation of Birds During the Natufian and Early Neolithic of the Southern Levant. *Archaeofauna* 2:121–143.

Tchernov, E., 1993c, From Sedentism to Domestication—A Preliminary Review for the Southern Levant. In *Skeletons in Her Cupboard: Festschrift for Juliet Clutton-Brock*, edited by A. Clason, S. Payne, and H. P. Uerpmann, pp. 189–233. Oxbow Monograph 34, Oxford.

Tchernov, E., 1994, *An Early Neolithic Village in the Jordan Valley II: The Fauna of Netiv Hagdud*. American School of Prehistoric Research Bulletin 44, Peabody Museum of Archaeology and Ethnography, Harvard University, Cambridge.

Tchernov, E., 1997, Are Late Pleistocene Environmental Factors, Faunal Changes and Cultural Transformations Causally Connected? The Case of the Southern Levant. *Paléorient* 23(2):209–228.

Tchernov, E., and Valla, F., 1997, Two New Dogs, and other Natufian Dogs, from the Southern Levant. *Journal of Archaeological Science* 24(1):65–95.

Turville-Petre, F., 1932, fig. List

Turville-Petre, F., 1932, Excavations in the Mugharet el-Kebarah. *Journal of the Royal Anthropological Institute* 62:271–276.

Uerpmann, H. P., 1987, *The Ancient Distribution of Ungulate Mammals in the Middle East*. Beihefte zum Tübinger Atlas des vorderen Orients, Reihe A, Nr. 27, Ludwig Reichert Verlag, Weisbaden.

Unger-Hamilton, R., 1989, The Epi-Palaeolithic of Southern Levant and the Origins of Cultivation. *Current Anthropology* 31(1):88–103.

Unger-Hamilton, R., 1991, Natufian Plant Husbandry in the Southern Levant and Comparison with That of the Neolithic Periods: The Lithic Perspective. In *The Natufian Culture in the Levant*, edited by O. Bar-Yosef and F. R. Valla, pp. 483–520. International Monographs in Prehistory, Ann Arbor.

Upham, S., 1990, Decoupling the Processes of Political Evolution. In *The Evolution of Political Systems*, edited by S. Upham, pp. 1–17. Cambridge University Press, Cambridge.

Valla, F. R., 1984, *Les industries de silex de Mallaha (Eynan) et du Natoufien dans le Levant*. Mémoires et Travaux du Centre de Recherche Français de Jérusalem 3, Association Paléorient, Paris.

Valla, F. R., 1987, Les Natoufiens Connaissaient—Ils L'Arc? In *La Main et L'Outil: manches et Emmanchements Préhistoriques*, edited by D. Stordeur, pp. 165–174. Maison de l'Orient, Lyon.

Valla, F. R., 1988, Les Premiers Sédentaires de Palestine. *La Recherche* 199:576–584.

Valla, F. R., 1990, Le Natoufien: Une Autre Façon de Comprendre le Monde? *Journal of the Israel Prehistoric Society* 23:171–175.

Valla, F. R., 1995, The First Settled Societies—Natufian (12,500–10,200 B.P.). In *The Archaeology of Society in the Holy Land*, edited by T. Levy, pp. 169–189. Leicester University Press, London.

Valla, F. R., 1996, L'Animal "Bon à Penser": La Domestication et la Place de l'Homme dans la Nature. In *Nature et Culture II*, edited by M. Otte, Vol. 68, pp. 651–668. ERAUL, Liège.

Valla, F. R., 1998, Natufian Seasonality: A Guess. In *Seasonality and Sedentism: Archaeological Perspectives from Old and New World Sites*, edited by T. R. Rocek and O. Bar-Yosef, Vol. 6, pp. 93–108. Peabody Museum Bulletin, Peabody Museum of Archaeology and Ethnology, Cambridge.

Valla, F. R., 1999, The Natufian: A Coherent Thought? In *Dorothy Garrod and the Progress of the Palaeolithic*, edited by W. Davies and R. Charles, pp. 224–241. Oxbow, Oxford.

Valla, F. R., Khalaily, H., Samuelian, N., March, R., Bocquentin, F., Valentin, B., Marder, O., Rabinovich, R., Le Dosseur, G., Dubreuil, L., and Belfer-Cohen, A., 2001, Le Natoufien Final de Mallaha (Eynan), Deuxième Rapport Préliminaire: Le Fouilles de 1998 et 1999. *Journal of the Israel Prehistoric Society* 31:43–184.

Valla, F. R., Le Mort, F., and Plisson, H., 1991, Les Fouilles en Cours sur la Terrasse d'Hayonim. In *The Natufian Culture in the Levant*, edited by O. Bar-Yosef and F. R. Valla, pp. 93–110. International Monographs in Prehistory, Ann Arbor.

van Zeist, W., and Bakker-Heeres, J. A. H., 1985, Archaeobotanical Studies in the Levant: Neolithic Sites in the Damascus Basin, Aswad, Ghoraife, Ramad. *Prehistoria* (1982) 24:165–256.

van Zeist, W., and Bakker-Herres, J. A. H., 1986, Archaeobotanical Studies in the Levant, III, Late Paleolithic Mureybet. *Palaeohistoria* 26:171–199.

van Zeist, W., and Bottema, S., 1991, *Late Quaternary Vegetation of the Near East*. Beihefte zum Tübinger Atlas des Vorderen Orients, Reihe A (Naturwissenschaft) Nr.18, Dr. Ludwig Reichert Verlag, Weisbaden.

Vita-Finzi, C., and Higgs, E. S., 1970, Prehistoric Economy in the Mount Carmel Area of Palestine: Site Catchment Analysis. *Proceedings of the Prehistoric Society* 36(1):1–37.

Voigt, M. M., 1990, Reconstructing Neolithic Societies and Economies in the Middle East: An Essay. *Archaeomaterials* 4:1–14.

Voigt, M. M., 2000, Çatal Höyük in Context: Ritual at Early Neolithic Sites in Central and Eastern Turkey. In *Life in Neolithic Farming Communities: Social Organization, Identity, and Differentiation*, edited by I. Kuijt, pp. 253–293. Plenum, New York.

Weinstein-Evron, M., and Belfer-Cohen, A., 1993, Natufian Figurines from the New Excavations of the El-Wad Cave, Mt. Carmel, Israel. *Rock Art Research* 10(2):102–106.

Weinstein-Evron, M., and Ilani, S., 1994, Provenance of Ochre in the Natufian Layers of El-Wad Cave, Mount Carmel, Israel. *Journal of Archaeological Science* 21(4):461–467.

Weinstein-Evron, M., Lang, B., Ilani, S., Steinitz, G., and Kaufman, D., 1995, K/AR Dating as a Means of Sourcing Levantine Epipalaeolithic Basalt Implements. *Archaeometry* 37:37–40.

Wobst, M. H., 1976, Locational Relationship in Palaeolithic Society. *Journal of Human Evolution* 5:49–58.

Wobst, H. M., 1999, Style in Archaeology or Archaeologists in Style. In *Material Meanings: Critical Approaches to the Interpretation of Material Culture*, edited by E. S. Chilton, pp. 118–132. Foundations of Archaeological Inquiry, University of Utah, Salt Lake City.

Wright, G. A., 1978, Social Differentiation in the Early Natufian. In *Social Archaeology: Beyond Subsistence and Dating*, edited by C. L. Redman, M. J. Berman, E. V. Curint, W. T. J. Langhorne, N. M. Versaggi, and J. C. Wanser, pp. 201–233. Academic Press, New York.

Wright, K. I., 1992, A Classification System for Ground Stone Tools from the Prehistoric Levant. *Paléorient* 18(2):53–81.

Yamada, S., 2000, Development of the Neolithic: Lithic Use-Wear Analysis of Major Tool Types in the Southern Levant, Ph.D. Dissertation, Harvard University.

Zohary, D., 1996, The Mode of Domestication of the Founder Crops of Southwest Asian Agriculture. In *The Origins and Spread of Agriculture and Pastoralism in Eurasia*, edited by D. Harris, pp. 142–158. UCL Press, London.

Zohary, D., Tchernov, E., and Kolska Horwitz, L., 1998, The Role of Unconscious Selection in the Domestication of Sheep and Goats. *Journal of Zoology* 245:129–135.
Zohary, M., 1973, *Geobotanical Foundations of the Middle East*. Springer Verlag, Stuttgart.
Züchner, C., 1996, The Scaliform Sign of Altamira and the Origin of Maps in Prehistoric Europe. In *"El Hombre Fósil" 80 Años Después*, edited by A. Moure Romanillo, pp. 325–343. Universidad de Cantabria, Fundación Marcelino Botín Institute for Prehistoric Investigations, Santander.

Part II

Microevolutionary Approaches to Long-term Hunter-Gatherer Settlement Change

Introduction to Part II

Microevolutionary Approaches to Long-term Hunter-Gatherer Settlement Change

Unlike the first part of this volume, the second and third parts depart more radically from the theoretical hearth of the forager/collector model and take the issue of hunter-gatherer settlement and subsistence change beyond the now classic cultural ecological paradigm that was nurtured in archaeology in large measure by Binford's writings. Of these next two parts, Part II is the most closely allied in ecological terms, but unlike the cultural ecological approach, authors in this part have turned in greater or lesser degree toward the more formal modeling approach of evolutionary ecology.

Evolutionary ecologists seek to account for adaptive variation in ecological context by referring explicitly to Darwinian evolutionary mechanisms (Winterhalder and Smith 1992). Evolutionary ecology (more specifically behavioral ecology) has influenced anthropology and archaeology since the 1970s (e.g., Dyson Hudson and Smith 1978; Yesner 1981). It has been most visibly represented in archaeology by optimal foraging studies of hunter-gatherer subsistence. Behavioral ecological approaches have nevertheless been developed for a much broader array of anthropological topics (e.g., collective action and political stratification), and for both hunter-gatherers and food producers. To date, the formal behavioral ecological models that have been most effective in archaeology are those that deal with issues of subsistence, where archaeological data are most easily assessed according to economic utility (in a subsistence currency such as calories, travel distance, processing difficulty). Formal models have been developed to evaluate strategic and political social behavior, but so far these have proved relatively coarse for empirical evaluation with archaeological data (and the reverse is also true that archaeological data are commonly too coarse to be used to test models of strategic behavior directly). Fortunately for the topic of this volume, behavioral ecological models that deal with mobility and subsistence behavior are much more refined and have received considerable empirical support, as contributions to this part demonstrate.

Many archaeologists who have gravitated to evolutionary ecology have done so out of an interest in using the more formal and more precise modeling tools that are a hallmark of this approach. These models, commonly developed from the study of animal behavior and ultimately from microeconomic principles, tend to be more biologically grounded and reductionist (see Winterhalder and Smith 1992 for a defense of reductionism). They often avoid explicit use of concepts such as "culture" and "ideology," which are notoriously difficult to define as variables in formal models. Many researchers see these "cultureless" models as good ways to delineate those aspects of human social behavior that can and those that cannot be explained according to reductionist biological principles. Others mix evolutionary ecological models with arguments from more anthropologically inspired constructs.

One key aspect of evolutionary ecology that has drawn converts from other strains of ecological anthropology is its continued insistence on a hypothetico-deductive scientific method. Though out of fashion in some social science circles, the approach of generating and testing explicitly deductive hypotheses, based on simple and formal models, has lasting appeal to many. Evolutionary ecologists approach the task of constructing an understanding of the world like that of erecting a building. They seek to establish firm foundations rooted in the common biological heritage of all living creatures before erecting an explanatory edifice for dimensions of sociocultural behavior and organization that have emerged from this foundation. Using a more solidly anchored foundation, higher stories will be better grounded.

One area where cultural ecological and evolutionary ecological models have converged is around the concept of human adaptations to risk and risky environments. The chapter here by Kipnis is an example of this convergence, where the "risk buffering" theme in cultural ecological research (e.g., Halstead and O'Shea 1989) has been joined with "risk-sensitivity" analyses from evolutionary ecology and microeconomics (e.g., Smith 1988; Winterhalder 1986; Winterhalder et al. 1999). This convergence has encouraged the development of more conceptually rigorous and ultimately more testable models that retain important insights of both sets of models. Though largely comparable, it may turn out that these two strains of "risk" modeling retain some yet unresolved and ultimately constructive contradictions (compare, for example, Bamforth and Bleed 1998 and Fitzhugh 2001).

Attempts to formalize issues of hunter-gatherer residential and logistical mobility, foraging range, prey selection, sharing, and group hunting have led to the development of models of central place foraging (Zeanah, this volume), prey selection/diet breadth, patch selection (Fitzhugh, this volume), hazard minimization (Winterhalder 1986), tolerated theft (Hawkes 1992), and collective action (Smith 1981). These models have shed light on the division of foraging labor by sex and made it possible to test cultural ecological assumptions that have long held sway in hunter-gatherer archaeology. In many cases, this approach has revealed significant, fine-grained variability in the environmental and archaeological records that changes the way we look at archaeological records and the explanation of sociocultural processes. (Eds.)

REFERENCES

Bamforth, D. B., and Bleed, P., 1998, Technology, Flaked Stone Technology, and Risk. In *Rediscovering Darwin: Evolutionary Theory in Archeological Explanation*, edited by C. M. Barton and G. A. Clark, pp. 109–139. Archeological Papers of the American Anthropological Association, No. 7. American Anthropological Association, Arlington.

Dyson Hudson, R., and Smith, E. A., 1978, Human Territoriality: An Ecological Reassessment. *American Anthropologist* 80(1):21–41.

Fitzhugh, B., 2001, Risk and Invention in Human Technological Evolution. *Journal of Anthropological Archaeology* 20(2):125–167.

Hawkes, K., 1992, Sharing and Collective Action. In *Evolutionary Ecology and Human Behavior*, edited by E. Smith and B. Winterhalder, pp. 269–300. Aldine de Gruyter, New York.

Halstead, P., and O'Shea, J., 1989, Introduction: Cultural Responses to Risk and Uncertainty. In *Bad Year Economics: Cultural Responses to Risk and Uncertainty*, edited by P. Halstead and J. O'Shea, pp. 1–7. Cambridge University Press, Cambridge.

Smith E. A.,1981, The Application of Optimal Foraging Theory to the Analysis of Hunter-Gatherer Group Size. In *Hunter-Gatherer Foraging Strategies: Ethnographic and Archaeological Analyses*, edited by B. Winterhalder and E. A. Smith, pp. 36–65. The University of Chicago Press, Chicago.

Smith, E. A., 1988, Risk and Uncertainty in the Original Affluent Society: Evolutionary Ecology of Resource Sharing and Land Tenure. In *Hunters and Gatherers 1: History, Evolution, and Social Change*, edited by T. Ingold, D. Riches, and J. Woodburn, pp. 222–252. Berg, Oxford.

Winterhalder, B. P., 1986, Diet Choice, Risk, and Food Sharing in a Stochastic Environment. *Journal of Anthropological Archaeology* 5:369–392.

Winterhalder, B. P., Lu, F., and Tucker, B., 1999, Risk-Sensitive Adaptive Tactics: Models and Evidence from Subsistence Studies in Biology and Anthropology. *Journal of Archaeological Research* 7(4):301–348.

Winterhalder, B. P., and Smith, E. A., 1992, Evolutionary Ecology and the Social Sciences. In *Evolutionary Ecology and Human Behavior*, edited by E. A. Smith and B. P. Winterhalder, pp. 3–23. Aldine de Gruyter, New York.

Yesner, D. R., 1981, Archaeological Applications of Optimal Foraging Theory: Harvest Strategies of Aleut Hunter-Gatherers. In *Hunter-Gatherer Foraging Strategies: Ethnographic and Archaeological Analyses*, edited by B. Winterhalder and E. A. Smith, pp. 148–170. University of Chicago Press, Chicago.

Chapter 6

Mobility, Search Modes, and Food-Getting Technology

From Magdalenian to Early Mesolithic in the Upper Danube Basin

LYNN E. FISHER

INTRODUCTION

The archaeology of Late Glacial and Early Postglacial central and western Europe offers evidence of long-term economic, technological, and demographic change in mobile hunter-gatherer societies at a time of dramatic environmental change. Late Glacial and Early Postglacial warming caused a rapid retreat of alpine glaciers and touched off a cascade of changes in natural landscapes, as first steppe-tundra and then woodland flora and fauna recolonized previously glaciated and periglacial areas. Climatic and landscape change north of the Alps took place in several episodes during a long period of unstable conditions between the first major glacier retreats about 16,000 years B.P. and the development of deciduous forests about 8000 years later. The archaeological record of this period shows clear evidence of changes in several aspects of hunter-gatherer spatial behavior, including migration into deglaciated and periglacial regions (Housley et al. 1997;

LYNN E. FISHER • Sociology/Anthropology Program, University of Illinois at Springfield, Springfield, Illinos 62794

Jochim et al. 1999) and changes in land use by local populations, as indicated by alterations in the number, size, character, and location of archaeological sites during this dynamic period (Eriksen 1991, 1996; Jochim 1991, 1998; Mellars 1994; Myers 1989; Straus 1991; Svoboda et al., eds. 1996).

Archaeological study of events at the Pleistocene–Holocene transition, therefore, provides opportunities for understanding change in land use as one way hunter-gatherers coped with alterations in a natural landscape. A number of different models have been proposed for understanding change in hunter-gatherer land use at the transition. Many anthropologists have addressed mobility as an aspect of subsistence systems and assign causal roles to Late Pleistocene population packing (Binford 1968; Vierra 1995) or to environmentally driven change in foraging tactics, emphasizing decline of the large-bodied migratory game animals that ranged Late Glacial steppe/tundra mosaics (Stewart and Jochim 1986), dispersal of prey in post-glacial woodlands (Binford 1990; Floss 1994; Holliday 1998; Myers 1989), or instability of resource distributions (Jochim 1991). The forager/ collector spectrum model (Binford 1980) has been influential in defining the terms of this debate. A long-standing debate about Late Pleistocene and Early Holocene subsistence and settlement systems has to do with relationships among mobility, environmental productivity, foraging efficiency, and the problems of searching for terrestrial game animals.

Binford's early formulation of the forager-collector model distinguished two different kinds of subsistence-related mobility: residential mobility, or movements of an entire residential group to a new camp, and logistical mobility, movements of task groups to resource locations and back to camp (1980). Binford illustrated these two types of mobility strategy by describing an idealized continuum from collector to forager settlement systems. While foragers, as Binford put it, "map on" to resources by moving consumers to resource locations in frequent residential moves and make only short logistical forays, collectors establish base camps at strategic locations and use longer logistic forays by task groups to bring in and store resources. Binford (1980, 1983) and Kelly (1983, 1995) analyzed ethnographic and environmental data to document and investigate global trends in the frequency of these strategies. As stated in a later reexamination of some of these ideas, "fully nomadic people" (roughly, foragers lacking a multiseason base camp) are rare at temperate and high latitudes (Binford 1990: 136).

Exactly why this is so has been of great interest to prehistorians of Pleistocene Europe because the colonization of Europe by hunter-gatherers involved movement into temperate and northern regions with relatively harsh glacial climates (Gamble 1986). The rarity of year-round nomadism among northern hunter-gatherers has been explained as a behavioral adaptation to the short northern growing season and consequent concentration of food-getting opportunities into narrow windows of resource availability (Binford 1980; Kelly 1983). Thus, a key underlying difference between collector and forager strategies had to do with resource distributions and efficient spatial behavior for locating concentrated/predictable or dispersed/unpredictable resources. This strong pattern and compelling explanation has led to a general expectation that groups that occupy cool climatic areas and those that have access to relatively concentrated resources such as reindeer herds or other gregarious game animals must have relied to some extent on logistic mobility.

One influential view, based on this ecological model of variation in mobility, draws a broad distinction between the mobility strategies of open country (steppe/tundra) and woodland or forest foragers and infers from this that reliance on residential mobility must have increased during the period of late glacial and postglacial warming (e.g., Floss 1994). A widely adopted working hypothesis based on this logic proposes that Magdalenian hunter-gatherers, as reindeer hunters on the tundra in relatively open country, may have been logistically organized to some degree, whereas Late Paleolithic (boreal forest) and Early Mesolithic (boreal/temperate forest) hunter-gatherers probably depended to a greater degree on residential mobility (Eriksen 1991, 1996; Floss 1994). This scenario interprets changes in number, size, character, and location of archaeological sites across Late Glacial and Early Postglacial times as a shift along the forager–collector spectrum.

In the upper Danube region of southwest Germany (see Fig. 6.1 and the chronological framework in Table 6.1), the archaeological evidence that supports these arguments consists primarily of differences in site size and artifact density between Late Upper Paleolithic (Magdalenian) and Early Mesolithic archaeological complexes and evidence for seasonal exploitation of reindeer in Magdalenian sites located in open country (Eriksen 1991, 1996; Jochim 1991, 1998). Specifically, some Magdalenian sites have large enough accumulations of artifacts and animal bones that they suggest repeated use of certain locations for hunting reindeer, the dominant large game animal on southern German Magdalenian sites (Albrecht and Hahn 1991; Eriksen 1996). These accumulations are associated with open tundra landscapes or their margins and in at least some cases, with favorable locations for intercepting reindeer (Albrecht and Berke 1987; Eriksen 1991; Hahn 1979, 1981; Pasda 1994; Schuler 1994; Weniger 1982, 1987, 1989, 1990). Magdalenian sites also have patterned variation in the locations of backed armature production, which suggests that some lithic tool-making took place in gearing up for concentrated bouts of hunting or other resource acquisition (Fisher 2000). The subsequent Late Paleolithic and

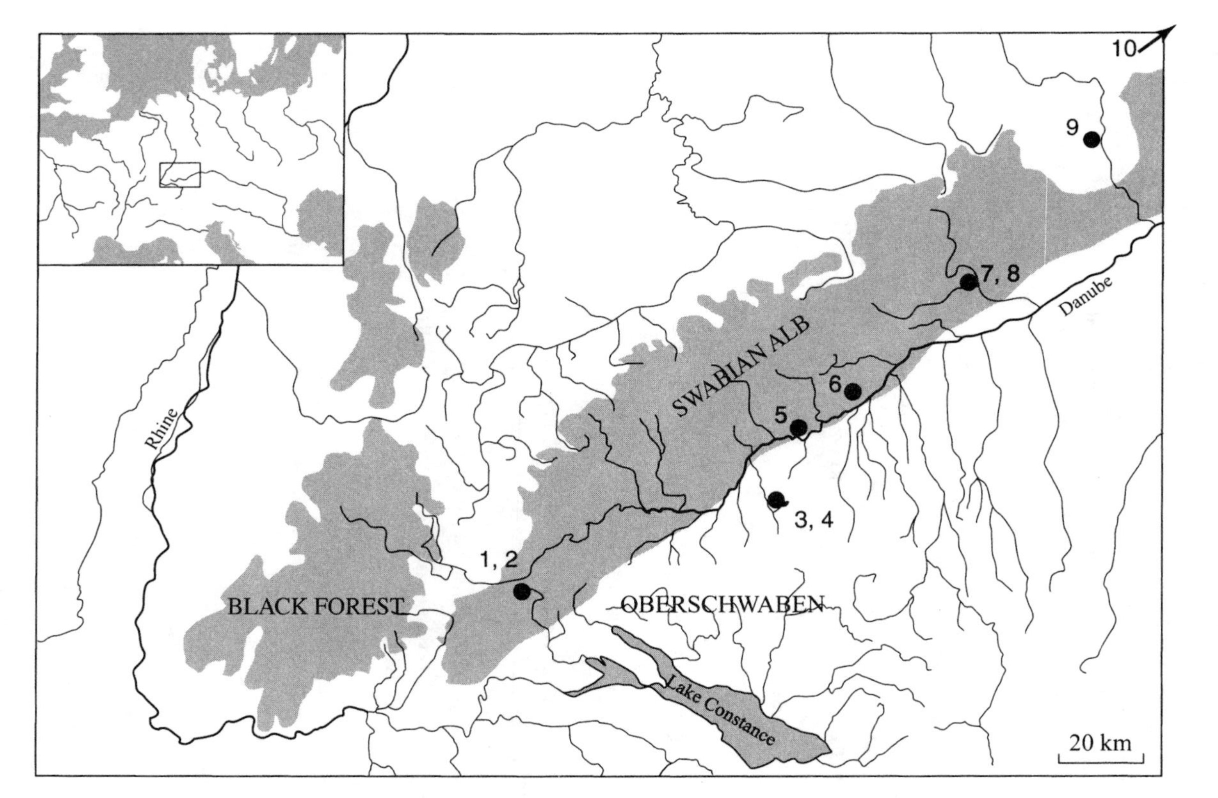

Figure 6.1. Map of the southern German study area in Europe, showing the location of sites mentioned in the text. Most Magdalenian sites are located in rock shelters in the limestone upland of the Swabian Jura (center). Few sites are known from open-air locations in the moraine lowlands of the alpine foreland to the south. More Late Paleolithic and Early Mesolithic sites are known from open-air locations in the lowlands. 1, 2: Petersfels, Gnirshöhle; 3, 4: Henauhof, Schussenquelle; 5: Felsställe; 6: Helga Abri; 7, 8: Malerfels, Spitzbubenhöhle; 9: Kaufertsberg; 10: Schräge Wand.

Table 6.1. Schematic Chronological Framework for Late Glacial and Early Postglacial Southern Germany

Years B.P.	Pollen Zone	Vegetation	Archaeological Complex
8000			
	Boreal	Northern deciduous forest with hazel, oak, elm	
9000	—		
	Preboreal	Open pine-birch forest	Early Mesolithic
10,000			
	Dryas III	Open birch-pine forest with willow scrub, park-tundra	
11,000	—		Late Paleolithic
	Allerød	Open pine-birch forest and scrub/steppe	
12,000	(Late Glacial Interstadial)		
	Bølling	Steppe/tundra mosaic with birch, juniper, grasses, herbs	Magdalenian
13,000	—		
	Dryas I	Tundra (dwarf birch, herbs)	
14,000			

Early Mesolithic industries in southern Germany lack this kind of evidence for seasonal ambush hunting of large game and more generally are characterized by smaller accumulations of material that suggest shorter and/or less spatially redundant occupations (Eriksen 1991, 1996; Fisher 2000; Jochim 1991, 1998; Stewart and Jochim 1986).

However, considerable uncertainty exists about the degree to which Magdalenian hunter-gatherers were logistically organized and further, whether a shift along the forager–collector continuum fully explains the diversity of Late Glacial and Early Postglacial subsistence and settlement systems. It is clear that Magdalenian hunter-gatherers could at some times and places target herds using ambush strategies, but it is less clear how important storage or transport of food resources from task locations to base camps may have been (cf. Pasda 1994). As has often been pointed out, it is difficult to infer logistic mobility (Lieberman 1993: 608) because the concept is based on a subtle contrast between individual and group mobility (Kelly 1992: 44). Binford suggested that the evolution of collector strategies at high latitudes may have no simple environmental cause but instead may be the result of the Late Pleistocene and Holocene "perfection of transport technologies, particularly water transport vessels and the use of pack and draft animals" (1990: 138). Fully nomadic groups, though rare, are not absent at high latitudes (Binford 1990: 137). These considerations, along with a growing awareness of the unique social and political context of multiseason base camps among modern foragers, have led many to question the assumption that ambush hunting of reindeer implies logistic mobility (Audouze and Enloe 1991, Pasda 1998).

An alternative approach sees this period as one marked by major adaptive shifts in foraging adaptations. Binford (1990) and Foley (cited in Lewin 1988) suggested that terminal Pleistocene foragers developed novel adaptive strategies in response to resource depletion brought about by some combination of climatic change and the impact of increasingly dense human populations. Expanding human populations moved into central and northern Europe, where they would have encountered lower biomass environments that had low densities of terrestrial game (Jochim et al. 1999). Binford (1990) points out that terrestrial hunters in these areas would have had to cover increasingly large areas to make a living off large mammal game. Therefore, high mobility costs and high search costs for preferred prey should be strong mechanisms of selection. Binford argues that this ecological setting would have favored strategies to reduce mobility costs in foraging (1990: 147). He predicts that foragers in continental environments should adopt highly portable housing and should experiment with transport technologies and facilities such as traps and snares to increase encounter rates with scarce game (1990: 147). The reverse is expected where aquatic resources are available, that is, foragers who face high mobility and search costs in terrestrial hunting should become increasingly "tethered" to locations from which aquatic resources can be harvested. This would have resulted in increased redundancy of occupation and more permanent housing (1990: 147). This kind of increased reliance on aquatic resources, it is argued, takes place as a result of increasing mobility costs, in the absence of any kind of density-dependent resource depletion (1990: 148). Thus, Binford places a key "habitat trade-off (between aquatic and terrestrial resources)" at the center of understanding variation in hunter-gatherer mobility strategies at temperate and higher latitudes (1990: 132). In a similar vein, Holliday (1998) argued that the low density of large game in developing woodlands in Late Glacial and Early Postglacial Europe may have formed a barrier to exploitation by terrestrial hunters, such that year-round residence in the forest was possible only with a mixed hunting/trapping/fishing economy or a focus on grassland/forest ecotones.

These arguments are based on resource geography but depend on technological innovation as an important mediating factor between people and their resource environments. The proposed novel adaptive strategies depend on technologies for transporting people and resources, as well as food-getting technologies for capturing aquatic animals and small mammals and birds (Binford 1990; Holliday 1998; Vierra 1995). Underlying these scenarios is the assumption that fishing and trapping are strongly localized activities that may be incompatible to some degree with the very high mobility required for hunting large terrestrial game (Holliday 1998). Here again, efficient spatial behavior for locating differently distributed resources is the central concept relied on to predict settlement patterns.

The notion of broad adaptive shifts at the Pleistocene–Holocene transition suggests that, rather than a shift from logistic to residential mobility, we may need to look for a variety of possible differences in subsistence and spatial behavior among highly mobile hunter-gatherers and perhaps for the first appearance of techniques and practices associated with modern logistical foraging. This raises questions about search behavior which have not yet been examined in detail. What is the nature of this proposed trade-off between high-ranked but scarce and unpredictable terrestrial resources and lower ranked but more stable aquatic resources? What kinds of variation in spatial behavior can be expected to accompany this trade-off? What role might nonaquatic resources such as small game or birds, so important in late Paleolithic and Early Mesolithic contexts, play in a similar trade-off in continental environments?

A recently developed model of search mode selection under conditions of partially directed search (Schmidt 1998) provides some guidance for examining the conditions under which foragers might increase foraging efficiency by targeting search effort toward lower ranked but stable, reliable, or abundant resources and taking higher ranked game opportunistically. This chapter outlines some implications of this model and presents a brief discussion of archaeological data from Magdalenian, Late Paleolithic, and Early Mesolithic southern Germany that, I suggest, can be used to evaluate the usefulness of a focus on variable search tactics for understanding the evolution of late Pleistocene and early Holocene subsistence and settlement systems. I propose that a theoretical focus on searching should provide a useful perspective on the evolution of hunter-gatherer settlement and subsistence systems.

MOBILITY AND SEARCH COSTS

It is common to assume that hunter-gatherer settlement dynamics are strongly determined by aggregation or dispersal of large mammal game, typically the highest ranked items in a hunter-gatherer diet. This accords with the classic diet breadth model (Pulliam 1974), which predicts that the choice of items to include in an optimal diet depends only on encounter rates with high-ranked prey, not on any changes in abundance or concentration of secondary prey. This makes it difficult to understand the role that abundant secondary resources might have in site location decisions without invoking density-dependent constraints on residential mobility (population packing) (Binford 1990).

The classic diet breadth model assumes that all prey are searched for at once by a forager who uses a single stereotyped search pattern. This undirected (or "fine-grained") search creates a single set of probabilities of encountering prey species. Based on these simplifying assumptions, diet choice is reduced to a single decision—whether or not to pursue an item once encountered. As is well known, this choice, it can be shown, depends only on the abundance of high-ranked prey, not on changes in abundance or concentration of low-ranked prey (Kelly 1995; Pulliam 1974; Schmidt 1998; Smith and Winterhalder 1992). Human foraging often contradicts this assumption because humans decide in which direction to go, which microhabitats to search, how quickly to travel, and which implements to take based on information about the abundance of a variety of resources (e.g., Smith 1991; Winterhalder 1981). Schmidt proposes that foragers face two choices—first, how and where to search, and second, what to take when encounter occurs (1998).

The choice of how and what to search for has been modeled by Kenneth Schmidt (1998) as an extension of the classic diet breadth model. He begins with the assumption that foragers may possess several different search modes, "each associated with a unique set of prey encounter probabilities", and supposes that foragers could choose from this repertoire "the search mode that maximizes ... foraging efficiency given the current distribution of prey" (Schmidt 1998: 263). Thus, the forager chooses a focal prey or microhabitat and biases the search toward that prey. The choice of a particular search mode involves a trade-off between an increased encounter with one prey species (the focal item) and a decreased encounter with others (nonfocal prey) (Schmidt 1998). Examples in animal and human foraging include a wide range of variations in the rate of travel, tactics of search, and attack distance that effectively position the forager to detect large or small prey, prey in different habitats, or prey in different parts of a single habitat (O'Brien et al. 1989, 1990; Robinson and Holmes 1982; Robinson 1984; Schmidt 1998: 264; Winterhalder 1981: 90–91).

Schmidt illustrates the model graphically using an isoleg model (Fig. 6.2) that analyzes the shape and dynamics of a hypothetical line in a state space defined by prey densities. Each line on the graph represents a set of points at which the energetic return of two different foraging strategies is equal (an isoleg). A well-known example concerns the diet breadth choice just

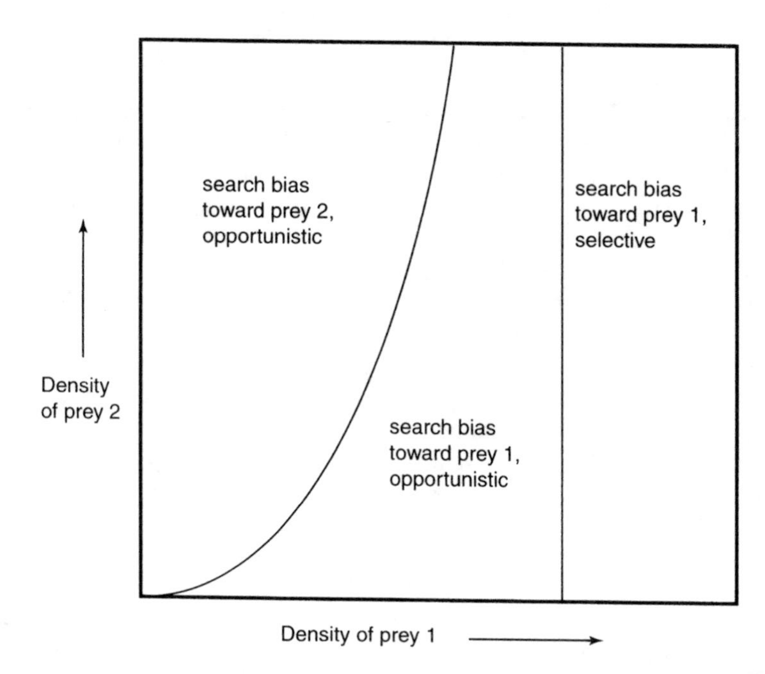

Figure 6.2. Schematic graph showing optimal strategy for a forager using unique search modes for each of two prey items or habitat types (after Schmidt 1998: Fig. 1, used with kind permission of Kluwer Academic Publishers).

mentioned, which appears as the vertical line on the right-hand side of Fig. 6.2. The line represents a set of points in this state space where a selective forager, who takes only prey one, will have exactly the same energy return as an opportunistic forager who takes both prey items. To the right of the isoleg, where the density of prey 1 is high, an optimal forager should always be selective; to the left, the same forager should always be opportunistic, that is, to the left of the vertical line, the reward per unit time for pursuing the less preferred prey exceeds the reward of taking only prey 1.

Schmidt added a consideration of search mode to this well-known model by assuming that rather than having a single, fixed set of probabilities of encountering prey items, the forager has two search modes, one biased toward the higher ranked prey, the other biased toward encounter with lower ranked prey (1998: 265). Schmidt's extension of the diet breadth model produces three possible foraging strategies. A forager may search for prey 1 and be selective, search for prey 1 and be opportunistic, or search for prey 2 and be opportunistic (Schmidt 1998: 265). The resulting isoleg that appears as the curved line in Figure 6.2 represents a set of points where the two search modes provide equal energetic returns. Note that it cannot cross the diet breadth isoleg. This model is based on the assumption that the prey types are perfect substitutes, that is, "the value of consuming an item of prey 2 is always a constant fraction ... of the value of consuming an item of prey 1" and that prey are nondepletable (Schmidt 1998: 265). Schmidt goes on to consider depletable resources (1998). This more complex model is beyond the scope of this chapter but produces similar general results that suggest a dynamic mix of search modes in foraging behavior where multiple search modes exist.

Two interesting results can be derived from this. First, as Figure 6.2 shows, the choice of the way to direct a search effort depends not only on abundance of the higher-ranked prey, as in the classic diet breadth model, but on the energy value, abundance, and pursuit and processing costs of both potential prey items (Schmidt 1998: 266). Second, in some cases, a forager can do better by biasing search toward nonpreferred or lower ranked prey and taking preferred prey opportunistically. Factors that tend to increase the area to the left of the curve (that is, the area of the state space in which a forager would do better by partially directing the search toward prey 2), include

1. an increase in the reward or decrease in the pursuit and processing costs of the less-preferred prey;
2. an increase in the probability of encountering either prey while using search mode 2; or
3. an increase in the pursuit and processing time of the preferred prey (Schmidt 1998: 266).

SEARCH MODES AND HUNTER-GATHERERS

This model can be used to derive qualitative predictions about the trade-off between alternate kinds of protein resources, including large game, small game, and fish, in hunter-gatherer economies. The point I want to focus on is the distinction between focal or target prey of a foraging trip and preferred or high-ranked prey. There is good evidence from the ethnographic record that a distinction between focal prey and preferred prey is likely to be important in understanding subsistence and locational decisions made by hunter-gatherers. Excellent examples come from Smith's analysis of Inuit foraging (1991). Smith introduces the concept of "hunt type." Hunt type is

> ... defined by a constellation of factors that ... lead to an expected foraging outcome that differs predictably from other types of foraging. A specific hunt type may be associated with particular prey species, a particular patch type

> (microhabitat) or set of patch types, specialized methods of search or capture, specialized transport or foraging technology, particular seasons or environmental conditions, or any combination thereof. (Smith 1991: 156)

Smith found broad agreement with a diet breadth model within hunt types (that is, foragers seemed to pass up low-ranked resources that would lower their overall foraging returns) and some tentative agreement with a patch choice model in the seasonal choice of hunt types (1991). His descriptive accounts of hunts, however, suggest that encounter probabilities with nonfocal prey are important in decisions about where to forage. Several hunt types, as for example the fall "canoe ptarmigan" hunt and the "lake ice jig hunt," targeted a focal prey species (ptarmigan, lake trout) that was not the highest ranked prey taken (these were ringed seal and ptarmigan, respectively) (1991: 227). An early application of the diet breadth model to anthropological data by Hames and Vickers (1982) also suggested that search biased toward small game (secondary prey) while opportunistically taking large game (high-ranked or preferred prey) was characteristic of depleted environments near horticultural villages where high-ranked prey were scarce.

Obviously, though high-ranked prey will always be taken when encountered, it is not always the target prey of hunter-gatherers because they may often be able to do better by making secondary or nonpreferred prey the target of search and taking the preferred prey opportunistically.

Case Study: Southern German Late Glacial—Early Postglacial

A distinction between focal prey or the target of search, and preferred or high-ranked prey may be useful in explaining change through time and variation in late glacial and early postglacial foraging tactics and settlement systems. Consider a simple two-prey model in which "large mammal game" (prey 1) is the preferred prey item and "small game" (prey 2) is a secondary prey item that substitutes perfectly for prey 1 (that is, the value of prey 2 is a constant fraction of the value of prey 1). This schematic scenario is one that can be applied to archaeofaunas throughout late glacial and early postglacial times in southern Germany.

In Magdalenian, Late Paleolithic, and Early Mesolithic assemblages, a shifting spectrum of large and small game is exploited, but large game remains the dominant element throughout (Eriksen 1991, 1996; Jochim 1998: 189; Jochim et al. 1999). Large game exploited by Magdalenian and Late Paleolithic foragers during late glacial times includes predominantly reindeer and wild horse, with rare bison, ibex, and chamois (Eriksen 1996: 113). Late glacial changes in the available spectrum of large game included most prominently the disappearance of the gregarious reindeer and horse by about the end of the Bølling (around 12,000 B.P.), and the establishment of woodland fauna, including red deer, roe deer, aurochs, elk (*Alces alces*), and brown bear. Species important to early Holocene foraging economies, especially the roe deer and wild boar, became established along with a deciduous forest only in the transition from the Preboreal to Boreal pollen zones at about 9000 B.P. Small game species reflect a similar sequence of replacement of arctic by boreal and finally temperate elements. Mountain hare, arctic fox, and carnivores (wolf) present in varying amounts in Magdalenian archaeofaunal assemblages are gradually replaced by wolverine, red fox, and lynx, polecat, ermine, and badger in Late Paleolithic assemblages and finally are supplemented by a forest fauna, including wild cat, pine marten, and otter in Early Mesolithic assemblages beginning about 9000 B.P. (Eriksen 1996: 113).

Game birds and fish appear in archaeofaunas throughout this sequence but also present a picture of species changing due to climatic change. Among fish species, the salmonids that dominate in cold early late glacial streams are augmented by warmer water species such as pike (abundant in Mesolithic contexts) and catfish (Jochim 1998: 131). Some Magdalenian assemblages include fairly abundant bones of grouse (*Lagopus lagopus*) and other game birds (e.g., *Perdix perdix*), but only rare examples of waterfowl (Mourer-Chauvire 1984). Some Mesolithic faunal assemblages in the wetland mosaic of the alpine foreland, on the other hand, include significant proportions of bone from migratory and resident waterfowl and some land birds such as grouse (e.g., Henauhof NW level 4, Jochim 1998: 189).

Table 6.2 relates patterns of late glacial and early postglacial environmental change to variables important to understanding a hypothesized search mode trade-off between large game and small game. Ecological conditions faced by late glacial foragers who expanded into southern Germany would have put a premium on exploiting animal resources under conditions in which rates of encounter with large game (presumed to be high-ranked resources under many conditions) would have been very low (Binford 1990). At the same time, rapid warming, glacial retreat, and forest recovery during the late glacial period in southern Germany produced an environment increasingly productive of fish, birds, and small game. Small mammal game (hare, squirrel) and some birds (e.g., spruce grouse) would have been able to exploit woodland or forest resources that were unavailable to most large mammals (Holliday 1998: 716), and a variety of fish, birds, and small game flourished in the developing wetlands of the alpine foreland. As the braided late glacial river systems began to stabilize during the Early Holocene (Boreal) into single-channel meanders surrounded by boggy floodplain oak forests, productive wetland resources would have developed in the Danube area as well (Frenzel 1995).

Table 6.2. Changing Aspects of Search Costs for Late Glacial/Early Postglacial Foragers in Southern Germany

Late Glacial/Early Postglacial Pollen Zones	Large Game ("Prey 1")	Small Game ("Prey 2")	Range Overlap Small/Large Game
Boreal	Sparse, solitary	Higher value forest (hare) and wetland (birds/fish) prey	High
Preboreal	Very sparse, solitary	Higher value forest prey (hare)	
Younger Dryas	Very sparse, solitary		?
Late glacial Interstadial (Bølling/Allerød)	Impoverished fauna. Gradual shift gregarious—solitary	? Arctic hare and grouse appear in archaeofaunas	Lower
Oldest Dryas	Gregarious		

Throughout this period, the challenges of searching for large mammal game would have been severe. During the earlier late glacial period, searching for game (particularly the wary horse) in open environments would have made stalking an arduous process in which the close approach to game required for accurate use of projectiles must have been difficult (Bartram 1997: 340; Bion 1997: 282; Hitchcock and Bleed 1997: 354). Later late glacial pine-birch forests, on the other hand, were probably characterized by very low-density, more solitary prey, in which large areas would have to be searched to encounter large game and ambush tactics were less likely to be useful. Early postglacial deciduous forests were more productive, similar to modern temperate forests, but were still characterized by dispersed and solitary large game.

As a result, southern German foragers from Magdalenian to Early Mesolithic times probably faced very low probabilities of encountering "prey 1", large terrestrial game. Populations of birds, small game, and fish (prey 2), on the other hand, probably became more abundant and stable through time. Developing postglacial forests and wetlands also may have increased the likelihood of encountering large game while directing a search toward relatively concentrated secondary prey, because the wetland mosaics that emerged during late glacial and early postglacial times, it can be argued, created resource "hot spots" that increased local densities of many types of resources (Nicholas 1988). This creates a context in which it may have been increasingly efficient for foragers to direct search efforts toward concentrated, predictable, and/or abundant secondary resources.

Several suggestions can be made about search modes in this scenario. First, from the presence of small game and birds on Magdalenian sites in southern Germany, it is clear that Magdalenian foragers who migrated into Germany already had technologies for capturing agile small prey (Hahn 1979, 1981; Jochim et al. 1999), though it is not known exactly what these were. At the same time, the continued importance of large mammal game in Mesolithic sites does suggest that large game remain preferred prey throughout this period. Second, the increasing scarcity and lower value of large game encounters, the increasing value and density of secondary prey, and the possible increase in habitat overlap of large and small game all suggest that the value of search directed toward secondary prey would have increased during this period. This can be expected to have changed preferences for site locations and mobility strategies, emphasizing search of habitats where small game (prey 2) is abundant and where the range of small and large game is likely to overlap.

We can examine this archaeologically by considering site locations, raw material economy, and changes in the design and maintenance of food-getting technologies. A focus on search modes, or (in Smith's terminology) hunt types, should provide useful concepts for understanding the ways in which associations between places, prey types, and technological assemblages change through time. Several long-noted archaeological patterns support the hypothesis that the role of small game in foraging economies changed over time in southern Germany. Here, I can only make three brief suggestions about the kinds of data that suggest that this model may help to explain archaeological patterns in late glacial and early postglacial southern Germany. In what follows, I am drawing on a very rich database of published analyses of Magdalenian, Late Paleolithic, and Early Mesolithic occupations in southern Germany, as well as my own comparative analyses of lithic technology from a sample of 17 assemblages that span this period (Fisher 2000).

Site Locations

First, consider a change in site locations. Examination of patterns in southern Germany suggests that the role of small game in foraging economies may have changed during the late glacial and early postglacial of southern Germany. The Magdalenian site of Petersfels, for example, was repeatedly occupied during the period from about 14,000–12,000 for reindeer hunting, with a lesser focus on horse (Albrecht 1984). It has been argued that the site, favorably located for reindeer to travel between upland and lowland grazing areas, was an excellent tactical base for intercept hunting of reindeer; the faunal assemblages at the site are, for the most part, reindeer-dominated and poor in species compared to regional averages (Albrecht 1976, 1984; Albrecht and Berke 1987; Hahn 1979, 1981; Weniger 1982, 1987,

1989, 1990). Lithic assemblages contain large numbers of backed bladelets, which show some association with open-country reindeer hunting in this region (Fisher 2000, in press). Some of the faunal assemblages at the site, though, are numerically dominated by hare and partridge (Albrecht 1976; Albrecht and Berke 1987). The tactical location of the site and its repeated use for reindeer hunting suggest that the site was visited for large game intercept hunting and small game were opportunistically taken as backup resources. This is similar to an interpretation that Enloe (2000) developed for Paris Basin Magdalenian sites.

Late Paleolithic and Early Mesolithic sites in the region do not show this kind of evidence for tactical use of places for ambush hunting (Stewart and Jochim 1986). Instead, the adjacent lowlands, now covered with forest and dotted with lakes and bogs, became a focus for settlement, where both large game and small game were taken. Early Mesolithic sites in the lowlands show a strong bias toward locations on the shores of lakes and streams (Jochim et al. 1998). These are likely to have been favorable locations for capturing small game, though most faunal assemblages remain numerically dominated by bones of large mammal game, including red deer and roe deer (Jochim 1976, 1998: 189). Although any conclusion must await a more detailed analysis, this suggests that small game, birds, and fish may have shifted from a backup resource to a focal, though secondary, resource during the Early Mesolithic. In this context, the preferred large game could have been opportunistically taken. Large game might also have been the focus of longer distance, more coarse-grained hunting forays in this and neighboring habitats.

Stone Raw Material Economy

Second, consider the use of local and transported stone raw material. Late glacial and early postglacial southern Germany shows a complex pattern of subtle changes in raw material economies, shaped by changes in the kinds of blanks and tools made (Fisher 2000, in press). The overall pattern that has long been noted, however, suggests that Early Mesolithic foragers scoured their environment more thoroughly to pick up larger quantities of dispersed local raw materials of variable quality from river gravels and moraine deposits, whereas Magdalenian and Late Paleolithic foragers targeted more concentrated raw material sources. This results in a slightly larger proportion of local raw material on Early Mesolithic sites in environments poor in stone raw material (Fig. 6.3). At the same time, however, Early Mesolithic sites continue to show very small amounts of stone raw materials drawn from the same range of nonlocal sources, up to a maximum of more than 200 kilometers, that show up in the earlier sites.

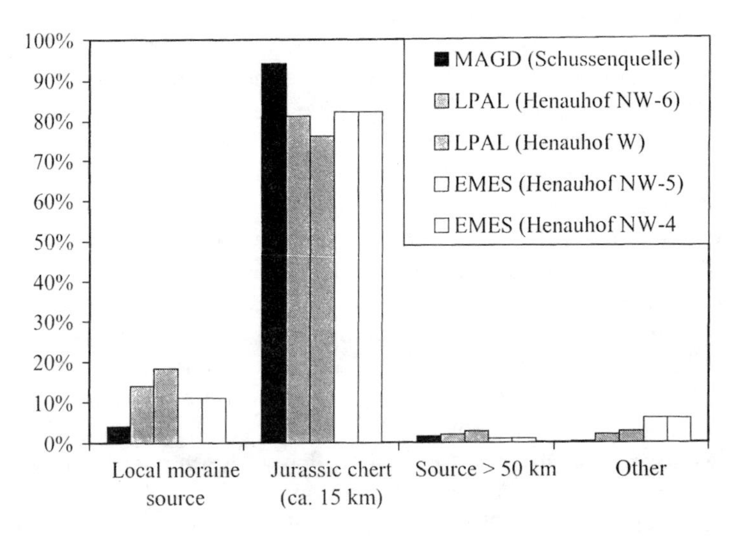

Figure 6.3. Relative frequency of stone raw materials in selected Magdalenian, Late Paleolithic, and Early Mesolithic assemblages in a setting poor in local raw material. These five assemblages derive from three excavated localities on or near the shoreline of the Federsee, a large late glacial and postglacial lake in the rolling moraine landscape of Oberschwaben south of the Swabian Jura limestone plateau. Sources: Fisher 2000; Jochim 1998; Schuler 1994.

Assuming that stone raw material transport provides at least a general indication of the spatial range of activities, one possible explanation of this pattern would be that foragers throughout this period continued to maintain access to large territories to pursue terrestrial large game, but that Early Mesolithic people were also using a more fine-grained search mode that, though tending to decrease encounters with large game, would tend to increase encounters with a wide variety of secondary game.

Design and Manufacture of Food-Getting Tools

Finally, trends in tool design might also indicate a shift toward focal exploitation of small game. Microlithic technologies, including Magdalenian backed bladelets and backed points, Late Paleolithic backed points, and Early Mesolithic geometric microliths and triangular micropoints, are likely have been multifunctional (Clarke 1976, 1978; Finlayson 1990a,b; Vierra 1995; Zvelebil 1985) but certainly included projectile armatures (Bachechi et al. 1997; Rozoy 1978; Ströbel 1959). If we assume that microliths are likely to be food-getting tools, in contrast to the blade and flake maintenance tools that predominate in many Paleolithic sites, it should be possible to examine trends in the design and manufacture of food-getting tools over time (Fisher 2000). Two broad trends are evident in the microlithic technologies of southern Germany. First, backed points, triangles, and triangular micropoints, all pointed microliths that are possible projectile armatures, decline in size beginning during Magdalenian times and continuing through the Early Mesolithic (Fig. 6.4). Second, Late Paleolithic and Early Mesolithic microliths are made on less regular bladelets, often on truncated flakes (Table 6.3).

Microlithic implements in all periods are typically made on slender blades. Magdalenian backed implements are almost entirely lamellar. Fragments had relatively parallel unretouched edges and dorsal arises, allowing confident identification of fragmentary pieces as bladelet segments. Late Paleolithic and Early Mesolithic backed points, backed bladelets, and microliths, on the other hand, include more pieces made on irregular blades or flakes that could not be identified as bladelet segments. These differences are significant as calculated by both chi-square (chi-square = 92.87, df = 3, $p < .0001$) and Kendall's tau (tau-b = .443, approx. significance < .0001).

These trends are part of a complex pattern undoubtedly caused by a number of different factors, but one overall result of these trends is probably a decrease in the value of individual projectile points. Ethnographic studies of projectile point use by modern hunters suggest that the cost of projectile points expended during pursuit is probably related to the value of prey (Bion 1997: 282). We might propose that the smaller, cheaper projectiles of Early Mesolithic times are more likely to have been fired at a combination of small and large game, rather than reserved for large game. Reducing pursuit costs of both small and large game would have been an important context of technological innovation during this period.

IMPLICATIONS FOR UNDERSTANDING VARIATION IN HUNTER-GATHERER MOBILITY

What does this suggest about mobility? The search mode model considered here has a number of interesting implications for improving our understanding of the relationships among mobility, environmental productivity, and the problems of searching for terrestrial game. First, the model suggests that adopting secondary resources as focal prey can be an efficient foraging option, even in the absence of population packing. Second, it suggests that locational decisions are likely to be affected by abundance of secondary as well as preferred prey. This provides a theoretical foundation for the argument that late glacial and early postglacial concentrations of

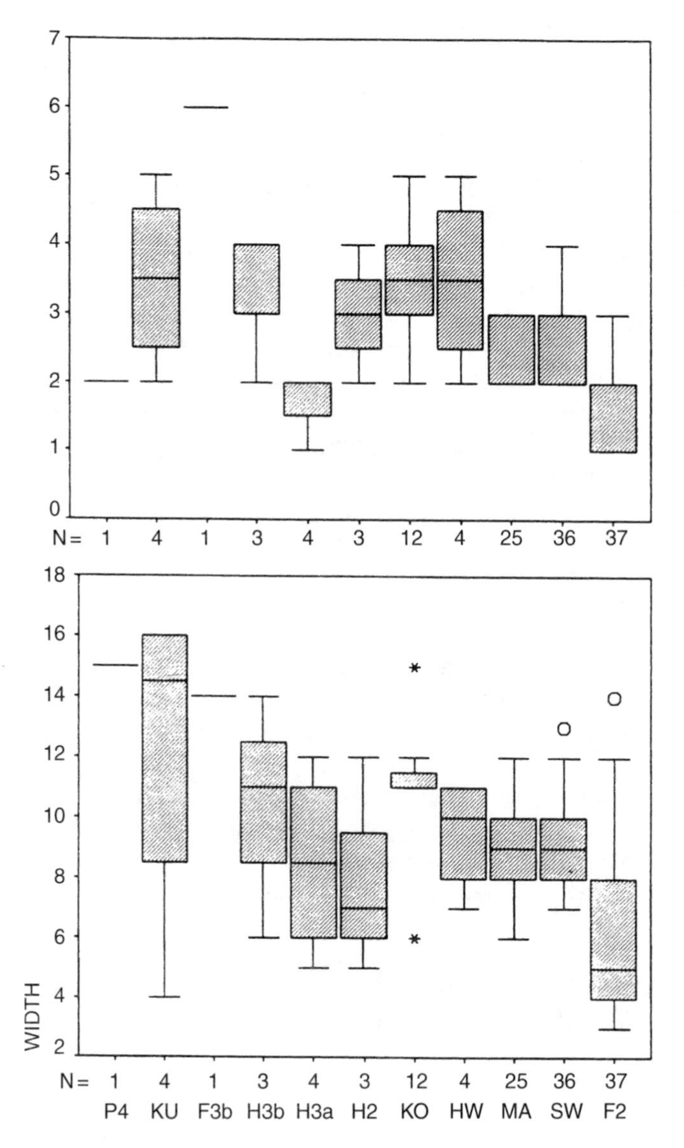

Figure 6.4. Width (mm) and thickness (mm) of whole and partial pointed microliths from samples of Magdalenian, Late Paleolithic, and Early Mesolithic assemblages in southern Germany. Assemblages are arrayed from earlier to later across the horizontal axis. Pointed microliths (including Magdalenian and Late Paleolithic backed points and Early Mesolithic triangular micropoints and geometric triangles) are grouped together as possible projectile armatures. Assemblage codes—Magdalenian: P4, Petersfels 1 level 4; KU, Kaufertsberg-Unten; F3, Felsställe 3a/b; H3, Helga Abri IIIa, b; Late Paleolithic: H2, Helga Abri IIF7; KO, Kaufertsberg-Oben; HW, Henauhof-West. Early Mesolithic: SW, Schräge Wand; MA, Malerfels I; F2, Felsställe 2a3. For details on these assemblages and analyses, see Fisher 2000.

Table 6.3. Frequency of Blank Types among Microlithic Tools in Assemblages Grouped by Period (excluding KO and KU)[a]

Period			Blank Type[b,c]		
			Blade	Flake/Frag	Total
	MA	Count	55	1	56
		% within Period	98.2	1.8	100
	LM	Count	161	4	165
		% within Period	97.6	2.4	100
	LP	Count	10	2	12
		% within Period	83.3	16.7	100
	EM	Count	66	50	116
		% within Period	56.9	43.1	100
Total		Count	292	57	349
		% within Period	83.7	16.3	100

[a]See Figure 4 caption for assemblage codes.
[b]Pearson chi-square = 92.87; df = 3; $p < .0001$. One cell has an expected count less than five.
[c]Kendall's tau-b = .443; approx. significance < .0001.

secondary prey were likely to have been a draw for human settlement, even as foragers were maintaining large territories to pursue terrestrial game. Third, it suggests that wide-ranging mobility and some localization of foraging patterns are not necessarily incompatible but instead might be linked into a single economy that had multiple search modes or hunt types. Finally, the search mode model provides a link between adoption of new food-getting technologies and decisions about the most efficient use of available resources. Specifically, weapons, facilities, or hunting strategies that tend to decrease handling costs of secondary prey increase the likelihood that foragers should direct their search toward nonpreferred prey.

The problems of searching for scarce large game and incorporating secondary game into foraging economies have been at the heart of a series of debates about the evolution of hunter-gatherer subsistence and settlement systems in the late Pleistocene and early Holocene. However, search tactics and spatial behavior, as it relates directly to searching for varying resources, have not been the focus of much theoretical attention in evolutionary ecology (but see Bell 1991; O'Brien et al. 1989, 1990; Robinson and Holmes 1982; Schmidt 1998). This brief exploration of one novel extension of the diet breadth model suggests that theoretical focus on searching for different broad classes of prey or in different habitat types should provide a useful perspective on the evolution of hunter-gatherer settlement and subsistence systems.

As many have pointed out, a dominant analytical focus on large game species such as reindeer or horse as the targets of Paleolithic economies fails to take into account a wide range of variation in tactics for exploiting small game, fish, birds, and a variety of secondary species (Bicho et al. 2000; Hahn 1979, 1981; Eriksen 1991; Stiner et al., 2000). Much debate has centered on increasing diet breadth as a major evolutionary trend in Paleolithic and Mesolithic subsistence. Less attention has been paid to the variety of roles these resources may have played in prehistoric hunter-gatherer subsistence systems. A distinction between focal prey and preferred prey should be widely applicable in archaeology as a conceptual device for distinguishing between changes in diet breadth and changes in search tactics as a component of hunter-gatherer spatial behavior.

REFERENCES

Albrecht, G., 1976, *Magdalenien-Inventare vom Petersfels*. Tübinger Monographien zur Urgeschichte Band 6, Verlag Archaeologica Venatoria, Tübingen.

Albrecht, G., 1984, Intensive Fall Hunting at Petersfels During the Magdalenian: Questions Concerning the Motives. In *Jungpaläolithische Siedlungsstrukturen in Europa*, edited by H. Berke, J. Hahn, and C. J. Kind, pp. 99–120. Archaeologica Venatoria, Tubingen.

Albrecht, G., and Berke, H., 1987, Das Brudertal bei Engen/Hegau—Beispiel für eine Arealnutzung im Magdalénien. *Mitteilungsblatt der Archaeologica Venatoria* 12:1–11.

Albrecht, G., and Hahn, A., 1991, *Rentierjäger im Brudertal: Die Jungpaläolithischen Fundstellen um den Petersfels und das Städtische Museum Engen im Hegau*. Fuhrer zu archaologischen Denkmalern in Baden-Wurttemberg 15, Konrad Theiss Verlag, Stuttgart.

Audouze, F., and Enloe, J. G., 1991, Subsistence Strategies and Economy in the Magdalenian of the Paris Basin. In *The Late Glacial of Northwest Europe: Human Adaptation and Environmental Change at the End of the Pleistocene*, edited by R.N.E. Barton, A. J. Roberts, and D. A. Roe, pp. 63–71. Council for British Archaeology, London.

Bachechi, L., Fabbri, P. F., and Mallegni, F., 1997, An Arrow-Caused Lesion in a Late Upper Palaeolithic Human Pelvis. *Current Anthropology* 38(1):135–140.

Bartram, L., 1997, A Comparison of Kua (Botswana) and Hadza (Tanzania) Bow and Arrow Hunting. In *Projectile Technology*, edited by H. Knecht, pp. 321–344. Interdisciplinary Contributions to Archaeology, Plenum, New York.

Bell, W. J., 1991, *Searching Behaviour: The Behavioural Ecology of Finding Resources*. Chapman and Hall, London.

Bicho, N. F., Hockett, B., Haws, J., and Belcher, W., 2000, Hunter-Gatherer Subsistence at the End of the Pleistocene: Preliminary Results from Picareiro Cave, Central Portugal. *Antiquity* 74:500–506.

Binford, L. R., 1968, Post-Pleistocene Adaptations. In *New Perspectives in Archeology*, edited by S. R. Binford and L. R. Binford, pp. 313–341. Aldine, Chicago.

Binford, L. R., 1980, Willow Smoke and Dogs' Tails: Hunter-Gatherer Settlement Systems and Archaeological Site Formation. *American Antiquity* 45:4–20.

Binford, L. R., 1982, The Archaeology of Place. *Journal of Anthropological Archaeology* 1:5–31.

Binford, L. R., 1983, Long Term Land Use Patterns: Some Implications for Archaeology. In *Lulu Linear Punctated: Essays in Honor of George Irving Quimby*, edited by R.C. Dunnell and

D.K. Grayson, pp. 27–53. Anthropological Papers No. 72, Museum of Anthropology, University of Michigan, Ann Arbor.

Binford, L. R., 1990, Mobility, Housing and Environment: A Comparative Study. *Journal of Anthropological Research* 46(2):119–152.

Bion G. P., 1997, Technology and Variation in Arrow Design among the Agta of Northeastern Luzon. In *Projectile Technology*, edited by H. Knecht, pp. 267–286. Interdisciplinary Contributions to Archaeology, Plenum, New York.

Clarke, D. L., 1976, Mesolithic Europe: The Economic Basis. In *Problems in Economic and Social Archaeology*, edited by G. de G. Sieveking et al., pp. 449–481. Duckworth, London.

Clarke, D. L., 1978, *Mesolithic Europe: The Economic Basis*. Duckworth, London.

Enloe, J. G., 2000, Le Magdalénien du Bassin Parisien au Tardiglaciaire: La Chasse aux Rennes Comparáeé à Celle d'Autres Espéces. In *Le Paléolithique Supérieur Recent: Nouvelles Données sur le Peuplement et l'Environnement. Actes de la Table Ronde de Chambéry*, edited by G. Pion, pp. 39–45. Société Préhistorique Française Mémoire XXVIII.

Eriksen, B., 1991, *Change and Continuity in a Prehistoric Hunter-Gatherer Society*. Archaeologica Venatoria Band 12, Verlag Archaeologica Venatoria, Tübingen, Germany.

Eriksen, B., 1996, Resource Exploitation, Subsistence Strategies, and Adaptiveness in Late Pleistocene–Early Holocene Northwest Europe. In *Humans at the End of the Ice Age: The Archaeology of the Pleistocene-Holocene Transition*, edited by L. G. Straus, B. V. Eriksen, J. M. Erlandson, and D. R. Yesner, pp. 101–128. Plenum, New York.

Finlayson, B., 1990a, The Function of Microliths: Evidence from Smittons and Starr, SW Scotland. *Mesolithic Miscellany* 11(1):2–6.

Finlayson, B., 1990b, Lithic Exploitation During the Mesolithic in Scotland. *Scottish Archaeological Review* 7:41–57.

Fisher, L., in press, Retooling and Raw Material Economies: Technological Change in Late Glacial and Early Postglacial Southern Germany. In *Lithic Raw Material Economy in Late Glacial and Early Postglacial Western Europe*, edited by L. E. Fisher and B. V. Eriksen, International Monographs in Prehistory, Ann Arbor.

Fisher, L., 2000, *Land Use and Technology from Magdalenian to Early Mesolithic in Southern Germany*. Unpublished Ph.D. Dissertation, Anthropology, University of Michigan, Ann Arbor.

Floss, H., 1994, *Rohmaterialversorgung im Paläolithikum des Mittelrheingebietes*. Monographien des Römisch-Germanischen Zentralmuseums 21, Rudolf Habelt, Bonn.

Frenzel, B. (ed.), 1995, *European River Activity and Climatic Change During the Late Glacial and Early Holocene*. Gustav Fischer Verlag, Stuttgart.

Gamble, C., 1986, *The Palaeolithic Settlement of Europe*. Cambridge University Press, Cambridge.

Hahn, J., 1979, Essai sur l'Écologie du Magdalénien dans le Jura Souabe. In *La Fin des Temps Glaciaires en Europe*, edited by D. de Sonneville-Bordes, pp. 203–211. Colloques Internationaux C.N.R.D. No. 271, Centre National de la Recherche Scientifique, Paris.

Hahn, J., 1981, Abfolge und Umwelt der Jüngeren Altsteinzeit in Südwestdeutschland. *Fundberichte aus Baden-Württemberg* 6:1–27.

Hames, R., and Vickers, W., 1982, Optimal Foraging Theory as a Model to Explain Variability in Amazonian hunting. *American Ethnologist* 9:358–378.

Hitchcock, R., and Bleed, P., 1997, Each According to Need and Fashion: Spear and Arrow Use among San Hunters of the Kalahari. In *Projectile Technology*, edited by H. Knecht, pp. 345–370. Interdisciplinary Contributions to Archaeology, Plenum, New York.

Holliday, T. W., 1998, The Ecological Context of Trapping among Recent Hunter-Gatherers: Implications for Subsistence in Terminal Pleistocene Europe. *Current Anthropology* 39(5):711–720.

Housley R. A., Gamble, C. S., and Pettitt, P. 1997, Radiocarben Evidence for the Late Glacial Human Recolonisation of Northern Europe. *Proceedings of the Prehistoric Society* 63:25–54.

Jochim, M. A., 1976, *Hunter-Gatherer Subsistence and Settlement: A Predictive Model*. Academic Press, New York.

Jochim, M. A., 1991, Archaeology as Long-Term Ethnography. *American Anthropologist* 93:308–321.

Jochim, M. A., 1998, *A Hunter-Gatherer Landscape: Southwest Germany in the Late Paleolithic and Mesolithic*. Plenum, New York.

Jochim, M. A., Glass, M., Fisher, L., and McCartney, P., 1998, Mapping the Stone Age: An Interim Report on the South German Survey Project. In *Aktuelle Forschungen zum Mesolithikum*, edited by N. J. Conard and C. J. Kind, pp. 121–132. Mo Vince Verlag, Tubingen.

Jochim, M. A., Herhahn, C., and Starr, H., 1999, The Magdalenian Colonization of Southern Germany. *American Anthropologist* 101(1):129–142.

Kelly, R. L., 1983, Hunter-Gatherer Mobility Strategies. *Journal of Anthropological Research* 39:277–306.

Kelly, R. L., 1992, Mobility/Sedentism: Concepts, Archaeological Measures, and Effects. *Annual Review of Anthropology* 21:43–66.

Kelly, R. L., 1995, *The Foraging Spectrum: Diversity in Hunter-Gatherer Lifeways*. Smithsonian Institution Press, Washington, DC.

Lewin, R., 1988, New Views Emerge on Hunters and Gatherers. *Science* 240:1146–1148.

Lieberman, D. E., 1993, The Rise and Fall of Seasonal Mobility among Hunter-Gathers. The Case of the Southern Levant. *Current Anthropology* 34(5):599–631.

Mellars, P., 1984, The Upper Paleolithic revolution. In *The Oxford Illustrated Prehistory of Europe*, edited by B. Cunliffe, pp. 42–78. Oxford University Press, Oxford.

Mourer-Chauviré, C., 1984, Die Vögel der Würmzeitlichen und Holozänen Fundstelle Spitzbubenhöhle. In *Der Steinzeitliche Besiedlung des Eselsburger Tales bei Heidenheim (Schwäbische Alb)*, edited by J. Hahn, pp. 80–83. Forschungen und Berichte zur Vor- und Frühgeschichte in Baden-Württemberg 17, Landesdenkmalamt Baden-Württemberg (Kommissionsverlag Konrad Theiss), Stuttgart.

Myers, A., 1989, Reliable and Maintainable Technological Strategies in the Mesolithic of Mainland Britain. In *Time Energy and Stone Tools*, edited by R. Torrence, pp. 78–91. Cambridge University Press, Cambridge, England.

Nicholas, G. P., 1988, Ecological Leveling: The Archaeology and Environmental Dynamics of Early Postglacial Land Use. In *Holocene Human Ecology in Northeastern North America*, edited by G. P. Nicholas, pp. 257–296. Plenum, New York.

O'Brien, W. J., Browman, H. I., and Evans, B. I., 1990, Search Strategies of Foraging Animals. *American Scientist* 78:152–160.

O'Brien, W. J., Evans, B. I., and Browman, H. I., 1989, Flexible Search Tactics and Efficient Foraging in Saltatory Searching Animals. *Oecologia* 80:100–110.

Pasda, C., 1994, *Das Magdalénien in der Freiburger Bucht*. Materialhefte zur Archäologie in Baden-Württemberg 24, Konrad Theiss, Stuttgart.

Pasda, C., 1998, *Wildbeuter im Archäologischen Kontext. Das Paläolithikum in Südbaden*. Archäologie im Südwesten II, Verlag Dr. G. Wesselkamp, Bad Bellingen.

Pulliam, H. R., 1974, On the Theory of Optimal Diets. *American Naturalist* 108:137–154.

Robinson, S. K., and Holmes, R. T., 1982, Foraging Behavior of Forest Birds: The Relationships among Search Tactics, Diet, and Habitat Structure. *Ecology* 63(6):1918–1931.

Robinson, S. K., 1984, Effects of Plant Species and Foliage Structure on the Foraging Behavior of Forest Birds. *Auk* 101:672–684.

Rozoy, J. G., 1978, *Les Derniers Chasseurs: L'Épipaléolithique en France et en Belgique. Essai de Synthèse.* Bulletin de la Société Archéologique Champenoise, Numéro Spécial Juin 1978.

Schmidt, K. A., 1998, The Consequences of Partially Directed Search Effort. *Evolutionary Ecology* 12:263–277.

Schuler, A., 1994, *Die Schussenquelle: Eine Freilandstation des Magdalénien in Oberschwaben.* Materialhefte zur Archäologie in Baden-Württemberg, Heft 27, Kommisionsverlag, Konrad Theiss Verlag, Stuttgart.

Smith, E. A., 1991, *Inujjuamiut Foraging Strategies: Evolutionary Ecology of an Arctic Hunting Economy.* Aldine de Gruyter, New York.

Smith, E. A., and Winterhalder, B., 1992, Natural Selection and Decision Making: Some Fundamental principles. In *Evolutionary Ecology and Human Behavior*, edited by E. A. Smith and B. Winterhalder, pages 25–60. Walter de Gruyter, Inc., New York.

Stewart, A., and Jochim, M. A., 1986, Changing Economic Organization in Late Glacial Southwest Germany. In *The End of the Palaeolithic in the Old World*, edited by L. G. Straus, pages 47–62. British Archaeological Reports, International Series 284. BAR, Oxford, England.

Stiner, M. C., Munro, N. D., and Surovell, T. A., 2000, The Tortoise and the Hare: Small-Game use, the Broad-Spectrum Revolution, and Paleolithic Demography. *Current Anthropology* 41(1):39–73.

Straus, L. G., 1991, Human Geography of the Late Upper Paleolithic in Western Europe. *Journal of Anthropological Research* 47:259–278.

Ströbel, R., 1959, Tardenoisspitze in einem Bovidenknochen von Schwanningen am Neckar (Kr. Rottweil). *Fundberichte aus Schwaben* N. F. 15:103–106.

Svoboda, J., Ložek, V., and Vlček, E. eds 1996, *Hunters between East and West: The Paleolithic of Moravia.* Plenum Press, New York.

Vierra, B. J., 1995, *Subsistence and Stone Tool Technology: An Old World Perspective.* Anthropological Research Papers No. 47, Arizona State University.

Weniger, G. C., 1982, *Wildbeuter und Ihre Umwelt: Ein Beitrag zum Magdalenien Südwestdeutschlands aus Ökologischer und Ethnoarchäologischer Sicht.* Archaeologica Venatoria, Tübingen.

Weniger, G. C., 1987, Magdalenian Settlement Pattern and Subsistence in Central Europe, In *The Pleistocene Old World: Regional Perspectives*, edited by O. Soffer, pp. 201–215. Plenum, New York.

Weniger, G. C., 1989, The Magdalenian in Western Central Europe: Settlement Pattern and Regionality. *Journal of World Prehistory* 3:323–372.

Weniger, G. C., 1990, Germany at 18,000 BP. In *The World at 18,000 BP: High Latitudes*, edited by O. Soffer and C. Gamble, pp. 171–192. Unwin Hyman, Boston.

Winterhalder, B., 1981, Foraging Strategies in the Boreal Forest: An Analysis of Cree Hunting and Gathering. In *Hunter-Gatherer Foraging Strategies: Ethnographic and Archeological Analyses*, edited by B. Winterhalder and E. A. Smith, pp. 66–98. University of Chicago Press, Chicago.

Zvelebil, M., 1985, Economic Intensification and Postglacial Hunter-Gatherers in North Temperate Europe. In *The Mesolithic of Europe*, edited by C. Bonsall, pp. 80–88. John Donald, Edinburgh.

Chapter 7

Long-term Land Tenure Systems in Central Brazil

Evolutionary Ecology, Risk-Management, and Social Geography

RENATO KIPNIS

INTRODUCTION

My goal here is to look at the land tenure system based on reciprocal access to foraging areas, as this relates to spatial and temporal resource variability in central Brazil. The theoretical perspective I employ is based on evolutionary ecology and risk-management theory. The approach takes into account both technoenvironmental and social constraints and was first suggested by Eric Smith (1991b).

In this chapter, I argue that we should look at hunter-gatherer settlement patterns and land use through models derived from evolutionary ecology theory. Specifically, I use the concepts and methodologies of optimal foraging to generate hypotheses about foragers' decision-making based on the assumption that foragers are risk minimizers rather than net-rate-of-acquisition maximizers. The application of such models can be used to formalize Binford's forager–collector continuum and thereby subject it to greater logical and empirical examination.

RENATO KIPNIS • Laboratório de Estudos de Evolutivos Humanos, Instituto de Biociências, Universidade de São Paulo, São Paulo, SP 05508-900 Brasil.

The human ecological approach to hunter-gatherer studies has shown that such societies use a broad range of ways to mitigate risk, including mobility, storage, logistical collecting, exchange, communal sharing, intensification, and diversification (e.g., Cashdan 1985; Colson 1979; Goland 1991a; Halstead and O'Shea 1989; Spielmann 1986; Wiessner 1982b, 1996; Winterhalder 1990). Combinations of these mechanisms have been used to define contrasting strategies: foragers/collectors (Binford 1980), immediate-return systems/delayed-return systems (Woodburn 1980), nomadic hunter-gatherers/sedentary hunter-gatherers (Testar 1982; 1988), travelers/processors (Bettinger and Baumhoff 1982), and generalized foragers/complex hunter-gatherers (Hayden 1990, 1996; Wiessner 1996). These strategies are culturally defined adaptive responses to specific sets of environmental conditions that vary spatially and temporally in a given environment.

These contrasting strategies have often been used in the ethnographic and archaeological literature as typological characterization rather than their original heuristic intent. From this classificatory perspective, the different systems, which in many cases represent a continuum, have been dichotomized. The most widely used and misused model has been the forager–collector model. Binford's spectrum of organizational strategies that hunter-gatherers use to adapt to various environments has been used to classify populations as either foragers, who move consumers to resources, or collectors, who move resources to consumers, when in reality different systems along the forager–collector continuum exist, as well as systems that mix different strategies.

Binford's model has been very influential for two main reasons: first, it explains the difference between foragers and collectors with reference to variability in the quantity and seasonal distribution of resources at their disposal measured by a simple variable (effective temperature); and secondly, it derives hypotheses about the material correlations of the strategies. The temptation to pigeonhole archaeological findings based on Binford's simple but elegant argument into one of the two strategies proved to be too much to be ignored by many scholars. This resulted in several studies where the variability in the hunter-gatherer settlement pattern was reduced to the two ends of the spectrum, that is, foragers or collectors, and where technoenvironmental variables were given priority to the detriment of other potential explanatory variables (i.e., political, social, and ideological).

But at the same time, Binford's work stimulated others to built on his model and suggest more inclusive approaches. One such study is Wiessner's work among the !Kung society (1982a,b, 1983, 1984, 1986, 1996). Wiessner argues that environmental variables alone do not suffice to explain the diversity of settlement pattern and land use of hunter-gatherers.

To construct a more robust model, she proposes to look at strategies used in organization around other persons in the social relationships of production. Specifically, she suggests that risk theory can be used to predict a much wider range of responses that can be associated with material remains generated by different strategies of organizations than Binford's forager/collector model (Wiessner 1982a,b).

Risk is an important concept in evolutionary ecology. In the past two decades, the explanation of past and present human behavioral variability from an ecological perspective has profited from approaching it through evolutionary ecology. The latter is an overarching theoretical approach to explain the variability of decision-making, in an environmental context based on the neo-Darwinian theory of natural selection and evolution. Specifically, evolutionary ecology in hunter-gatherer studies aims to establish a micro ecological approach to choices of production and consumption in an ecological setting (Smith and Winterhalder 1992; Winterhalder 1981, 1990).

Evolutionary ecology provides a robust body of theory for understanding behavioral diversity in relation to an organism's environment. This theory is a powerful basis for deriving explanatory hypotheses, and it has been successfully applied in anthropology. This theoretical approach is very useful for pursuing some of the basic archaeological goals, namely, the explanation of cultural diversity and the understanding of the processes of cultural evolution.

Evolutionary ecology applies optimization and game theory models to construct and test evolutionary explanations. Models derived from optimal foraging theory are used as heuristic devices to generate expected values for comparison with empirical data. The goal is to understand the sources of observations that deviate from modeled expectations, not to show that observations fit the models. In other words, the role of optimization theories in evolutionary ecology is to understand the diversity of life, not to demonstrate that organisms optimize.

THEORETICAL BACKGROUND

Risk and Land Tenure

The unpredictable nature of short- (e.g., day-to-day) and long-term success in the quest of food is a critical component of decision-making in foraging societies. The resources of foragers and farmers as well are subject to episodes of scarcity due to drought, flood, frost, epidemic, and other irregular calamities. Foragers and farmers require adaptations that mitigate the effects of unpredictable subsistence shortfalls.

Short-term temporal variation in food supply (i.e., variability from season to season or from year to year) may cause foragers to starve. Variation in the amount of food eaten during some period may also affect fertility and mortality, even though the likelihood of starvation is low (Kaplan and Hill 1992: 188). Human (Kaplan et al. 1990; Winterhalder 1986a) and animal foragers (Stephens and Krebs 1986) are not indifferent to the variability in foraging returns and in the mean rate of capture or, in more technical terms, they are risk sensitive.

Risk and uncertainty refer to situations where decision-makers cannot always uniquely predict the outcomes of their actions. Both are associated with stochastic processes (i.e., variation in outcomes that cannot be controlled by the decision-maker). Risk refers to the probability of falling below a minimum requirement for survival due to production shortfalls that may result from resource fluctuations. These fluctuations can be described by three variables predictability, frequency, and severity. A general definition of uncertainty is that it refers to an individual's lack of knowledge about the state of the world (Cashdan 1990: 2). Uncertainty refers to the lack of perfect information that afflicts decision-makers (Smith 1991b: 231). But if the individual can assign some probability to each outcome, the situation is one of risk, and no longer one of uncertainty. In risk, there is an objective probability associated with possible outcomes. In contrast, in the case of uncertainty, the objective probabilities are unknown. Risk pertains to situations in which outcomes are variable, regardless of whether the actor has complete information about the probability distribution. Uncertainty can be reduced through information acquisition (i.e., learning), but risk cannot (Clark 1990: 48).

Based on experimental and theoretical work, it has been suggested that a forager's decision-making to manage risk follows the expected energy budget (e.g., daily caloric returns); if positive avoid risk, but if the expected energy budget is negative, then prefer risk (Stephens and Krebs 1986: 134–137). The technical term in evolutionary ecology for the two different strategies is risk-averseness (avoidance of risk), and risk-proneness (preference for risk). Then, risk is better defined as the probability of falling below a minimum requirement for survival, for example, production shortfalls that may result from resource fluctuations.

In its most general form, the problem of decision-making under risk is a problem of choosing among (or evaluating) probability distributions. Risk, as a probabilistic variation, is a normally distributed variable (e.g., return rates from collecting palm nuts) with variance around the mean. A risk-prone strategy seeks to increase variance; if one expects to obtain less food than needed, choose a large variance. A risk-averse strategy seeks to reduce variance; if one expects or acquires more food than needed, choose a small variance.

The risk-proneness/risk-averseness concept is better visualized by the *Z*-score model (Bettinger 1991: 118–124; Stephens and Krebs 1986: 137–141). The model assumes a normally distributed variable (e.g., return rate for a given behavior) that has mean μ, standard deviation σ, and a fixed threshold n which the forager must meet for survival. Z is the ratio between the amount the forager requires for survival (n) minus the mean (μ) of the return rate's distribution, and the standard deviation (σ) of the return rate's distribution. As always, a mathematical algorithm expresses the relationship and the model more elegantly:

$$z = \frac{n - \mu}{\sigma}$$

The model (Fig. 7.1) seeks to minimize Z, the probability of falling below the threshold. In other words, minimizing Z maximizes the probability of survival. If the threshold (n) is smaller than the mean (μ), Z will be negative; conversely, if n is greater than the mean (μ), Z will be

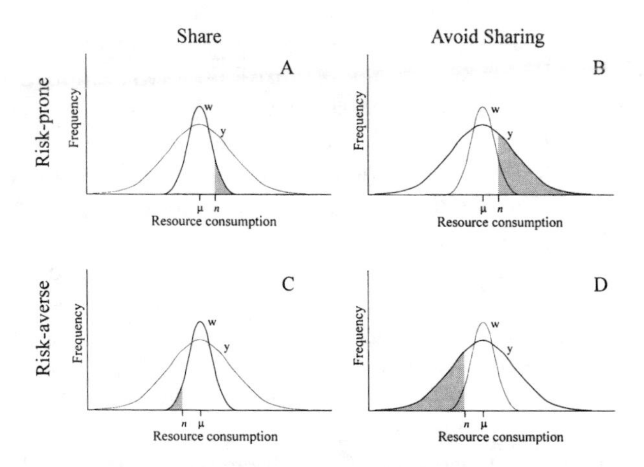

Figure 7.1. The Z-score model, showing two distributions of resource consumption with same mean, but with different standard deviations. One low-variance resource consumption (w-distribution curve), and one high-variance resource consumption (y-distribution curve). For the risk-prone forager (A/B) an increase in the variance increases the probability of falling on the right side of the distribution tail (shaded area), thus increases the probability of survival. It is a logical outcome, if the forager expects to acquire less than needed, the best strategy is the one that increases the chances of getting enough food, and at the same time decreasing the probaility of not getting enough (left side of the tail). On the other hand, if the risk-averse forager (C/D) expects to acquire more than the threshold, then the best strategy is the one that minimizes the chances of getting less than necessary (shaded area), regardless of what happens on the rightside of the distribution tail.

positive. Thus, for a risk-averse forager, for whom the expected μ is greater than n, a decrease in σ (i.e., variance) will result in greater probability of survival (i.e., Z decreases). A risk-prone forager, for example, one who expects to acquire less food than needed, will have a value of μ smaller than n, and an increase in σ (i.e., variance) will result in a decrease in Z.

One important realization from the early experimental work on risk sensitivity is that it suggests a simple currency of optimization for the decision-making of foragers: minimization of energetic shortfall (Stephens 1990: 26). Anthropologists who work within the evolutionary ecology framework quickly saw the potential of the risk-minimization concept for explaining human decision-making. Using the same concepts and methodologies of optimal foraging, the diet breadth model, for example, could be used to generate hypotheses about hunter-gatherer decision-making based on the assumption that they are risk minimizers rather than net-rate-of-acquisition maximizers (Winterhalder 1990).

Winterhalder (1986a,b, 1990) has examined, through computer simulations, the different foraging strategies that the diet breadth model would predict for hunter-gatherer societies when the goal is to minimize risk. Based on the results of the ecological models, he suggests two fundamental components of hunter-gatherers' economy: (1) production decisions will emphasize high harvest rates, while (2) distribution will emphasize sharing to achieve low consumption variance (Winterhalder 1990: 82). Thus, looking at social interaction and specifically systems of reciprocal exchange (i.e., sharing), the results of risk theory holds true. Foragers should be risk-averse, and by sharing food they even out fluctuations in food supply and avoid the threat of starvation

Sharing of independently acquired resources among foragers is an effective mechanism to reduce variance (Clark 1990; Stephens 1990; Winterhalder 1986a, 1990). Reduction in variance can be achieved with quite small numbers of foragers, especially when the interforager return-rate correlation has negative or low to medium positive values. For example, 60% of the potential risk reduction from sharing can be gained from only six or fewer cooperating foragers, whatever the interforager correlation (Winterhalder 1990: 79).

Sharing of food is a prevalent characteristic of egalitarian societies based on a foraging economy, it has been observed among hunter-gatherer groups living in many different environments (Bicchieri 1972; Kelly 1995; Lee and DeVore 1968; Wiessner 1996). Foraging societies that have no resource storage mechanisms manage a wide range of risks by pooling (e.g., sharing) over the long term through reciprocal relationships. In these situations, cultural mechanisms (e.g., leveling mechanisms; Clastres 1989; Wiessner 1996) are established to reduce the desire for immediate return from reciprocation, which would balance the relationship between sharers. The reciprocation is delayed for a later time (Sahlins 1972).

Changes in diet breadth (i.e., generalizing or specializing) as risk management strategies are effective only when the forager's minimum requirement is much smaller or larger than the expected value of the return rate. This is based on recognizing that each foraging choice has a mean (i.e., expected value) and variation about the mean. The variability in the net acquisition rate is due to fluctuations in the resource encounter rate or other stochastic factors that affect foraging success over time.

Winterhalder (1986a, 1990) shows that, if the minimum threshold is much smaller than the average return rate, the forager should sacrifice efficiency for reduced variance (i.e., lowest probability of failure) by adding more resources to the diet (see Fig. 7.1 c,d). On the other hand, if the minimum threshold is much larger than the average net return rate, the model predicts that the forager will contract diet breadth by taking only the highest ranked resources (see Fig. 7.1 a,b). In the latter case, the forager loses some foraging efficiency for high variance (i.e., greatest chance of success). But in situations where the minimum threshold does not differ much (above or below) from the net acquisition rate, the optimal risk-minimizing diet choice converges with the rate-maximizing choice (Winterhalder 1990: 75–76).

Although food sharing among hunter-gatherer societies is very common, the application of risk theory accounts for the variation in social interaction. In a fashion similar to the diet-breadth model, the risk model predicts the context in which sharing and nonsharing should occur, or in other words, it explains the degree of variation in sharing. For example, in situations when resources are extremely scarce and starvation might be expected (i.e., average resource consumption is much smaller than the minimum threshold), the foragers will share less or not share at all, thus increasing variance (i.e., greater chance of success; see Fig. 7.1 a,b).

So far I have discussed the technoenvironmental constraints that favor systems of reciprocal access between foragers of different local bands. Now, I will briefly consider the social constraints. Reciprocal access between bands raises issues of monitoring and enforcing reciprocity and of avoiding the material and evolutionary costs of indiscriminate altruism (Smith 1991b). It also raises issues that involve coordinating of two sets of foragers. In particular, we need to pay attention to costs incurred by a host group that allows visitors access to the local foraging area. Two primary costs are (1) the possibility that visitors allowed access to resources will not reciprocate, and hence the need to maintain controls against possible deception; and (2) the effects of the visitor's foraging on the host's-foraging efficiency.

These problems can be explained by viewing them in terms of uncertainty and information and by employing the strategic logic of evolutionary game theory. The main point is that evolutionary game theory specifies the conditions required to perpetuate a system of land tenure based on reciprocal access in a social group. In technical terms, it specifies the conditions in which an evolutionarily stable strategy (ESS) can develop (Smith 1991b: 250):

1. Residents possess more information about the location and abundance of local resources and the recent and current allocation of foraging effort than visitors.
2. Uncoordinated search exacts a penalty of interference and inefficiency by overcrowding foraging effort above the equilibrium that would result from greater information.
3. Today's visitors are likely to be tomorrow's hosts.
4. Residents can impose sanctions against those who do not reciprocate.

Control over social boundaries and information flow is an effective mechanism for reducing uncertainty that would arise if movement and resource exploitation were uncoordinated. Smith (1991b: 251) predicts that "the lower the correlation in resource availability between adjacent regions, the more demand for reciprocal access, and hence the more likely elaborate devices will evolve to control this access and reduce uncertainty it could bring." Smith suggests a continuum between areas without borders and regions of total control over territory where the two ends of the spectrum would be extremely rare.

Information

It is probably evident by now that information about resource distribution and yields is a critical variable for decision-making. For example, the risk-sensitive model assumes that the forager knows which alternative is risky and which is certain. Information acquisition and the effects of incomplete information are important issues in optimal foraging theory, particularly relevant for human foragers, who rely extensively on learning and communication to acquire, assess, and store information about resource distribution, abundance, and profitability.

As a simplifying assumption, information constraints are generally relaxed in optimal foraging models. Usually, optimal foraging models assume that the forager has complete information about resource distributions and yields and is concerned with the average payoff of different choices (Stephens and Krebs 1986: 34–35). But incomplete information is more realistic. Even among humans, with our substantial capacities and information storage mechanisms, information is not perfect.[1] To make an informed decision about some matter, a forager relies on information. The level of knowledge of the facts relevant for decision-making defines the level of uncertainty. For example, a forager can be uncertain about present or future conditions of the environment, which might be critical for decision-making. Uncertainty can be reduced by information gathering, which leads to a problem of costs. The basic idea is that information provides benefits, but its acquisition is costly. The problem is at what point the costs of obtaining information exceed the benefits. Or put differently, What is the rate-maximizing behavior when a forager must acquire both information and forage?

The degree to which resources can be predicted requires that more or less effort be put into information gathering; thus a trade-off is made between collecting information to increase efficiency or minimize risk and performing another activity that might have a more immediate or direct return (i.e., hunting, child care, etc.). The task of generalizing how incomplete information affects the economics of decision-making is difficult; most of the efforts have been empirical case studies (Stephens and Krebs 1986: 75 and references therein). One of the few attempts to incorporate information variability in optimal foraging models is by Stephens (1990; Stephens and Davis 1986). He suggests that types (recognizable classes of resource items or patches) may be composed of many indistinguishable subtypes that make types ambiguous (Stephens 1990: 31–32). For example, for a hunter, a tapir may be a recognizable type, and the indistinguishable subtypes might be a lean tapir versus a fat tapir or, female tapir versus a male tapir. Ambiguity (the existence of more than one subtype) is the basic property of incomplete information problems.

The diet breadth model assumes a repetitive cycle of search-encounter of resource items (types) and a decision to pursue or not pursue the resource. It is usually assumed in optimal foraging models that foragers recognize types upon encounter (i.e., complete information). It is also assumed that, upon encounter of resource items, foragers recognize types (i.e., complete information), but don't recognize subtypes.

Furthermore, in these optimal foraging models, the forager is not allowed to change her or his assessment of resource quality as the resource is being pursued. But in reality, foragers do use their experience to guide their behavior (i.e., decisions). In other words, in reality there is uncertainty

[1]Smith (1991a: 58–61) distinguishes between perfect and complete information. Based on perfect information, a forager can predict the outcome of a particular strategy exactly, whereas with complete information (Stephens and Krebs 1986: 11), the forager can predict with certain probability.

about types (i.e., incomplete information, lack of knowledge about subtypes), and foragers do use information to reduce this uncertainty. Then, the problem, according to Stephens and Davis (1986: 76), is how a long-term rate maximizer should treat recognizable types when it knows that they are ambiguous (i.e., divided into indistinguishable subtypes).

Stephens approaches this problem by using three models: (1) the recognition cost model (resource discrimination between 'good' and 'bad subtypes'); (2) the environmental tracking model (tracking environmental changes when subtypes can be readily distinguished); and (3) the patch sampling model (decision about how long to forage in a given patch as information about the return rate is acquired while foraging). These are baseline models, which allow incremental adjustments in complexity and realism (in the search for greater explanatory power).

Results from all three models suggest that information is of greatest value at intermediate levels of variation, whether synchronous (subtype discrimination and patch sampling) or temporal (environmental tracking). For example, because of environmental instability, tracking environmental changes (i.e., acquiring information about where resources are and when they are available) is necessary. But information gained about unstable attributes of the environment (types that change frequently) is not as valuable as information gained about stable attributes of the environment (Stephens and Davis 1986: 81–90). For example, getting information about the exact location of prey animals is not as valuable as the same information about trees with fruits. If a forager goes to a location where an animal prey had been sighted earlier, it is likely that the prey will not be there, but if the forager goes to a location where a tree with fruits was found, the tree will be there.

A fourth model examined by Stephens (1990: 38–42) takes learning into account (the ability of decision-makers to use experience to modify their behavior). The value of such a strategy depends on environmental persistence (i.e., whether types and subtypes are likely to remain the same through time). Stephens suggests that environmental persistence is comprised of two types: within-generation persistence and between-generation persistence. When within-generation persistence is low, information is not of much value. The same is true for the opposite situation. When within-generation and between-generation persistence is high, it is not worthwhile to invest in information gathering. Information is expected to be extremely valuable when between-generation persistence is low and within-generation persistence is high.

Behavioral ecologists treat risk and information separately though they interact. Risk and value problems (how a forager values food or mates, etc.) make assumptions about how well informed the decision-maker is (Stephens 1990: 43). Unfortunately, the interaction of the two variables (risk and information) in optimal foraging models has not been explored. But, there are good reasons to assume that hunter-gatherers have more or less complete information about the outcome of alternative strategies. Foragers often engage in behaviors that provide information that increases long-term gain (Kaplan and Hill 1992; Kelly 1995). The assumption of complete information makes the models mathematically tractable and more easily interpreted (Smith 1991a).

Having discussed foraging strategies within an evolutionary ecology framework, I will turn now to practical applications of the theoretical concepts discussed earlier in explaining human behavioral strategies in coping with environmental fluctuations.

Human Foragers and Risk Management

The acquisition of information about resource quality, quantity, and distribution is an important dimension of mobility. Monitoring of information about the spatial and temporal distribution of a particular resource is a key component in mobility strategies (Binford 1980, 1983; Flannery 1986; Hegmon and Fisher 1991; Kelly 1995; Moore 1981; Smith 1991a). Mobility allows the acquisition of comparative information from diverse locales that helps foragers to make appropriate choices. When resources are scarce and unpredictable, foragers may respond to information on resource scarcity or unpredictability by increasing their foraging range. Larger foraging ranges are exploited in order to ensure adequate returns. For instance, the Hadza have well-developed knowledge of resources across a large area so that their movements frequently take place before a food shortage develops. This knowledge is acquired by frequent movement of individuals and groups of people, and constant change in band size and composition (Woodburn 1968).

When critical resources are scarce and unpredictable in certain seasons due to the fact that local resources are quickly exhausted, increased monitoring of information, as well as higher mobility, are strongly expected. For instance, with greater frequency of moving or visiting friends and relatives, foragers will receive more up-to-date information on the spatial and temporal distribution of resources (Hegmon and Fisher 1991).

Risk, consciously perceived or not by individuals, is a daily reality. Even today, with our advanced technological methods to acquire, store, and disseminate information, which increases our ability to predict (spatially and temporally) short and long-term environmental variations accurately, widespread famines triggered by environmental factors are still common.[2] Other less acute types of food stress (chronic stress) afflict a large percentage of today's population daily.

[2]Although modern widespread famines might be triggered by environmental factors (e.g., flood, drought), they can be explained only by including socioeconomic and historic agencies.

Oscillation in food resources abundance associated with environmental fluctuations (i.e., predictability, frequency, and severity) is a widespread phenomenon with which all societies, past and present, have to cope. As mentioned earlier, the different classifications of hunter-gatherer societies are based on variables that deal with risk and environmental uncertainty. The fact that the different attempts to classify hunter-gatherer societies according to subsistence strategies and degree of social complexity have not been successful, though they provide heuristically and analytically useful distinctions, reflects the variability of the risk-managing strategies that those societies employ.[3]

Through time, human societies have developed a gamut of cultural mechanisms for buffering environmental variability. Two of those strategies are diversification and mobility, which, it has been suggested, are effective mechanisms for mitigating variation in food supply. Oscillation in food resources abundance is one of the main adversities with which human societies have to cope, especially drastic nonaverage fluctuations. Anthropologists have argued that a society's socioeconomic organizational responses to risk follow a ranked order. Minnis (1985), following Slobodkin and Rapoport's (1972) suggestion that selection of strategies to cope with environmental perturbations results in a hierarchical set of responses, proposes a model of hierarchical economic and social organizational responses to food stress by nonstratified societies.[4] The sequence of responses starts with low level, more ephemeral ones (i.e., those that are quickly activated and easily reversed), and if they fail to cope with the problem, a higher level, more permanent response (i.e., less reversible and less quickly activated) is adopted.

The criterion used by Minnis to order the responses is social inclusiveness, and the definition of food stress that he uses (Minnis 1985: 5) is any type of food shortage or perceived food shortage, regardless of severity, that invokes a reaction by the human group. Assuming that mobility is restricted, he argues that responses that involve greater numbers of social groups will become more frequent after less inclusive responses are tried. The smaller the number of social units involved in a single response, the more ephemeral the response is, and conversely, the greater the number of social units, the more permanent the response. The levels of inclusive

[3]Smith and Winterhalder (1981: 3–4) remark that the attempts to dichotomize hunter-gatherer societies "share the assumption that the hunter-gatherer adaptation is a uniform one. The possibility that diverse behavioral forms characterize hunter-gatherer societies and that the range of variation is the correct subject for explanation is generally unrecognized."

[4]Nonstratified society, as defined by Fried (1967), is one which lacks mechanisms that limit individual members' access to basic resources that sustain life.

responses are: (1) household (e.g., adhering resource types to the diet); (2) kin group (e.g., expansion of the kinship system); (3) community (e.g., aggregation of small foraging units into more permanent social groups); and (4) extracommunity (e.g., partnership networks and trading partnerships). Various levels of responses can occur simultaneously. For example, responses can facilitate dispersal within one's territory and exchange of food and goods (Spielmann 1986; Wiessner 1982b, 1986, 1996).

Minnis argues that adopting of more inclusive responses in an increasingly stressful situation increases the probability that a cultural system will survive; if more inclusive responses are not employed with increased stress, we would expect increased probability that the cultural system will dissolve. The prolongation of more inclusive responses involves greater changes in the system. This is due to the problems posed by increasing and/or intensifying the social network and due to changes in risk management.

Halstead and O'Shea (1989) argue that because risk-management behaviors are related to many aspects of society, changes in those mechanisms will trigger a chain reaction in other aspects of the social system, which can lead to social change. Therefore, responses to increased environmental risk that include change in buffering activities together with increased inclusiveness can eventually lead to permanent system change. These changes can be economic (e.g., domestication of plants, increased storage), or sociopolitical (e.g., emergence of hierarchical decision-making structures), or a combination of both. Thus, environmental variability exercises a powerful selective pressure on human behavior, especially during fluctuations that result in severe and unpredictable depletion of resources.

Colson (1979) suggested five risk-buffering mechanisms based on the ethnographic literature and on her own work among the Makah of the Pacific Northwest Coast and the Gwenbe Tonga of Zambia. These mechanisms have been used as a framework for other similar risk-management models (Halstead and O'Shea 1989; Minnis 1985; Miracle 1995; Redding 1988). The five coping devices to buffer risk that Colson (1979: 21) suggested are (1) diversification of subsistence activities, rather than specialization or reliance on a few plants or animal; (2) storage of foodstuffs; (3) storage and transmission of information on "famine" foods; (4) conversion of surplus food into durable valuables which could be stored and traded for food in an emergency; and (5) maintenance of a social network to allow tapping in food resources of other regions.

Diversification, as the inclusion of undesirable or "famine" food types, or the exploitation of broader and more varied areas (Colson 1979; Minnis 1985), is an expansion of diet breadth. Diversification is an effective mechanism for buffering spatial and temporal environmental fluctuations.

Hunter-gatherers are limited in using food storage as a long-term buffering mechanism because storing food for more than a single season is impractical. The ecological conditions under which long-term storage would be an effective buffering mechanism against periodic scarcity are also fairly restricted for foragers (Goland 1991b). Thus, storage is probably the least important strategy used to cope with long-term environmental fluctuations. On the other hand, storage as a short-term, seasonal buffering strategy is effective especially in temperate zones. In the tropics, it is difficult to store food for more than a relatively short period (Minnis 1985: 34). Among the !Kung, for example, food is consumed within 48 hours. Occasionally strips of meat may be cut from large animals for drying into biltong. "In 1 to 3 days, the meat is fully dry and can be kept in this state for 2 months or more. Drying meat attracts attention as well as flies, and most of the biltong is distributed to relatives and neighbors within a few days. What is left is eaten by the household, and all of that is usually gone within a month of the kill" (Lee 1979: 156).

A different kind of storage, information storage, is probably a more ubiquitous mechanism despite the scarcity of studies among foraging societies designed to look at information as a risk-management strategy. One mechanism used by nonliterate societies for preserving survival knowledge is incorporating it in oral tradition. Information which is not regularly reinforced, such as environmental cycles of more than a life span, is encoded in myths, story telling, folktales, songs, and histories and is passed along from generation to generation (Colson 1979; Minc 1986; Sobel and Bettles 2000). For example, the information can be some type of cue for identifying crises at a relatively early stage of development, or for the location and/or existence of alternate resources (Miracle 1995: 24), or it can signal special situations when food taboos can be broken.

Conversion of surplus food into durable valuables that can be traded for food in the future buffers spatial and temporal fluctuations. The surplus can work in two different ways to support specialized craft production of goods that can be used later for trade/exchange, or the surplus itself can be traded/exchanged for goods that again, will be used later for exchange/trade. For this to occur, an exchange network and/or reciprocity system must be in place. Basically, surplus conversion is a way to circumvent food storage limitations other than the existence of a surplus. It converts a currency (i.e., food resources, energy), whose values diminish with time (e.g., food decomposes with time and/or is consumed by other organisms like rodents, fungus, etc.), into a more stable currency (i.e., valuable material goods). During bad years, the latter can be converted back into food.

Colson (1979: 23) attributes the maintenance of social networks as the ultimate insurance against failure to acquire food. Mechanisms such as trading patterns (Wiessner 1982b, 1996), extended kinship, and fictive kinship ties allow one to tap the resources of other regions. This buffering strategy can be very effective when the social network is composed of groups that live in different environments, especially when fluctuations in food resources are not synchronic.

Although Colson does not include food processing in her list of buffering strategies, she does mention food processing in her description of the Tonga's responses to bad years (Colson 1979: 25; also see Lee 1979: 181). Intensification of food processing at the household level is one of the low-level responses, and it buffers temporal and spatial fluctuations. It can be as simple as consumption of marrow from small animal bones that would not be explored in normal years, or more complex as processing roots, nuts and other botanical resources by detoxification, grinding, cooking, etc.

Mobility, a coping mechanism mentioned by Halstead and O'Shea (1989) and Redding (1988), buffers against periodic resource stress. During bad years, groups can move to another place, they can disband, and members can disperse along affinal or consanguineal kin lines to other regions. When resource failure affects the entire range of a foraging society, the solution on a regional scale is temporary movement to areas of greater abundance. Usually, this residential mobility depends on reciprocal access (Smith 1991b). For foraging societies, colonizing a new, previously uninhabited area, there are no major restrictions on mobility,[5] and long-term displacement or migration is an option. But as the population grows, constraints arise, access to neighboring areas will depend on social networks, and long-term mobility or migration will be more costly, especially if the "neighborhood" is characterized by high resource predictability.[6]

Each buffering mechanism addresses particular dimensions of variability (e.g., spatial and temporal scales, predictability, etc.). For example, food storage is effective on more limited temporal scales and in environments

[5]Having uninhabited adjacent areas might facilitate mobility, but group movement, especially during food crises, will be only effective if people know that the place into which they are moving is not under the same severe conditions. Under these conditions, an information network would be highly advantageous. Another important variable to take into consideration in this situation is maintaining the regional social network to ensure reproducing the population. Therefore, groups that colonize unexplored regions are very likely to colonize a whole region relatively fast, in order to insure information about resources from a wide area. This relatively rapid colonication will result in low levels of population density in its initial stages.

[6]The land tenure model suggested by Dyson-Hudson and Smith (1978) predicts that foragers who live in areas characterized by high resource predictability and high resource densities will be territorial, and even if the resource density is low, the model predicts "passive territoriality." Thus, the groups that live in those areas would be willing to pay higher costs to defend their territory from other foragers, especially when population density and competition increase.

where periodic abundance is predictable. Exchange (e.g., mutualistic relationships; Spielmann 1986) requires contiguous ecozones, which regularly provide different types of subsistence resources. Mobility is a strategy that enables populations and communities to adjust to resource availability across regional spatial scales. Diversification functions along both temporal and spatial scales, and that is why it is considered to be the most fundamental response to variability (Colson 1979; Goland 1991a). The pervasive aspect of diversification is also due to the fact that this response can be intensified and/or incorporated into the risk management strategy at a lower level of inclusiveness. For example, new species can be added to the diet at the household level. On the other hand, storage might require a higher level of inclusiveness once it is based on the seasonal abundance of certain resources. To harvest those resources efficiently in a short period of time a community-level response might be necessary.

Some buffering mechanisms are naturally complementary, whereas others are, in practice, incompatible (Halstead and O'Shea 1989). For example, mobility, which is an effective risk buffering mechanism for small foraging groups with expansive territories, may preclude effective use of storage. Storage often co-occurs with greater sedentariness.[7] The ability to store is limited by several factors: (1) the capacity to produce a surplus, which is most common in environments where periodic abundance is predictable; (2) the existence of the technology for preservation; and (3) the degree of mobility (Goland 1991a). On the other hand, mobility and diversification seem to be complementary strategies.

The scale of the response is associated with the severity of environmental variability. Low-level mechanisms are the most efficient and the most reliable but have limited scope. High-level strategies are effective in buffering large-scale (temporal and/or spatial) resource shortages, but they are more costly, less efficient, and unreliable (Halstead and O'Shea 1989: 4). Constraints such as population density, intensity of subsistence exploitation, and the level of technological sophistication together with different buffer-mechanism costs and buffer mechanism effectiveness in relation to the severity of environmental fluctuations results in an hierarchical sequence of responses (Table 7.1). Less costly strategies will be used more often and earlier than more costly ones, thus forming a hierarchy of cultural responses to increasingly severe resource crises.

Intra-band pooling and sharing of resources is a very effective mechanism for reducing risk (e.g., short-term asynchrony in food acquisition) among hunter-gatherers. It is not included in Table 7.1 because it is assumed to be a ubiquitous characteristic among human foraging societies. On the

[7]Storage and mobility are not always incompatible strategies (e.g., Eskimos).

Table 7.1. Different Buffering Strategies, Mechanisms, and Constraints. Strategies are Listed in Hierarchical Order

Strategy	Mechanisms	Buffers	Constraints
Mobility	Individual, family, or group migration. Group fissioning.	Buffers both spatial and temporal environmental variability. Enables populations and communities to adjust to resource availability across regional spatial scales. Permanent migration keeps population density at low levels.	Access to neighboring areas. Information to monitor environmental variability across a large area beyond the normal annual range of a population. Population density. Transport.
Diversification	Diet breadth expansion (addition of low-ranked, undesirable, "famine" foods). Increase range of habitats exploited by foraging new areas or marginal environments.	Buffers both spatial and temporal environmental variability.	Access to new areas in the case of expanding foraging range. Knowledge of resources.
Food extraction that maximize/conserves food value	Food processing (e.g., marrow extraction, grinding, cooking).	Buffers both spatial and temporal environmental variability.	Food processing technology.
Food storage	Stabilization of available food to be consumed in the future. It can be as simple as keeping grains in a pot, salting fish/meat, transforming manioc root into manioc flour, etc.	Buffers short-term temporal fluctuations and is most effective in environments where periodic abundance is predictable and where storage is possible.	Capacity to produce a surplus, technology for preservation, degree of mobility, storable resources. Social interactions. Theft of stored food.
Conversion of surplus food into goods	Exchange, reciprocity, partnerships, craft specialization.	Buffers primarily spatial variability, but also temporal variability through its dependence on stored goods.	Social network. Theft of goods.
Regional social networks	Exchange, affiliation, kinship, partnerships.	Buffers spatial resource variability and also temporal fluctuations.	Requires contiguous ecozones where fluctuations in subsistence resources are not synchronous.

other hand, inter-band food sharing is rare. In many places, asynchrony in the availability of resources increases with increasing distance (Smith and Boyd 1990). Therefore, increasing asynchrony in resource fluctuation as a function of distance implies that risk buffering through inter-band sharing should be greatest if conducted over longer distances. Also, prolonged asynchrony of resource availability may often be greater among members of different bands. The fitness effects of interband sharing will be more pronounced on those long-term fluctuations than the short-term variability managed by intraband sharing (Smith 1991b; Smith and Boyd 1990). But interband sharing where it would be the most advantageous (i.e., long distance) has a high cost of transport, information acquisition, and social-network maintenance. Excluding situations of low-cost travel by watercraft, it implies long-distance movement on foot during stressful situations, a permanent flow of information from distance places so that environmental fluctuations can be monitored, and preservation of long-distance networks. The obvious coping alternatives to interband sharing are (1) local storage and (2) movement of people between local groups rather than movement of goods.

As already mentioned, storage, the first alternative, is usually very limited as a long-term coping strategy because storing food for more than a single season is almost impractical in the tropics, and even relative short-term storage is very costly. Thus, we can expect that under the conditions of the risk reduction model there would be a potential selective advantage for members of different bands to develop systems of reciprocal access to foraging areas.

In sum, the structure of risk-management mechanisms must match the frequency, duration, and severity of stresses. As localized periods of resource stress become longer and more intense, greater amounts of space and larger numbers of people must be incorporated into the response. The problem posed by resource fluctuations may be solved through either temporal or spatial averaging of resource availability.

Risk Management, Information, and Rock Art

As discussed earlier, information significantly influences decision-making by risk managers. Ethnographic studies indicate that hunter-gatherer societies employ a wide range of cultural mechanisms to acquire, process, share, and store information about resource location and availability (e.g., Hegmon and Fisher 1991; Minc 1986; Mithen 1990; Smith 1991a; Sobel and Bettles 2000; Wiessner 1982a,b). For example, among Nambiquara communities of Central Brazil, visiting and feasts are very common activities, during which information about resource availability and abundance in a much larger area is acquired (Setz 1983: 127–128).

Culturally mediated information is usually divided into two general types, long- and short-term. Short-term information concerns social and environmental attributes that change with some frequency, such as current resource abundance and distribution, the location of a particular foraging group, and current composition of exchange networks. Mobility is probably the main strategy employed by foragers to acquire and monitor short-term information. In contrast, long-term information is defined as unchanging during the life of an individual. Long-term information includes knowledge of environmental cycles longer than a human life span, distant topography, distant locations of potential resources, kinship networks, and land tenure rights. Long-term information is acquired, stored, and transmitted by symbolic communication such as myths, stories, folktales, songs, and histories, and is passed along from generation to generation.

Acquisition, maintenance, and transmission of both short- and long-term information are based on cultural mechanisms largely embedded in social organization and occur in a variety of social contexts that correspond to different geographic and social distances (Hegmon and Fisher 1991: 129). Information pooling and sharing among and between groups of people, it has been shown, reduces the high costs of information (Moore 1991). Information pooling can result from interaction among coresident members of a household or camp, frequent visiting and residential moves over a wide geographic area, and occasional aggregation of larger groups (Hegmon and Fisher 1991: 129; Kelly 1995: 181–189).

Short- and long-term information processes are intimately related to the two risk-management strategies discussed before—mobility and regional social networks (see Table 7.1). Both strategies include, to varying degrees, short- and long-distance movement of people and groups and social interactions; mobility is facilitated by interpersonal and intergroup relationships, and social networks are facilitated by movement among individuals and groups. The latter involves interpersonal cohesion among individuals and the varying social units of which they are a part and which are facilitated by mobility. Information gathering and transmission can be achieved and intensified while foraging, moving camp, visiting; as well as by aggregations for ritual and other communal activities; and by kinship and exchange networks that facilitate contact between distant groups. This sociospatial perspective of hunter-gatherer societies can be approached through the concept of social geography.

Social geography is a dynamic phenomenon that involves patterned distributions of humans in their social space and landscape, their knowledge of these patterns, and the potential for this knowledge to have a material form and to constitute social geometries, that is, lattices of individual and social group arrangements within space (Conkey 1984, 1997, also see

Myers 1986 for an excellent ethnographic example). Thus, information concerning social units (e.g., individual members, nuclear family, group, etc.) and the resulting lattices of their interactions, as variables in hunter-gatherer decision-making, are as critical as information about the environment. In addition, the ways information about social geography is acquired, maintained, and transmitted should be similar to those described before. Moreover, hunter-gatherer societies structure the environment in ways to produce and reproduce mechanisms of information flow (e.g., totemic landscape, rock art, residential location and layout).

A material manifestation of information processes that consequently, becomes part of the archaeological record (e.g., occupation sites, art forms, artifacts, clothing, etc.), enables us to understand and explain prehistoric social geographies. Specifically, we can define technoenvironmental and social constraints (see discussion before) and attributes (e.g., artifacts, cave paintings, site locations, etc.) to construct and test explanatory models of past hunter-gatherer societies.

For example, based on the previous discussion of land tenure and risk management, we should expect to find evidence for broad communication networks where reciprocal access between members of land-owning groups is highly developed and where transfer of memberships, which grant property rights, are relatively easily negotiated. On the other hand, in a context of strong controls on local-group membership and restricted reciprocal access (e.g., territoriality or private ownership), we should expect to find evidence for more closed and exclusive networks. Furthermore, we should expect the material evidence of symbolic activities to be continuous or clinal for open networks and relatively discontinuous and discretely clustered for closed or exclusive networks. At the same time, we should find spatial and temporal variations that correspond to variability in risk-management strategies.

Wiessner (1982b, 1983) found that among !Kung members who pool risk (i.e., members who participate in *hxaro* exchange networks), an effort was made to repress individual variation so as to deemphasize household or band identity. For example, !Kung stylistic variation in projectile points is suppressed within groups whose members share a sense of identity and commitment to group membership and who interact in cooperative activities (e.g., hunting and food sharing), in ritual activities, and belong to the same breeding population.

Jochim (1983: 216) has argued that the continuous distribution of many cultural traits, including stone tools and personal ornaments, during the earlier part of the Upper Paleolithic, suggests of broad communication networks. He also suggests that the regionalization of cave art style during the late Solutrean and early Magdalenian in Southwestern France and Northern

Spain indicates the formation of closed communication networks as a result of increased sedentism and territoriality.

Based on ethnographic analogy with the art and ceremony of Central Australia, Gamble (1982, 1991) has argued that the stylistic similarities of the "Venus figurines" of Europe indicate the existence of open social networks in the harsh periglacial environments of Northern Europe. Changes from stylistic uniformity towards increasing regionalism in rock art, have been viewed as evidence of changes from open to closed networks in Australian prehistory (David and Lourandos 1998; Flood 1997; Layton 1991, 1992; Lewis 1988; Morwood, and Hobbs 1995; Rosenfeld 1997; Taçon 1993; Taçon and Brockewell 1995).

Rock art (paintings and engravings on cave and rock shelter walls/ceilings), as a purposive, symbolic way of socializing the landscape, has the immediate potential of signifying and marking place (Conkey 1984). In Australia, the location of these symbolic markers often became centers of ritual activity for individuals, families, clans, and sometimes aggregates of groups that spoke more than one language. Secular and/or sacred activities were practiced at those sites. With time, marked locations became linked to mythologies that described the origin and power of larger landscapes (Taçon 1994).

In the context described above, rock art can have a dual function. It can serve as a landmark to the landscape for asserting group rights and interrelationships, such as macroband affiliation, and as a marker of ritualized locations critical for acquiring, transmitting, and maintaining of information. Although the sociospatial function is an important component, we should keep in mind that rock art "expresses and mediates social relations rather than explicitly territorial affiliation" (Rosenfeld 1997: 296).

In sum, I suggest here that we can derive models based on risk management and evolutionary ecology to understand why and under what circumstances different prehistoric social geographies developed and test these models with the archaeological record.

MODELING SETTLEMENT SYSTEMS FOR CENTRAL BRAZIL[8]

Paleoecological Context of Central Brazil

Anthropologists have generally assumed that tropical environments are spatially and temporally homogeneous (Binford 1980; Jochim 1981; Whitelaw

[8]Central Brazil is a geographical area very closely associated with the spatial distribution of *cerrado* and *caatinga* vegetations (Fig. 7.2).

1991). However, tropical environments are very complex and highly diverse ecosystems that have important local variations in resource availability, seasonality, and interannual fluctuations (Bourlière 1983; Gentry 1990; Prance 1982; Sarmiento 1984; Whitmore 1998).

It has been assumed that climatic changes during the Terminal Pleistocene affected the whole lowland region of South America equally and that environmental changes during this period were not significant in the tropics. However, in the last decade, new paleoclimatic studies from different areas in Brazil, specifically central Brazil, have produced a more detailed picture. Moreover, the dramatic environmental changes associated with deglaciation at the end of the Pleistocene (Wright et al. 1993) have also been recorded in tropical regions (Guilderson et al. 1994).

Although the degree of environmental change in the tropics, and particularly in Brazil, might not have been as pronounced as it was in high latitude or high altitude regions, general paleoclimatic fluctuations in the tropics were synchronous with oscillations elsewhere in the world.

Paleoenvironmental data from Greenland high-resolution ice cores show that Pleistocene climates were extremely dynamic and unstable. Conditions during the warm Bölling/Alleröd period and the cold Younger Dryas were much more variable and unpredictable than earlier, stadial or later, interglacial climates. The onset of the Bölling interstadial occurred around 12,500 radiocarbon years B.P. and was marked by extremely rapid climatic changes (Dahl-Jensen et al. 1998; Stager and Mayewski 1997; Taylor et al. 1997; Thompson 2000). Changes in past continental climate and coastal sea level seem to be associated with changes in the global ocean surface circulation (Mörner 1996), and there is increasing evidence that the high-amplitude climatic swings observed in the Greenland ice cores during the last glacial maximum also occurred in the Southern Hemisphere.

The existence of a Younger Dryas event in South America between 11,000 and 10,000 radiocarbon years B.P. is not clear. Available data from South America show either no change or numerous paleoenvironmental fluctuations during the transition period from late-glacial to postglacial times (Markgraf 1993). Recent glaciochemical analysis of a high-resolution ice core from east Antarctica suggests that rapid climate change events occurred more frequently in Antarctica than previously demonstrated. The evidence indicates that events similar in variability to those seen in the Greenland ice-cores during the Bölling/Alleröd and the cold Younger Dryas periods do exist in Antarctic ice core records (Mayewski et al. 1996). Two other high-resolution ice cores from North-Central Andes of Peru also suggest the occurrence of high-amplitude climatic changes and a Younger Dryas equivalent in South America (Thompson et al. 1995; Thompson 2000).

The high-resolution records from the Southern and Northern Hemispheres have also shown that the Holocene, though relatively more stable than the late Pleistocene, was also subject to rapid change. At least one climatic change between 8200 and 7800 years ago was widely distributed and abrupt in both the tropics and the southern regions (Meese et al. 1994; Stager and Mayewski 1997).

In South America, the climatic instability resulted in increased seasonality. Generalized extreme low amounts of rainfall and climatic instability between 25,000 and 20,000 years B.P. and around 12,500 years B.P. in South America resulted in the extension of savannas[9] into tropical humid forests and in the reduction of savanna woodland in the areas now covered with savannas (Markgraf 1993; van der Hammen 1983; van der Hammen and Hooghiemstra 2000). In central Brazil, dry conditions with increased seasonality and precipitation values lower than evaporation persisted until the beginning of the Holocene (ca. 8000 years B.P.).

Although we lack high-resolution paleoclimatic records in central Brazil from which we can directly measure climatic fluctuations, it is possible to infer that temperature, and more importantly, dry/humid fluctuations had an important impact on the environment. Savanna ecosystems seem to react rapidly to every (even to a minor) change in effective precipitation (caused directly by a change in the amount of annual precipitation, by a change in the average annual temperature, or by increased or decreased seasonality) and to adjust themselves to the new circumstances by incorporating or eliminating woody elements, closing or opening the herb stratum, displacing of herds of herbivores, etc. (van der Hammen 1983: 33).

With increased seasonality and longer dry seasons during the late Pleistocene and early Holocene, arid habitats expanded at the expense of rain-forest areas. The savanna vegetation (*cerrado*) of Central Brazil extended north/northwest into the tropical rain forest and south/southwest into the Atlantic pluvial forest. In the northeast, parts of the savanna were probably replaced by thorn scrub vegetation (*caatinga*) under xeric conditions. Fluctuations in the climate from dry to moist conditions during the last glacial maximum, together with colder temperatures resulted in the expansion of pine forest (*Araucaria* forest) into the savanna in the southeast region during moist periods, which retracted when dry conditions

[9]The extension of savanna vegetation into the Amazon rain forest is currently a matter of much contention (Absy et al. 1991; Colinvaux et al. 1996, 2000; De Oliveira 1996; Hooghiemstra and van der Hammen 1998; van der Hammen and Hooghiemstra 2000). The paleoenvironmental record (e.g., pollen sequence from lake sediments) suggests that although there is evidence of savanna vegetation in the Amazon during the late Pleistocene, it is not as widely dispersed as previously thought.

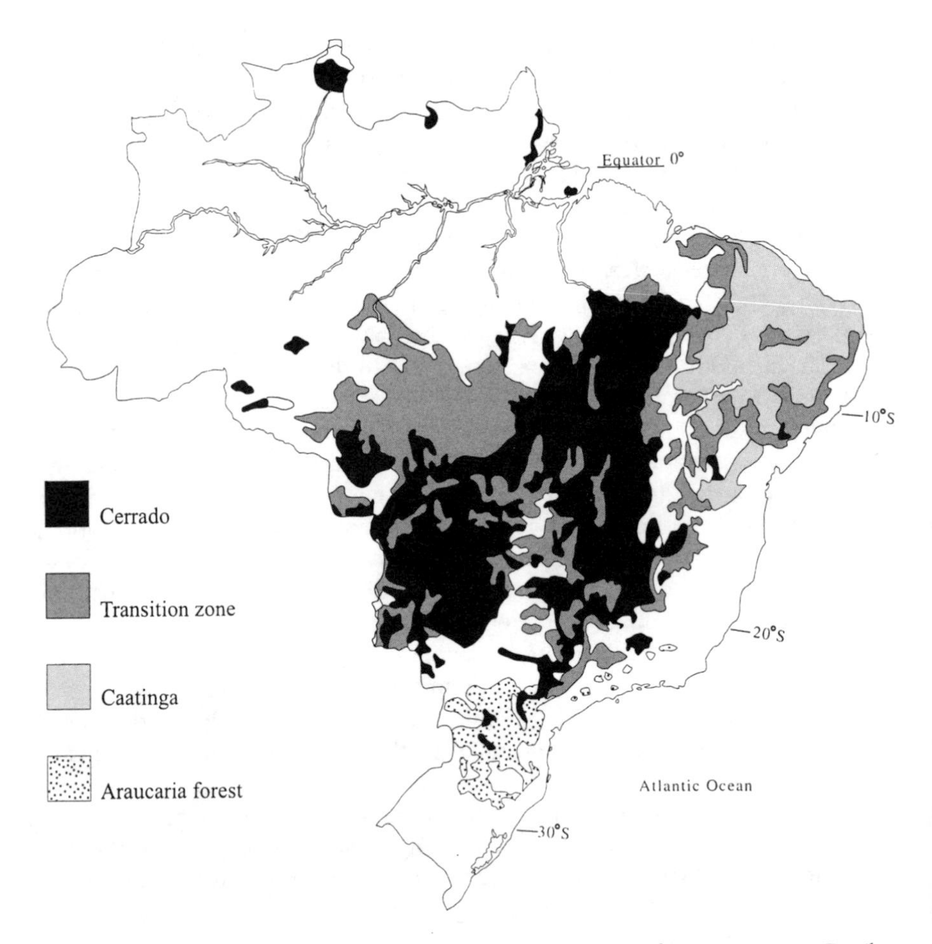

Figure 7.2. Modern distribution of *cerrado*, *caatinga* and *araucaria* forest vegetaion in Brazil.

returned. After ca. 10,000 years B.P., in warm conditions, the pine forest became restricted to the south (Fig. 7.2).

The expansion of the *caatinga*, associated with semiarid conditions, into the northern *cerrado* region very likely resulted in a decrease of large and medium size mammal biomass. The present mammal fauna of the *caatinga* avoids the environmental effects of aridity and climatic unpredictability during harsh periods by using the numerous mesic enclaves scattered throughout the region. Even during periods of elevated precipitation, most mammal species reach their highest density in these relatively restricted areas (Mares et al. 1985). The increased aridity and seasonality of the late Pleistocene and early Holocene reduced in the number and size of these mesic enclaves, and consequently, reduced the mammal population, especially mammals that have relatively large territorial ranges and low density (e.g., deer, peccaries, tapir). Tapir (*Tapirus terresris*) and capybara (*Hydrochaeris hydrochaeris*) are adapted to areas with a permanent supply of water and are rarely found in the region today. Mares and colleagues (1981, 1985) published a list of mammals collected in northeastern Brazil between 1975 and 1978. The list also includes a representative sample of mammals collected throughout the *caatingas* that are housed in the National Museum in Rio de Janeiro and the Museum of Zoology of the University of São Paulo. Tapir, peccary, and capybara are not found on the list.

Mares et al. (1985: 65) hypothesize that due to an anomalous shifts in precipitation between extreme drought and severe flooding, small mammals had a "boom and bust" pattern of population expansion and retraction in the past. The reason for the "boom and bust" pattern is that the *caatinga* mammal fauna is neither endemic nor arid adapted. None of the *caatinga* mammals shows any pronounced physiological adaptations to aridity, nor does the *Kerodon rupestris* (rock cavy), the only *caatinga* endemic. Thus, these animals rely on the availability of mesic vegetation with in the *caatinga* for survival. Populations of small mammals die back during the droughts, and during wet years mammals are abundant (Mares 1997; Mares et al. 1985). Most of the *caatinga* species also occur in the *cerrado* (Mares et al. 1981, 1985). We can reasonably assume then, that the cycles of mammal population expansion and retraction also increased with increased unpredictability and seasonality during the late Pleistocene and early Holocene.

In other *cerrado* regions, savanna expansion to the detriment of rain forests together with the retraction of gallery forests, and palm swamps, has been documented in several regions (Behling 1997; Behling and da Costa 1997; Behling and Lichte 1997; De Oliveira 1996; Ferraz-Vicentini et al. 1996; Ledru 1993; Ledru et al. 1996, 1998; Salgado-Labouriau 1997; Salgado-Labouriau et al. 1998) and may have caused a sharp decrease in the number of species more adapted to humid forests. A decrease in the continuous canopy may also have caused a decline in the primate populations.

A Model of Hunter-Gatherer Responses in Central Brazil

One of the key climatic variables in central Brazil, specifically in the northeast regions, is precipitation. Recent severe droughts occurred in 1942, 1951, 1953, 1970, 1983, 1992, and 1993. The latter two followed dry years since the late 1980s and were extremely intense. This limited recharge of depleted soil moisture levels (Rao et al. 1986, 1995) and severely compromised agriculture in the region. In 1993, mass migration was being considered as the only drought-avoidance strategy available for millions of people living in the area (McGregor and Nieuwolt 1998: 262). Similar climatic occurrences, probably more severe, occurred during the late Pleistocene and early Holocene when the general climate was drier than present, as discussed before.

Based on current paleoecological and archaeological data available for central Brazil, I define six main assumptions for developing a foraging model: (1) during the late Pleistocene and early Holocene period, the region was characterized by increased seasonality and dry conditions; (2) the climate was unstable; (3) the magnitude, frequency, duration, and spatial extent of climatic instability and change affected the availability of critical food resources; (4) the first colonization of the region by humans occurred at the end of the Pleistocene period; (5) when the first foraging societies settled in eastern central Brazil, ca. 12,000 years B.P., they encountered an environment impoverished of large fauna, particularly in the northeast region, and (6) the local megafauna was already extinct.[10]

When the first Terminal Pleistocene hunter-gatherers settled in central Brazil, some of the high-ranking items (e.g., large fauna) were scarce already. Because of the rarity and sometimes total absence of several of the high-ranking resources, and based on optimal foraging model and archaeological data (Kipnis 2002), I showed that those foraging societies had a generalized diet breadth based on small animals and plant resources.

Late Pleistocene/early Holocene hunter-gatherers living in central Brazil had to cope with both climatic instability and local paleoecological variation. Critical variables were interannual rainfall fluctuation, rainfall unpredictability, and long-term environmental change. Economic pursuits must have been affected by the strong seasonality in the region, and it was likely that frequent prolonged and unpredictable drought affected water supplies. Although localized food shortages due to strong seasonality, long dry seasons, and prolonged droughts could have been partially alleviated by expanding diet breadth by adding low-ranked resources, I have shown elsewhere (Kipnis 2002) that this latter alternative was very limited.

In this context, I suggest that a system of reciprocal access to local foraging areas by members of different local bands evolved as a risk-management mechanism in the region. As the human population increased, societies responded to environmental fluctuations by increasing their social network, which would have facilitated movement within and between territories and made access to critical environmental and social information more reliable and efficient.

I do not predict that coping mechanisms such as storage and/or exchange of food as responses were employed. Storage and/or exchange

[10]To date, there has been no clear association between extinct megafauna and human occupation in central Brazil.

of food as responses to stress due to environmental stochasticity depend on the existence of surplus (see Human Foragers and Risk Management section earlier). It is unlikely that societies with a foraging economy in central Brazil could have relied on surplus availability during the late Pleistocene and early Holocene to maintain a food exchange strategy or a conversion of surplus strategy. It is more likely that after the climatic optimum (ca. 5000 years B.P.), when the climate became more stable and domesticated plants were adopted, conditions were more suitable for developing of exchange relationships, probably mutualistic relationships.[11] Storage also depends on technology and conditions for preservation, which make this strategy very costly and inefficient for hunter-gatherer societies that live in the tropics.

THE ARCHAEOLOGICAL RECORD

Archaeology of Central Brazil

Well-documented evidence of human occupation in central Brazil dates back to the Terminal Pleistocene; the earliest radiocarbon dates are at about 11,500 B.P. (Dillehay 2000; Kipnis 1998; Prous and Fogaça 1999; Schmitz 1987). The principal archaeological sites are in the states of Minas Gerais (several sites in the Peruaçu Valley and Lagoa Santa/Serra do Cipó region), Mato Grosso (Abrigo Santa Elina), Piauí (several sites at São Raimundo Nonato), and Goiás (GO-JA-01). The vast majority of the sites are rock shelters and caves. The rich archaeological record of the late Pleistocene/early Holocene hunter-gatherer societies of central Brazil include lithic assemblages, subsistence remains (fauna and flora), perishable wooden and reed artifacts, human burials, and rock art.

Lithic Industries

The lithic assemblage of the first human settlers of central Brazil is known as the *Itaparica* tradition. It was first defined on the basis of the

[11]Agriculture allows better control of the spatial and temporal distribution of resources compared to foraging. The practice of agriculture together with more stable climate gives conditions for producting surplus food. According to Spielmann (1986), intersocietal subsistence exchange can become an efficient strategy for low-density hunter-gatherer societies when two conditions are met: (1) the presence of horticultural societies which supply carbohydrate resources, and (2) availability of small but regularly harvestable supplies of protein or predictable, periodic large concentrations of protein resources. The presence of horticultural societies and small regularly harvestable supplies of protein might have been the case in central Brazil after the climatic optimum.

Figure 7.3. Main archaeological research regions in Central Brazil. (1) Serra da Capivara/Piauí, (2) Peruaçu valley/Minas Gerais, (3) Lagoa Santa-Serra do Cipò region/Minas Gerais, (4) Seridó/Rio Grande do Norte, and (5) Serranópolis-Caiapônica/Goiás.

material from sites in the Serranópolis region of Goiás State (Fig. 7.3). The tradition spans from ca. 11,500 to 8500 B.P. and is found throughout central Brazil. This industry is noteworthy for its variety of large flake tools that grade into one another and into smaller groups of edge-trimmed flakes.

The *Itaparica* tradition is characterized by a unifacial and expedient lithic industry, that had few multifunctional tools, many scrapers, drills, very rare bifacial points, and few curated tools. One of the most typical and most formalized tools is the *limaces* (slugs). These artifacts are usually found associated with cores, flakes, and *débitage*. These are predominantly symmetrical elongated thick unifacial blades with steeply, semiabrupt, retouched edges, made from parallel-sided blades from a prepared core, and usually present a dorsal crest. This core technology is a specialized development of the unidirectional type of core (Barbosa 1992; Barbosa and Velasco 1994; Prous 1992; Prous and Fogaça 1999; Schmitz 1987; Schmitz et al. 1989).

As already mentioned, bifacial points are rare, and usually isolated tanged points appear in the later stage of the *Itaparica* tradition between 9000 and 8500 years ago (Barbosa 1992). In western central Brazil, the *Itaparica* tradition has one phase (*Paranaíba* 10,750 to 9000 B.P.) which is characterized by thick blades and polished and pecked lithic tools used for grinding. The most common lithic tool of the *Paranaíba* phase is the *limace*. Also characteristic of this phase are asymmetrical elongated unifacial tools made of thin narrow blades that are relatively common and diverse in terms of worked edges. Several irregular tools used flakes (for scraping, drilling, and cutting) are found in this industry as well. The great majority of the raw material (98%) is found locally (Barbosa et al. 1994).

The *Serranópolis* phase industry (9000 to 2000 B.P.) that immediately follows the *Paranaíba* in western central Brazil, is completely different. The well-finished unifacial thick blades disappear from the archaeological record. What appears is an "ill"-defined lithic industry with irregular flakes with burins, awls, drills, and small scrapers. The polished and pecked tools (probably anvils and grinding stones) from the previous phase continue in this one. *Débitage* flakes are bigger and thicker (Barbosa et al. 1994; Schmitz et al. 1989).

Work by Wesley Hurt (1960, 1964) at Cerca Grande in the Lagoa Santa region (Fig. 7.3) revealed evidence of human occupation dating back to 9028 ± 120 B.P. The lithic industry associated with the Early Holocene occupation is composed predominantly of quartz crystal flakes. Hurt suggests that it is improbable that all of the abundant flakes and fragments found at several sites represent the spoils and rejects from tool manufacture, but seem to be tools themselves (but see Prous 1992/93: 383 for a different interpretation). Only a few artifacts have deliberately fabricated forms or retouched edges. Formalized artifacts made of quartz or quartz crystal include a few barbed arrow points, square-stemmed arrow points, end scrapers, semilunar scrapers or "spokeshaves," triangular scrapers, and ovoid scrapers. A few scrapers and knives were also made of jasper. Another common artifact is the pitted hammerstone (*quebra-cocos*[12]).

[12]*Quebra-cocos* (pitted hammerstone) usually are simple blocks that have one or more flat or concave surfaces with small circular depressions or more often circular smooth areas with a dark stain. As its Portuguese name indicates, they were used for cracking palm nuts.

Grande Abrigo de Santana do Riacho is a rock shelter located in the Serra do Cipó (Lagoa Santa region) where evidence of human occupation dates back to 11,960 ± 250 B.P. and is continuous up to ca. 3000 B.P. (Prous 1991a). The main lithic industry associated with the early occupations at Santana do Riacho is characterized by an expedient flaked lithic assemblage mainly of quartz, which is locally available within a 1- to 2-km radius from the site (Prous et al. 1991). Other raw materials far less used at Santana do Riacho are quartzite, flint, chalcedony, and jasper. Quartzite is found at the site and surroundings, but flint, jasper, and chalcedony are not. The closest known source of flint is 40 km from Santana do Riacho. Jasper and chalcedony are also exotic to the area and are not found less than 60 km from Serra do Cipó.

Formalized artifacts include side scrapers, end scrapers, nosed scrapers, borers, and *limaces*. A few fragments of bifacial artifacts (10 quartz pieces), probably projectile points, were also found. Several used flakes were also recovered from the excavation. Quartz tools, cores, and *débitage* were present in relatively great quantities (Prous 1991c). The flint assemblage, rare in the most recent occupations, increases in quantity as one goes back in time. Quartzite artifacts are all from older occupations (Prous et al. 1991: 190).

Pecked and polished stone artifacts were also recovered from the excavations at Santana do Riacho. The mortar fragments were found distributed throughout all periods. Complete and semifinished, polished and pecked axes were found at all archaeological levels (Prous 1991d: 212). Local igneous basic rocks, as well as exotic material like sillimanite, hematite, and silexite were used to manufacture the polished artifacts. The closest source of hematite is 30 km from Serra do Cipó. Sillimanite occurs at Serra do Cipó as very small pieces. Sizable pieces can be found 15 km from Santana do Riacho.

Use-wear analysis of a sample of the lithic collection showed conclusive evidence of woodworking on 4 of 69 analyzed quartz scrapers, and two other scrapers showed evidence of hide or pigment processing. Twenty-five other scrapers also had evidence of use, but the use-wear did not show a distinguishable pattern. The other 35 analyzed tools did not have clear indications of use (Lima 1991: 277).

Lapa do Boquete in the Peruaçu region (Fig. 7.3) contains a rich sequence of remains from the Terminal Pleistocene (12,000 ± 300) to the nineteenth century. At Lapa do Boquete, the lithic material is very abundant, mainly flint of different colors and knapping quality. Some chalcedony, jasper, and quartz *débitage* was also recovered. A few *limaces* and *débitage* associated with their production were found in levels dating between 10,000 and 12,000 B.P. (Prous 1991b; Prous et al. 1992; Prous and Fogaça 1999).

At Lapa do Boquete, the *limaces* are made of metamorphic sandstone. Microwear analysis indicates that *limaces*, thick scrapers, and unmodified flakes of various size and shape were used as woodworking tools (Alonso 1994; Fogaça and Lima 1991; Prous et al. 1992). Out of 62 retouched pieces analyzed from different levels, 28 showed microwear, and 9 pieces were used for scraping wood, 1 piece for cutting wood, and the other 18 pieces did not show a clear functional microwear pattern. Five nonretouched flakes found at the surface of the site were analyzed, and one had positive microwear woodcutting pattern (Alonso 1994).

The lithic products associated with levels dated between 6000 and 9000 B.P. are composed of very small flakes, on average 2 cm, and a maximum length of 5.5 cm. Rare small blades with prepared striking platforms are present. Cores and tools suggested by the *débitage* are very rare. Note here, that the industry from older occupations is better known than those from younger levels (Prous 1991b).

Lapa Pequena is a limestone cave excavated by Alan Bryan and Ruth Gruhn (1978). The site is 200 km southeast of the Peruaçu region (Fig. 7.3). In Unit B, which spans from 8240 ± 160 to 7030 ± 100 B.P., used small flakes (length mode = 25 mm), cores, *quebra-cocos*, and fragments of bone points were recovered. According to the authors, the edges of the flakes and cores were designed or used for scraping not for cutting. At Lapa Pequena, there was no evidence of bifacial flaking. Only eight tools were classified as formalized artifacts. Three core scrapers were classified as end scrapers that had steep or stepped flaking around an excurvate to nearly straight edge. Two other pieces and four *becs* (nosed scraper) had invasive flaking. The principal raw material used to produce lithic tools during that period was chert (79%, including jasper and chalcedony); crystalline quartz (19%) and limestone (2%) comprised the rest. All of the raw material is readily available in the region (Bryan and Gruhn 1978).

More recently, Bryan and Gruhn (1993) excavated a series of sites in a karstic area in the semiarid outback region (*caatinga* environment) of Bahia state, approximately 500 km north of the Peruaçu region. Only Abrigo do Pilão, a limestone rock shelter, yielded a radiocarbon date in the early Holocene (9610 ± 90 B.P.) in association with human occupation. The radiocarbon dates suggest that intensive occupation of the cave lasted less than a millennium before the site was abandoned (8790 ± 80 B.P.). It was not reoccupied until the ceramic horizon (860 ± 60 B.P.). The earliest lithic assemblage at Abrigo do Pilão features large jasper or quartzite flakes with edge trimming. Bryan and Gruhn associate the large flake tools from Abrigo to Pilão with the *Itaparica* tradition. *Limaces*, the marker for the *Itaparica* tradition, are absent from the Abrigo do Pilão assemblage, and the lithic industry is characterized by very few formally shaped artifacts. Pebbles, flakes, or fragments of quartz, quartzite, chert, or limestone were modified by use or minimal unifacial retouching of working edges or surfaces. The traditional lithic tools in the central region were always simple, essentially formless according to the authors. The lithic artifacts were used mainly to work vegetal materials, to shape wooden tools, weapons, and utensils; and to process plant foods.

Rock Art

Very rich and widespread prehistoric rock art (paintings and engravings) is found throughout central Brazil (Fig. 7.4). Eastern and northeastern

Figure 7.4. Main archaeological sites and regions presenting rock art (paintings and/or engravings).

central Brazil, in particular, have a high density of such sites, very often associated with karstic areas and perennial rivers. Although documentation and analysis is still very incomplete and absolute chronology poorly known, recent studies and syntheses, together with a relative chronology based on superimposed paintings and engravings, make possible a regional approach for some specific areas (Albano 1979/80; Guidon 1985, 1992, 1995; Martin 1995, 1996; Pessis 1995, 1999; Prous 1989, 1992, 1994, 1995, 1999; Schmitz 1997; Schmitz et al. 1986, 1997).

The occurrences of rock art in central Brazil have been divided into several distinct and relatively discrete broad stylistic regions associated with the *Itacoatiara*, *Geométrica*, *Nordeste*, *Agreste*, *São Francisco*, and *Planalto* traditions (Fig. 7.4). The *Nordeste* and *Agreste* are two of the most important and better studied stylistic units in northeast central Brazil, and I will focus my discussion on these two traditions.

The main characteristics of the early paintings in central Brazil, known as the *Nordeste* and *Agreste* traditions, are the narrative representations of collective hunting, ceremonial activities, dancing, intercourse, and individual and collective acts of violence (Figs. 7.5–7.7). The thematic drawings consist of anthropomorphic, zoomorphic, phytomorphic, and rare geometric figures. The core area of these two traditions is the semiarid outback region (*caatinga* environment) of Pernambuco, Piauí, and Rio Grande do Norte states. Paintings associated with the *Nordeste* tradition have also been found in northern Minas Gerais, southern Goiás, Bahia, and Mato Grosso

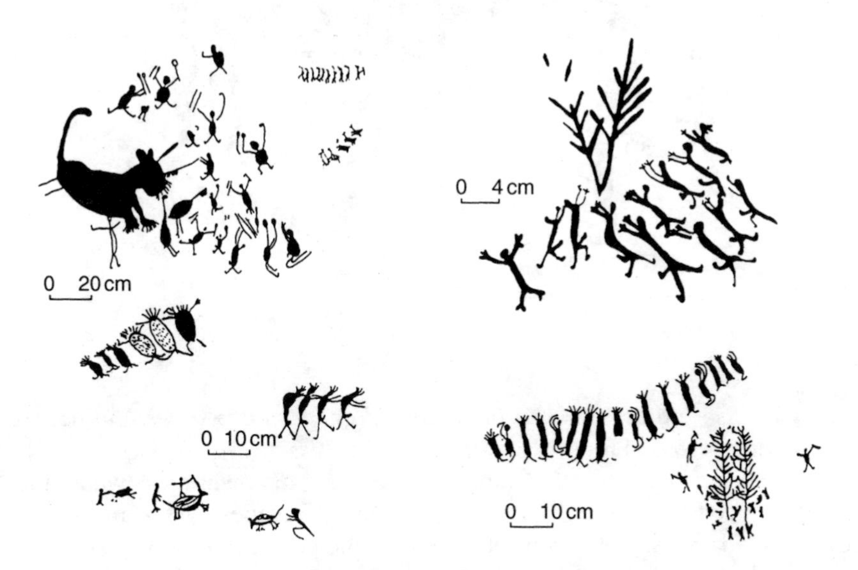

Figure 7.5. Northeast Tradition (adapted from Guidon 1995).

Figure 7.6. Northeast Tradition (adapted from Guidon 1995).

Figure 7.7. Agreste Tradition (adapted from Guidon 1995).

states in association with *caatinga*, *cerrado* and *caatinga/cerrado* transition (Guidon 1992; Prous 1992, 1994; Schmitz 1987).

Southeast Piauí and Rio Grande do Norte are the main areas where the *Nordeste* tradition is found. Most of the paintings in these areas are monochrome (red) but sometimes biochrome (red and gray). Anthropomorphic figures predominate (26–60%), followed by representations of cervids, rhea, and armadillo. Other animals may also be depicted. Zoomorphs account for 24–50% of all figures at in a given site. Non figurative geometric drawings (often stick lines) are usually less than 15%, and sometimes completely absent. Anthropomorphs are often depicted holding clubs, spears, or stick throwers.

Within this broad figurative tradition, considerable regional differences exist. The *Nordeste* tradition is divided into three subtraditions. The *Várzea Grande* and *Salitre* subtraditions are found mainly in southeast Piauí, and the *Seridó* subtradition is present in Rio Grande do Norte state (Guidon 1992, 1995; Martin 1995, 1996; Pessis 1995). The *Várzea Grande*, the most studied and better known of the three subtraditions, is comprised of three main stylistic units: *Serra da Capivara* style, *Serra Talhada* stylistic complex, and *Serra Branca* style. These stylistic units, arranged in chronological order, are not a sequence of clear distinct and discrete stylistic units, but rather reflect a slow and continuous development (Guidon 1995: 122). The different stylistic units are found in neighboring sites, or often in the same site side by side and/or superimposed.

The *Serra da Capivara* style is characterized by figures primarily painted in red, with closed contours that are often filled with a solid color. Anthropomorphic figures are small, and often smaller than the zoomorphic depictions. The zoomorphs are usually placed in a very visible location, and they dominate the composition. The main themes are dancing, sex acts, ceremonial rites, and individual hunting of small animals. Graphic representations concern groups rather than individuals (Pessis 1995: 118).

Following the *Serra da Capivara* is the *Serra Talhada* stylistic complex, characterized by an increase in the diversity of representations, the miniaturization of figures, the use of several colors (red, yellow, white, gray, and brown), and the frequent presence of bichrome and three-color figures. Rows of sticks also appear. The usual *Nordeste* themes are represented, but shift from individual to collective hunting. In addition, acts of sex become ritually collective, and new themes are incorporated, especially collective and individual acts of violence (Pessis 1995: 118). The *Serra Talhada* stylistic complex is widespread throughout Piauí, Pernambuco, and Rio Grande do Norte states. Prous (1994) has identified evidence of this stylistic complex in northern Minas Gerais (Fig. 7.8) and has suggested that some of the designs (Fig. 7.9) described by Schmitz et al. (1986) for southern Goiás and by Vialou and Vialou (1986) for Mato Grosso belong to the *Serra Talhada* stylistic complex.

The *Serra Branca* style is composed of anthropomorphic figures with a very particular body shape, that is adorned with vertical lines or geometric designs. The zoomorphic depictions are often not closed, are filled with geometric designs similar to those adorning the anthropomorphic figures, or are filled with a solid color. The representation of material objects increases.

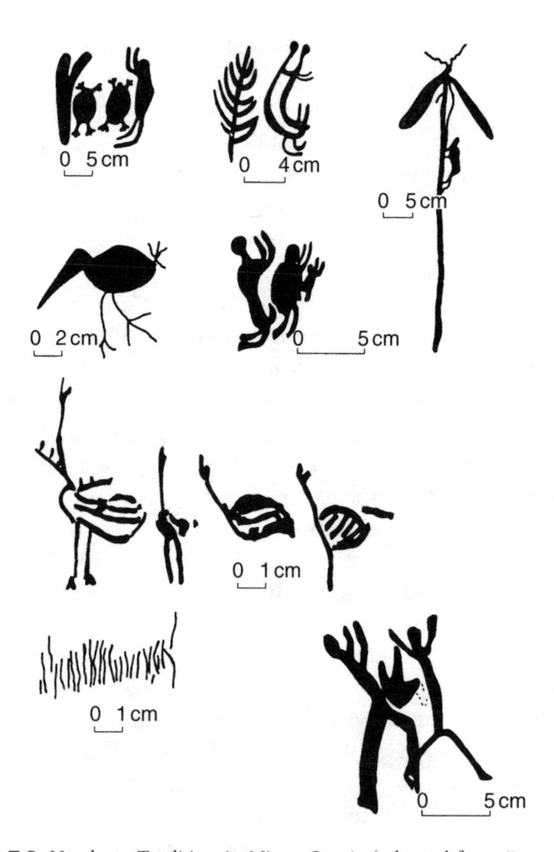

Figure 7.8. Nordeste Tradition in Minas Gerais (adapted from Prous 1995).

Although the number of themes decreases, there is an increase in the formality and complexity of representations. The decreasing depiction of sex acts is replaced by the representation of pregnancy. The theme of collective and individual acts of violence develops with much complexity. Individual figures are more numerous than collective arrangements (Pessis 1995: 118).

One of the main characteristics of the *Seridó* subtradition (Fig. 7.10) is the dearth of zoomorphs and the almost complete absence of nonfigurative geometric designs (Prous 1994; Martin 1995). This subtradition is also characterized by ubiquitous drawings that have been interpreted as pirogues with boatmen and oars (Martin 1996). Most of the representations are anthropomorphs (more than 80%) and objects (bows and axes) that they carry. The representations are carefully delineated by fine lines, mainly red, but also white, yellow, or black. Though cervids were predominant zoomorphs in the *Várzea Grande* subtradition, the rhea and other birds

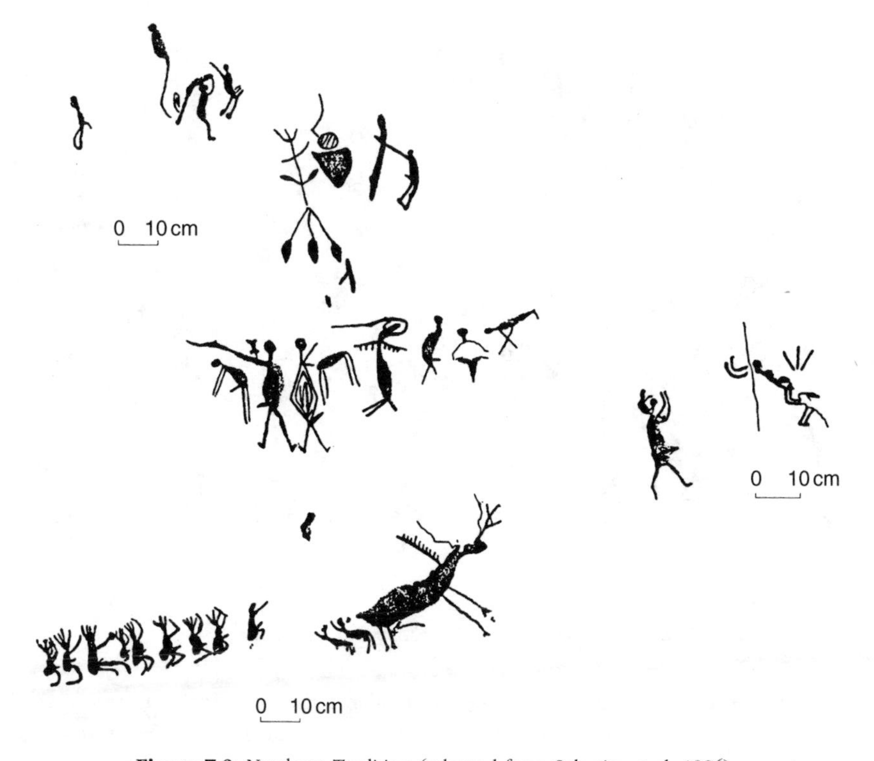

Figure 7.9. Nordeste Tradition (adapted from Schmitz et al. 1986).

(e.g., toucan and macaw) are the most frequent zoomorphs depicted in the *Seridó* subtradition. The theme of aggression appears frequently and includes a significant number of compositions in which only two figures are involved (Martin 1995: 130).

The spatial distribution of the *Agreste* tradition is similar to that of the *Nordeste* tradition, but it also occurs in areas where there is no evidence of *Nordeste* tradition such as in the northern region of Piauí and the southeast area of Pernambuco. The *Agreste* in northern Piauí is characterized by anthropomorphs, few zoomorphs, and the absence of phytomorphs and object depictions. The paintings are large figurative drawings, particularly the anthropomorphs. The rare thematic representations, when they do appear, are always of hunting activities. The presence of nonfigurative designs is abundant, and there is a clear morphological diversity (Guidon 1992, 1995; Prous 1994). In the Serra da Capivara region, the *Agreste* tradition is divided into three subtraditions: *Serra do Tapuio*, *Extrema*, and *Gerais*. Unfortunately, descriptions of these subtraditions have not been published so far.

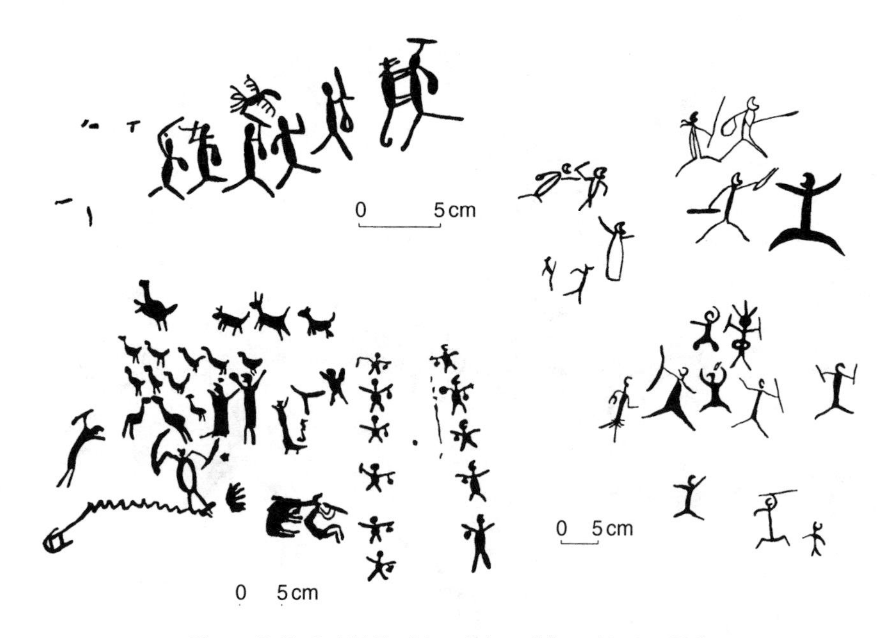

Figure 7.10. Seridó Tradition (adapted from Martin 1995).

One of the best known archaeological regions in central Brazil is the Serra da Capivara, Piauí State. In this region, the largest concentration of sites associated with the *Nordeste* tradition is found. The strongest evidence of dated rock art for this region comes from the terminal Pleistocene and early Holocene. Large fragments of painted wall broken by erosion were found in association with occupational floors dating to 11,000 years ago. Radiocarbon dating of deposits immediately overlying paintings at one of the sites (*Toca do Baixão do Perna I*), also yielded a date of 10,500 B.P. (Guidon 1992; Lage 1999; Pessis 1999). These early dates are associated with the *Serra da Capivara* style of the *Nordeste* tradition and the *Agreste* tradition. The *Serra Branca* style appears around 7000 B.P. (Martin 1995). The evidence of the *Nordeste* tradition in southeast Piauí is continuous up to 6000 B.P. Martin (1995) produced a date of 9410±100 B.P. for the *Seridó* subtradition at the Mirador site in the Seridó region. In the Serra da Capivara region, the *Agreste* tradition is peripheral between 10,500 and 6000 B.P. At the end of the *Nordeste* tradition around 6000 B.P., the *Agreste* tradition becomes dominant in the region until between 4000 and 3000 B.P. when it disappears from the region (Guidon 1992).

In northern Minas Gerais, the *Nordeste* tradition (Fig. 7.8) is characterized by gesticulating anthropomorphs that show ritual and/or sexual intercourse. The miniaturized, monochromatic drawings are primarily done in black, with rare occurrences of white, yellow, or red paintings. Rare zoomorphs (rhea and other small animals) and numerous stick lines appear in horizontal sequence. The general characteristics of the paintings are similar to the *Serra Talhada* stylistic complex. In this region, the *Nordeste* tradition drawings are often placed at the periphery of the panels or of the shelter walls in a very discrete position. They are found in sites where no other tradition occurs and in sites associated with other traditions. In the latter case, the paintings attributed to the *Nordeste* tradition are sometimes superimposed on the *São Francisco* tradition (Lima et al. 1989; Prous 1994, 1999). Based on this context, Prous suggests a much more recent date for the *Nordeste* tradition in northern Minas Gerais than its occurrence in northeast central Brazil.

Several rock shelters in southern Goiás have paintings associated with the *Serra Talhada* stylistic complex (Figure 9). The themes are similar to those found in the Peruaçu Valley. The depiction of rows of small animals is a common characteristic of the paintings. Anthropomorphs are less frequent than in other regions where this stylistic complex is present (Prous 1994).

DISCUSSION

The lithic assemblage associated with Terminal Pleistocene and early Holocene dates in central Brazil is characterized by the lack of intentionally shaped formal tools with rare bifacial points; the more common *limaces* are the most formalized lithic artifacts manufactured during this period. The more informal tools, based on flakes struck from unstandardized cores, were probably the basis for manufacturing a much more elaborated material culture made from wood and bone. The abundance of lithic raw material found locally at most sites and easy access to hard, durable organic material that can be converted into lightweight portable tools eliminate the need for a standardized core technology among mobile foragers (Parry and Kelly 1987), and this seems to be the case for central Brazilian prehistoric foragers.

At the same time, the presence of exotic raw materials from quarries as far as 60 km from where they were discarded (e.g., Santana do Riacho) suggests long-distance ties. So far we do not have enough information to discern the processes that explain the flow of exotic materials (i.e., residential or logistical movements or trade), but it does indicate linkages (direct or indirect) between distant locations. This point is important because independently of the specific ways in which the materials were being moved around, the flow of exotic raw materials was based on social relationships. In other words, if access to exotic materials was through trade, it implies the presence of trade networks, and if the material was acquired directly by movement of people, it implies access to neighboring regions.

Despite the difficulties in reconstructing mobility strategies from prehistoric technology (Kelly 1992), the lithic assemblages from central Brazil suggest a high degree of mobility during the Terminal Pleistocene and early Holocene. They also indicate the presence of social networks. Foraging activity in itself in *cerrado* seems to already imply a higher degree of mobility when compared to foraging in more forested regions. Setz (1983, 1991) has compared the subsistence activity of two Nambiquara groups, one located in a forested environment and the other in *cerrado*. Hunting seems to contribute equally (i.e., same percentage) to both communities, with differences in fishing, plant collecting, and agriculture. But the hunting territory associated with the *cerrado* group is four times greater than the one used by the society living in the forest. Probably, the strongest evidence for the development and maintenance of social networks underlying systems of reciprocal access is the rich and widespread prehistoric rock art paintings and engravings found throughout central Brazil.

The geographically widespread distribution of the earliest paintings, dating to the late Pleistocene/early Holocene period (*Nordeste* and *Agreste* traditions), suggests the development of a system that served to align a low-population-density social network during unstable climatic conditions in a predominantly semiarid environment. The degree of variability within each major stylistic unit (*Várzea Grande*, *Salitre*, *Seridó*, *Serra do Tapuio*, *Extrema*, and *Gerais* subtraditions) might indicate the development of preferential interlocking networks of personal interactions, but paintings and/or engravings of each localized area share essential characteristics.

The regionally dispersed and relatively homogenous rock art of the *Serra da Capivara* style, with distinct local variations, but with narratives that emphasize collective actions, suggests a cultural mechanism that facilitates interband mobility and aggregation of groups. The long-distance interactions would have maintained a degree of uniformity within this stylistic unit, would have reinforced the sociocultural networks, and would have facilitated the flow of information. The better watered canyons and uplands where rock art is found are logical locations for the aggregation of dispersed hunter-gatherer groups during the dry season, and the frequent drawings of ritual ceremonies seem to be depicting these interactions.

The significance of the lack of marked cultural discontinuities in a relatively vast area, which would suggest large regional systems, and movement of people over long distances, may not be fully apparent to Western observers. What seems to us today as long distances might not be perceived as constraints by non-Western societies. For example, contact between distant communities as far apart as 500 km, structured through ceremonial gatherings, has been documented for the Pintupi people in Australia (Myers 1986: 45). Long-distance movement is critical for developing of ritual and ceremonial congregations at "sacred" locations that facilitate interband interaction, exchange of information, renewal of kinship/partnership ties, and marriage arrangements.

However, interaction between individuals and/or between groups was not static. There is strong evidence from the *Serra Talhada* stylistic complex that social interactions were not always peaceful. The appearance of collective and individual acts of violence depicted in the *Serra Talhada* complex could be associated with punitive actions against individuals and/or groups who failed to reciprocate, or it could be associated with periods of widespread drought that triggered risk-seeking behavior. This does not indicate the breakdown of networks, but rather a change in social relationships.

One could argue that the increase in the formality and complexity of the graphic display of violence, together with a shift toward representation of individuals to the detriment of collective figures in the *Serra Branca* style, indicates a change in land tenure in the direction of more restricted networks. But there is no discrete and distinct regionalization of representations to suggest close networks, or the development of territoriality, or private ownership. The development of such systems is predicted if increases in return rates due to climatic amelioration were sufficient to justify the costs of defending one's territory and constricting access by other groups (Dyson-Hudson and Smith 1978; Smith 1991b).

The more important aspect of this study is that changes in social geography might reflect the constant negotiation of social relationships of risk managers. The co-occurrence of different stylistic units at the same site might indicate changing sociocultural relationships. It is very likely that the shared characteristics present in each major area (e.g., *Planalto*, *Nordeste*, *São Francisco* traditions) originated early in each region. Through time, particular representations of social interaction were produced and reproduced within each interlocking network, different interactions are associated with different sets of shared symbols.

Some of the sites where ceremonial gatherings occurred in the past were probably important locations for long periods. Lee (1989, 1993) reports that bands from different regions of the !Kung territory traditionally gathered at /Xai/ Xai in the dry season to trade, dance, and arrange marriages. /Xai/ Xai had been a meeting place far more than a century and probably much longer. The constant use of these preferred locations through time accumulates the representations of ongoing negotiation of group membership and social interactions. Thus, the co-occurrences and changes of stylistic compositions we see in the rock art of central Brazil reflect changing social geographies.

In this perspective, the superimposition or not of paintings and/or engravings is itself meaningful, and we should not be surprised that superimposition at different sites might show different relative chronologies. The fact that the *Serra Talhada* stylistic complex is present in northern Minas Gerais after it had ended in the core area of southeast Piauí, together with its peripheral location within the panels, might be related to social relationships within a large interlocking network, rather than actual replacement of one population by another, as is the more common interpretation (Prous 1995).

In short, I suggest that a system of reciprocal access to local foraging areas by members of different local bands developed in central Brazil during the terminal Pleistocene/early Holocene period. Furthermore, the land tenure system was structured by a visual system with two main roles: one as landmarks to the landscape for asserting group rights and relationships such as macroband affiliation and the other as a marker of ritualized locations critical for acquiring, transmitting, and maintaining of information.

CONCLUSION

Eric Smith (1995: 224) has suggested that by paying greater attention to the impact of risk and uncertainty on individual choice, one is compelled to pay explicit attention to social interaction and structural constraint. I have suggested earlier that we can use such an approach to generate more inclusive ecological models and formalize expectations about prehistoric social geographies that can be tested with the archaeological record.

Based on risk-management and evolutionary ecology theory and on paleoenvironmental reconstruction, I suggested that a system of reciprocal access to local foraging areas by members of different local hunter-gatherer groups evolved as a risk-management mechanism in central Brazil. As human population increased and consequently, the region was, colonized societies responded to environmental fluctuations by increasing their social network, which would have facilitated movement within and between territories and made access to critical environmental and social information more reliable and efficient.

The archaeological record from central Brazil, specifically the rich paintings and engravings found in thousands of rock shelters and caves, suggest that late Pleistocene/early Holocene hunter-gatherers in this region invested in developing and maintaining large and extended networks of interaction and integration. Furthermore, it indicates that the composition, in time and space, of this social geography was constantly being negotiated.

Although we lack more detailed regional archaeological and paleoenvironmental data to better substantiate the particular case of central Brazil, I think that the interpretation of the archaeological record proposed here is open to greater logical and empirical examination than the more traditional interpretation. Also, I hope that the theoretical approach described in this paper will be an important contribution to guide future research on the settlement patterns of hunter-gatherer societies in general.

ACKNOWLEDGMENTS

I am grateful to Junko Habu and Ben Fitzhugh for inviting me to participate in this volume and for commenting on drafts of the manuscript. Elizabeth Sobel and an anonymous reviewer made very useful comments that significantly improved this chapter. This research was possible thanks to a graduate grant received from National Research Council of Brazil (CNPq—Proc. 203022/89.0), research grants awarded by National Science Foundation (Grant 9510523), Wenner-Gren Foundation For Anthropological Research (Grant 5833), Karstic Research Fellowship, and a post-doctoral fellowship from The State of São Paulo Research Foundation (FAPESP—Proc. 01/06881-1).

REFERENCES

Absy, M. L., et al., 1991, Mise en évidence de Quatre Phases d'Ouverture de la Forêt Dense dans le Sud-Est de l'Amazonie au Cours des 60,000 Dernières Années. Première Comparaison avec d'Autres Régions Tropicales. *Comptes Rendus de l'Academie des Sciences, Serie 2* 312:673–678.

Albano, R., 1979/80, Bibliografia Sobre Arte Rupestre. *Arquivos do Museu de História Natural da Universidade Federal deMinas Gerais* 4/5:185–189.

Alonso, M., 1994, *Estudo Traceológico de Indústria Líticas de Alguns Sítios do Vale do Rio Peruaçu*. Unpublished manuscript.

Barbosa, A. S., 1992, A Tradição Itaparica: Uma Compreensão Ecológica e Cultural do Povoamento Inicial do Planalto Central brasileiro. In: *Prehistoria Sudamericana: Nuevas Perspectivas* edited by B. J. Meggers and C. Evans, pp. 145–160. Taraxacum, Washington, DC.

Barbosa, A. S., and Velasco, N. I., 1994, Processos Culturais Associados à Vegetação de Cerrado. In *Cerrado: Caracterização, Ocupação e Perspectives* edited by Pinto, M. N. pp. 155–170. Editora Universidade de Brasília, Brasília.

Behling, H., 1997, Late Quaternary Vegetation, Climate and Fire History from the Tropical Mountain Region of Morro de Itapeva, SE Brazil *Palaeogeography, Palaeoclimatoloty, Palaeoecology* 129(3–4):407–422.

Behling, H., and da Costa, M. L., 1997, Studies on Holocene Tropical Vegetation, Mangrove and Coast Environments in the State of Maranhão, NE Brazil. *Quaternary of South America and Antarctic Peninsula* 10:93–118.

Behling, H., and Lichte, M., 1997, Evidence of Dry and Cold Climatic Conditions at Glacial Times in Tropical Southeastern Brazil. *Quaternary Research* 48:348–358.

Bettinger, R. L., 1991, *Hunter-Gatherers: Archaeological and Evolutionary Theory*. Plenum, New York.

Binford, L. R., 1980, Willow Smoke and Dog's Tails: Hunter-Gatherer Settlement Systems and Archaeological Site Formation. *American Antiquity* 45(1): 4–20.

Binford, L. R., 1983, *In Pursuit of the Past*. Thames and Hudson, London.

Bicchieri, M. G. (ed.), 1972, *Hunters and Gatherers Today*. Waveland Press, Prospect Heights, Ill.

Bourlière, F., 1983, The Savanna Mammals: Introduction. In *Tropical Savannas*, edited by F. Bourlière, pp. 359–361. Elsevier, Amsterdam.

Bryan, A. L., and Gruhn, R., 1978, Results of Test Excavation at Lapa Pequena, MG, Brazil. *Arquivos do Museu de História Natural da Universidade Federal de Minas Gerais* 3:261–326.

Bryan, A. L., 1993, *Archaeological Research at Six Cave or Rock shelter Sites in Interior Bahia, Brazil*. Center for the Study of the First Americans, Oregon State University, Corvallis, OR.

Cashdan, E., 1985, Coping with Risk: Reciprocity Among the Basarwa of Northern Botswana. *Man, New Series* 20(3):454–474.

Cashdan, E., 1990, Introduction. In *Risk and Uncertainty in Tribal and Peasant Economies*, edited by E. Cashdan, pp. 1–16. Westview Press, Boulder.

Clark, C. W., 1990, Uncertainty in Economics. In *Risk and Uncertainty in Tribal and Peasant Economies*, edited by E. Cashdan, pp. 49–63. Westview Press, Boulder.

Clastres, P., 1989, *Society Against the State*. Zone Books, New York.

Colinvaux, P. A., De Oliveira, P. E., Moreno, J. E., Miller, M. C., and Bush, M. B., 1996, A Long Pollen Record from Lowland Amazonia: Forest and Cooling in Glacial Times. *Science* 274:85–88.

Colinvaux, P. A., De Oliveira, P. E., and Bush, M. B., 2000, Amazonian and Neotropical Plant Communities on Gacial Time-Scales: The Failure of the Aridity and Refuge Hypotheses. *Quaternary Science Reviews* 19:141–169.

Colson, E., 1979, In Good Years and in Bad: Food Strategies of Self-Reliant Societies. *Journal of Anthropological Research* 35(1):18–29.

Conkey, M. W., 1984, To Find Ourselves: Art and Social Geography of Prehistoric Hunter Gatherers. In *Past and Present in Hunter-Gatherer Studies*, edited by C. Schire, pp. 253–276. Academic Press, New York.

Conkey, M. W., 1997, Beyond Art and Between the Caves: Thinking About Context in the Interpretive Process. In *Beyond Art: Pleistocene Image and Symbols*, edited by M. W. Conkey et al., pp. 343–367. California Academy of Sciences, San Francisco.

Dahl-Jensen, D. K., et al., 1998, Past Temperatures Directly from the Greenland Ice Sheet, *Science* 282(5387):268–271.

David, B., and Lourandos, H., 1998, Rock Art and Socio-Demography in Northeastern Australian Prehistory. *World Archaeology* 30(2):193–219.

De Oliveira, P. E., 1996, Glacial Cooling and Forest Disequilibrium in Western Amazonia. *Academia Brasileira de Ciênicas*, supplement 1 68:129–138.

Dillehay, T. D., 2000, *The Settlement of The Americas*. Basic Books, New York.

Dyson-Hudson, R., and Smith, E. A., 1978, Human Territoriality: An Ecological Reassessment. *American Anthropologist* 80:21–41.

Ferraz-Vicentini, K. R., C. F. and Salgado-Labouriau, M. L., 1996, Palynological Analysis of a Palm Swamp in Central Brazil. *Journal of South American Earth Science* 9(3/4):207–219.

Flannery, K. V., 1986, Ecosystem Models and Information Flow inthe Tehuacán-Oaxaca Region. In *Guilá Naquitz: Archaic Foraging and Early Agriculture in Oaxaca, Mexico* edited by K. V. Flannery, pp. 19–228. Academic Press, New York.

Flood, J., 1997, *Rock Art of The Dreamtime: Images of Ancient Australia*. Angus & Robertson, Australia.

Fogaça, E., and Lima, M. A., 1991, L'Abri du Boquete (Brésil): Les Premières Industries Lithiques de l'Holocène. *Journal de la Société deAméricanistes, N.S.* 77:111–123.

Fried, M. H., 1967, *The Evolution of Political Society: An Essay in Political Economy*. Random House, New York.

Gamble, C., 1982, Interaction and Alliance in Paleolithic Society. *Man* 17:92–107.

Gamble, C., 1991, The Social Context for European Paleolithic Art. *Proceedings of the Prehistoric Society* 57:316.

Gentry, A. H. (ed.), 1990, *Four Neotropical Rainforests*. Yale University Press, New Haven, CT.

Goland, A. C., 1991a, *Cultivating Diversity: Field Scattering as Agricultural Risk Management in Cuyo, Cuyo, Dept. of Puno, Peru*. Unpublished Ph.D. Dissertation, The University of Michigan, Ann Arbor.

Goland, A. C., 1991b, The Ecological Context of Hunter-Gatherer Storage: Environmental Predictability and Environmental Risk. In *Foragers in Context*, edited by P. T. Miracle, L. E. Fisher, and J. Brown, pp. 107–125. Michigan Discussion in Anthropology, No 10, Department of Anthropology, The University of Michigan, Ann Arbor.

Guidon, N., 1985, A Arte Pré-Histórica da Área Arqueológica de São Raimundo Nonato: Sintese de Dez anos de Pesquisa. *Clio* 2(7):3–81.

Guidon, N., 1992, As Ocupações Pré-Históricas do Brasil (Excetuando a Amazônia). In *História dos Índios no Brasil*, edited by M.C. da Cunha, pp. 37–52. FAPESP/SMC/Companhia das Letras, São Paulo.

Guidon, N., 1995, Traditions rupestres de l'Aire Archeologique de São Raimundo Nonato, Piauí, Brésil. In *Rock Art Studies in the Americas*, edited by J. Steinbring, pp. 121–128. Oxbow Monograph 45, Oxbow Books, Oxford.

Guilderson, T. P., Fairbanks, R. G., and Gubenstone, J. L., 1994, Tropical Temperature Variations Since 20,000 Years Ago: Modulating Interhemispheric Climate Change. *Science* 263:663–665.

Halstead, P., and O'Shea, J., 1989, Introduction: Cultural Responses to Risk and Uncertainty. In *Bad Year Economics: Cultural Responses to Risk and Uncertainty*, edited by P. Halstead and J. O'Shea, pp. 1–7. Cambridge University Press, Cambridge.

Hayden, B., 1990, Nimrods, Piscators, Pluckers, and Planters: The Emergence of Food Production. *Journal of Anthropological Archaeology* 9(1):31–69.

Hayden, B., 1996, Feasting in Prehistoric and Traditional Societies. In *Food and the Status Quest*, edited by P. Wiessner and W. Schiefenhövel, pp. 127–146. Berghahn Books, Providence, RI.

Hegmon, M., and Fisher, L. E., 1991, Information Strategies in Hunter-Gatherer Societies. In *Foragers in Context*, edited by P. T. Miracle, L. E. Fisher, J. Brown, pp. 127–145. Michigan Discussions in Anthropology, Vol. 10, Department of Anthropology, The University of Michigan, Ann Arbor.

Hillel, B., Leaman, A., Stansall, P., and M., Bedford, 1978, Space Syntax. In Social Organization and Settlement, edited by D. Green et al., *British Archaeological Reports, International Series* (suppl.) 47(11): 343–381.

Hooghiemstra, H., and van der Hammen, T., 1998, Neogene and Quaternary Development of the Neotropical Rain Forest: The forest Refugia Hypothesis, and a Literature Overview. *Earth-Science Reviews* 44:147–183.

Hurt, W. R., 1960, The Cultural Complexes from the Lagoa Santa Region, Brazil. *American Anthropologist* 62(4):569–585.

Hurt, W. R., 1964, Recent Radiocarbon Dates for Central and Southern Brazil. *American Antiquity* 29:25–33.

Jochim, M. A., 1981, *Strategies for Survival: Cultural Behavior in an Ecological Context*. Academic Press, New York.

Jochim, M. A., 1982, Paleolithic Cave Art in Ecological Perspective. In *Hunter-Gatherer Economy in Prehistory: A European Perspective* edited by G.N. Bailey, pp. 212–219. Cambridge University Press, Cambridge.

Kaplan, H., and Hill, K., 1992, The Evolutionary Ecology of Food Acquisition. In *Evolutionary Ecology and Human Behavior*, edited by E. A. Smith and B. Winterhalder, pp. 167–201. Aldine de Gruyter, Chicago.

Kaplan, H., Hill, K., and M., Hurtado, 1990, Risk, Foraging and Food Sharing among the Ache. In *Risk and Uncertainty in Tribal and Peasant Economies*, edited by E. Cashdan, pp. 107–143. Westview Press, Boulder.

Kelly, R. L., 1992, Mobility/Sedentism: Concepts, Archaeological Measures, and Effects. *Annual Review of Anthropology* 21:43–66.

Kelly, R. L., 1995, *The Foraging Spectrum: Diversity in Hunter-Gatherer Lifeways*. Smithsonian Institution Press, Washington, DC.

Kipnis, R., 2002, *Foraging Societies of Eastern Central Brazil: An Evolutionary Ecological Study of Subsistence Strategies During the Terminal Pleistocene and Early/Middle Holocene*. Unpublished Ph.D. Dissertation, The University of Michigan, Ann Arbor.

Kipnis, R., 1998, Early Hunter-Gatherers in the Americas: Perspectives from Central Brazil. *Antiquity* 72(277):581–592.

Lage, M. Conceição Soares Meneses, 1999, Dating of the Prehistoric Rock Paintings of the Archaeological Area of the Serrada Capivara National Park. In *Dating and the Earliest Known Rock Art*, edited by M. Strecker and P. Bahn, pp. 49–52. Oxbow Books, Oxford.

Layton, R., 1991, Trends in the Hunter-Gatherer Rock Art of Western Europe and Australia. *Proceedings of the Prehistoric Society* 57:163–174.

Layton, R., 1992, *Australian Rock Art: A New Synthesis*. Cambridge University Press, Cambridge.

Ledru, M., 1993, Late Quaternary Environmental and Climatic Changes in Central Brazil. *Quaternary Research* 39:90–98.

Ledru, M., Behling, H., Fournier, M., Martin, L., and Servant, M., 1994, Localisation de la Forêt d'*Araucaria* du Brésil au Cours de l'Holocène. Implications Paléoclimatiques. *Comptes Rendus de l'Academie des Sciences, Série 2* 317:517–521.

Ledru, M., Braga, P. I. S., Soubiès, F., Fournier, M., Martin, L., Suguio K., and Turcq, B., 1996, The Last 50,000 Years in Neotropics (Southern Brazil): Evolution of Vegetation and Climate. *Palaeogeography, Palaeoclimatology, Palaeoecology* 123:239–257.

Ledru, M., Salgado-Labouriau, M. L., and Lorscheitter, M. L., 1998, Vegetation Dynamics in Southern and Central Brazil During the last 10,000 yr B.P., *Review of Palaeobotany and Palynology* 99(1):131–142.

Lee, R. B., 1993, *The Dobe Ju/'Hoansi*, 2nd ed. Harcourt Brace, New York.

Lee, R. B., 1979, *The !Kung San: Men, Women, and Workin a Foraging Society*. Cambridge University Press, Cambridge.

Lee, R. B., and DeVore I. (eds.), 1968, *Man The Hunter*, Aldine, New York.

Lewis, D., 1988, *The Rock Paintings of Arhnem Land, Australia*. BAR International Series 415, Oxford.

Lima, M. A., Guimarães, C. M., and Prous, A., 1989, Os Grafismos Tipo Nordeste no Vale do Peruaçu, MG. *Dédalo, Publicação Avulça* 1:313–321.

Lima, M. A., 1991, Indústria Lítica de Santana do Riacho: Análise Funcional de Microtraceologia. *Arquivos do Museu de História Natural da Universidade Federal de Minas Gerais* 12:275–284.

Markgraf, V., 1993, Climatic History of Central and South America Since 18,000 BP: Comparison of Pollen Records and Model Simulations. In *Global Climates*, edited by H. E., Wright, Jr., J. E., Kutzbach, T., Webb III, W. F., Ruddiman, F. A., Street-Perrott, and P. J. Bartlein, pp. 357–385. University of Minneapolis Press, Minneapolis.

Mares, M. A., 1997, The Geobiological Interface: Granitic Outcrops as Selective Force in Mammalian Evolution. *Journal of the Royal Society of Western Australia* 80:131–139.

Mares, M. A., Willig, R., and Lacher, T. E. Jr., 1985, The Brazilian Caatinga in South America Zoogeography: Tropical Mammals in a Dry Region, *Journal of Bioeography* 12:57–69

Mares, M. A., Willig, M. R., Streilein, K. E., and Lacher, T.E. Jr., 1981, The Mammals of Northeast Brazil: A Preliminary Assessment. *Annals of Carnegie Museum* 50:81–137.

Martin, G., 1995, The Seridó Sub-Tradition of Prehistoric Rock Painting in Brazil. In *Rock Art Studies in the Americas* edited by J. Steinbring, pp. 129–135. Oxbow Monograph 45, Oxbow Books, Oxford.

Martin, G., 1996, The Rock Art Sites of Seridó in Rio Grande do Norte (Brazil) in the Context of the Peopling of South America, *FUNDHAMENTOS Revista da Fundação Museu do Homem Americano* 1(1):340–346.

Mayewski, P.A., et al., 1996, Climate Change During the Last Deglaciation in Antarctica. *Science* 272(5268):1636–1638.

McGregor, G. R., and Nieuwolt, S., 1998, *Tropical Climatology,* 2nd ed. John Wiley & Sons, Chichester, England.

Meese, D. A., et al., 1994, The Accumulation Record from the GISP2 Core as an Indicator of Climate Change Through the Holocene. *Science* 266(5191):1680–1682.

Minc, L. D., 1986, Scarcity and Survival: The Role of Oral Tradition in Mediating Subsistence Crises. *Journal of Anthropological Archaeology* 5:39–113.

Minnis, P. E., 1985, Social *Adaptation to Food Stress*. The University of Chicago Press, Chicago.

Miracle, P. T., 1995, *Broad-Spectrum Adaptations Re-Examined: Hunter-Gatherer Responses to Late Glacial Environmental Changes in the Eastern Adriatic*. Unpublished Ph.D. Dissertation, The University of Michigan, Ann Arbor.

Mithen, S. J., 1990, *Thoughtful Foragers: A Study of Prehistoric Decision Making*. Cambridge University Press, Cambridge.

Moore, J. A., 1981, The Effects of Information Networks in Hunter-gatherer Societies. In *Hunter-Gatherer Foraging Strategies* edited by Winterhalder, B. and Smith, E. pp. 194–217. University of Chicago Press, Chicago.

Mörner, N., 1996, Global Change and Interaction of Earth Rotation, Ocean Circulation and Paleoclimate. *Academia Brasileira de Ciênicas* 68:77–94, Supplement 1.

Morwood, M. J., and Hobbs, D. R., 1995, Themes in the Prehistory of Tropical Australia. *Antiquity* 69(265):747–768.

Myers, F. R., 1986, *Pintupi Country, Pintupi Self: Sentiment, Place, and Politics among Western Desert Aborigines*. Smithsonian Institution Press, Washington, DC.

Parry, W. J., and Kelly, R. L., 1987, Expedient Core Technology and Sedentism. In *The Organization of Core Technology*, edited by J.K. Johnson and C.M. Morrow, pp. 285–304. Westview Press, Boulder, Colorado.

Pessis, A., 1995, Graphic and Social Representation in the Nordeste Tradition of Rock Painting in Brazil. In *Rock Art Studies in the America*, edited by J. Steinbring, pp. 117–119, Oxbow Monograph 45, Oxbow Books, Oxford.

Pessis, A., 1999, The Chronology and Evolution of the Prehistoric Rock Paintings in the Serra da Capivara National Park, Piauí, Brazil. In *Dating and the Earliest Known Rock Art*, edited by M. Strecker and P. Bahn, pp. 41–47. Oxbow Books, Oxford.

Prance, G. T., 1982, Forest Refuges: Evidence from Woody Angiosperms. In *Biological Diversification in the Tropics* edited by G. T. Prance, pp. 137–156. Columbia Press, New York.

Prous, A., 1989, Arte Rupestre Brasileira: Uma Tentativa de Classificação. *Revista de Pré-História* 7:7–31.

Prous, A., 1991a, Santana do Riacho—Tomo I. *Arquivos do Museu de História Natural da Universidade Federal de Minas Gerais* 12:3–384.

Prous, A., 1991b, Fouilles de l'Abri du Boquete, Minas Gerais, Brésil. *Journal de la Société de Américanistes, N.S.* 77:77–109.

Prous, A., 1991c, Indústria Lítica de Santana do Riacho: Os Instrumentos Lascados. *Arquivos do Museu de História Natural da Universidade Federal de Minas Gerais* 12:229–274.

Prous, A., 1991d, Indústria Lítica de Santana do Riacho: Tecnologia, Tipologia e Traceologia: Os Instrumentos Polidos e Picoteados. *Arquivos do Museu de História Natural da Universidade Federal de Minas Gerais* 12:211–228.

Prous, A., 1992, *Arqueologia do Brasil.* Editora UnB, Brasília.

Prous, A., 1992/93 Santana do Riacho- Tomo II. *Arquivos do Museu de História Natural da Universidade Federal de Minas Gerais* 13/14:3–420.

Prous, A., 1994, L'art rupestre au Brésil. *Bulletin de la Société Préhistorique Ariège-Pyrénées* 49:77–144.

Prous, A., 1995, Stylistic Modification and Economic Changes in Peruaçu Valley (Brazil). In *Rock Art Studies in the Americas* edited by J. Steinbring, pp. 143–149. Oxbow Monograph 45, Oxbow Books, Oxford.

Prous, A., and Fogaça, E., 1999, Archaeology of the Pleistocene-Holocene Boundary in Brazil. *Quaternary International* 53/54:21–41.

Prous, A., Moura, M. T. T., and Lima, M. A., 1991, Indústria Lítica de Santana do Riacho: Tecnologia, Tipologia e Traceologia: Matérias Primas. *Arquivos do Museu de História Natural da Universidade Federal de Minas Gerais* 12:187–198.

Prous, A., Lima, M., Fogaça, E., and Brito, M. E., 1992, Indústria Lítica da Camaca VIII da Lapa do Boquete, Vale do Peruaçu, MG, Brasil. *Anais do 3° Congresso da Associação Brasileira de Estudos do Quaternário*, ABEQUA/UFMG, Belo Horizonte, pp. 342–362.

Rao, V. B., Hada, K., and Herdies, D. L. 1995, On the Severe Drought of 1993 in North-East Brazil. *International Journal of Climatology* 15:697–704.

Rao, V. B., Satyamurty, P., and de Brito, J. I. V., 1986, On the 1983 Drought of 1993 in North-east Brazil. *Journal of Climatology* 6:43–51.

Redding, R. W., 1988, A General Explanation of Subsistence Change: From Hunting and Gathering to Food Production. *Journal of Anthropological Archaeology* 7:56–97.

Rosenfeld, A., 1997, Archaeological Signatures of the Social Context of Rock Art Production. In *Beyond Art: Pleistocene Image and Symbols*, edited by M. Conkey, O. Soffer, D. Stratmann, and N. G. Jablonski, pp. 289–300. California Academy of Sciences, San Francisco, California.

Sahlins, M., 1972, *Stone Age Economics.* Aldine, Chicago.

Salgado-Labouriau, M. L., 1997, Late Quaternary Vegetation and Climatic Changes in Cerrado and Palm Swamp from Central Brasil. *Palaeo* 128:215–226.

Salgado-Labouriau, M. L., Barberi, M., Ferraz-Vicentini, K. R., and Parizzi, M. G., 1998, A Dry Climatic Event During the late Quaternary of Tropical Brazil. *Review of Palaeobotany and Palynology* 99:115–129.

Sarmiento, G., 1984, *The Ecology of Neotropical Savannas.* Harvard University Press, Cambridge.

Schmitz, P. I., 1987, Prehistoric Hunters and Gatherers of Brazil. *Journal of World Prehistory* 1(1):53–125.

Schmitz, P. I., 1997, *Serranópolis II— As Pinturas e Gravuras dos Abrigos.* Instituto Anchietano de Pesquisas/UNISINOS, São Leopoldo/RS.

Schmitz, P. I., Barbosa, A. S., Jacobus, A. L., and Ribeiro, M. B., 1989, *Arqueologia nos Cerrados do Brasil Central, Serranópolis I.* Pesquisas, Série Antropologia N° 44. Instituto Anchietano de Pesquisas/UNISINOS, São Leopoldo, Rio Grande do Sul.

Schmitz, P. I., Barbosa, M. O., and Riberio, M. B., 1997, *As Pinturas do Projeto Serra Geral.* Instituto Anchietano de Pesquisas/UNISINOS, São Leopoldo/RS.

Schmitz, P. I., Riberio, M. B., Barbosa, A. S., Barbosa, M. O., and de Miranda, A. F., 1986, *Caiapônica.* Instituto Anchietano de Pesquisas/UNISINOS, São Leopoldo/RS.

Setz, Eleonore Zulnara Freire, 1983, *Ecologia Alimentar em um Grupo Indigena: Comparação Entre Aldeias Numbiquara de Floresta e de Cerrado.* Unpublished Master Thesis, Instituto de Biologia da Universidade Estadual de Campinas, Campinas.

Slobodkin, L., and Rapoport, A., 1972, On the Inconstancy of Ecological Efficiency and the Form of Ecological Theories. In *Growth by Intussusception: Ecological Essays in Honor of G. Evelyn Hutchinson* edited by E.S. Deevey, pp. 293–305. The Connecticut Academy of Arts and Science, New Haven, CTO.

Smith, E. A., 1991a, *Inujjuamiut Foraging Strategies: Evolutionary Ecology of an Artic Hunting Economy.* Aldine de Gruyter, New York.

Smith, E. A., 1991b, Risk and Uncertainty in the "Original Affluent Society": Evolutionary Ecology of Resource-Sharing and Land Tenure. In *Hunters and Gatherers 1: History, Evolution, and Social Change*, edited by T. Ingold, D. Riches, and J. Woodburn, pp. 222–251. Berg, Oxford.

Smith, E. A., 1992, Natural Selection and Decision-Making: Some Fundamental Principles. In *Evolutionary Ecology and Human Behavior*, edited by E. A. Smith and B. Winterhalder, pp. 25–60. Aldine de Gruyter, New York.

Smith, E. A., and Boyd, R., 1990, Risk and Reciprocity: Hunter-Gatherer Socioecology and the Problem of Collective Action. In *Risk and Uncertainty in Tribal and Peasant Economies* edited by E. Cashdan, pp. 167–191. Westview Press, Boulder.

Smith, E. A., and Winterhalder, B., 1981, New Perspectives on Hunter-Gatherer Socioecology. In *Hunter-Gatherer Foraging Strategies*, edited by B. Winterhalder and E. A. Smith, pp. 1–12. University of Chicago, Chicago.

Sobel, E., and Bettles, G., 2000, Winter Hunger, Winter Myths: Subsistence Risk and Mythology among the Klamath and Modoc. *Journal of Anthropological Archaeology* 19:276–315.

Spielmann, K. A., 1986, Interdependence Among Egalitarian Societies. *Journal of Anthropological Archaeology* 5:279–312.

Stager, J. C., and Mayewski, P. A. 1997, Abrupt Early to Mid-Holocene Climatic Transition Registered at the Equator and the Poles. *Science* 276(5320):1834–1836.

Stephens, D. W., 1990, Risk and Incomplete Information in Behavioral Ecology. In *Risk and Uncertainty in Tribal and Peasant Economies*, edited by E. Cashdan, pp. 19–46. Westview Press, Boulder.

Stephens, D. W., and Krebs, J. R., 1986, *Foraging Theory.* Princeton University Press, Princeton.

Taçon, P. S. C., 1993, Regionalism in the Recent Rock Art of Western Arnhem Land, Northern Territory. *Archaeology in Oceania* 28:112–120.

Taçon, P. S. C., 1994, Socializing Landscapes: The Long-Term Implications of Signs, Symbols and Marks on the Land. *Archaeology in Oceania* 29:117–129.

Taçon, P. S. C., and Brockewell, S., 1995, Arnhem Land Prehistory in Landscape, Stone and Paint, *Antiquity* 69:676–695.

Taylor, K. C., et al., 1997, The Holocene-Younger Dryas Transition Recorded at Summit, Greenland. *Science* 278(5339):825–827.

Testar, A., 1982, The Significance of Food Storage among Hunter-Gatherers: Residence Patterns, Population Densities, and Social Inequalities. *Current Anthropology* 23:523–530.

Testar, A., 1988, Food Storage among Hunter-Gatherers: More or Less Security in the Way of life? In *Coping with Uncertainty in Food Supply*, edited by I. de Garine and G. A. Harrison, pp. 170–174. Oxford University Press, Oxford.

Thompson, L. G., 2000, Ice Core evidence for Climate Change in the Tropics: Implications for our Future. *Quaternary Science Reviews* 19:19–35.

Thompson, L. G., et al., 1995, Late Glacial Stage and Holocene Tropical Ice Core Records from Huascaran, Peru. *Science* 269(5220): 46–50

Van der Hammen, T., 1983, The Palaeoecology and Palaeogeography of Savannas. In *Tropical Savannas*, edited by F. Bourlière, pp. 19–35. Elsevier, Amsterdam.

Van der Hammen, T., and Hooghiemstra, H., 2000, Neogene and Quaternary History of Vegetation, Climate, and Plant Diversity in Amazonia. *Quaternary Science Reviews* 19:725–742.

Vialou, D., and Vialou, A., 1986, Mato Grosso: Préhistoire au Coeur du Brésil. *Archeologia* 213:36–48.

Whitelaw, T., 1991, Some Dimensions of Variability in the Social Organization of Community Space among Foragers. In *Ethnoarchaeological Approaches to Mobile Campsite*, edited by C. Gamble and W.A. Boismier, pp. 139–188. International Monographs in Prehistory, Ann Arbor.

Whitmore, T. C., 1998, *An Introduction to Tropical Rain Forests*. Oxford University Press, Oxford.

Wiessner, P., 1982a, Beyond Willow Smoke and Dog's Tails: A Comment on Binford's Analysis of Hunter-Gatherer Settlement Systems. *American Antiquity* 47:171–178.

Wiessner, P., 1982b, Risk, Reciprocity and Social Influences on !Kung San Economics. In *Politics and History in Band Societies*, edited by E. Leacock and R. Lee, pp. 61–84. Cambridge University Press, Cambridge.

Wiessner, P., 1983, Style and Social Information in Kalahari San Projectile Points. *American Antiquity* 48(2):253–276.

Wiessner, P., 1984, Reconsidering the Behavioral Basis for Style: A Case Study among the Kalahari San. *Journal of Anthropological Archaeology* 3:190–234.

Wiessner, P., 1986, !Kung San Networks in a Generational Perspective. In *The Past and Future of !Kung Ethnography: Critical Reflections and Symbolic Perspectives*, edited by M. Biesele, R. Gordon, and R. Lee, pp. 103–136. Helmut Buske Verlag, Hamburg.

Wiessner, P., 1996, Leveling the Hunter: Constraints on the Status Quest in Foraging Societies. In *Food and the Status Quest*, edited by P. Wiessner and W. Schiefenhövel, pp. 171–191. Berghahn Books, Providence, RI.

Winterhalder, B., 1981, Optimal Foraging Strategies and Hunter-Gatherer Research in Anthropology: Theory and Models. In *Hunter-Gatherer Foraging Strategies*, edited by B. Winterhalder and E. A. Smith, pp. 13–35. The University of Chicago Press, Chicago.

Winterhalder, B., 1986a, Diet Choice, Risk, and Food Sharing in a Stochastic Environment. *Journal of Anthropological Archaeology* 5:369–392.

Winterhalder, B., 1986b, Optimal Foraging: Simulation Studies of Diet Choice in a Stochastic Environment. *Journal of Ethnobiology* 6:205–223.

Winterhalder, B., 1990, Open Field, Common Pot: Harvest Variability and Risk Avoidance in Agricultural and Foraging Societies. In *Risk and Uncertainty in Tribal and Peasant Economies*, edited by E. Cashdan, pp. 67–88. Westview Press, Boulder.

Woodburn, J., 1968, An Introduction to Hadza Ecology. In *Man The Hunter*, edited by R. Lee and I. DeVore, pp. 49–55. Aldine, New York.

Woodburn, J., 1980, Hunters and Gatherers Today and Reconstruction of the Past. In *Soviet and Western Anthropology*, edited by E. Gellner, pp. 93–117. Columbia University Press, New York.

Wright H.E., Jr., et al., 1993, *Global Climates Since the Last Glacial Maximum*. University of Minnesota Press, Minneapolis.

Chapter 8

Central Place Foraging and Prehistoric Pinyon Utilization in the Great Basin

David W. Zeanah

INTRODUCTION

Great Basin archaeologists eagerly incorporated Binford's (1980) forager–collector model into their settlement pattern analyses because Julian Steward's (1933, 1938, 1941) work on the cultural ecology of Great Basin hunter-gatherers predisposed them to think of the influence of resource distributions on foraging and mobility strategies (Rhode 1999; Zeanah and Simms 1999).[1] Thomas' epistemologies for Monitor Valley and the Carson Desert of Nevada (Thomas 1983a, 1985) stood out as exemplary applications of the model (Bettinger 1991a: 70–73) because they demonstrated that ethnographic Great Basin bands that shared the same culture, language, and technology ran the gamut from pure foragers (i.e., Kawich Mountain Shoshone), through seasonally mixed foragers and collectors (i.e., Reese River Valley Shoshone and Carson Desert Paiute) to full-time

[1]Steward's contribution to the forager–collector model has been discussed elsewhere (Rhode 1999; Thomas 1983).

DAVID W. ZEANAH • Department of Anthropology, California State University, Sacramento, California 95819

collectors (i.e., Owens Valley Paiute). The dilemma posed by Thomas was that although the forager–collector model captures adaptive diversity among Great Basin hunter-gatherers, it fails to explain why such variability occurred in a region that lacked the global-scale differences in effective temperature posed by Binford as driving the forager–collector continuum. Almost 20 years ago, Thomas noted that "we currently lack the theoretical models to explain this variability" (1983a: 39), although he was optimistic that archaeological field research dedicated to development and application of "mid-range" models would eventually yield a theoretical understanding of variability among Great Basin settlement systems.

Substantial headway has been made in formulating the theoretical models that Thomas found lacking in 1983, but the progress results from a research tack different from that he anticipated (Zeanah and Simms 1999). Theoretically inspired by behavioral ecology, explorations of the cost–benefits of foraging in the Great Basin have, piece by piece, simplified the forager–collector model into constituent economic choices (i.e., which resources to harvest, which resources to transport, when to field process resources, and where to live). The contribution of field research has been to identify spatial and temporal variability in hunter-gatherer behavior not anticipated in Thomas' application of the forager–collector model to ethnographic cases. Archaeological research into prehistoric Great Basin pinyon (*Pinus monophylla*) procurement strategies illustrates this case.

GREAT BASIN PINYON PROCUREMENT STRATEGIES

Steward (1938: 27–28, 232) emphasized the importance of pinyon seeds as a storable food, whose productivity determined the size, permanence, and dispersion of winter villages among many ethnographic Great Basin groups. For this reason, a primary research goal of many subsistence-settlement studies was to assess the antiquity of pinyon procurement in the Great Basin (Bettinger 1976, 1977; Thomas 1973; Thomas and Bettinger 1976; Wells 1983). Site distributions discerned from probabilistic sample surveys of biotically defined strata (see Binford 1964), were analyzed under the assumption that statistically significant associations of camp assemblages (ground stone tools, rock-ring dwellings, and storage features) with pinyon-juniper woodlands reflect pinyon usage. The antiquity of pinyon procurement strategies was inferred from temporally diagnostic artifacts, radiocarbon dates, and obsidian hydration readings associated with pinyon camps.

Findings of these studies revealed that pinyon procurement strategies were geographically and temporally variable across the Great Basin (Fig. 8.1). For example, pinyon camps appeared in the forests above Reese

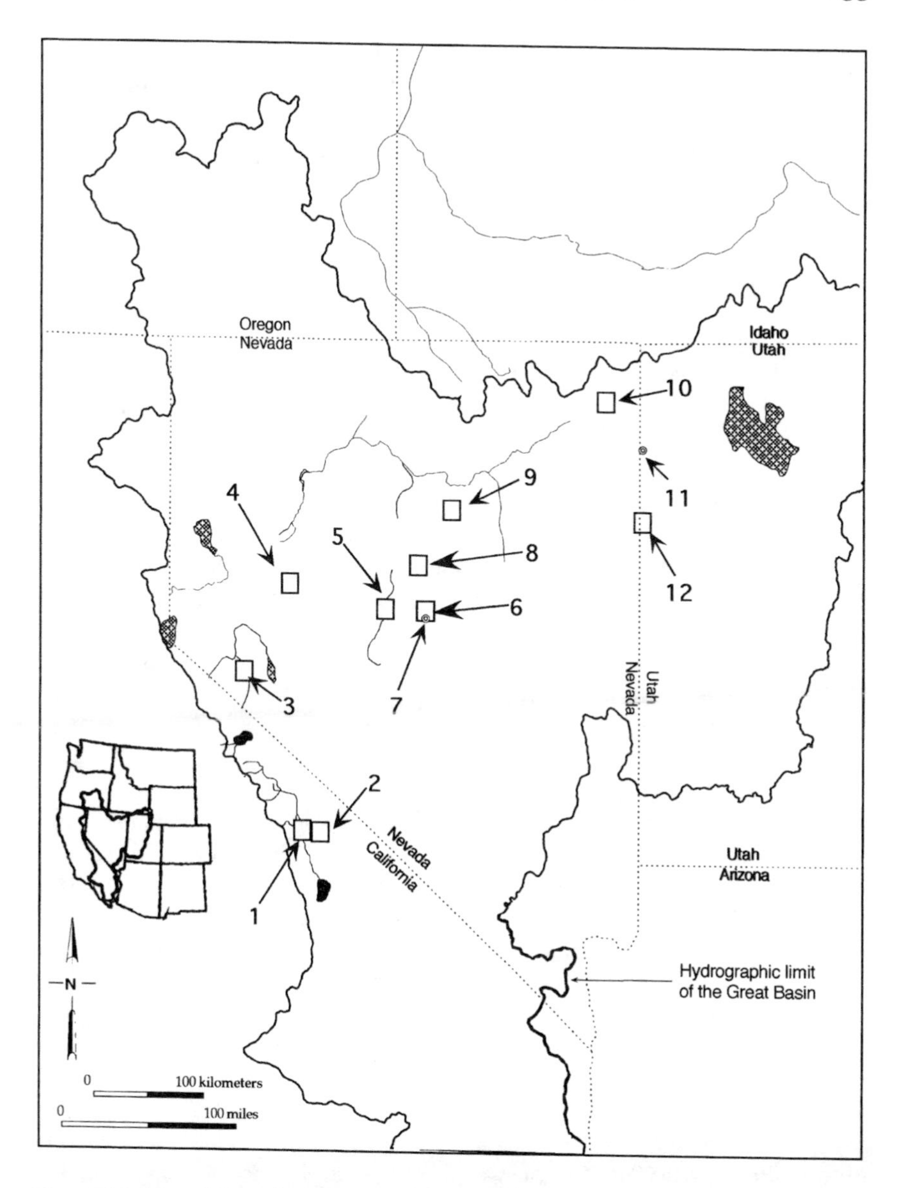

Figure 8.1. Map of Great Basin showing locations of pinyon-juniper surveys and sites mentioned in text. Key to locations:1. Owens Valley Survey (Bettinger 1976, 1977); 2. Deep Springs Valley Survey (Delacorte 1990); 3. Walker Lake Uplands Survey (Rhode 1990a); 4. Stillwater Survey (Kelly 2001); 5. Reese River Valley Survey (Thomas 1973; Thomas and Bettinger 1976); 6. Monitor Valley Survey (Thomas 1988); 7. Gatecliff Shelter (Thomas 1983b); 8. Grass Valley Survey (Wells 1983); 9. Cortez Survey (Delacorte et al. 1992); 10. Toano Draw Survey (Zeanah 1992); 11. Danger Cave (Rhode and Madsen 1998); 12. Deep Creek Survey (Lindsay and Sargent 1979).

River Valley as early as 6000 B.P., implying that the use of pinyon extends back to the Middle Holocene (Thomas 1973; Thomas and Bettinger 1976; Grayson 1993: 257). In contrast, woodlands of Owens Valley remained sparsely occupied until about 1350 B.P., indicating that hunter-gatherers bypassed pinyon as a food resource until that time (Bettinger 1977; Delacorte 1990). The results of these two surveys have different implications for interpreting Great Basin prehistory; the Reese River Valley survey suggests that the ethnographic pattern of pinyon procurement operated throughout the Holocene, whereas the Owens Valley survey indicates considerable variability in the role of pinyon over time. Subsequent surveys of pinyon-juniper woodlands elsewhere in the Great Basin failed to identify any pinyon camps whatsoever comparable to those of the Reese and Owens River Valleys (Delacorte et al. 1992; Kelly 2001; Lindsay and Sargent 1979; Zeanah 1992).

Inferences about the antiquity of pinyon procurement drawn from survey data were criticized because they lacked direct evidence in the form of macrofossils retrieved from dated contexts that pinyon was exploited as a dietary item from camps in pinyon zones. This was troubling because pinyon achieved its modern distribution only in the last few millennia (Madsen 1986; Grayson 1993), leaving open the possibility that associations of "pinyon camps" with pinyon-juniper woodlands were fortuitous and unrelated to pinyon procurement (Madsen 1981; see also Thomas 1981; Bettinger 1981; Delacorte et al. 1992).

In response to these criticisms, more concerted efforts to recover pinyon macrofossils from excavated contexts bolstered inferences drawn from site distributions (Bettinger 1989; Rhode 1980; Rhode and Thomas 1983; Wells 1983). In addition, paleoenvironmental research has emphasized tracing the Holocene expansion of pinyon so that its availability could be compared with local archaeological records. Many investigators expected that the development of economically exploitable pinyon groves in the central and northern Great Basin intensified occupation of those regions (Thomas 1982; Simms 1985; Grayson 1993). If so, variability in pinyon-juniper settlement patterns simply reflects local differences in the time that pinyon achieved its modern distribution.

Paleoenvironmental studies have revealed that the Holocene spread of pinyon through the Great Basin was a more complex process than the simple northward expansion initially expected (Lanner 1983; Madsen 1986). For example (Fig. 8.2), pinyon pine arrived in the vicinity of Danger Cave as early as 6700 B.P. (Rhode and Madsen 1998) but may not have appeared in the Stillwater Range until after 1250 B.P. (Kelly 2001: 36; Wigand and Nowak 1992). Nevertheless, this unanticipated complexity does not correlate

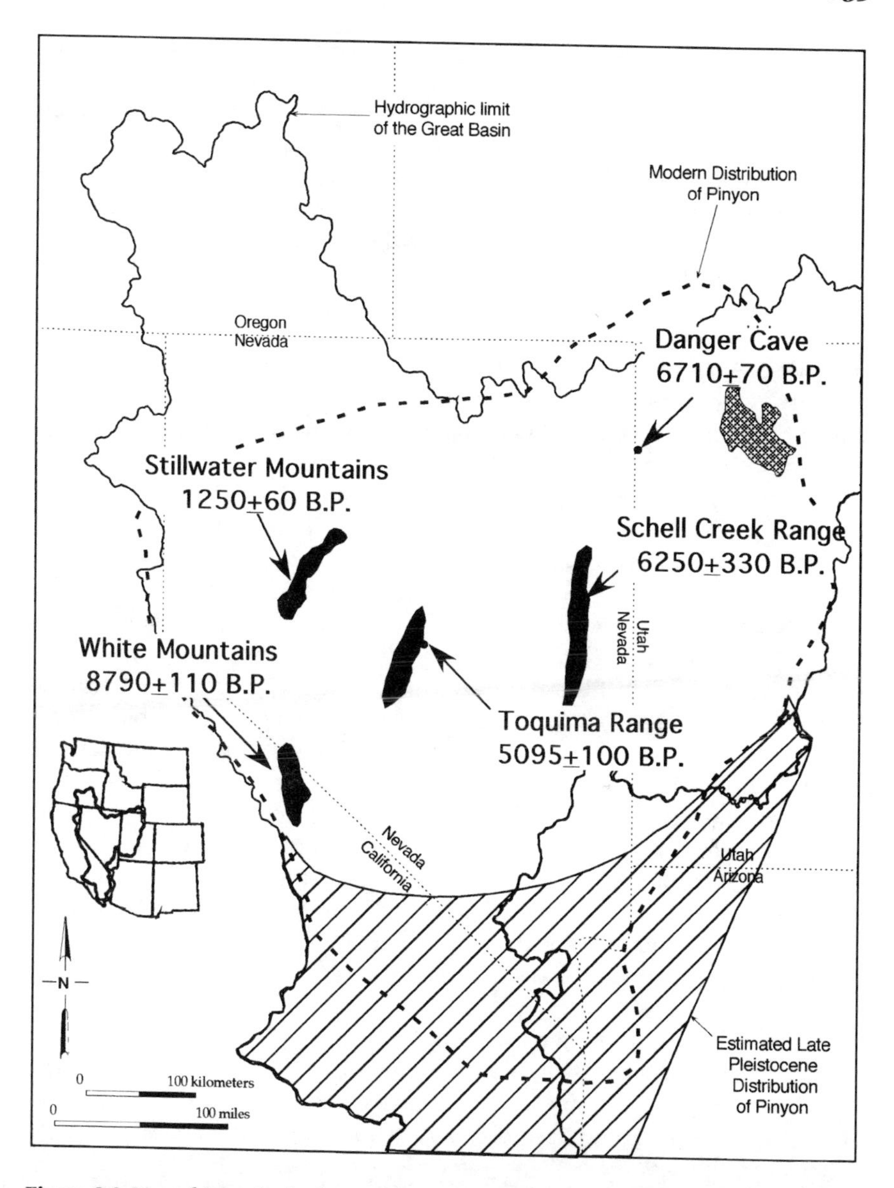

Figure 8.2. Map of Great Basin showing the modern distribution of pinyon (West et al. 1998), estimated Late Pleistocene distribution of pinyon (Madsen 1986), and radiocarbon dates from packrat middens and archaeological sites documenting the Holocene spread of pinyon (Jennings and Elliot-Fisk 1993; Rhode and Madsen 1998; Thompson and Hattori 1983; Wigand and Nowak 1992).

with variability among archaeological pinyon-juniper settlement patterns. Pinyon arrived in the Toquima Range about 6000 B.P. about when occupation of Gatecliff Shelter began (Thompson and Hattori 1983) and roughly contemporaneous with the inception of pinyon camp occupation in nearby Reese River and Grass Valleys (Thomas 1973, 1982; Wells 1983). In contrast, pinyon was present in the White Mountains by 8800 B.P. (Jennings and Elliot-Fisk 1993), many millennia before the association of sites with the pinyon woodlands of adjacent Owens and Deep Creek Valleys (Bettinger 1977; Delacorte 1990).

Another criticism of assessments of the antiquity of pinyon procurement derived from settlement pattern analyses concerned the effects of the mobility strategy employed for pinyon procurement on pinyon-juniper settlement patterns. McGuire and Garfinkel (1976) suggested that the 1350 B.P. appearance of pinyon camps in Owens Valley represented a local intensification of previously existing pinyon collection strategies, not the incorporation of a previously bypassed resource into hunter-gatherer diets. This inference was derived from inventory (although not a probabilistic sample survey) and test excavation of pinyon camps along the Pacific Crest Trail of the southern Sierra Nevada, south of Owens Valley (Garfinkel et al. 1980; McGuire and Garfinkel 1980). Drawing on pinyon collection strategies of the ethnographic Tubatulabal and Kawaisu Paiute as analogies, Garfinkel and McGuire proposed that pinyon camp assemblages could be categorized as either pinyon bases or temporary camps, suggesting that the 1350 B.P. appearance of pinyon camps in northern Owens Valley represented a shift from logistic to residential usage of pinyon woodlands. Rhode (1980) developed expectations for the deposition of pinyon macrofossils designed to distinguish the two types of pinyon camps. Pinyon should have been fully processed at base camps leading to the deposition of cones scales and hulls at these sites. In contrast, only initial stages of pinyon processing should have occurred at temporary pinyon camps, leaving only pine cones and scales.

Investigations of the Pacific Crest Trail sites failed to produce radiocarbon dates of pinyon macrofossils older than 1350 B.P., and it was not found that pinyon macrofossils varied as expected by pinyon camp type. However, their insightful consideration of pinyon camp variability clearly anticipated Binfords' (1980) forager–collector model and Thomas' (1983a, 1985) application of the model to Great Basin ethnographic cases. They pointed out how logistic and residential mobility strategies for pinyon procurement could affect pinyon-juniper settlement patterns in ways that had not been anticipated in previous considerations of the ethnographic record.

Nevertheless, the forager–collector model offered no satisfactory explanation for the change of pinyon procurement strategies over time or for anticipating circumstances where logistic or residential pinyon procurement should occur. For example, Thomas could not explain why some ethnographic groups residentially "mapped onto" pinyon (i.e., Kawich Mountain and Reese River Valley Shoshone), whereas others resided elsewhere and logistically collected pinyon (i.e., Owens Valley and Carson Sink Paiute) in a manner that allowed predicting the mode of prehistoric pinyon procurement in prehistoric Monitor Valley (Thomas 1983a: 156–165; 1983b: 514–516).

Despite 30 years of archaeological research devoted to procuring direct subsistence evidence, paleoenvironmental data, and additional surveys of pinyon-juniper woodlands, the regional variability in pinyon-juniper settlement patterns remains unexplained in any satisfying way. Neither ethnographic descriptions nor the forager–collector model offer robust, testable expectations of the reasons that such variability should occur.

CONTRIBUTIONS OF BEHAVIORAL ECOLOGY APPROACHES BASED ON FORAGING THEORY

It is in this research milieu that various scholars began to consider the economic cost–benefits of pinyon use from the theoretical framework of behavioral ecology. In the earliest of these, Simms (1985) applied general principles of the diet breadth model to predict how prehistoric hunter-gatherers modified their subsistence-settlement strategies in response to the Holocene expansion of pinyon. Harvesting experiments demonstrated that pinyon nuts yield higher caloric return rates than many seeds recovered from archaeobotanical contexts that predate the expansion of pinyon into the central and northern Great Basin (Table 8.1). Thus, according to the prediction of the diet breadth model that optimal foragers raise their overall foraging return rate by taking high-ranked resources whenever they come across them, Simms reasoned that Great Basin hunter-gatherers should have added pinyon to their diet as soon as nuts were locally available. Based on experimental postencounter rates, it seemed unlikely that the variability of pinyon-juniper settlement patterns reflected simple use or nonuse of pinyon as a dietary item, an assessment subsequently supported by recovery of 6700-year-old pinyon hulls from Danger Cave (Rhode and Madsen 1998). Instead, such variability must pertain to the intensity and organization of pinyon usage. Initiating the research tack followed in this chapter, Simms posed the costs necessary to transport pinyon to residential bases as an economic constraint that caused the variation observed among archaeological site distributions in various pinyon woodlands.

Jones and Madsen (1989) devised a measure of the relative transportability of different resources for the maximum transport distance (MTD) that a burden carrier could fill and carry a standardized volume of a resource before incurring a net caloric loss (Table 8.1). They found that MTD ranged from 829 to 0 km for various Great Basin resources; pinyon with a high MTD (812 km) may be economically procured logistically, whereas resources such as pickleweed that has a low MTD (0 km) are better candidates for acquisition by foraging (Jones and Madsen 1989).[2]

In Table 8.1, calculations of the MTD have been modified to reflect uphill and downhill transport costs.[3] Madsen and Jones assume a constant uphill gradient of 3% that, though suitable for a general comparison of resources, is misleading when applied to the Great Basin where the elevation of vegetation communities places resources in consistent topographic relationships with each other. As can be seen in the table, grade greatly affects the caloric costs of travel and transport (Brannon 1992; Zeanah 2000). Because it is unlikely that Great Basin hunter-gatherers ever faced the prospect of transporting pinyon nuts uphill to a shadscale patch, a more realistic ranking of the relative portability of pinyon and shadscale would be the downhill MTD of pinyon (2272 km) and the uphill MTD of shadscale (297 km).

Barlow and Metcalfe (1996; Metcalfe and Barlow 1992) modeled the extent of field processing for resources necessary to obtain and transport the resource optimally, dependent on the round-trip distance between home bases and pinyon patches. Their model assumes that central place foragers maximize the utility of packages returned home, compared with effort expended in field processing and transport. Resources consist of high utility (the edible seed in the case of pinyon) and low utility parts (i.e., cones, cone

[2]Note that these figures are not meant to estimate the actual distance that Great Basin hunter-gatherers carried resources. Obviously, the distance that a collector should logistically procure a distant resource is constrained by the return of resources close at hand (Rhode 1990b) and the return for simply moving to the distant resource patch (see Kelly 1990).

[3]Calculations for uphill and downhill transport assume the following caloric cost constants derived from MacDonald (1961).

Percent Grade	Caloric Cost of Walking Per Km	Caloric Cost of Carrying 1 Kg Per Km
−10	36.6	0.42
10	115.2	1.32

Table 8.1. Caloric Return Rate (kcal/h) and Maximum Transport Distance (km) of Selected Great Basin Resources[a]

[a]Simms 1987; Jones and Madsen 1989; Zeanah 2000.

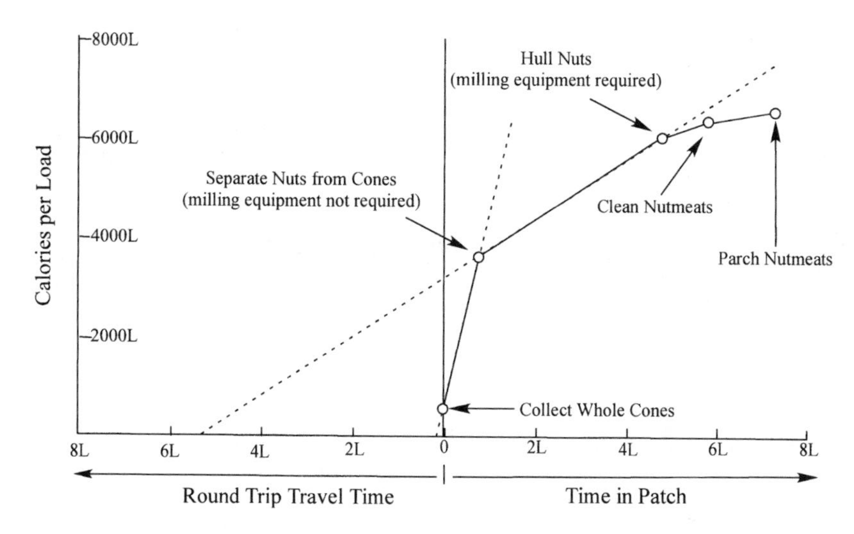

Figure 8.3. Changes in the utility of pinyon with field processing time. The dashed lines drawn through the x axis indicate hours of round-trip travel time between a central place and the pinyon grove when removing pine nuts from cones and hulling pine nuts (Barlow and Metcalfe 1996).

scales, and hulls) that can be discarded either at home or at the field processing location. The goal of field processing is to increase the utility of a transported load by culling low utility parts, but too much field processing reduces the number of trips that foragers can make to and from the resource patch. A central place forager must trade off the number of trips and the utility of each load to optimize the return rate of a resource transported home.

The round-trip distance between the central place and the patch determines the extent of field processing worthwhile, more processing is expected as the distance increases. Barlow and Metcalfe (1996) used their model to predict field processing decisions for pinyon nuts, as illustrated in Figure 8.3. The y axis illustrates the utility of pinyon (in calories per kilogram) samples at each processing stage. The x axis shows time spent procuring and field processing pinyon to the right of the y axis, and round-trip travel time is to the left. Based on calculations for 15-kg loads, Barlow and Metcalfe predict that hunter-gatherers can economically transport cones and cones scales back to residential bases that are no more than 2.5 km away from the pinyon grove. In contrast, they can profitably hull and clean nut meats in the field only if the transport distance back home exceeds 120 km.

The Barlow and Metcalfe field processing model mathematically formalized Rhode's (1980) intuitive expectations for the deposition of pinyon

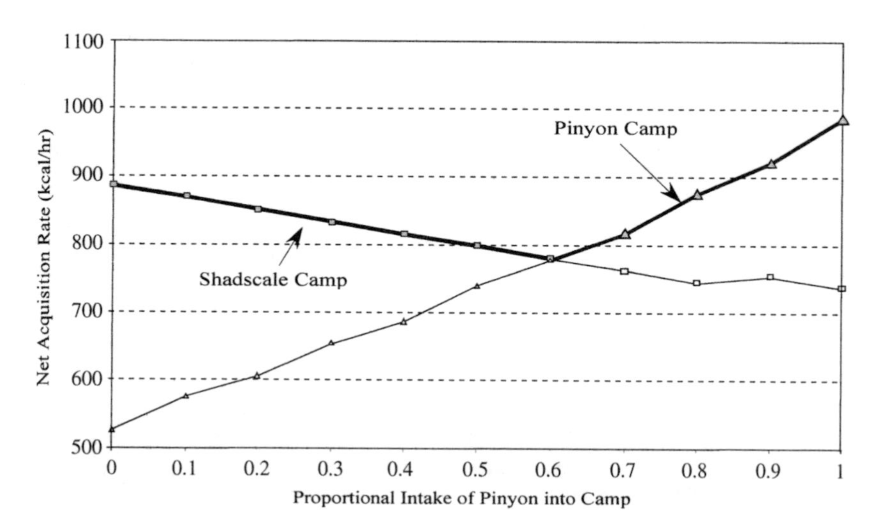

Figure 8.4. Central place settlement model for pinyon-juniper and lowland shadscale base camps based on proportional intake of pinyon into camp (Zeanah 2000).

macrofossils and grounded them firmly within the theoretical framework of behavioral ecology. Model predictions received strong empirical support from the frequent recovery of pinyon seed hulls, but not scales, from archaeobotanical contexts outside of pinyon woodlands (Basgall and McGuire 1988; Madsen 1979; Rhode and Madsen 1998; Scharf 1992; Wells 1983). The field processing model also holds implications for the optimal location of central place base camps. When spatially discrete resources are used from the same central place, transport costs should tether residential camps to patches of resources that cannot be field processed into high utility loads (Barlow and Metcalfe 1996; see Bettinger et al. 1997).

Zeanah (1996, 2000) developed a transport cost model that predicts where central place hunter-gatherers should reside when they provision camp from two spatially discrete resource patches and have the option to camp at one resource patch and logistically use the other. In Figure 8.4, the model is cast for two hypothetical winter villages where food stores are filled with pinyon and shadscale. Because harvesting experiments indicate that pinyon is a resource ranked slightly higher, the model assumes that hunter-gatherers should supply the winter village with as many nuts as the abundance of the crop will allow and make up any deficit by procuring shadscale seed. Then, central place foraging return rates can be calculated for both locations under different scenarios of pinyon procurement by subtracting procurement and transport costs obtained from each camp from a gross caloric requirement and dividing by the time necessary for procurement and transport.[4] Hunter-gatherers should choose the camp location with highest net acquisition rates after transport. Three important things are apparent about this simulation.

First, the model closely reproduces the winter village location decision described by Steward (1938: 28, 52–53, 65, 118, 142, 157); Great Basin hunter-gatherers chose to overwinter in woodlands if the pinyon harvests were sufficiently large but shifted camp elsewhere and logistically transported pinyon back to camp when harvests were small. Second, the portability of pinyon strongly influences net acquisition rates obtainable from different camps. Although pinyon and shadscale have comparable postencounter caloric return rates (1400 and 1200 kcal/h, respectively), the greater portability of pinyon makes it necessary for the pinyon harvest to exceed 60% of the caloric requirement to make residence at the pinyon camp more economical than the shadscale camp. If pinyon comprises less than 60% of the caloric intake of the camp, central place foragers achieve a higher net acquisition rate by residing near shadscale and logistically transporting pinyon. Finally, the central place foraging model directly links the forager–collector model with issues of diet breadth and subsistence intensification. As diet breadth expands to include spatially dispersed resources, the economics of central place foraging pull residential bases to less portable

[4] The formula for calculating net acquisition rates is as follows:

$$\frac{R + \sum_{i=1}^{2}\left[FN_i * C_h + \left(\sum_{s=1}^{n} D_{is} * NN_i * W_s\right) + \left(\sum_{s=1}^{n} D_{if} * NN_i * L_i * T_s\right)\right]}{\sum_{i=1}^{2}\left(FG_i + \frac{\sum_{s=1}^{n} D_{is} * NG_i}{V}\right)},$$

where
R = net caloric requirement during period of camp occupation;
FN_i = handling time (hours) for total loads of resource i comprising net caloric requirement;
FG_i = handling time (hours) for total loads of resource i comprising net caloric requirement plus additional loads required to cover caloric costs;
C_h = caloric cost of handling resources (300 kcal/h);
D_{is} = distance of slope s to and from nearest patch of resource i (km);
D_{if} = distance of slope f from nearest patch of resource i (km);
NN_i = number of loads of resource i transported comprising net caloric requirement;
NG_i = number of loads of resource i transported comprising net caloric requirement plus additional loads required to cover caloric costs;
W_s = caloric costs of walking across grade s (kcal/km);
L_i = total weight of one load of resource transported (max. = 25 kg);
T_s = caloric costs of carrying load across grade s (kcal/km);
V = walking speed (3 km/h).

resources that contribute significantly to the dietary intake of the camp. Highly portable resources, such as pinyon, are likely to witness a shift from residential to logistic usage as diet breadth expands (Zeanah 2000).

APPLICATION OF MODELING INSIGHTS

One aspect of spatial and temporal variability among prehistoric pinyon procurement strategies that has vexed Great Basin archaeologists for decades is the abundance of groundstone milling equipment in pinyon-juniper woodlands. Milling slabs and handstones are ubiquitous in pinyon-juniper forests of Owens Valley (Bettinger 1976, 1977; Delacorte 1990). However, in many other parts of the Great Basin, the rarity of ground stone tools is not in keeping with the ethnographically documented importance of pinyon as a food resource in the same regions (Thomas 1973; Thomas and Bettinger 1976; Simms 1985; Kelly 2001; Bettinger 1999b). Contending explanations for the phenomenon are either that pinyon usage developed relatively recently in forests where ground stone tools are rare (Bettinger 1999b), or that tool curation, scavenging, and site looting have artificially depressed the quantities of ground stone tools in some woodlands (Thomas and Bettinger 1976; Simms 1985).

Central place foraging models suggest a new explanation for the differential abundance of milling equipment. Ethnographically (Chamberlin 1911; Coville 1892; Dutcher 1893; Wheat 1967), ground stone tools were used to remove hulls from pinyon seeds and grind seeds into flour (Fig. 8.5). The Barlow and Metcalfe field processing model (Fig. 8.3) suggests that these processing steps should be economically undertaken only at a home base or when the pinyon is to be transported for distances that far exceed those typical of ethnographic Great Basin foragers (see Rhode 1990b). The transport cost and central place location models suggest that pinyon harvests must be relatively high to make residing in pinyon zones more economical than in lowland residence. If these inferences are correct, then the rarity of milling equipment in many Great Basin woodlands simply reflects different decisions of whether to reside in pinyon zones or to procure and transport pinyon logistically.

Such regional differences in the tendencies to reside in pinyon-juniper woodlands might reflect local variability in the quantity of harvested nuts.[5] One regional classification of Great Basin pinyon-juniper woodlands

[5]The local presence of sufficient quantities of resources higher ranked or more portable than pinyon would also lead to a decision not to camp in pinyon woodlands (see Delacorte et al. 1992; Kelly 1995), irrespective of the quantity of the local pinyon crop.

Figure 8.5. Two-handed huller being used to shell pine nuts on a wooden metate. Ground stone tools are likely to have been used to process pinyon only at base camps or when logistic transport distances were exceptionally large. Photograph by Margaret Wheat, 1958. Margaret Wheat Collection, Special Collections, University of Nevada-Reno Library.

reveals that the composition of pinyon-juniper woodlands varies across the Great Basin in ways that correlate with archaeological patterns (West et al. 1998). In the study, 463 woodland plots 20×50 m in 66 Great Basin mountain ranges were compared for the proportional composition of pinyon and juniper.[6] The study documented tremendous regional variability both within and between mountain ranges but revealed one consistent regional

[6]At a minimum, plots had to contain at least 25 pinyon or juniper trees with at least one fully mature tree, and no evidence of recent cutting, chaining, or burning. These criteria ensured sampling of comparable stands at least 50 years old.

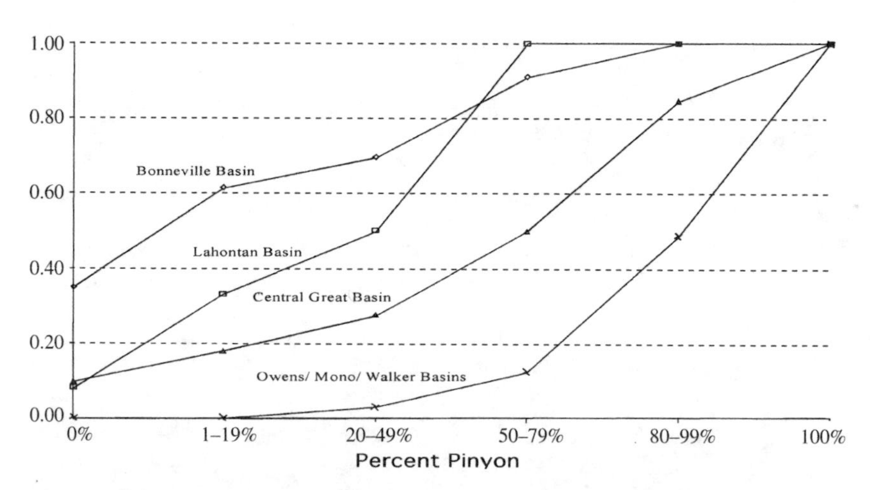

Figure 8.6. Percent pinyon in woodlands of four geographic regions of the Great Basin (West et al. 1998).

trend. Woodlands in the western, central, and southern portions of the Great Basin are richer in pinyon trees than comparable stands in the northern and eastern Great Basin (Fig. 8.6). For example, 56% of the stands sampled in the Walker, Mono, and Owens River Basins of the western Great Basin are pure pinyon woodlands. In central Great Basin ranges, 16% of the stands are pure pinyon, but a further 57% are at least 50% pinyon. In the Lahontan and Bonneville/Upper Humboldt basins, juniper dominates woodlands (respectively, 50% and 71% of the stands bear less than 50% pinyon). These regional trends appear to result from the role of Pacific winter storm fronts in inhibiting pinyon growth (Beeson 1972; West et al. 1978).

Obviously, differences in the proportional representation of pinyon in woodlands will not translate directly to the productivity of pinyon harvests; a pure but sparse stand of pinyon might produce smaller crops than a dense stand dominated by juniper. However, available data support a relationship; in a poor cone production year, Jordan (1974) inventoried an average yield of 681 cones per acre in a woodland tract of pure pinyon but only 553 cones per acre in a comparable tract of 77% pinyon trees. Figure 8.7 shows estimated favorable year productivity of pine nut harvests for 29 Nevada ecological sites bearing pinyon (Soil Conservation Service 1992a,b,c,d, 1993). The estimates reveal considerable variability in the size of the seed crop (ranging from 30 to more than 330 kg/ha) and reveal a significant positive correlation between crop size and percentage coverage by pinyon.

Figure 8.8 illustrates the effect of pinyon density on pinyon-juniper settlement decisions. Ethnographic accounts indicate that a family of Great

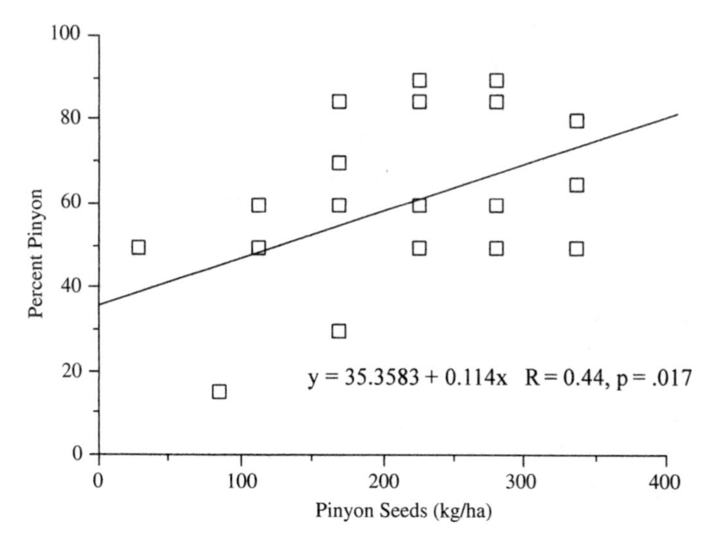

Figure 8.7. Pine nut productivity by percentage Pinyon for 29 Nevada ecological sites.

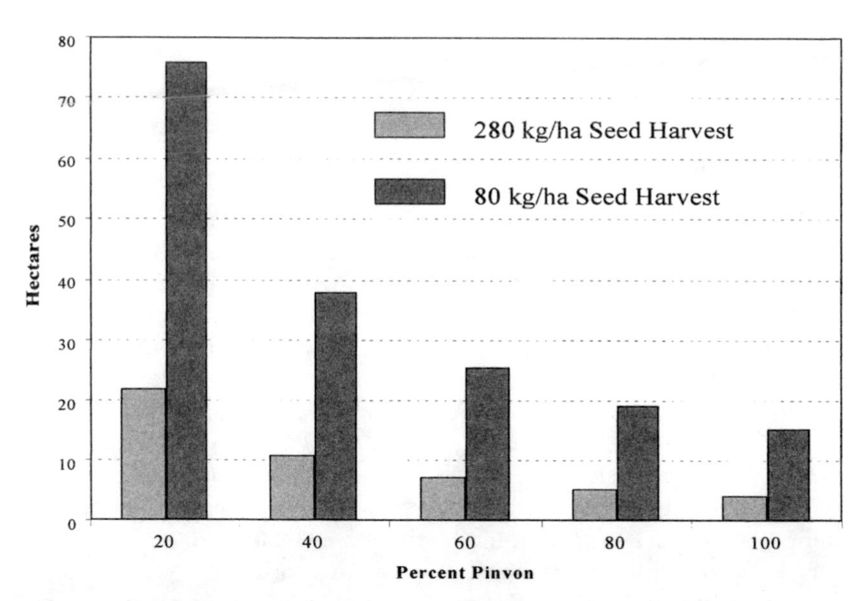

Figure 8.8. Area (hectares) required to harvest 680 kg of clean pinyon seeds with 280 kg/ha and 80 kg/ha crop yields.

Basin hunter-gatherers could store about 680 kg of pinyon nuts in a good year (Cook 1941; Price 1962; Steward 1938), but harvests could range from 45 to 275 kg in less productive years (Cook 1941: 54). A typical pinyon nut harvest in a favorable year is about 280 kg/ha (Fischer and Montano 1977; Jeffers 1994). Figure 8.8 indicates the area required to harvest 680 kg of clean pinyon seed for maximum harvests of 280 kg/ha and 80 kg/ha of unhulled nuts in pure pinyon woodlands, scaling the harvest size by the proportion of pinyon in the woodland and assuming that prehistoric hunter-gatherers lose none of the crop to spoilage or competitors. For favorable pinyon harvests, less than twenty hectares can provide an entire winter food supply, even in forests of 20% pinyon. However in years of a more moderate harvest of 80 kg/ha,[7] 15 hectares of pure pinyon woodlands could supply the winter larder, but more than 75 hectares would be necessary in woodlands of 20% pinyon.

More diffuse pinyon crops in hilly terrain encourage placement of residential bases in lowland areas below the pinyon-juniper woodland and make it less likely that camp spots in pinyon woodlands will recurrently be the most economic locations for winter villages. If so, differences in the relative richness of pinyon will correlate with the density of ground stone tools in woodlands; ground stone will be relatively rare in woodlands that have low densities of pinyon because those woodlands are less likely to serve recurrently as suitable winter occupation locations.

Table 8.2 arrays data on ground stone tool distributions in nine surveys of Great Basin woodlands. The span of time that pinyon has been available in these areas does not account for ground stone tool densities. Pinyon has probably grown in the northeastern (Deep Creek, Toano, and Pequop Ranges) and central Great Basin (Toquima, Shoshone, Toiyabe, and Monitor Ranges) for the last 6000 years and arrived in the northern (Cortez Range) and northwestern (Stillwater Mountains) area only in the last few millennia. Yet ground stone tool densities are higher in the Stillwater and Cortez Mountains than in any of the central or northern ranges. Neither does the antiquity of pinyon procurement inferred from archaeological evidence explain ground stone densities. Pinyon camps in Owens and Deep Springs Valleys postdate 1350 B.P., yet ground stone occurs in much higher densities there than in the Reese River Valley (where pinyon camps, it is inferred, are much older) or in the northeastern Great Basin ranges (where hulls from Danger Cave prove that pinyon was consumed 6700 years ago).

Data on the proportion of pinyon in plots sampled in pinyon-juniper forests adjacent to eight of the surveys (West et al. 1998) are presented in Table 8.3 and ranked for overall pinyon density. Comparison of these data

[7]An 80-kg/ha harvest would be about the maximum yield of 2500 cones per acre observed in a 5 year study of a sample plot in northern Utah (Lanner 1983).

Table 8.2. Groundstone Tool Densities in Pinyon Juniper Woodland Surveys

Survey	Ground Stone Tools	Survey Area (hectares) in Pinyon-Juniper Woodlands	Groundstone Tools per Hectare
Owens Valley	60	775	0.0774
Deep Springs	32	750	0.0427
Reese River	10	1000	0.0100
Stillwater	22	968	0.0227
Toano Draw	15	1984	0.0076
Deep Creek	3	448	0.0067
Monitor Valley	36	4775	0.0075
Cortez	23	1152	0.0200
Walker Lake Uplands	4	192	0.0208
Grass Valley	29	1000	0.0290

with ground stone densities reveals a strong, significant correlation between the proportion of pinyon and the number of ground stone tools per hectare (Kendall's tau = .74, $p = .0133$). Consistent with expectations based on the central place foraging models, pinyon productivity accounts for ground stone tool densities in pinyon-juniper woodlands better than the span of pinyon availability or the antiquity of pinyon procurement.

The Owens Valley Case

Despite the regional correlation between ground stone and pinyon densities, variations in pinyon productivity cannot account for all spatial and temporal diversity among pinyon-juniper settlement patterns. Local differences in the availability of alternative resources, paleoenvironmental change, and long-term trends in population growth and resource intensification must also influence the pinyon use strategy in any particular time and place. However, a theoretical understanding of the economics of central place foraging provides a context from which to compare local trends in the archaeological record against particular resource distributions. This is best illustrated by Owens Valley where some of the richest pinyon forests of the Great Basin contain high densities of ground stone, yet lack pinyon camps until 1350 B.P. (Bettinger 1977, 1989; Delacorte 1990). What factors account for the relatively late but intensive development of pinyon camps in this region?

The intensive post-1350 B.P. usage of pinyon suggested by high ground stone tool densities is consistent with the ethnographic record, which shows that Owens Valley supported one of the highest population densities in the

Table 8.3. Relative Ranking of Pinyon Densities by Adjacent Mountain Range (West et al. 1998)

Survey	Mountain Range	Total Inventoried Plots	100% Pinyon	80–99% Pinyon	50–79% Pinyon	20–49% Pinyon	1–19% Pinyon	No Pinyon %	Relative Rank
Owens Valley/ Deep Springs	White Mountains	7	71	14	14	0	0	0	1
Walker Lake Uplands	Wassuk and Pine Grove Mountains	17	47	35	12	6	0	0	2
Reese River	Toiyabe and Shoshone Ranges	27	48	33	11	4	0	4	3
Grass Valley	Toiyabe Range and Simpson Park Mountains	13	8	76	0	8	8	0	4
Monitor Valley	Toquima and Monitor Mountains	14	14	36	50	0	0	0	5
Toano Draw	Toana and Pequop Ranges	9	0	11	56	11	11	11	6
Deep Creek	Deep Creek Mountains	8	0	25	38	0	38	0	6
Stillwater	Stillwater Mountains	No Data	–	–	–	–	–	–	–
Cortez	Cortez Mountain	No Data	–	–	–	–	–	–	–

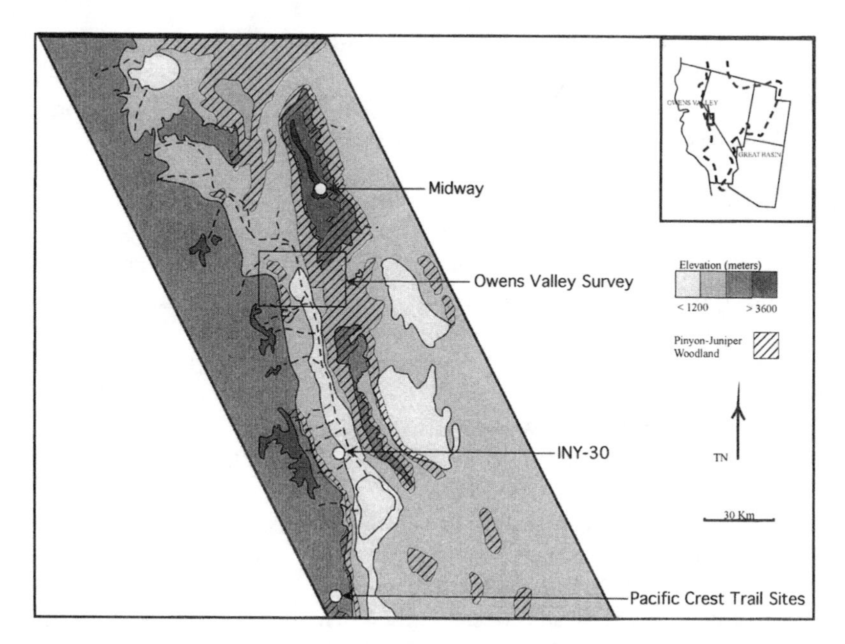

Figure 8.9. Map of Owens Valley and surrounding areas showing the locations of Bettinger's Owens Valley survey and sites yielding evidence of pre-1350 B.P. pinyon usage.

Great Basin and that pinyon was an important overwinter food resource (Steward 1933, 1938). Yet archaeological research in Owens Valley shows that the ethnographic pattern developed only 1350 years ago as a consequence of subsistence-settlement intensification in response to population growth (Basgall and McGuire 1988; Bettinger 1977, 1989; Delacorte 1990, 1999). Figure 8.9, a map of Owens Valley and adjacent Mono Basin and Coso Range regions, shows the approximate location of Bettinger's (1977) Owens Valley survey, which is about the size of an ethnohistoric territory (Steward 1933: map 1). Analyses of obsidian distributions and assemblage composition and diversity reveal that pre-1350 B.P. settlement systems probably encompassed the entire region illustrated on the map (Basgall 1989; Basgall and McGuire 1988; Bettinger 1999a; Delacorte 1999). Clearly, the Owens Valley survey was not designed to sample site variability adequately in the earlier settlement pattern. Confronted with differences in the mobility of this scale, the possibility must be considered that archaeologists have simply not looked in the right place for evidence of pre-1350 B.P. pinyon usage.

Seasonality data from a limited set of pre-1350 B.P. faunal and floral assemblages are evidence that earlier hunter-gatherers overwintered near Owens Lake and the Coso Range (Basgall and McGuire1988: 321, 348) and traversed the central and northern regions in summer and early fall (Scharf

1992). If so, evidence for pre-1350 B.P. pinyon use should occur in the southern portion of the region, but studies that support a 1350 B.P. inception of pinyon use have been done predominantly in central and northern Owens Valley (Bettinger 1976, 1977, 1989; Delacorte 1990).

Figure 8.9 also indicates the locations of the Pacific Crest Trail sites that have been proposed to represent pre-1350 B.P. pinyon camps (Garfinkel et al. 1980; McGuire and Garfinkel 1980) and two sites that have yielded archaeobotanical pinyon hulls from pre-1350 B.P. contexts (Basgall and McGuire 1988; Scharf 1992). Certainly, this suggests that some of the best evidence for pre-1350 B.P. pinyon usage (the Pacific Crest Trail pinyon camps and site CA-INY-30) comes from southern Owens Valley. Also note that the two sites bearing pre-1350 B.P. pinyon hulls (Midway and CA-INY-30) are located outside of pinyon woodlands, suggesting that pre-1350 B.P. hunter-gatherers in Owens Valley logistically procured and transported to camps outside of pinyon-juniper woodlands.

Post-1350 B.P. pinyon camps in central Owens Valley characteristically contain habitation structures, storage features, and ground stone tools (Bettinger 1977, 1989) that reflect the preparation, storage, and consumption of pinyon at residential bases, rather than field processing of pinyon at logistic camps. If so, the 1350 B.P. appearance of pinyon camps in central Owens Valley marks the beginning of residential occupation of pinyon woodlands in central Owens Valley, not necessarily the inception of pinyon usage in the entire region. However, a shift from logistic to residential pinyon procurement strategies appears to contradict central place foraging models. Pinyon should be a more economical target for logistic exploitation when less transportable resources, such as lowland seeds, must also be garnered from the same central place.

One possible explanation for the discrepancy pertains to the intensity and organization of pinyon usage in Owens Valley after 1350 B.P. Known pre-1350 B.P. occupation sites occur only on the valley floor and usually appear to be short-term, transient camps (Basgall and Giambastiani 1995; Bettinger 1977, 1991b; Bettinger et al. 1984; Delacorte 1990, 1999). Residential sites bearing rock-ring habitation structures that suggest prolonged, seasonal occupation have been found only in southern Owens Valley (Basgall and McGuire 1988). In contrast, post-1350 B.P. residential sites always contain high investment features, such as rock-ring floors, storage features, plant processing facilities, and refuse middens. These sites also contain a more eclectic array of plant, animal, and tool stones than their earlier counterparts, but they frequently originate from within a smaller catchment radius of each base camp (Basgall and McGuire 1988; Bettinger 1989, 1991b; Delacorte 1999). Post-1350 B.P. base camps occur throughout Owens Valley region and appear in upland vegetation communities for the first time

(Basgall and Giambastiani 1995; Bettinger 1977, 1989, 1991b; Delacorte 1990). Therefore, the 1350 B.P. appearance of pinyon camps in central Owens Valley represents only one element of a regional subsistence-settlement shift that marked an expansion of residential site locations into multiple vegetation communities, a constriction of the catchment basins exploited from residential sites, and an intensification of subsistence strategies.

This transition ultimately led to the semisedentary and territorial settlement patterns observed ethnographically in Owens Valley (Steward 1933; Thomas 1983a: 32–34). The regional archaeological record suggests that growing populations quickly filled every habitable location in Owens Valley, constraining both logistic and residential mobility (see Steward 1933: map 1). In this context, the 1350 B.P. appearance of residential occupation sites in the pinyon-juniper woodlands of central Owens Valley may reflect the choice of hunter-gatherers to map onto resource patches that could be more economically used logistically because better base camp locations were already occupied and logistic collecting opportunities constrained.

The data presented in Figure 8.8 shows that small patches of rich pinyon groves can supply an entire winter harvest of nuts, even in years with a relatively poor crop, so long as hunter-gatherers minimize losses to spoilage and competing foragers. One effective ethnographically documented strategy for maximizing the seed crops yielded from a grove of pinyon trees was to harvest green, unopened cones (Bettinger 1994; Bettinger and Baumhoff 1983). Procurement of green cones, amassing large overwinter pinyon caches, territorial ownership of pinyon groves, and using alternative, high-cost upland resources are means by which post-1350 B.P. hunter-gatherers intensified their subsistence strategies to make up for the costs of camping in suboptimal base camp locations (Bettinger 1991a; Grayson 1991). Therefore, the 1350 B.P. shift from logistic to residential usage of pinyon woodlands results from circumscription of residential base catchments and restriction of logistic mobility that would result from demographic packing in the central Owens Valley region (Bettinger 1991b; Zeanah 2000).

CONCLUSIONS

Theoretical models of the costs and benefits of central place foraging in the Great Basin have provided the theoretical tools necessary to investigate variability in logistic and residential mobility strategies in the Great Basin. These models reveal no theoretical reason to expect any single strategy of pinyon exploitation as typical of Great Basin hunter-gatherers. Instead, they follow Thomas' lead to model formally how the logistic and residential strategies of hunter-gatherers, equipped with similar technologies and procuring similar resources, might vary in different local landscapes.

The development of transport models in Great Basin subsistence studies illustrates the important role that general explanatory theory can play in such considerations. These models derive from evolutionary theory and bear clear implications for understanding variability in pinyon procurement strategies. They also have test implications for site distributions and assemblage composition that clarify earlier disputes about the relative validity of survey data versus archaeobotanical remains as evidence of prehistoric subsistence. Therefore, they qualify as middle range theory at its best by explicitly subsuming mid-range issues under higher order theory.

REFERENCES

Barlow, K. R., and Metcalfe, D., 1996, Plant Utility Indices:Two Great Basin Examples, *Journal of Archaeological Science* 23:351–371.

Basgall, M. E., 1989, *Obsidian Acquisition and Use in Prehistoric Central Eastern California*, edited by R. E. Hughes. Contributions of the University of California Research Facility No. 48.

Basgall, M. E., and Giambastiani, M. A., 1995, *Prehistoric Use of a Marginal Environment:Continuity and Change in Occupation of the Volcanic Tablelands, Mono and Inyo Counties California*. Center for Archaeological Research at Davis Publication No. 12, University of California, Davis.

Basgall, M. E., and McGuire, K. R., 1988, *The Archaeology of CA-INYO-30:Prehistoric Culture Change in Southern Owens Valley, California*. Far Western Anthropological Research Group, Inc., Davis, CA.

Beeson, C. D., 1972, *The Distribution and Synecology of Great Basin Pinyon-Junipers*, M.S. Thesis, University of Nevada, Reno.

Bettinger, R. L., 1976, The Development of Pinyon Exploitation in Central Eastern California. *Journal of California and Great Basin Anthropology* 3(1):81–95.

Bettinger, R. L., 1977, Aboriginal Human Ecology in Owens Valley:Prehistoric Change in the Great Basin. *American Antiquity* 42(1):3–17.

Bettinger, R. L., 1981, Settlement Data and Subsistence Systems. *American Antiquity* 46:640–643.

Bettinger, R. L., 1989, *The Archaeology of Pinyon House, Two Eagles, and Crater Middens:Three Residential Sites in Owens Valley*. Anthropological Papers 67. American Museum of Natural History, New York.

Bettinger, R. L., 1991a, *Hunter-Gatherers:Archaeological and Evolutionary Theory*. Plenum, New York.

Bettinger, R. L., 1991b, Aboriginal Occupation at High Altitude:Alpine Villages in the White Mountains of Eastern California. *American Anthropologist* 93(3):656–679.

Bettinger, R. L., 1994, How, When, and Why Numic Spread. In *Across the West:Human Populations Movement and the Expansion of the Numa*, edited by D. B. Madsen and D.R. Rhode, pp. 44–55. University of Utah Press, Salt Lake City.

Bettinger, R. L., 1999a, From Traveler to Processor:Regional Trajectories of Hunter-Gatherer Sedentism in the Inyo-Mono Region, California. In *Settlement Pattern Studies in the Americas; Fifty Years Since Viru*, edited by B. R. Billman and G. M. Feinman, pp. 39–55. Smithsonian Institution Press, Washington.

Bettinger, R. L., 1999b, What Happened in the Medithermal. In *Models for the Millennium:Great Basin Anthropology Today*, edited by C. Beck, pp. 62–74. University of Utah Press, Salt Lake City.

Bettinger, R. L., and Baumhoff, M. A., 1983, Return Rates and Intensity of Resource Use in Numic and Prenumic Adaptive Strategies. *American Antiquity* 48:830–834.

Bettinger, R. L., Delacorte, M., and McGuire, K. R., 1984, Archaeological Excavations at the Partridge Ranch Site (CA-INY-2146), Inyo County, California. Submitted to California Department of Transportation, Sacramento.

Bettinger, R. L., Malhi, R., and McCarthy, H., 1997, Central Place Models of Acorn and Mussel Processing. *Journal of Archaeological Science* 24(10):887–899.

Binford, L. R., 1964, A Consideration of Archaeological Research Design. *American Antiquity* 29:425–441.

Binford, L. R., 1980, Willow Smoke and Dog's Tails:Hunter-Gatherer Settlement Systems and Archaeological Site Formation. *American Antiquity* 45(1):4–20.

Brannon, J. A., 1992, On Modeling Resource Transport Costs:Suggested Refinements. *Current Anthropology* 33(1):56–60.

Chamberlin, R. V., 1911, The Ethno-Botany of the Goisute Indians of Utah. *American Anthropological Association Memoirs* 2:331–404.

Cook, S. F., 1941, The Mechanisms and Extent of Dietary Adaptation Among Certain Groups of California and Nevada Indians. *Ibero-Americana* 18:1–59.

Coville, F. V., 1892, The Panamint Indians of California, *American Anthropologist* 5:351–361.

Delacorte, M. G., 1990, *Prehistory of Deep Springs Valley, Eastern California:Adaptive Variation in the Western Great Basin*, Ph.D. Dissertation, University of California, Davis.

Delacorte, M. G., 1999, The Changing Role of Riverine Environments in the Prehistory of the Central-Western Great Basin:Data Recovery Excavations at Six Prehistoric Sites in Owens Valley, California, Far Western Anthropological Research Group, Inc. Submitted to CALTRANS District 9.

Delacorte, M. G., Gilreath, A., and Hall, M. C., 1992, A Class II Sample Survey for the Cortez Cumulative Effects Study Area, Lander and Eureka Counties, Nevada, Far Western Anthropological Research Group. Submitted to U.S. Department of Interior, Bureau of Land Management, Battle Mountain and Elko Districts, Nevada.

Dutcher, B. H., 1893, Pinyon Gathering Among the Panamint Indians, *American Anthropologist* 6:376–380.

Fischer, J. T., and Montano, J. M., 1977, Management of Pinyon for Ornamentals, Christmas Trees, and Nut Production. In *Ecology, Uses, and Management of Pinyon-Juniper Woodlands*, edited by E. F. Aldon and T. J. Loring, pp. 35–40. General Technical Report 39. USDA Forest Service Rocky Mountain Forest and Range Experimental Station, Albuquerque.

Garfinkel, A. P., Schiffman, R. A., and McGuire, K. R., 1980, Archaeological Investigations in the Southern Sierra Nevada:The Lamont Meadow and Morris Peak Segments of the Pacific Crest Trail. Bureau of Land Management (California) Cultural Resources Publications in Archaeology, Bakersfield.

Grayson, D. K., 1991, Alpine Faunas from the White Mountains, California, Adaptive Change in the Late Pleistocene Great Basin. *Journal of Archaeological Science* 18:483–506.

Grayson, D. K., 1993, *The Desert's Past. A Natural Prehistory of the Great Basin*. Smithsonian Institution Press, Washington, DC.

Jeffers, R. M., 1994, Piñon Seed Production, Collection, and Storage. In Desired Future Conditions for Pinyon-Juniper Ecosystems, edited by D. W. Shaw, E. F. Aldon, and C. Losapio, General Technical Report 258. USDA Forest Service Rocky Mountain Forest and Range Experimental Station, Ogden, UT.

Jennings, S. A., and. Elliott-Fisk, D. L., 1993, Packrat Midden Evidence of Late Quaternary Vegetation Change in the White Mountains. *Quaternary Research* 39:214–221.

Jones, K. T., and Madsen, D. B., 1989, Calculating the Cost of Resource Transportation, *Current Anthropology* 30:529–534.

Jordan, M., 1974, *An Inventory of Two Selected Woodland Sites in the Pine Nut Hills of Western Nevada*, M.A. Thesis, University of Nevada, Reno.

Kelly, R. L., 1990, Marshes and Mobility in the Western Great Basin. In *Wetland Adaptations in the Great Basin*:Papers from the Twenty-First Great Basin Anthropological Conference, Museum of Peoples and Cultures Occasional Papers No. 1. Brigham Young University, Provo, pp. 258–276.

Kelly, R. L., 1995, Hunter-Gatherer Lifeways in the Carson Sink:A Context for Bioarchaeology. In *Bioarchaeology of the Stillwater Marsh:Prehistoric Human Adaptation in the Western Great Basin*, edited by C. S. Larson and R. L. Kelly, Anthropological Papers, Vol. 58, Part 1. American Museum of Natural History, New York.

Kelly, R. L., 2001, *Prehistory of the Carson Desert and Stillwater Mountains:Environment, Mobility, and Subsistence in a Great Basin Wetland.* University of Utah Anthropological Papers Number 123, Salt Lake City.

Lanner, R. M., 1983, The Expansion of Singleleaf Piñon in the Great Basin. In *The Archaeology of Monitor Valley 2. Gatecliff Shelter*, edited by D. H. Thomas, pp. 167–171. Anthropological Papers, Vol. 59, Part 1, American Museum of Natural History, New York.

Lindsay, L. W., and Sargent, K., 1979, *Prehistory of the Deep Spring Mountain Area, Western Utah.* Antiquities Section Selected Papers No. 114, Utah Division of State History, Salt Lake City.

MacDonald, I., 1961, Statistical Studies of Recorded Energy Expenditure of Man. *Nutrition Abstracts and Reviews* 31(3):739–757.

Madsen, D. B., 1979, Prehistoric Occupation Patterns, Subsistence Adaptations, and Chronology in the Fish Springs Area, Utah. In Archaeological Investigations in Utah, edited by D. B. Madsen and R. E. Fike. Cultural Resource Series No. 12, Bureau of Land Management, Salt Lake City.

Madsen, D. B., 1981, The Emperor's New Clothes. *American Antiquity* 46:637–540.

Madsen, D. B., 1986, Great Basin Nuts. In *Anthropology of the Desert West:Essays in Honor of Jesse D. Jennings*, edited by C. J. Condie and D. D. Fowler, pp. 21–41. University of Utah Anthropological Papers No 110, Salt Lake City.

McGuire, K., and Garfinkel, A. P., 1976, Comment on "The Development of Pinyon Exploitation in Central Eastern California", *Journal of California Anthropology* 3(2):83–85.

McGuire, K., and Garfinkel, A. P., 1980, Archaeological Investigations in the Southern Sierra Nevada:The Bear Mountain Segment of the Pacific Crest Trail. Bureau of Land Management (California) Cultural Resources Publications in Archaeology, Bakersfield.

Metcalfe, D., and Barlow, K. R., 1992, A Model for Exploring the Optimal Trade-off between Field Processing and Transport. *American Anthropologist* 94:349–356.

Price, J. A., 1962, *Washoe Economy.* Nevada State Museum Anthropological Papers 6, Carson City, NV.

Rhode, D., 1980, Plant Macrofossils. In Archaeological Investigations in the Southern Sierra Nevada:The Lamont Meadow and Morris Peak Segments of the Pacific Crest Trail, edited by A. P. Garfinkel, R. A. Schiffman, and K. R. McGuire, Bureau of Land Management (California) Cultural Resources Publications in Archaeology, Bakersfield.

Rhode, D., 1990a, Settlement Patterning and Residential Stability at Walker Lake, Nevada:The View From Above. In *Wetland Adaptations in the Great Basin*:Papers from the Twenty-First Great Basin Anthropological Conference, Museum of Peoples and Cultures Occasional Papers No. 1, Brigham Young University, Provo, pp. 107–120.

Rhode, D., 1990b, On Transportation Costs of Great Basin Resources:An Assessment of the Jones–Madsen Model. *Current Anthropology* 31:413–419.

Rhode, D., 1999, The Role of Paleoecology in the Development of Great Basin Archaeology and Vice Versa. In *Models for the Millennium:Great Basin Anthropology Today*, edited by C. Beck, pp. 29–52. University of Utah Press, Salt Lake City.

Rhode, D., and Madsen, D. B., 1998, Pine Nut Use in the Early Holocene and Beyond:The Danger Cave Archaeobotanical Record. *Journal of Archaeological Science* 25: 1199–1210.

Rhode, D., and Thomas, D. H., 1983, Flotation Analysis of Selected Hearths. In *Archaeology of Monitor Valley:2. Gatecliff Shelter*, edited by D. H. Thomas, pp. 151–157. Anthropological Papers, Vol. 59, American Museum of Natural History, New York.

Scharf, E. R., 1992, *The Archaeobotany of Midway:Plant Resource Use at a High Altitude Site in the White Mountains of Eastern California*, M.A., University of Washington.

Simms, S. R., 1985, Pine Nut Use in Three Great Basin Cases:Data, Theory, and a Fragmentary Material Record. *Journal of California and Great Basin Anthropology* 7:166–175.

Soil Conservation Service, 1992a, Major Land Resource Area 25:Owyhee High Plateau, Nevada Site Descriptions. U.S. Department of Agriculture, Reno.

Soil Conservation Service, 1992b, Major Land Resource Area 26:Carson Basin and Mountains, Nevada Site Descriptions. U.S. Department of Agriculture, Reno.

Soil Conservation Service, 1992c, Major Land Resource Area 27:Fallon-Lovelock Area, Nevada Site Descriptions. U.S. Department of Agriculture, Reno.

Soil Conservation Service, 1992d, Major Land Resource Area 29:Southern Nevada Basin and Range, Site Descriptions. U.S. Department of Agriculture, Reno.

Soil Conservation Service, 1993, Major Land Resource Area 28B:Central Nevada Basin and Range, Site Descriptions. U.S. Department of Agriculture, Reno.

Steward, J. H., 1933, Ethnography of the Owens Valley Paiute. In *University of California Publications in American Archaeology and Ethnology*, Berkeley, Vol. 33, pp. 423–438.

Steward, J. H., 1938, *Basin-Plateau Aboriginal Sociopolitical Groups* Bureau of American Ethnology Bulletin No. 120, Smithsonian Institution, Washington, DC.

Steward, J. H., 1941, Culture Element Distributions: XIII. Nevada Shoshone. In:*University of California Anthropological Records*, Berkeley, Vol. 4, pp. 209–259.

Thomas, D. H., 1973, An Empirical Test for Steward's Model of Great Basin Settlement Patterns. *American Antiquity* 38(1):155–176.

Thomas, D. H., 1981, God's Truth in Great Basin Archaeology? *American Antiquity* 46:644–648.

Thomas, D. H., 1982, An Overview of Central Great Basin Prehistory. In *Man and Environment in the Great Basin*, edited by D. B. Madsen and J. F. O'Connell, pp. 156–171. Paper No. 2, Society for American Archaeology, Washington, DC.

Thomas, D. H., 1983a, *The Archaeology of Monitor Valley 1:Epistemology*. Anthropological Papers 58, Part 1, American Museum of Natural History, New York.

Thomas, D. H., 1983b, *The Archaeology of Monitor Valley 2:Gatecliff Shelter*. Anthropological Papers 59, Part 1, American Museum of Natural History, New York.

Thomas, D. H., 1985, *The Archaeology of Hidden Cave, Nevada*. Anthropological Papers of the American Museum of Natural History 61, New York.

Thomas, D. H., 1988, *The Archaeology of Monitor Valley 3:Survey and Additional Excavations*. Anthropological Papers of the American Museum of Natural History 66, Part 2, New York.

Thomas, D. H., and Bettinger, R. L., 1976, *Prehistoric Piñon Ecotone Settlements of the Upper Reese River Valley, Central Nevada*. In Anthropological Papers, Vol. 53, Part 3. American Museum of Natural History, New York.

Thompson, R. S., and Hattori, E. M., 1983, Packrat (Neotoma) Middens from Gatecliff Shelter and Holocene Migrations of Woodland Plants. In *Archaeology of Monitor Valley:2. Gatecliff Shelter*, edited by D. H. Thomas, pp. 157–169. Anthropological Papers. Vol. 59. American Museum of Natural History, New York.

Wells, H. F., 1983, *Historic and Prehistoric Pinyon Exploitation in the Grass Valley Region, Central Nevada:A Case Study in Cultural Continuity and Change*, Ph.D. Dissertation, University of California, Riverside.

West, N. E., Tausch, R. J., Rea, K. H., and Tueller, P. T., 1978, Phytogeographical Variation Within Juniper-Pinyon Woodlands of the Great Basin. *Great Basin Naturalist* Memoirs 2:119–136.

West, N. E., Tausch, R. J., and Tueller, P. T., 1998, A Management-Oriented Classification of Pinyon-Juniper Woodlands of the Great Basin. General Technical Report 12, USDA Forest Service, Rocky Mountain Research Station, Ogden UT.

Wheat, M. M., 1967, *Survival Arts of the Primitive Paiutes*. University of Nevada-Reno, Reno.

Wigand, P. E., and Nowak, C., 1992, Dynamics of Northwestern Nevada Plant Communities During the Last 30,0000 Years. In *The History of Water:Eastern Sierra Nevada, Owens Valley, White-Inyo Mountains*, Vol. 4, edited by C. A. Hull, V. Doyle-Jones, and B, Widawski, pp. 40–62. White Mountain Research Station Symposium, University of California Press, Berkeley.

Zeanah, D. W., 1992, *A Transport Model of Optimal Pinyon Camp Location in the Great Basin*. Paper Presented at the 23rd Biennial Meeting of the Great Basin Anthropological Conference, Boise, ID.

Zeanah, D. W., 1996, *Predicting Settlement Patterns and Mobility Strategies:An Optimal Foraging Analysis of Hunter-Gatherer Use of Mountain, Desert, and Wetland Habitats in the Carson Desert*, Ph.D. Dissertation, University of Utah, Salt Lake City.

Zeanah, D. W., 2000, Transport Costs, Central-Place Foraging, and Hunter-Gatherer Alpine Land-Use Strategies. In *Intermountain Archaeology*, edited by D. B. Madsen and M. D. Metcalf, pp. 1–14. University of Utah Anthropological Papers No. 122, Salt Lake City.

Zeanah, D. W., and Simms, S. R., 1999, Modeling the Gastric: Great Basin Subsistence Studies Since 1982 and the Evolution of General Theory. In *Models for the Millennium:Great Basin Anthropology Today*, edited by C. Beck, pp. 118–114. University of Utah Press, Salt Lake City.

Chapter 9

Residential and Logistical Strategies in the Evolution of Complex Hunter-Gatherers on the Kodiak Archipelago

BEN FITZHUGH

INTRODUCTION

My goal in this chapter is to model and evaluate evolutionary change in hunter-gatherer settlement and subsistence strategies on the North Pacific coast, specifically the southeastern portion of the Kodiak Archipelago in the Gulf of Alaska. The North Pacific forms an interesting case with reference to Binford's forager–collector model because it is characterized by a diverse and seasonally productive maritime environment (in summer) punctuated by seasonal impoverishment (in winter). A combination of oceanographic, geological, and climatic factors encourages high primary productivity, and a variety of resident and migratory species can be found in the near shore and littoral zones. In many areas, complex coastlines are characterized by tight ecological packing of exposed rocky shorelines, deep fjords, open embayments, and lagoons. This compression encourages considerable

BEN FITZHUGH • Department of Anthropology, University of Washington, Seattle, Washington 98195.

ecological diversity across a limited distance. These are conditions that Binford and others have claimed should encourage logistical mobility and a reliance on seasonal storage (Binford 1980, 1990; Steffian et al. 1998; Yesner 1980, 1998).

Discussing these and other factors of coastal environments, Yesner (1980) points out that maritime environments typically have greater stability (fluctuations of lower amplitude) and are supplemented by *unearned resources*, migratory species that increase the local carrying capacity but derive a large part of their nourishment elsewhere. From this, it is commonly argued that North Pacific hunter-gatherers would have rapidly developed residential sedentism (or a stable set of seasonal residential sites: see Mitchell 1983) and the kinds of logistical procurement strategies that are famously documented in North Pacific Rim ethnographies. Although not everyone sees the North Pacific Rim as a "Garden of Eden" unaffected by unpredictable variance in subsistence returns, most tend to view subsistence, settlement, and land use as largely invariant during the last 6000 or more years (see Maschner 1999; Steffian et al. 1998; Yesner 1998).

Yesner (1998) has published the most comprehensive model of North Pacific maritime adaptation, suggesting that fully maritime lifeways formed only after 6000 B.P., when an increasingly cold neoglacial climate decreased the productivity of terrestrial resources and human populations shifted gradually toward the coasts. Earlier coastal occupation is recognized (to at least 10,000 B.P.), but Yesner views these cases as only marginally maritime and interprets them as incremental shifts toward coastal life following a reduction of terrestrial productivity in the terminal Pleistocene. This compelling model follows a logic developed by Binford (1990), relating to shifting trade-offs between the opportunity costs of dispersed terrestrial foraging and increasingly logistical but geographically focal littoral and maritime foraging. Fisher, in this volume, presents a similar model for shifts between large and small game in the late European Paleolithic.

Yesner's model makes reasonable use of the published environmental and archaeological data (meager as they are), and it is clearly testable with the addition of new data on early coastal lifeways and the nature and timing of environmental changes through the Holocene. For example, emerging evidence from the Kodiak Archipelago indicates that fully maritime hunter-gatherers were in place by at least 6600 uncal. B.P./ca. 7500 cal. B.P. This suggests an earlier maritime focus on the adjacent mainland coasts of the Gulf of Alaska and/or Aleutians (Fitzhugh 2002). Rising sea level may have obscured earlier occupations. And of course models for the origins of the first migrations to America continue to leave open the possibility of Pleistocene maritime adaptations along the North Pacific Rim (Dixon 1993; Easton 1992; Fladmark 1979; Mandryk 1993).

Notwithstanding the murky picture surrounding the origins of maritime adaptations in the North Pacific, the models of Yesner and others for the relative productivity and stability of mid- to late Holocene marine and littoral environments leave little expectation of significant systemic change in subsistence and settlement patterns during the past 6000 years. This expectation has not, however, been rigorously examined archaeologically. In contrast to the presumed stasis in subsistence and settlement, the evolution of political complexity on the North American northwest coast is usually seen developing only during the late Holocene, variously around the North Pacific between 3000 and 1000 B.P. (Ames and Maschner 1999; Fitzhugh 2002; Matson and Coupland 1995). Given arguments by Kelly (1995) and others that politically complex hunter-gatherers owe their status to a chain of events triggered by reduced residential mobility, renewed examination of the historical development of settlement and land use patterns should help to clarify our understanding of social and cultural evolution in maritime contexts. The argument that I will make in this chapter is first that maritime hunter-gatherers of the North Pacific have not always organized subsistence and settlement according to the ethnographically documented modes that govern our preconceptions of North Pacific maritime organization. Second, I seek to demonstrate that an understanding of the variables that structure maritime settlement is critical to an understanding of cultural evolution in this region.

The Semantics of "Evolution"

Evolution is, of course, a loaded term full of ambiguity in recent archaeological literature. In this chapter, I refer to what might be called "structural evolution" or "systemic evolution," which I define as change in relationships within and between social/environmental parameters and individual strategic behaviors, where this change results in unprecedented organizations (of systems, modes of interrelationship, etc.). Although this is not the place to outline an epistemology of evolution more fully, I recognize two very different and ultimately compatible uses of the term in archaeological writing. The first refers to structural or systemic evolution, as just described, and the second relates to the innovation and differential persistence of heritable variance by Darwinian (or analogous) processes. The two concepts are certainly interrelated and presumably Darwinian processes underlie many social and systemic processes, but ignoring the difference between the concepts has been problematic[1] and has opened ecological anthropology to charges of teleological functionalism and

[1]One of the problems to which I refer is the argument, especially popular in 1970s and 1980s anthropological archaeology, that a particular cultural *system* evolved because it, as a unit,

scientific vacuousness (see Dunnell 1980; contributions in O'Brian 1996). Many ecological archaeologists, myself included, have sought to escape unjustified group functionalism in social evolution by seeking explanatory mechanisms in the cumulative effects of individual behavior in a socioecological context and by drawing inspiration from behavioral/evolutionary ecological theory (Boone and Smith 1998; Broughton and O'Connell 1999; O'Connell 1995). Notwithstanding these trends, retaining the systemic notion of social evolution on a scale above that of the behavioral mechanisms that innervate it saves us from counterproductive denial of macroevolutionary outcomes and allows us to continue to pursue a suite of questions about the development of social diversity and complexity at one level while we continue to pursue the microfoundations of systemic operation and change at another. Searching the forager–collector model for insights into macroevolutionary processes is an example of the attempt to use behavioral scale mechanisms to understand cumulative change.

I make no apologies to those who, following currently fashionable trends, disavow the reality or relevance of social evolution in favor of historic particularism. It is, however, important to acknowledge that any specific cultural trajectory is produced through a combination of patterned and potentially predictable processes, interlaced with unpredictable events and unique combinations of forces (historic contingencies). The emphasis placed on one or the other will be dictated by the analytical goals of analysis. In this case, I am interested in generalities that will benefit our understanding of similar contexts beyond the case under consideration. It was, of course, in the same vein that Binford developed his forager–collector model and that he directed much of his prior and subsequent work.

THE FORAGER–COLLECTOR MODEL AND MARITIME HUNTER-GATHERERS

Binford's original formulation (1980, 1982) was derived from a theoretical consideration of terrestrial hunter-gatherers stimulated by ethnographic comparisons, and in it he traced the implications of spatiotemporal

was more adaptive or fit than some alternative or preceding system. Though this kind of argument need not be incorrect, it has proven analytically ambiguous because social systems rarely operate as sufficiently integrated units for group selection to operate on this scale. Internal contests between group members are often more directly implicated as mechanisms of change than group "adaptation." The mechanisms of adaptation are also analytically vague at this level. I do not wish to imply, however, that group-selection arguments are necessarily wrong; only that, given the lack of clearly developed theoretical units and mechanisms, arguments of this sort have inevitably assumed what they should have sought to explain, namely, that cultural practices arise through a process of group selection.

resource variation for residential and logistical mobility and storage. Spatial incongruity between simultaneously available resources, it was argued, encourages logistical over residential means of bringing resources and consumers together. Thus, increasingly patchy environments should generate more logistical procurement strategies and decreased residential mobility. Temporal variation on the appropriate scales should, on the other hand, encourage storage. And storage in turn increases the effective patchiness of the environment by increasing the costs of residential mobility (storage transport). Because patchy environments also tend to be the most seasonal (shorter growing seasons), logistical collecting and storage are often highly correlated in ethnographic samples. We infer from this that when hunter-gatherers were "pushed" or "pulled" into increasingly extreme and patchy environments, residential mobility would have diminished, whereas logistical production and storage would have increased. The reverse is expected as groups occupy less patchy landscapes.

Of course, environmental conditions change in response to cyclical and stochastic processes or as populations move from one location to another, and Binford was careful to point out that logistical and residential variability represent strategic alternatives "which may be employed in varying mixes in different settings" (Binford 1980:355). Unlike our goals in this volume, Binford was also not explicitly considering long-term changes in hunter-gatherer settlement and mobility patterns in his 1980 article. He did, however, mention that long-term change could be expected and that density-dependent factors that limit residential mobility should lead to increased logistical production (Binford 1980). Factors such as increased population density are expected to have this effect.

Population density is but one of several variables that impact the foragers' experience of resource patchiness in space and time and, subsequently, the relative desirability of residential mobility compared to logistical production and storage. And as Binford (1980) and Kelly (1995) have pointed out, other factors besides subsistence resource distributions can limit residential mobility (such as the location of fresh water, fuel, shelter, or defensive locations). Technological capacity (for both transport and subsistence procurement and processing) also strongly conditions the effective patchiness of an environment. We must add to these the constraints and opportunities of the social environment for accessing resources, mediating resource variability, and ensuring reproductive opportunities (e.g., Wiessner 1982; Wobst 1974).

Maritime Hunter-Gatherers

Maritime hunter-gatherers face a number of geographically and technologically unique conditions that affect their residential and logistical options (e.g., Yesner 1980). In addition to higher productivity and packed ecological diversity, maritime populations are typically tethered residentially to the coastline (bringing them as close as physically possible to the resources that they most depend on). By using boats, residential flexibility should be inversely proportional to the complexity of coastal geography, all else being equal. Occupants of linear coastlines are more limited residentially (and to a lesser extent logistically) than those living in archipelagos of closely spaced islands or along deeply embayed landscapes. Linear rocky coastlines afford fewer residential opportunities, and we would expect maritime hunter-gatherers to be more tethered to accessible locations (river mouths, spits, and bays).

One set of assumptions that is commonly made by archaeologists who study the development of North Pacific hunter-gatherer societies is that ecological conditions would have favored residential stability, primarily logistical subsistence organization, and seasonal scale storage from the earliest phases of coastal occupation (Maschner 1999; Steffian et al. 1998). This follows Binford's (1980, 1990) proposal that hunter-gatherers should *tend* to solve seasonal scale imbalances in food availability through storage and Yesner's (1980) observation that maritime hunter-gatherers in relatively high latitudes (in ethnographic samples) have tended to be organized in just these ways. Minimal archaeological data have supported these beliefs (e.g., Aigner and Del Bene 1982; see also Cannon this volume; Ames and Maschner 1999), but there has, not yet been sufficient treatment of this problem. Drawing on socioecological theory and data derived from an intensive archaeological survey of the southeastern region of the Kodiak Archipelago, I will evaluate what I will refer to as the stasis model of maritime settlement/subsistence organization.

BACKGROUND

I spent the past several years studying the evolution of settlement and land use in the Gulf of Alaska's Kodiak Archipelago (Fig. 9.1) with the overall goal of understanding better the evolutionary dynamics that give rise to hunter-gatherer social and political complexity (Fitzhugh 1996, 2002). To this end, I conducted an intensive archaeological survey and several excavations in the Sitkalidak region to understand better the relationship between economic, technological, social, and political factors in the evolution of "complex hunter-gatherers." Detailed description of the Sitkalidak Archaeological Survey project can be found in Fitzhugh (1996). My examination of the Kodiak data here will follow from a discussion of Kodiak environmental variability and the formulation of a model that relates to change in

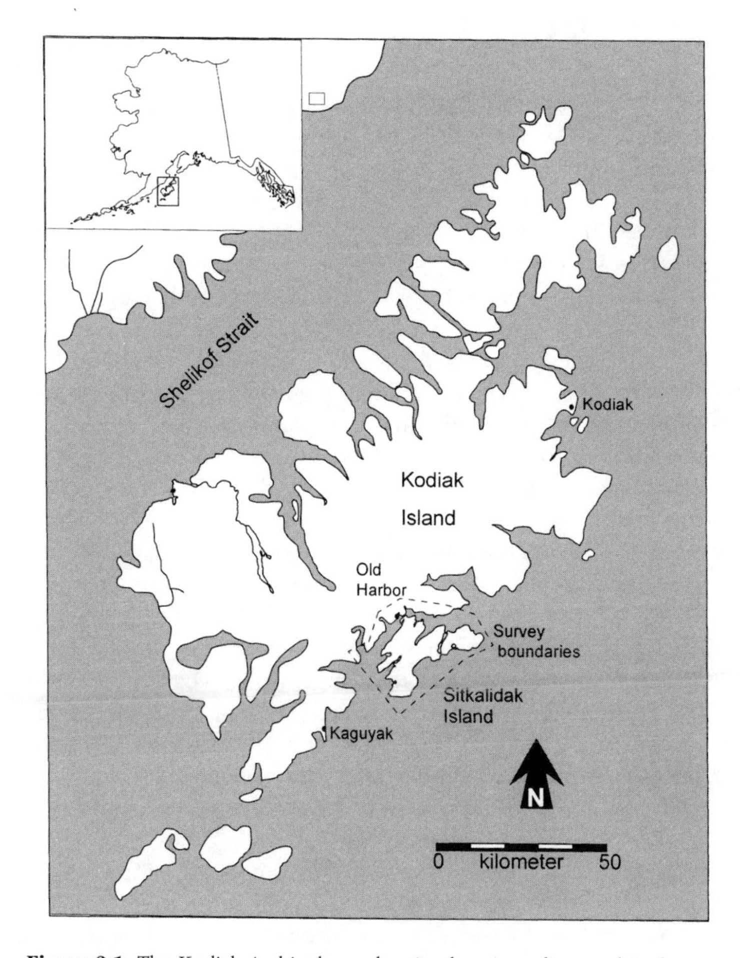

Figure 9.1. The Kodiak Archipelago, showing locations discussed in the text.

mobility strategies, settlement, and land use. I will specifically examine three lines of evidence: site distributions, site function, and artifact variability.

Seasonal Variation

As part of the American Pacific Northwest ecozone, Kodiak is located in a relatively productive coastal environment and experiences high seasonal variation in resource availability (Hood 1986). The vast majority of this productivity is concentrated in marine and littoral zones, whereas the terrestrial environment is comparatively impoverished. Prehistorically important resources include a variety of sea mammals, fish, birds, eggs, shellfish, and plant products, whose availability fluctuates dramatically between seasons.

Summer is the season of the highest resource productivity, followed by spring and fall. Seals congregate in haul-outs on the rocky outer coasts. Seal haul-out zones and year-round sea lion rookeries are more accessible in summer due to typically better boating weather. Herring aggregate and spawn close to shore in late spring/early summer, and salmon amass in spawning rivers from summer into fall. At the same time, mature halibut move into bays and estuaries from their winter grounds out on the continental shelf. In spring and fall, whales swim close to shore, and migratory waterfowl congregate where they can find open water in lakes and lagoons.

By any measure, winter is the season of scarcity for hunter-gatherers, as it is throughout the subarctic and arctic latitudes. Nevertheless, several resources continue to be available. These include sea birds, shellfish, seals and sea lions, and resident near-shore fish (cod, juvenile halibut, flounder, sculpin, etc.). Of these, sea birds are relatively lean and poor-tasting, shellfish are potentially lethal (Fitzhugh 1995) and also lean (Erlandson 1988), seals disperse and are less predictably located, and resident near-shore fish are small and occur in relatively low densities. Within the range of written records, unique locations such as the Karluk River have supported anadromous fish and waterfowl well into the winter (Barsch 1985; Patrick Saltonstall, personal communication 2000), but such locations are rare, and none is located in the area to be examined in this chapter. Overall, winter is a season of both reduced productivity and reduced patchiness. The latter follows from the elimination of concentrated summer resources such as seal haul-outs and salmon runs (see Fig. 9.2).

In addition to variability in seasonal subsistence, significant differences are documented in weather patterns that effect mobility opportunities (Wilson and Overland 1986; Steffian, in prep). At least in modern climatic regimes, fierce storms occur roughly every four days in winter. Strong wind and waves, icing, and fog are also difficulties faced regularly by winter boaters. Paddling skin boats in these conditions would be extremely taxing. Seasonally foreshortened daylight also limits the distances that could be safely traveled, reducing foraging opportunities for many resources. Nevertheless, boat-based foraging opportunities remain.

The potential for overland travel is facilitated in winter by frozen ground and snow, and this could have served as a form of mobility not often considered by Kodiak archaeologists (Saltonstall, personal communication, 2000). Nevertheless, overland mobility would have been qualitatively different from boating that has greater capacity for transporting people and resources. Foraging theory suggests that the ranges of terrestrial foraging would be smaller than boating ranges (see also Ames, this volume). Add to

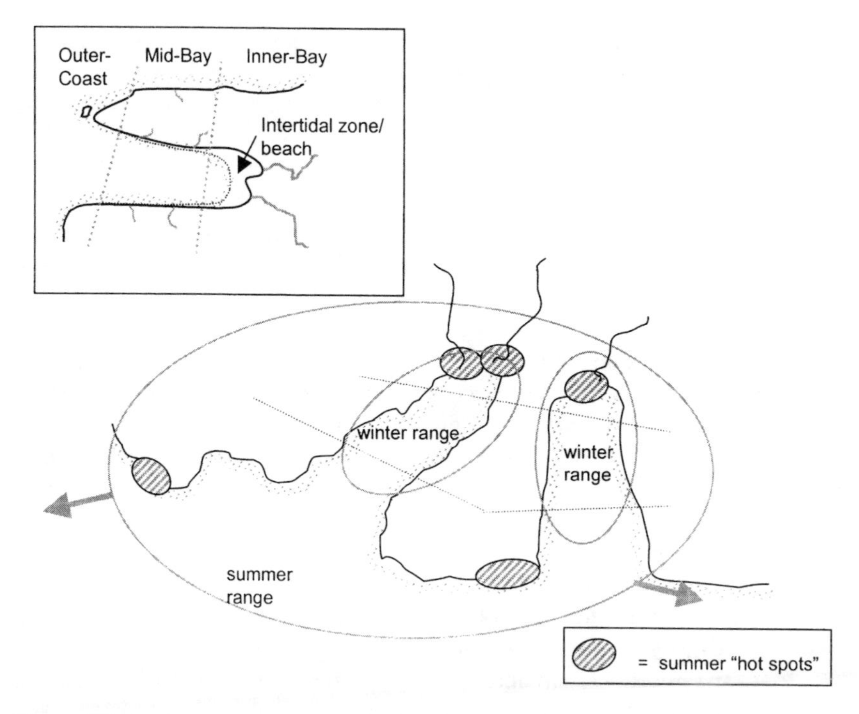

Figure 9.2. A schematic representation of ecological distributions and their implications for maritime foragers living on the reticulated Kodiak coast. See text for discussion.

this a consideration of the mountainous and deeply fjorded terrain and the lower terrestrial productivity, especially in winter, and we are left with little reason to expect significant amounts of terrestrial foraging. This expectation is generally supported by analyzed zooarchaeological assemblages that report only small proportions of terrestrial species, such as bear, fox, and river otter (Amorosi 1987, 1988; Clark 1974, 1997; Yesner 1989). Because of the greater difficulty involved in transporting resources overland, terrestrial mobility should have been primarily residential, moving people to new coastal patches following local resource depletion; and social, linking individuals or families into larger networks for informational, reproductive, social, and spiritual purposes. Even these activities were probably facilitated by boat travel more often than by land travel.

It is reasonable to divide Kodiak's convoluted coastline into three zones on the basis of resource distributions and the relative hazards of boating in each (Haggarty et al. 1991). Each bay or estuary on Kodiak, where the coastline is exposed between bays, is a more or less redundant ecological unit (Fig. 9.2). The zone outside of the bay and at the bay mouth is exposed to high-energy surf. As a result, the coast tends to be steep and rocky and often supports seal haul-outs, sea lion rookeries, sea bird colonies, and shellfish adapted to high-energy coasts (e.g., blue mussel, sea urchin, chiton, and limpet). Inside of this zone is a more protected midbay portion. In this section, hills tend to rise steeply on both sides, and the subtidal zone drops sharply from the shores. The relatively steep slopes make for contracted intertidal and subtidal zones and increase the diversity of resources available close to shore. Short freshwater streams punctuate the midbay coast, and steep gravel beaches offer good boat landing. The inner portion of the bay commonly receives discharge from one or more larger streams that drain one or more substantial inland valleys. High sediment outwash and low coastal energy combine in the inner bay zone to generate a shallower coastline that has significant horizontal tidal displacement. The broad intertidal zone, composed of fine sediments, supports burrowing shellfish but can be hazardous to pedestrians and boaters. Rivers that drain the heads of bays commonly support the larger runs of anadromous fish, although some species spawn even in the brackish discharge of the steeper streams that line the midbay regions. Lower topography and prograded beach berms in this zone provide ample locations for situating dwellings. Lagoon systems are variants on the inner bay theme, shallow with extensive mud and clay intertidal systems, although larger lagoons can include substantial "midbay" sections.

KODIAK PREHISTORY

Following convention (Clark 1984, 1992a, 1994), I divide the Kodiak cultural sequence into three prehistoric periods and one Historic period: the Ocean Bay period (5500–1500 B.C.), the Kachemak period (1500 B.C.–A.D. 1200), the Koniag period (A.D. 1200–1784), and the Historic period (after A.D. 1784). These *periods* are often subdivided into tighter chronological *phases* on the basis of modest changes in material culture. These are commonly identified as Ocean Bay I and II, Early and Late Kachemak, Early and Developed Koniag, and Russian and American Historic phases. The larger period designations are sufficient for our purposes. The Historic period is identified generically to avoid confusion over the ethnic or chronological attribution of sites derived from native Kodiak Alutiiq (Qikertarmiut), Russian, and Euro-American source populations. Substantial summaries of these periods and phases can be found elsewhere (Clark 1974, 1979, 1984, 1996, 1997; Fitzhugh 1996; Jordan and Knecht 1988; Knecht 1995; Steffian 1992).

The oldest accepted period on the Kodiak Archipelago is the Ocean Bay period. This relatively long-lasting tradition is characterized by core and blade and bifacial technologies supplemented and largely supplanted late in the period by ground slate technologies (Clark 1979, 1982). Occupations commonly contain floor deposits of red ocher, which are interpreted as reflecting heavy use of ocher as a treatment for skin tents and other hide preparations (Fitzhugh 2002; see Jewitt 1987; Philibert 1994). Late in the Ocean Bay period, by 2500 B.C., semisubterranean sod houses become common and ocher floors disappear. The origins of the Ocean Bay tradition are unclear, although it probably arose from the original maritime settlement of the Gulf of Alaska. The earliest known Ocean Bay sites share some technological affinities with the terminal Pleistocene/ early Holocene Paleoarctic tradition and with the unique assemblage from the Anangula site in the eastern Aleutian Islands (Fitzhugh 2002; Jordan 1992). As a technological tradition, Ocean Bay assemblages are identified throughout Kodiak on the adjacent shores of the Alaska Peninsula (G. Clark 1977; Dumond 1987; Jeanne Schaaf, personal communication 1998), in the Kachemak Bay region (Workman 1998), and possibly in Prince William Sound (Linda Yarborough, communication 1998).

The Kachemak period follows Ocean Bay on Kodiak and appears to develop from a late Ocean Bay base (Clark 1996, 1997). This period is characterized by new tool forms and (late in the period) the development of elaborate ornamentalism (labret-wear, decorated stone lamps) and mortuary ritual (Heizer 1956; Workman 1992). It is notably different from Ocean Bay in the use of mass harvesting and processing techniques (nets and semilunar knives) and the establishment of substantial residential sites recognized today as the earliest villages of the area (Crowell 1986; Erlandson et al. 1992; Haggarty et al. 1991). Toward the end of this period (at about A.D. 800), local competition is evident in the construction of defensive facilities and complex mortuary treatment (Fitzhugh 2002; Simon and Steffian 1994; Urcid 1994).

The Koniag period occupies the last 600 years before Russian contact in the late eighteenth century. This period witnessed a number of technological and social changes, including the use of composite fish harpoons, gravel tempered pottery, and multiroom houses large enough to enclose extended families (see Jordan and Knecht 1988 or Knecht 1995 for a more complete accounting of Koniag material culture). Whale hunting becomes a major social and economic activity (Fitzhugh 2002). Villages become larger, defensive sites expand to service entire communities, and social inequality is evident in archaeological house-size variation (Fitzhugh 1996, 2002). Cultural continuity between Kachemak and Koniag periods has been debated since the 1930s (Hrdliçka 1944; Clark 1988, 1992b, 1994; Dumond 1988a,b, 1994, 1998; Jordan and Knecht 1988; Knecht 1995). Clearly, the Kodiak Archipelago of the Koniag period became part of a much larger socioeconomic network of trade and warfare than it had been previously, which led to an increase in the transmission of off-island material culture and practices onto the islands. What is also increasingly clear is that there was no cultural hiatus or hostile takeover of the Kodiak region. Instead the archaeological evidence demonstrates continuity and gradual change in most material characteristics through the transition (Jordan and Knecht 1988; Knecht 1995; see also Fitzhugh 1996).

In contrast, the Russian takeover of Kodiak in the years following A.D. 1784 was certainly hostile and devastating to the native (Qikertarmiut or Kodiak Alutiiq) populations living there (Black 1992; Crowell 1997). Able-bodied males were conscripted to hunt sea otters as far away as the Kuril Islands and California Channel Islands (Ogden 1941, 1991; Shubin 1994). Others were forced into service supplying subsistence products to Russian outposts or as domestic servants. In the years 1837–1839, a massive smallpox epidemic claimed thousands of lives. Following that event, an imposed resettlement consolidated the Qikertarmiut population into a small number of villages (the common estimate of seven villages may, however, be a slight underestimate; Aron Crowell, personal communication 2000). The depopulation and resettlement are believed to have impacted traditional land use practices, among other effects (Knecht and Jordan 1985). The combination of harsh treatment and exposure to introduced diseases during the Russian occupation reduced the Qikertarmiut population to less than one-quarter of its size at contact (Clark 1984; Erlandson et al. 1992; Haggarty et al. 1991), despite humanitarian interference by Orthodox missionaries beginning in the 1790s. The American purchase of Alaska in 1867 further dispossessed Alaskan Natives, who subsequently came under the jurisdiction of the American Bureau of Indian Affairs and were subjected to the assimilationist policies of the United States. Only since the 1971 Alaska Native Claims Settlement Act have the Qikertarmiut and other Alaskan Natives been given limited legal rights to administer territory or derive economic advantage from their traditional homeland. Through most of these dramatic changes, Qikertarmiut maintained the use of traditional subsistence foods, technologies of food capture, and house construction techniques into the early twentieth century, and to this day they hold onto many aspects of traditional culture, especially in the area of subsistence. Let us now turn back to the issue of subsistence-settlement change during this 7500 year period of maritime hunting and gathering on Kodiak.

A MODEL OF SUBSISTENCE, SETTLEMENT, AND LAND USE CHANGE

A simple observation concerning the nature of population density on subsistence seasonality goes against the common assumption that hunter-gatherers of the North Pacific have always been semisedentary, logistically oriented foragers committed to seasonal storage practices to survive lean winters. Small populations are far more likely than large ones to be able to survive lean winters without seasonal scale storage. In the Kodiak Archipelago today, winter is a much more difficult period in which to secure food than spring, summer, or fall. The most productive resources (salmon, adult halibut, migratory waterfowl, and sea mammals) leave the region or are at least less accessible during this period (minimally November to April). Combine this with a tendency for much more extreme weather in winter (Wilson and Overland 1986), and foragers would be forced to subsist in winter primarily on a handful of locally available bottom fish, sea birds, and shellfish. Assuming rough climatic and ecological comparability between today's environment and that of the archaeological past (a problematic assumption at best; see Grayson 1984:174–177), small groups could have subsisted throughout the winter without storage, so long as they could move occasionally when local fish, bird, and shellfish stocks became diminished. This would be possible only as long as foraging groups were small and intergroup density low enough that alternative residential locations (foraging ranges) could be reached without much competition from other groups.

To make this argument somewhat more formally, we can use the patch choice model of optimal foraging theory (Kaplan and Hill 1992; Stephens and Krebs 1986). According to that model, a forager (or group of foragers) who experiences diminishing returns should choose to leave a patch when the foraging return rate is reduced to the point that the energetic and logistical costs of relocation are less than the costs of remaining in the diminished patch. Increased patchiness (heterogeneity) amplifies the cost of relocation, whereas reduced patchiness (homogeneity) decreases it, all else being equal, and as a result, foragers are expected to persist in increasingly diminished patches in proportion to the spatial patchiness of the environment. It turns out that overwintering resources are more evenly distributed around the Kodiak coastal environment than the more productive and more concentrated summer resources (see Fig. 9.2). Overwintering resources are also less productive and would be more easily diminished by foraging pressure.

Kelly (1995) has argued that residential mobility is usually motivated by diminishing returns of near-camp resources, regardless of the productivity of more logistically targeted resources that can be pursued from a number of alternative camp sites. In this hypothetical case of winter foraging on Kodiak, most winter resources could be captured profitably only close to camp due to the increased energetic costs and greater hazards of winter movement (both accentuated by cold weather, reduced daylight, and increased storminess). This means that winter foraging ranges should have been smaller than summer ranges, winter foraging would have more easily led to diminishing returns, and winter would be the season when periodic residential mobility would have improved the energetic return rate for central-place foraging. However, given the assumed difficulty of travel in winter (for foraging or relocating camp), we would not expect people to move very far in any single move. The same sets of longer distance logistical targets would commonly be retained following a move.

If winter is a time of near-camp foraging and occasional residential moves, what of summer? Again, assuming rough ecological equivalence between past and present, summer was a time of enhanced ecological productivity and comparatively greater scheduling conflicts. Seals and sea lions aggregating in haul-outs and rookeries along the exposed outer coast were attractive as high-yield, proteinaceous, fat-rich resources with valuable raw materials such as hides and bone for tents, boats, and tool production. Inner bays with larger rivers support high concentrations of anadromous fish, and availability ranges from a few days to several months depending on the characteristics of the stream. Small streams along midbay sections also supported short fish runs and would be nearest to good halibut and cod fishing. Residential camps with the greatest access to all ecological zones would be most advantageous, and these camps would be located where distances between outer coast and inner bay foraging could be minimized. Therefore, opportunistic and logistically optimal residential locations should lie in the midbay areas. Foraging in summer would be central-placed and logistical (in the sense of moving resources to camp, as opposed to the inverse). The main differences between summer and winter foraging activities should be the range of territory exploited and the focal productivity of outer coast and inner bay zones.

From these seasonal predictions, can we derive predictions pertaining to measurable archaeological parameters? This model has led me to expect that early hunter-gatherer populations (associated especially with the earliest Ocean Bay I phase on Kodiak) would most often have situated themselves in midbay base camps that would be occupied throughout the productive months and then moved one or more times in the winter. Such a settlement pattern would tend to encourage the use of portable technologies and minimal investment in settlement permanence. Site deposits should be thin and structures/features minimally "built-up." These predictions are largely met for the earliest phase of Kodiak prehistory, as discussed elsewhere (see Fitzhugh 1996, 2002). My interest here is in contrasting the predicted pattern of mobility and settlement variation with predictions for subsequent phases, once higher populations and mass-harvesting and storage techniques became popular.

From the above scenario, it follows that archaeological intersite variability of Ocean Bay sites should be low with a series of base camps, in similar (midbay) locations throughout the year, but moved occasionally. Binford (1982) argued that along the forager–collector continuum, residentially mobile foragers "mapping onto" resources should produce more homogenous archaeological assemblages (based on intersite comparisons) compared to more logistically organized collectors. A mix of logistical and residential strategies is predicted for the Ocean Bay period. However, because populations were low, competition minimal, and foraging largely boat-based (Ames, this volume), archaeological patterns should reflect a more fully residential (forager) pattern, but should have greater accumulations of debris on camp sites than one typically finds for more residentially mobile terrestrial foragers. When population aggregations occurred, they were most likely during the more productive summer months and, if archaeologically visible, should be reflected by larger than average sites close to predictably productive summer resource locations. This pattern of subsistence and settlement is expected to persist until its effectiveness is compromised.

The Ocean Bay period, then, should be characterized by a central-place foraging pattern, where residential camps are in logistically optimal midbay locations. The duration of base camp use is unpredicted, but the general character of these sites should be that of temporary camps, lacking in the durable structures that are made by people anticipating future use. Extraction locations should be minimally represented archaeologically because most foraging excursions would be boat-based. Resource search, pursuit, initial processing, and field discard would occur on the water or at its edge where dynamic processes severely limit archaeological preservation. Therefore, site types should be minimally differentiated. Artifacts should indicate minimal investment in processing relative to procurement, and artifact assemblages should reflect low variability compared to later (more specialized and differentiated) tool kits.

Crowding, Technological Intensification, Mass Harvesting, and Storage Effects on Settlement and Land Use

The Kachemak tradition represents a significant change in subsistence technology from that of the Ocean Bay. Net-sinkers and semilunar (*ulu*) knives indicate the development of mass-harvesting strategies for capturing and processing fish (Fitzhugh 2002). This change, present from the beginning of the Kachemak period, represents the best evidence (along with changes in site location characteristics, see below) that seasonal subsistence practices were redirected at this time to include the production of winter food stores. Schooling fish (especially salmon and herring) occur in large numbers for short periods during summer. They can be caught in nets stretched out from shore and across spawning rivers or in weirs built across rivers. Once caught, fish spoil rapidly before processing, making a large labor force of processors prerequisite to bulk harvesting of these resources. Salmon and herring, unlike most other subsistence targets, are also highly resilient to predatory pressure and could withstand massharvesting more effectively than sea mammals, birds, shellfish, and most bottom fish (Fitzhugh 2002). How can we understand this strategic and technological change and what predictable effects can we anticipate in settlement and land use?

As many scholars have pointed out, storage fundamentally changes the relationship between people and the productive landscape (Binford 1980, 1990; Goland 1991; Testart 1982; Winterhalder and Goland 1997; Woodburn 1981, 1982). All societies practice *some degree* of storage (Binford 1990; Kelly 1995), but it would be a mistake to assume that significant resource "overproduction" and storage were always desirable or even possible on the North Pacific coast. Seasonal or interannual storage often involves large labor investments in preserving and protecting stores. Bettinger (1999:52) calls this investment "storage time" to distinguish it from time spent preparing a stored resource for consumption, or "handling time." He argues that storage preparation can involve major trade-offs in time taken away from foraging for higher ranked resources that have greater benefits in immediate consumption. When population densities are low and periodic residential mobility can compensate for locally depleted patches (as predicted for the Ocean Bay period discussed above), investing in storage production and the accompanying increased processing activities may be perceived as disadvantageous. It is also possible that social values associated with sharing in immediate-return economies add an element of conservatism, further decreasing the appeal of a delayed-return economy (Bettinger 1999:54; Boehm 1993). On the other hand, behavioral ecological arguments have been proposed that are more consistent with the general line of reasoning developed in this chapter (Blurton Jones 1987; Hawkes 1992; Winterhalder 1986, 1996, 1997). According to these views, the high variability of hunting success in large, high-ranked prey would encourage sharing these items, whereas gathered products are typically small and available to all foragers simultaneously. Gatherers would rarely collect more than they felt was necessary for consumption by themselves and their dependents. As the importance of gathered foods relative to hunted foods (especially large game) shifted, storage could become a more prominent economic activity.

Another impediment to a storing economy might be the difficulty of forming cooperative task groups large enough to concentrate storable resources efficiently. Efficient harvesting of aggregated resources, be they caribou herds, beluga pods, bison, or salmon, often requires the collaboration of several hunters and processors (e.g., Lucier and Van Stone 1995;

Spiess 1979). Extensive cooperation in resource production would be more difficult for dispersed populations that deal with seasonal variability by "immediate-return" foraging (*sensu* Woodburn 1981) and residential mobility. The "economies of scale" involved in shifting to a "delayed-return" strategy might pay off only when large enough populations find themselves in relatively close quarters and are motivated to cooperate in the production process. Finally, the absence of appropriate technology could play a role in delaying adoption of a storage-based economy (Fitzhugh 2001).

Tipping the balance from a mobility-mediated to a storage-mediated subsistence strategy should involve a change in the perceived opportunity costs of these two strategies. Part of the trade-off would entail reduced ability to escape local resource depression and a decrease in the set of potential foraging patches. Given the reduced flexibility such a shift entails, foragers are not expected to change to a storage-based strategy without compelling motivation. I follow many others in believing that population packing would provide at least proximate circumstances to trigger this change (Bettinger 1999; Price and Brown 1985).

It is not unreasonable to expect external factors, such as climatic deterioration, to compel the adoption of previously unappealing alternatives to lean-season survival. Interestingly, the development of intensive storage production on Kodiak appears to coincide with the onset of the neoglacial cold period, which would have imposed greater constraints on mobility and subsistence production than would have been the case in the previous Hypsithermal (Mann et al. 1998). It is also reasonable to expect that population growth would occur in the absence of climatic variability until foraging bands began to encroach on the foraging ranges of others, thereby reducing the effectiveness of either residential or logistical mobility to compensate for spatiotemporal resource variation. Before "filling" of the landscape, population growth should occur at a slow rate commensurate with rather severe seasonal fluctuations in dietary sufficiency and attendant fertility effects (Ellison 1994), but otherwise unencumbered.

Once hemmed in, the costs of interpatch (residential) mobility would increase, and foragers would remain longer in local resource patches, tolerating higher levels of resource depletion (especially of slowly reproducing species). Diet breadths are expected to expand as foraging return-rates decline, resulting in the dietary inclusion of more resources requiring higher processing inputs. Strategic and technological innovations, such as assembly of larger cooperative foraging and processing parties (perhaps drawing collaboration from those uncomfortably close neighboring bands) and improvements in the technologies of mass harvesting and processing would substantially improve foraging efficiencies in these previously low-ranked species or minimally storable resources. In the absence of strategic and technological innovations, the decrease in per capita foraging efficiency expected with expanding diet breadth should preclude any significant population growth and might even compel population decline. No necessary relationship is supposed between population pressure and technological change (but see Fitzhugh 2001).

As Binford (1980) notes, major storage systems will fundamentally alter settlement strategies by increasing the costs of residential mobility while simultaneously decreasing its utility. And as I have noted before, for maritime systems on the Kodiak Archipelago, residential mobility and storage are primarily alternatives to winter scarcity (see Halstead and O'Shea 1989). Therefore, we would expect the development of storage to lead more directly to an increase in residential permanence.

From the foregoing, it should be clear that a proportional increase in storage strategies on Kodiak can be seen as a density-dependent development. It arises from decreased mobility opportunities with climatic deterioration and population packing and an increase in the proximity of potential collaborators for the harvesting and processing of time-limited (economy of scale) resources. The expectation of a density-dependent cause for storage and sedentism directly contradicts the prediction that high latitude maritime hunter-gatherers should be obligate storers (Binford 1990) or residentially stable (Yesner 1980) (although the predictions may still hold in comparison to most terrestrial foragers). My view is justified by the realization that small populations could survive winters on Kodiak without storage during the Ocean Bay period. One probable reason why few high latitude maritime hunter-gatherer groups have been observed ethnographically without seasonal scale storage or residential permanence may relate to the fact that all such environments had been occupied for thousands of years before ethnographic documentation. Density-dependent effects had already altered the structure of seasonal residence patterns. Thus, we are reminded that ethnographic analogs are incomplete, and evolutionary change is likely to contain variants not presently observed (Wobst 1974).

Interestingly, a second by-product of the development of a storage economy is the formation of *corporate groups*—teams of individuals who regularly collaborate in producing and consuming stored foods. These groupings should form along kinship lines and would be encouraged by the mutual desire to claim access to shares of collaborative production. Thus, shift to a processing-intensive storage strategy is also expected to generate a revolutionary transformation in the nature of settlement, leading to the first relatively sedentary, aggregated settlements that have residentially cohesive corporate units. Significant inequalities are not expected at this point—prospective subordinates could still leave—and corporate groups should be roughly equal in size.

In the patchy North Pacific environment, the emergence of a storage-based, processing economy should simultaneously stimulate *increased sedentism* (defined as long-term, permanent, site occupation) and more logistically organized *residential mobility*. Extraction locations that had previously attracted opportunistic foragers on daily excursions from their base camp would become targets for more intensive residential outposts (bulk extraction), bringing larger numbers of producers to resources for efficient production during periods of availability. We might characterize this as the development of more intensive land use, in which residential sites would be used repeatedly by larger groups, under a more structured set of seasonal activities. This is also the development of a semisedentary or seasonal mobility pattern structurally similar to that observed in ethnographic contexts around the North Pacific in recent centuries.

This land use pattern brings to mind John Murra's Andean "vertical archipelago" model (1980), in which distinct ecological zones are integrated through intensive resource extraction and transport to a "colonial" center.[2] In our loosely analogous "horizontal archipelago" model, resource extraction camps are used seasonally by task groups who "colonize" patches during their productive season to bring the resources back to a centralized village for storage and consumption. The result is both intensified land use and a reversal of the seasonality of residential movement. Whereas immediate-return groups would be most likely to move home base in winter, delayed-return groups are most like to move in spring, summer, or fall. The result is the formation of the "winter village," known ethnographically as the focal point of residential and ceremonial life and the stable hub of regional subsistence activities. We might then say that resource extraction camps at this point are logistical from an annual perspective but residential from a seasonal perspective. Stores are produced at residential camps and returned to the winter village for storage and consumption. This pattern is similar in some respects to Bettinger's (1999) account of terrestrial hunter-gatherer settlement patterns in the Middle Period of Nevada's Owens Valley, calling into question the notion that maritime hunter-gatherers are qualitatively different from terrestrial ones in broad ecological and evolutionary dimensions (but see Yesner 1980). This pattern varies around the North Pacific, depending on the relative productivity of extraction locales. Entire villages relocate seasonally around major fishing rivers such as the Columbia and Fraser Rivers of the North American northwest coast and the Karluk and Ayakulik Rivers of the Kodiak Archipelago (Ames this volume;

[2]This comparison is limited by the much greater scale of Andean economic systems subsumed under Murra's model and by the permanence of Andean "colonists" integrated economically by trade rather than seasonal mobility.

Jordan and Knecht 1988; Mitchell 1983). In other areas, villages disaggregate in the productive season and reconvene in the fall.

Under this new regime, we can expect certain resource extraction locations to attract residential camps (or transplanted villages), whereas others would be targeted individually and as part of daily logistical forays. In this volume, Ames notes that boat-based hunter-gatherers commonly travel large daily distances and bring resources back to central villages for processing. I would argue that this form of modified logistical mobility is most appropriate for resources that can withstand delayed processing, that are captured away from suitable landing spots, and/or that are predictable in space and time. I am also convinced that, despite the greater range of daily travel with boats, travel time is sometimes significantly costly to encourage more logistical use of the landscape (where processing is conducted away from the residential base). Resources that might be more easily returned to a residential base for processing include birds, sea mammals, plant products, and shellfish. Around the North Pacific Rim, anadromous fish, especially various species of salmon, are the most concentrated, spatially predictable, and time-limited resource. These products might also be returned to the village or central camp if availability is constant and predictable during a significant portion of the season or if the residential base is relatively close to the extraction location.

We should see more durable construction and on-location processing where fish runs are of relatively short duration (days or weeks), unpredictably timed, and critical for winter storage, so that missing a day or more of the run could dramatically reduce the size of winter stockpiles. At the more productive rivers, where fish are available for months, on-location processing would be less necessary, as Ames' ethnographic descriptions indicate. Around Kodiak, anadromous fish tend to concentrate seasonally in residentially appropriate inner bay zones and rivers and in smaller and shorter interval runs in midbay streams. With an intensive focus on seasonal storage, residential outposts should be formed at inner bay locations where resource diversity is low but seasonal productivity is high.

By adding storage to replace or supplement foraging near camp during winter and through the continued availability of logistical targets, populations could expand. Whereas small, immediate-return groups would have used several bays to compensate for low winter productivity in the initial (Ocean Bay) phase, delayed consumption of summer returns would support residentially stable aggregated settlements in adjacent bays, or even in the same bay, increasing the density of contemporaneous settlements. As long as opportunities exist for establishing new settlements, economic competition would be minimal, even though it would become more likely with collective production and consumption of seasonal stores.

In summary, following the development of an intensive storage economy, we can expect the establishment of the first sedentary villages, "built-up" and marked by the construction of durable features in anticipation of future use. A second site type should be developed for seasonal production of stored fish. These sites should be located close to fish streams and should be dominated by artifacts specifically targeted to fish capture and processing (specialized technologies). Other mass harvesting would be possible around the midbay village, but these activities would be less likely to influence settlement patterns (although they would affect artifact variability in village assemblages). Greater diversity (richness and evenness) is expected in faunal debris and artifact classes at the winter villages, whereas fish extraction sites should have lower faunal diversity and fewer artifact classes. There should be little evidence of sustained conflict in the form of military artifacts, defensive sites, or violent trauma. These conditions could persist peacefully for a considerable interval, supporting Binford's (1990) expectation that neither sedentism nor storage are sufficient to generate significant political complexity by themselves, although the greater diversity of site types and increased nature of artifact specialization qualify as increased complexity in social and technological dimensions.

Kachemak period assemblages, as is already known, contain unprecedented mass-harvesting and processing technologies (net sinkers and *ulu* knives) and the earliest clearly documented village sites (Clark 1984; Erlandson et al. 1992; Fitzhugh 2002; Haggarty et al. 1991). This period should also include greater intersite variability in site types and artifact assemblage composition (relating to site function). The largest classes of sites (villages) should contain the highest artifact diversity compared to more specialized logistical extraction sites. Smaller extraction sites are expected to occur in inner bay locations and at midbay fish streams. Moving seasonally to extraction facilities compromises logistical flexibility (maximal diversity) for efficiency in the production of a single resource class, and except for fishing camps, most residential activity should continue to be centered in midbay areas in the Kachemak period.

Density-Dependent Competition

In contrast to issues of sedentism and storage, ecologically minded anthropologists more commonly agree that the emergence of sociopolitical inequality and related aspects of social complexity are tied to density-dependent effects (Binford 1980; Boone 1992; Carniero 1970; Dyson-Hudson and Smith 1978). At issue are the extent to which potential subordinates can leave to find a better life and the degree to which potential elites can offer incentives/sanctions to make subordination an acceptable or least distressing option.

Unencumbered population expansion would eventually encounter environmental limits, which people would experience as increased spatial and temporal resource variability, reduced residential and logistical flexibility, and an increased tendency to impact more vulnerable patches. Increased exposure to subsistence variability (risk) would lead to heightened levels of tension within corporate groups and competition over resource extraction sites among these groups. In a filled environment, competitors who control high-quality, stable resource patches would have a competitive advantage (Boone 1992). Once the environment fills up with corporately owned extraction facilities, vulnerable/marginal groups would find few alternatives to selling labor for security. Those who control more stable extraction sites would find it advantageous to support the newly subordinated population (see Fitzhugh 2002 for a more detailed discussion of these dynamics).

This last phase of the model represents a significant divergence from the economic logic of the previous phases. Ecological variables continue to drive the predictions, but socioecological contexts are significantly different from what they were when foragers could escape subordination through mobility or when resource patches were uncontrolled. Under despotic conditions (see Boone 1992), strategies for leveling inequity yield to negotiations over the value of labor and commodities and the obligations of patronage. Subordinates seek to minimize subordination by making themselves valuable to the emerging elites and by simultaneously retaining options to shift allegiance to other patrons. Meanwhile, would-be elites compete for labor, productive resource patches, and trade in material and symbolic resources (see Fitzhugh 2002).

Settlement patterns should indicate higher site densities in both villages and extraction locations. Competition between emerging elites is expected to develop over access to the best resource extraction sites, long-distance trade routes, and nonelite labor (distant kin and slaves). Extraction sites should be more heavily constructed as a form of ownership claim and defense of property. Intensified production would lead to greater functional diversity in site classes. Military equipment and defensive facilities should appear. Social inequality should emerge and find expression in tangible archaeological variability, such as interhousehold comparisons of residential and entertainment space, storage capacity, and proportional abundance of expensive implements (made of rare materials, highly decorated, or otherwise uncommon). I expect heterogeneous distributions of nonlocal commodities, elaborate crafts, and facilities for hosting visiting dignitaries.

SITKALIDAK DATA

The predictions enumerated above and in Table 9.1 are investigated here using data from 150 archaeological sites documented between 1993 and 1995 during the Sitkalidak Archaeological Survey. I conducted this survey to gather comprehensive data on settlement pattern change through time to evaluate the nature of economic, technological, social, and political change in southeastern Kodiak from the earliest prehistoric occupation into the centuries of cultural contact with militarily dominant Europeans. Data collected that will be used in the following analysis include dated site components, locational parameters, functional site classifications, and artifact classifications. A brief description of each data category follows.

Chronology

Site components were initially dated by typological methods, using established temporal diagnostic characteristics to place components into

Table 9.1. Model Predictions

Phase 1: Ocean Bay
—Redundant site locations and undifferentiated site types—midbay positioning
—Larger (aggregation) sites should be located in particularly productive locations
—Larger (aggregation) sites should emphasize summer (spring to fall) fauna (including short shelf-life taxa)
—Expedient camp construction, anticipating future residential moves
—Multibay foraging range (summer)—greater raw material variability
—Artifacts should exhibit low intersite variability and low diversity (specialization)

Phase 2: Kachemak
—Evidence of labor-intensive processing activities
—First durable features and "built-up" sites, planning to reuse houses, features, etc. (semi-sedentism)
—First bulk fish extraction camps (adding inner bay/river locations to settlement pattern)
—Specialized technologies (especially mass harvesting and processing)
—Increased technological and faunal diversity (richness and evenness) in winter village sites (midbay)
—Relatively low technological and faunal diversity in extraction camps
—Absence of defensive sites, military artifacts, wealth differentials

Phase 3: Koniag
—Higher site densities
—Larger village sites
—Defensive sites
—Increased variability in site types
—Greater proportional abundance of status-oriented material culture

chronological periods and, if possible, into more precise temporal phases. Twenty-nine radiocarbon dates from 26 sites were assayed to provide finer chronological resolution and to corroborate typological attribution, which they did (see Fitzhugh 1996). Though 150 site locations were documented in the survey, 58 sites could not be placed in any chronological period. Of the remaining 92 sites, many with multiple components, 145 components have been assigned minimally to one of four archaeological periods (Ocean Bay: 18; Kachemak: 38; Koniag: 58; and Historic: 31; including Alutiiq, Russian, and Euro-American sites). Some attributions are uncertain (45) and additional unidentified components must surely exist, but the data set provides a robust sample from which to draw quantitative comparisons.

Of the radiocarbon dated sites, there is a significant gap in data between 4600 cal. B.P. and 2300 cal. B.P. (Fitzhugh 1996). Otherwise, the spread of dates suggest continuous occupation of the Sitkalidak region. The meaning of this gap remains unclear and is the target of future investigation. Sites dated to the Ocean Bay to Kachemak transition (ca. 4000 to 3500 cal. B.P.) are more scarce than sites before and after this interval (Erlandson et al. 1992; Mills 1994), but the Sitkalidak pattern is striking in the duration and apparent cultural sterility of this gap. Two alternative hypotheses include (1) regional emigration to more productive riverine systems with the shift to seasonal storage of fish or (2) geological processes that have systematically eliminated coastal sites dating from this period. Geomorphologist Gary Carver has geological evidence from several locations on Kodiak suggesting a significant modification of sea level during this interval (Carver, personal communication 1999). Without better resolution of this problem, the Sitkalidak data are less than ideal for tracing evolutionary processes through the Ocean Bay to Kachemak phase but are still suitable for comparisons across chronological periods.

Locational Parameters

Sites were classified according to their location on outer coasts, midbay sections, or inner bay/river banks. In some cases, discrimination was straightforward, whereas in others, subjective judgments were necessary. The principles underlying this classification were the bipolar considerations of exposure to open ocean and proximity to significant inner bay river systems. Any site that would have been unprotected from wind and swells off the open ocean was classified as "outer coast." Any site that was located close to or on a large stream was classified as "inner bay/riverine" or just "inner bay." Midbay sites were those that did not fall in either of the other two categories. In all cases, these locations had more accessible (steeper) beaches than inner bay sites and less exposure to ocean waves compared to outer coast sites (see Fig. 9.2). Note that 12 sites from the Ocean Bay Valley were left unclassified due to uncertainties in the geomorphic history of this infilled estuary/stream system.

Functional Site Classifications

Site components are classified according to a subjective multivariate procedure that takes into account the component area, the number of identifiable structures, the character of occupation, and the geographic setting. Following a derivative of Binford's (1980) hunter-gatherer functional classification that is consistent with ethnographic variability in maritime hunter-gatherer settlement and land use (Mitchell 1983), these sites types include locations, camps, settlements, villages, and refuges. Locations potentially include lookouts, kill sites, butchery stations, signal fires, or any other archaeological manifestation of short interval nonresidential land use. Camps, settlements, and villages are all residential site types that differ in their size, intensity of occupation, and nature of construction (with or without pit-houses). Camps are minimally built up without semisubterranean houses (or at most, one) and are generally smaller than 1000 m^2. Villages are large residential sites that have numerous semisedentary structures (visible at ground surface), deep middens, and otherwise thick deposits. Settlements are an intermediate category that includes larger versions of camps (sites larger than 1000 m^2 without pit-houses and pit-house sites with fewer than five structures). Refuges include any site whose residential features are situated on defendable landforms, such as cliff-faced islets and promontories. Generally these sites lack a convenient source of fresh water and would be cumbersome bases for logistical foraging, given the difficulty of launching boats from their exposed steep and rocky shores. More detailed discussion of this classification scheme can be found in Fitzhugh (1996).

Artifact Classification

The artifact classification presented in this chapter is newly developed for this analysis. Artifacts were collected whenever found in surveyed sites. Collections were made from controlled test excavations, as well as from surface and erosion profiles. Artifacts were identified according to conventional descriptive categories (projective point, nonprojectile biface, scraper, blade, hammerstone, pottery shard, etc.). In the following analyses, this variability is reduced to a set of six functional groupings, as presented in Table 9.2. These include ornamental objects, hunting tools, fishing tackle, processing tools (subsistence related), site maintenance tools, and evidence of tool production (manufacturing tools, performs, and debris). Except for

Table 9.2. Artifact Classification

Functional Grouping	Artifact Class
Ornamental	Beads
	Painted pebbles
	Labrets (lip plugs)
	Other ornaments
Hunting	Projectile points
	Lances
	Harpoons
Fishing	Grooved cobbles
	Notched shingles
Processing	Bifaces (not projectile point)
	Scrapers
	Prismatic blades/microblades
	Semilunar knives (*ulus*)
	Slate knives (nonulu).
	Utilized flakes/flake tools
	Pottery
Site maintenance	Ocher grinders
	Axes
	Nails
	Post remnants
	Bone wedges
Tool production	Abraders
	Adzes
	Debitage and other production debris
	Possible burins
	Burnishing stones
	Chisels
	Cores
	Gravers
	Grindstones
	Unsuccessful blades/microblades
	Tool preforms
	Worked slate
	Worked bone
	Small hammerstones
Hammerstones	Hammerstones

very small hammerstones (included in "tool production"), the category of hammerstones was eliminated from most analyses because of the generalized utility of this tool in processing, site maintenance, and tool production (and probably other activities, also). Because the grouping variables subsume a widely differing number of artifact classes, analyses are valid only for comparing proportional changes in that category relative to others across some other axis (e.g., chronological period or site function).

RESULTS

Site Positioning

It was predicted that there would be a change in the positioning of residential sites from Ocean Bay to Kachemak periods as mass harvesting and storage changed the nature of seasonal subsistence and as positioning strategies became more logistical. Table 9.3 and Figure 9.3 confirm the predicted change: over 90% of Ocean Bay components are located in midbay locations compared to only 61% of Kachemak and 45% of Koniag components. Non-mid-bay components in the Kachemak period are slightly skewed toward inner bay/riverine locations (21%). The six outer coast sites include exposed refuges, camps, and villages of the terminal Kachemak phase (after A.D. 800). Along with the Koniag sample, these sites indicate a

Table 9.3. Site Components by Geographical Position

Period	Outer coast	Mid Bay	Inner Bay	Total
Ocean Bay	0	13	1	14
Kachemak	6	20	7	33
Koniag	20	25	10	55
Historic	11	16	4	31

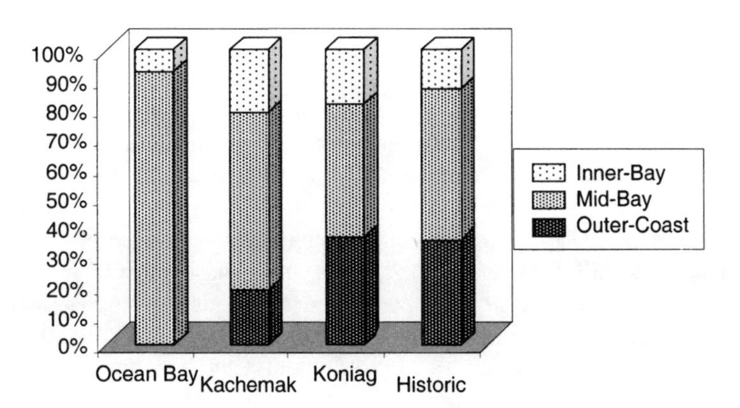

Figure 9.3. Proportional representation of site locations in inner bay, midbay, and outer coast zones by cultural historical period (derived from Table 9.3).

second gravitational change in settlement toward the outer coast that began just before the onset of the Koniag period. In general, the one major change from Ocean Bay to Kachemak appears to be the addition of inner bay site locations, as predicted.

Outer coast sites actually exceed inner bay sites in the Koniag sample, which is partly due to the fact that villages tend to be located near or in the outer coast zone in this period. Whale hunting and military strategy (visibility of enemy approaches and proximity to defensive sites) appear to explain this shift in settlement (Fitzhugh 1996, in press). Finally the Alutiiq period plot mixes native sites with Russian and American settler sites from the past 200+ years. In spite of this, trends established in the earlier (prehistoric) periods continue, dominated as they are by Koniag village sites that continue to be occupied following contact. Though it is not possible to categorize most nonvillage sites firmly by time or ethnicity, we know from historic records that people in the Sitkalidak area were resettled after the 1837–39 smallpox epidemic into only one or two villages. Old Harbor and Kaguyak were the descendant communities of these resettled populations (Fig. 9.1), although the exact nineteenth century locations of these villages are currently unresolved (see Clark 1974:7–8). Kaguyak was abandoned following the 1964 tsunami that leveled that small village. After 1840, I would expect a fall-off in land use away from resettled villages because postepidemic populations would not have needed to exploit the full potential of the landscape and because travel costs to more distant extraction locations would be higher. I would also predict greater use of short-term residential camps at greater distances from villages. These expectations seem to fit the evidence subjectively, but the historic sample here is too small and too poorly dated to provide quantitative confirmation.

Land Use Changes

Looking next at site types plotted against time period, we can tease out an additional dimension of changing settlement and land use (Table 9.4; Fig. 9.4). As predicted, portable camps by far dominate the Ocean Bay record, and several "settlement" sites (n = 4) may represent aggregation sites or particularly popular camping spots that were reused in slightly different areas at different times (Fitzhugh 1996). The season in which these sites were used is yet to be determined, but qualitative assessment of deposit character and occupation debris suggests that the difference between Ocean Bay "camps" and "settlements" has less to do with structural differentiation of site types and settlement and more with particularly popular residential locations. On the other hand, the differences between camps and settlements in Kachemak and Koniag samples is more marked

Table 9.4. Site Classification by Period

Site Class	Ocean Bay	Kachemak	Koniag	Alutiiq	Total
Location	1	0	0	1	2
Camp	8	1	1	2	12
Settlement	4	18	20	8	50
Village	0	7	14	10	31
Refuge	0	2	3	1	6
Total	13	28	38	22	101

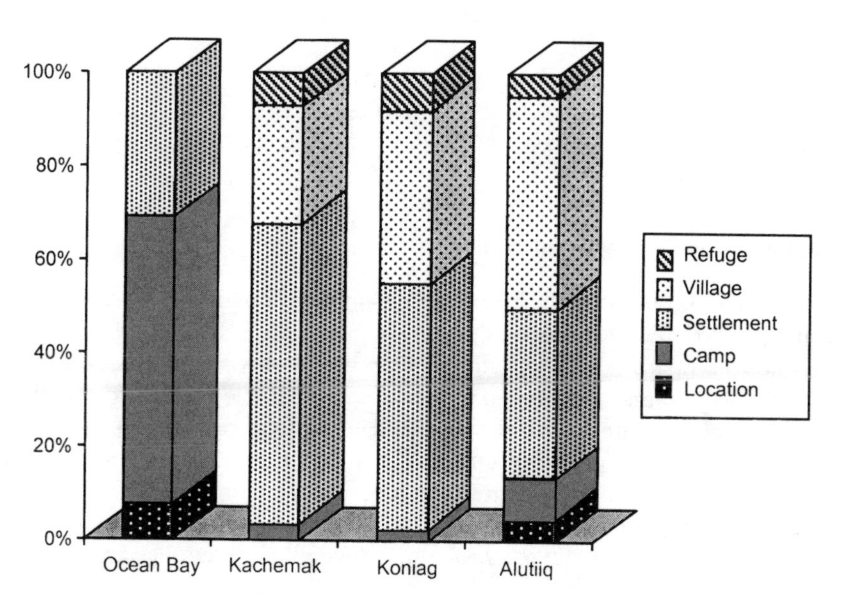

Figure 9.4. Proportional representation of functional site types by cultural historical period (derived from Table 9.4)

because camps represent comparatively ephemeral sites with no evidence of an investment in planned reuse, whereas settlements represent infrastructural investments in the form of sod house construction.

There is an ordinal shift apparent between Ocean Bay and Kachemak land use because the proportional importance originally placed on camps and settlements shifts to settlements and villages, respectively. This relates to a dramatic increase in the intensity of land use and in population density. It also masks the qualitative shift from homogenous base-camp organization to structured seasonal settlement cycle, as reflected in the change in positioning strategies already discussed (in other words, this is not just a change in the size and permanence of two classes of seasonal site types).

Following the rather exceptional changes in the character of land use from Ocean Bay to Kachemak, the subsequent changes are minimal and of degree more than kind. The Koniag settlement is similar to that of the Kachemak, except that Koniag winter villages expand to unprecedented sizes and are typically located in areas of greater visibility and exposure (outer coasts).

Compared to the bulk of the residential sites, special function sites are few and thus the trends are more speculative. Refuge sites are first found in the late Kachemak (A.D. 800 and later). This class takes on added importance in the Koniag period, when refuges expand from one or two family sites to encompass entire village populations (ca. 50 houses) and shift from semiprotected waters to outer coastal areas along with villages. "Locations" are an insignificant aspect of this data set, supporting the claim that maritime hunter-gatherers leave minimal archaeological evidence of nonresidential activities.

Variation in Tool Assemblages

Artifact groupings can be used to estimate the nature and diversity of activities at different kinds of sites. Figure 9.5a pools sites of all time periods and compares the proportional representation of tool classes by site types. This pooling approach provides a larger sample of sites for comparing site types. Figure 9.5b breaks these data by time period, but groups all site types together, allowing temporal comparisons. Figures 9.5c through 9.5e illustrate artifact trends across site types for each time period. These can be used for finer scale comparisons, but with smaller sample sizes. My discussion will center on Figures 9.5a and b. Several observations can be made:

1. Tool production and processing activities appear to be inversely related before the Historic period. The dominance of these classes derives from the higher volume of artifacts produced in these activities compared to others (e.g., debitage in the tool production case and expedient tools such as used flakes or unrefined bifaces in the processing case).

2. Tool production declines in relative importance as one moves from the most ephemeral site types to the least (locations and camps to villages). As shown in Figure 9.5b, this is clearly a product of change through time. One explanation for this pattern is the decline of the importance of flaked tools following the innovation of slate grinding in the late Ocean Bay period (Clark 1979, 1982), even though flint knapping continues throughout prehistory (Steffian et al. 1998). Slate tool production generates considerably less waste material than flint knapping. Interestingly, this trend does NOT support the argument that more tool production should be

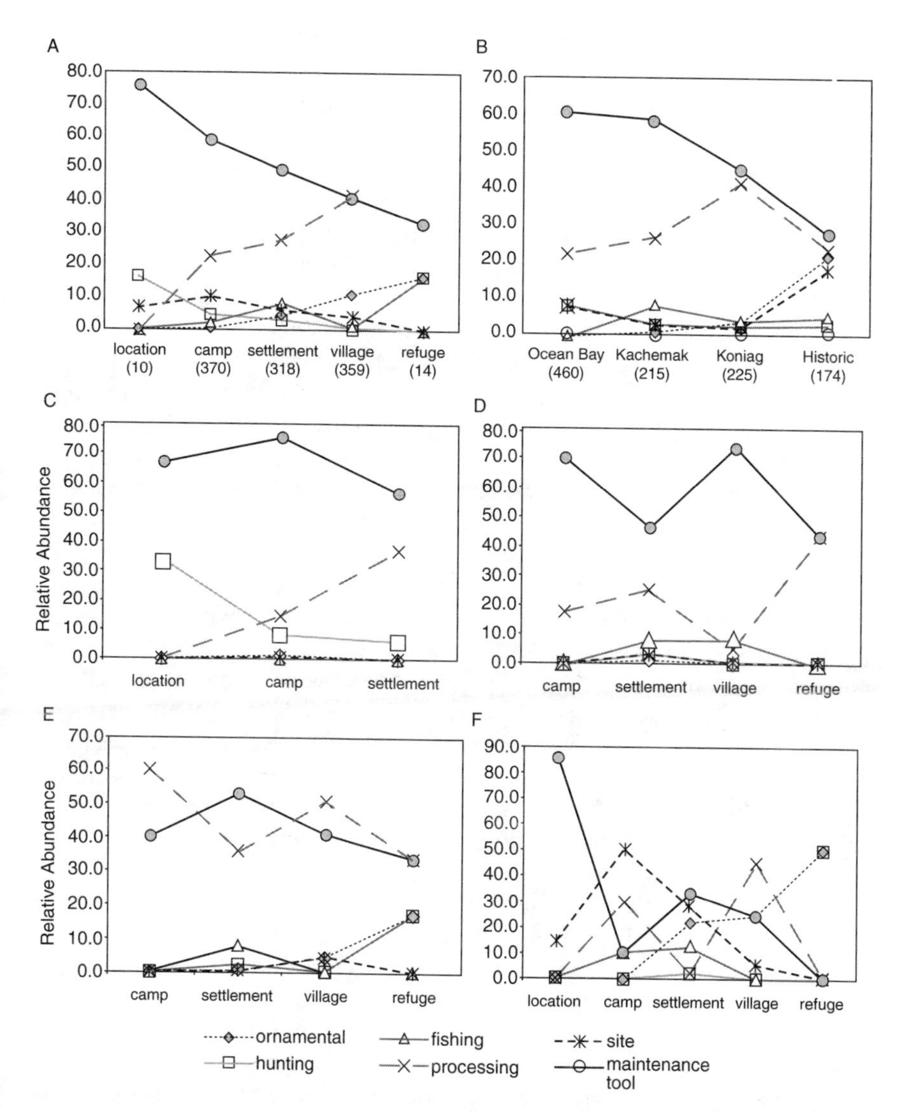

Figure 9.5. (a–f). Variation in functional classes of artifacts from the Sitkalidak Archaeological Survey by (a) site type, (b) culture historical period, (c–f) site type in (c) Ocean Bay, (d) Kachemak, (e) Koniag, and (f) Historic periods. (The same legend applies to all graphs in this figure.)

occurring at villages, assuming that villages were occupied in winter months when foraging activities were minimized and storage consumption important. From this, we might infer that lithic tools were produced during the seasons of greatest use (especially in camps and settlements).

Perishable products that required greater manufacturing time and effort (clothing, nets, boats, etc) might still have been produced disproportionately in winter villages, but they are not represented in any obvious way in these data.

3. Processing tools take on increasing importance along the same gradient of site durability and size (camps to villages) and from Ocean Bay to Koniag. This trend supports the prediction that processing-intensive practices increased in importance through time (following the shift to an economy based on seasonal food storage). In the light of the highly significant shift in settlement patterns already noted between Ocean Bay and Kachemak, it is somewhat surprising that the jump in processing tools is most pronounced between the Kachemak and Koniag periods. Nevertheless, these data support the predictions of increased processing activities through time.

4. If the shift to processing-intensive activities paralleled an increase in the use of mass-capture devices (e.g., nets), we would expect a simultaneous decrease in the importance of hunting tools. The hunting category measures change in the relative importance of projectile points. As predicted, this category falls off dramatically from Ocean Bay to Kachemak periods. These tools also decrease in relative abundance steadily from camp to settlement to village, supporting the notion that less specialized sites should have less specialized technologies (Binford 1980). Hunting implements are strongly represented in "locations" as would be expected if these were lookouts, but a small sample cautions against making too much of this site type. The refuge samples are also small but show a trend in a predictable direction. Hunting tools again increase proportionally, which would be in line with the reasonable conclusion that hunting tools also served a defensive function and that defense was particularly important at these refuge sites. This pattern is particularly significant because the refuge sites are otherwise most comparable to villages (which have the lowest representation of hunting tools).

5. Notched stones and grooved cobbles make up the bulk of the fishing category, which peaks in the Kachemak period at the settlement sites. Notched stone artifacts are commonly made of flat beach pebbles or shingles and occur in high numbers on the shoreward eroding sides of many Kachemak sites. Clark (1974: 67) notes that notched pebbles/shingles disappear from Kodiak assemblages in the transition to the Koniag period (they are also nearly absent from Ocean Bay sites), and this class was one of a variety of characteristics used in the typological dating of the Sitkalidak components. Therefore, it is not surprising that these implements are found predominantly in Kachemak contexts. There is some disagreement about the use of these stones (see Clark 1974:68, Knecht 1995:180–184), but I am

convinced that they were used as weights on nets in the capture of herring, salmon, birds, sea mammals (see Davydov 1976:221, 228), whereas notched or grooved cobbles were more typically weights for line fishing for capturing cod and halibut (Knecht 1995:186). In the Kachemak subsample (Fig. 9.5d), fishing appears to be most important in villages and secondarily in settlements, whereas in the Koniag period, fishing implements occur exclusively in settlements (most are grooved cobbles, indicating line fishing).

6. It is interesting to note that Kachemak notched stone pebble/shingles in this sample are found primarily in midbay sites not associated with significant fishing streams, despite the already discussed increase in inner bay/riverine use in this period. This may indicate that net fishing was more heavily oriented toward near-shore rather than riverine or river-mouth fishing. Other methods such as weir and spearfishing may have been practiced in rivers, where fish are already naturally aggregated. Kachemak and Koniag sites are commonly found associated with stone weirs on the Karluk and Ayakulik rivers (Jordan and Knecht 1988; Rick Knecht, personal communication 1993). Nonetheless, net-sinkers were used in Kachemak riverine contexts, as recent research at the Outlet site in northern Kodiak (Saltonstall, personal communication 1999) and in the Koniag Karluk One site (Knecht 1995: 180) attest.

7. Site maintenance is a small category (20 artifacts). Although we would predict greater emphasis on site maintenance as residential mobility shifts to semisedentary life following the Ocean Bay-Kachemak transition, this trend is not seen. In fact, if anything, site maintenance tools are less important as we move from Ocean Bay to Kachemak and again to Koniag, and similarly from camp to settlement to village. This category is actually misleading, because it is dominated by red ocher grinding stones from the Ocean Bay period Tanginak Spring site and iron from Historic sites, both artifact types absent in the intervening periods.

8. Finally, ornamental items increase in proportional importance through time and from less to more intensive site types (locations to villages). The highest proportion of ornamental objects occurs in the refuge sites investigated, possibly illustrating a link between warfare and ritual or status symbolism. The high representation of ornaments in historic contexts relates directly to the high frequency of trade beads found in some historic components. Just as flakes elevate the importance of tool production in Ocean Bay contexts, beads may artificially elevate the importance of ornamentation after Russian contact. This factor aside, the increase in ornamentation from Ocean Bay to Koniag is consistent with the emergence of heightened social competition and ritual late in the Kachemak period and the development of institutionalized ranking and slavery and a thriving exchange in prestige wealth by Koniag times (Fitzhugh 2002). Spatial

differences (camps to villages) support the prediction from ethnographic analogy that socializing, feasting, ritual, and ceremonial activities were focal activities in winter villages. Regalia decorated with beads and bangles are expected to be concentrated in villages, and it is in these sites that ornaments would have most often been manufactured, stored, used, and lost.

9. Finally, considering the temporal series (Fig. 9.5 c–e), with minor variation, the general spatial trends already discussed hold. The only deviation is in the Historic sample, which bears little relationship to the prehistoric sequence. This suggests that something very different was going on in historic settlement patterns. The most significant principle here is the mixture of sites from different ethnic groups with different land use practices. Add to this the rapid pace of social and economic change, introduced technologies and raw materials, and competing demands on productive labor. A full untangling of the impacts of these different forces on this distributional pattern would be fascinating, but must await better data (high-resolution dating and more secure ethnic attribution).

Assemblage Richness as a Measure of Activity Diversity

The measure of faunal diversity using assemblage richness and evenness is a common practice in the archaeological study of subsistence (Grayson 1984). Less often, similar measures are applied to artifact assemblages to assess the degree of activity specialization and settlement "grain" (Binford 1980) and hence the variety of activities attributable to a particular site or site component (Jones et al. 1983, 1989; but for cautions, see Binford 1982). Richness measures the number of different categories or classes present in a given assemblage, and evenness measures the degree of numerical constancy across classes within an assemblage. Richness and evenness are often used together, because each measures a different aspect of assemblage variation. However, both measures are vulnerable to sample size problems, which must be controlled if they are to be useful. Samples derived from a single statistical population can demonstrate that larger samples yield higher richness values until richness reaches "saturation," or the greatest number of possible classes in the population (Grayson 1984; Jones et al. 1989). This means that samples of different size cannot be compared accurately for richness without some way of correcting for sample size. Evenness handles comparisons between samples of different sizes, as long as each sample is large enough to yield statistically significant quantities in each class (Grayson 1984).

Given the generally small samples retrieved from sites in the Sitkalidak survey, evenness is an inappropriate measure of diversity. Richness, on the other hand, can be a highly effective measure of class diversity, even with small samples, as long as richness is presented as a function of sample size. As Jones and colleagues (1983, 1989; see also Grayson 1984) have shown, simple linear regression analysis of richness (total number of classes) against total sample size can yield a comparable statistic of class diversity, which can be used to compare samples of varying size. Depending on the nature of the sample class distribution, data transformation is often necessary to produce the best regressions (minimizing residuals). In the typical case in which most samples are very uneven, log/log transformations produce better regressions.

Figure 9.6 shows the results of richness analysis for the Sitkalidak data. In this case, $\log_{10}/\log_{10}$ transformation produced the best-fit regression lines (measured as a function of minimal R^2 values) for samples divided into Ocean Bay, Kachemak, Koniag, and Historic periods (Table 9.5). Given the linear relationship expected between taxonomic richness and sample size, it follows that regressions of equal slopes and similar alignment are statistically similar. Divergent regressions, on the other hand, indicate significant

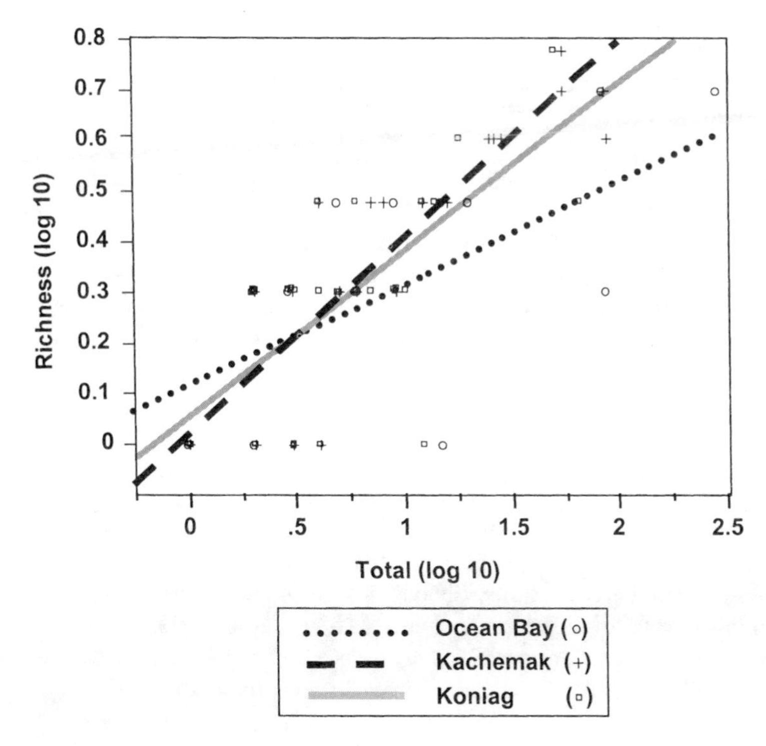

Figure 9.6. Logarithmic (log 10/log 10) regression analysis of artifact richness by culture historical period. See text for explanation.

Table 9.5. Regression Statistics Comparing Artifact Richness against Sample Size for Sitkalidak Assemblage Samples by Chronological Period

Sample Group	*N*	*r*	R^2	Regression Equation[a]
Ocean Bay	10	0.53	0.28	$Y = 0.10706 + 0.21452X$
Kachemak	23	0.89	0.79	$Y = 0.01055 + 0.37928X$
Koniag	37	0.74	0.54	$Y = 0.04766 + 0.31671X$
Historic	22	0.87	0.76	$Y = 0.08157 + 0.31812X$

[a] Y = number of classes ($\log_{10}$); *X* = number of tools ($\log_{10}$).

differences in richness diversity. Steeper slopes indicate more highly diverse assemblages (more classes in a given sample size), whereas shallower slopes represent less diverse assemblages (fewer classes for a given sample size). From this it follows that there is significantly more diversity in the Kachemak, Koniag, and Historic samples compared to the Ocean Bay sample. This pattern strongly supports the view, already demonstrated in other analyses, that the change from Ocean Bay to Kachemak was profound compared to later changes.

Extending this procedure to investigate artifact richness among different site types in each period, we are faced with a less expected but nevertheless consistent result (Fig. 9.7 a–d). Following Binford's (1980) prediction that more intensively occupied base camps or settlements should have more diverse assemblages, I would expect that richness values would increase from locations to camps to settlements to villages. In fact, the reverse trend is observed in each period plot except the Kachemak, for which richness appears approximately equal.[3] Among the Ocean Bay sample (Fig. 9.7a), camps yield more diverse assemblages than settlements. Among the Koniag and Historic samples, settlements are richer than villages.

This consistent discrepancy from expectations is difficult to fathom, especially given the high degree of consistency noted among all other measures of settlement, land use, and artifact variability. It may be instructive to note that Jones and colleagues (1989:77) found a similarly anomalous pattern in their study of artifact variability from Steens Mountain, Oregon. They were investigating artifact richness between "sites" and in "off-site" surface collections, divided into groups based on their proximity to high-density artifact deposits (sites). They predicted that artifact diversity would drop as one moves away from site locations into areas of reduced activity. Their results were the reverse of their predictions and could not be

[3]Regressions for samples of less than four site components are excluded because they yield highly irregular and unrepresentative regressions.

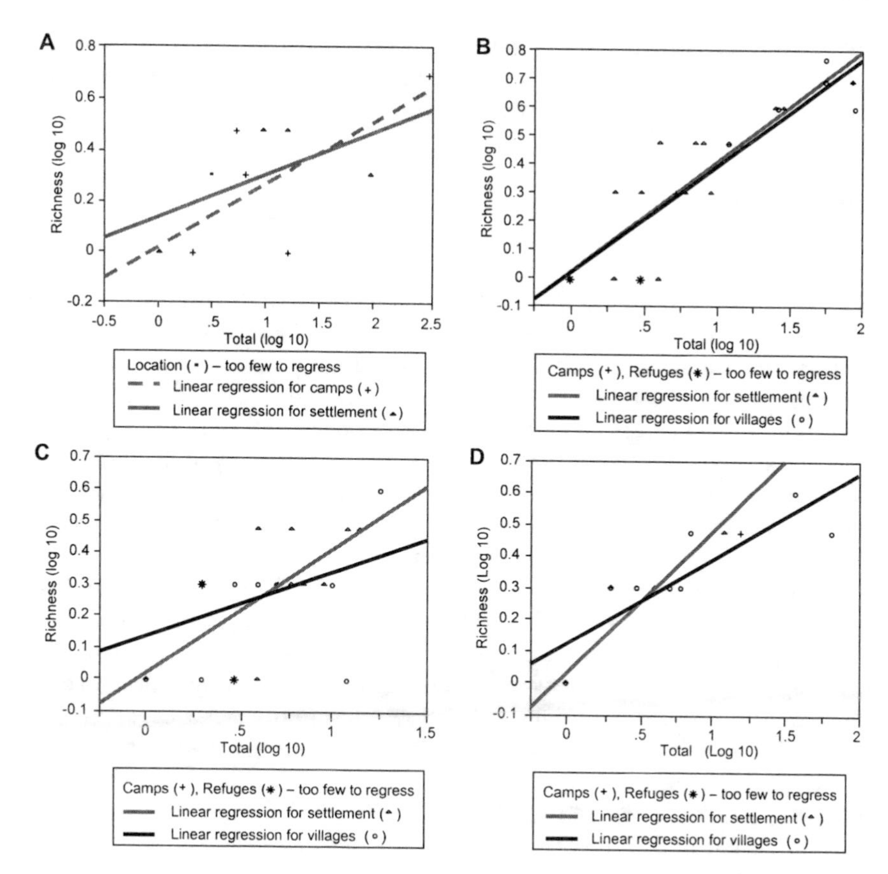

Figure 9.7. (a–d). Logarithmic (log 10/log 10) regression analysis of artifact richness by site type for (A) Ocean Bay, (B) Kachemak, (C) Koniag, and (D) Historic Alutiiq periods. See text for explanation.

satisfactorily explained. Taken together, these results call into question the "middle range theory" that underlies the link between increased specialization (in activities) and assemblage diversity (in artifacts). A theoretically informed alternative hypothesis is clearly needed at this stage to make sense of these discrepancies. One possible explanation might be that hunter-gatherers often use a more diverse set of technologies for procurement and processing in the field than they do at their home bases, or at least that they are more likely to break and discard/lose parts of these technologies at procurement/processing sites. Alternatively, it is possible, as Binford cautions (1982), that assemblages themselves may be poor predictors of activities and may instead represent accumulations of unconnected activities. Accordingly, we might expect residential camps to have been used differently at different times, leading to richer assemblages than the individual activities themselves would predict.

DISCUSSION AND CONCLUSION

In the preceding analysis, I have examined a series of theoretically generated predictions about changes in settlement and land use from the analysis of positioning strategies, site types, proportional artifact abundance, and artifact diversity. The results are largely consistent with the predictions outlined in Table 9.1. Settlement and land use reorganized with the advent of mass harvesting and storage to become more logistical compared to earlier times, and significant sociopolitical complexity followed much later. Interestingly, several measures of settlement and land-use organization remain fairly stable following the dramatic reorganization of the Ocean Bay to Kachemak transition. This suggests that the emergence of social inequality, extensive political alliance, warfare, and exchange during the past 1500 years had relatively little effect on the underlying subsistence-settlement system, which was established as a consequence of the shift to a storage economy as early as 3500 years ago.[4] In a rather important exception to this generalization, village locations shift to the outer coast zone in what appears to be a socially motivated settlement pattern adjustment, in which visibility and defense are optimized. This shift occurred as competition and conflict escalated in tandem with the emergence of social inequality toward the end of the Kachemak period and especially in the Koniag. The association of defensive sites close to several of the largest outer coast Koniag villages strongly suggests that economic efficiency is no longer the guiding principle of village positioning, even though open-water whale hunting does appear to become a significant economic addition around this time (Fitzhugh 1996; see also Crowell 1994). Not surprisingly, the Koniag period witnesses the greatest diversity of site types as central-place foraging yields to a seasonal cycle of dispersal for subsistence extraction and aggregation for social, ceremonial, and defensive functions.

How does this 7500-year trajectory fit or alter our understanding of hunter-gatherer adaptation and social evolution? I mentioned at the outset of this chapter that a popular model of the development of "complex hunter-gatherers" starts with the initial triggers of reduced residential mobility,

[4]Unfortunately, the gap in dated sites from the Sitkalidak Survey makes it impossible to know whether the settlement-subsistence stability measured between the Kachemak and Koniag periods in these data pertains to the interval of missing data (4600–2300 B.P.).

population growth, social circumscription, intensified resource production, resource hoarding, boundary defense, and warfare (Kelly 1995:310 et ff). From this self-reinforcing cycle, Kelly predicts that some individuals will be able to take control of intergroup networks and trade, monopolize the labor of disenfranchised group members to increase production and military strength, and use a combination of generosity and intimidation to attract and retain followers. Of course, if North Pacific maritime hunter-gatherers had established residentially sedentary occupations as early as 7000 years ago, this model would be hard pressed to account for roughly 5000 years of minimal development toward increased sociopolitical complexity.

Assuming that I am correct in judging that semisedentary winter village sites are an unprecedented phenomena of the Kachemak phase, we might expect to see indications of social adjustments and increasing social competition and tension during this period. Evidence for the development of an elaborate artistic and ritualistic tradition around Kodiak and the greater Gulf of Alaska supports this prediction. Particularly in the Late Kachemak phase, we see the development of an elaborate art tradition, including labor-intensive and visually stunning stone lamp relief carvings and the use of body ornamentation (labrets) made of high-quality, nonlocal materials (Steffian and Saltonstall 1995). Simultaneously, we see the use of a discriminatory mortuary tradition, in which some skeletons appear to have been held together long after death in a form of mummification, and others were relieved of their heads, hands, and feet after death (Simon and Steffian 1994; Urcid 1994; Workman 1992).

An analysis of Sitkalidak survey data on Kachemak and Koniag house measurements suggests that these phenomena did not succeed in transforming society into a ranked or stratified society during the Kachemak period, at least to the extent that variability in house sizes monitors such differentiation (for the supporting analysis, see Fitzhugh 1996 and 2002). On the other hand, analysis of the difference between Kachemak and Koniag period houses indicates the emergence of significant social differentiation and an enlargement of the scale of coresidential (corporate kin) groups by Koniag times. Multiroomed houses are used for the first time, and they appear to combine the function of housing extended families and serving as facilities for large social gatherings. These houses are extremely variable in size and the number of side rooms (used as bedrooms in ethnohistoric accounts).

The delay in the material manifestation of social and political reorganization represented by changes in intrasite/household residential patterns may be a case of cultural conservatism followed by relatively rapid ideological change, as both Cannon and Aldenderfer (this volume) suggest for other cultural sequences. It is plausible that social competition would escalate considerably in the behavioral domain (including emblems of status competition or political power) before it would be sanctified and codified in coresidence patterns and represented in architecture. On the other hand, the escalation of defensive fortification and increasingly regional focus of social conflict from Late Kachemak to Koniag suggest that the political system continued to expand and intensify in parallel with changes in residential architecture. We might conclude then that this change was not the latent effect of competition for social status on ideology but instead a product of still-emerging social complexities.

Whatever the proximate causes of social change in the Kodiak case, a more ultimate cause appears to have been population expansion and crowding. Elsewhere, I have presented a version of Boone's (1992) contest competition model of social inequality to suggest that social inequality on Kodiak developed after the most stable and productive resource patches had been claimed by corporate groups capable of defending their claims (see also Kelly 1995). Under this scenario, disenfranchised individuals find it necessary to sell their labor in exchange for access to resources, especially during periods of environmental crisis. High population densities make it increasingly difficult to exercise flexible strategies to ensure access to resources during crises, leading to greater indebtedness between those controlling stable and productive patches and those who do not. Unlike a standard argument for population pressure, this model recognizes the emergence of at least two categories of people, some of whom (subordinates) could starve to death without necessarily affecting the structure of the social system and another (elites) who might never suffer resource stress but for whatever reason choose to exclude desperate subordinates. Of course, under this scenario, starving subordinates are expected to put pressure on elites to be generous, and elites have to value surplus resources highly to pay the costs of defense. An emerging system of social competition between elites can provide the motivation for this situation to escalate into one of significant social and economic inequality.

These models of emerging complexity follow from a complex series of socioecological considerations; nevertheless, all suggest that we should see an expansion of population density in tandem with emerging complexity and inequality. Proxy evidence of population change throughout the past 7000 years on Kodiak strongly supports the notion that population growth accelerated consistently during the emergence of inequality, just as Kelly (1995) predicts (Fig. 9.8). Environmental change during this interval would not have made the environment any more productive—if anything, it would have become more difficult as the Little Ice Age began. This suggests that the late prehistoric emergent complexity is not a simple product of resource "richness" or abundance (see Hayden 1994, 1995), and it supports the ecological proposition that social complexity/inequality is a

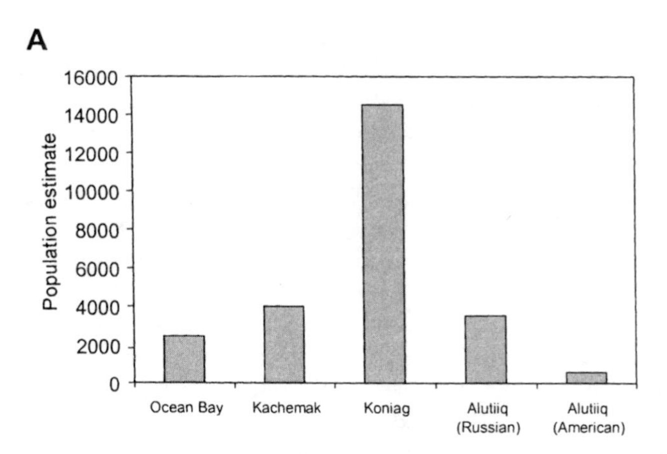

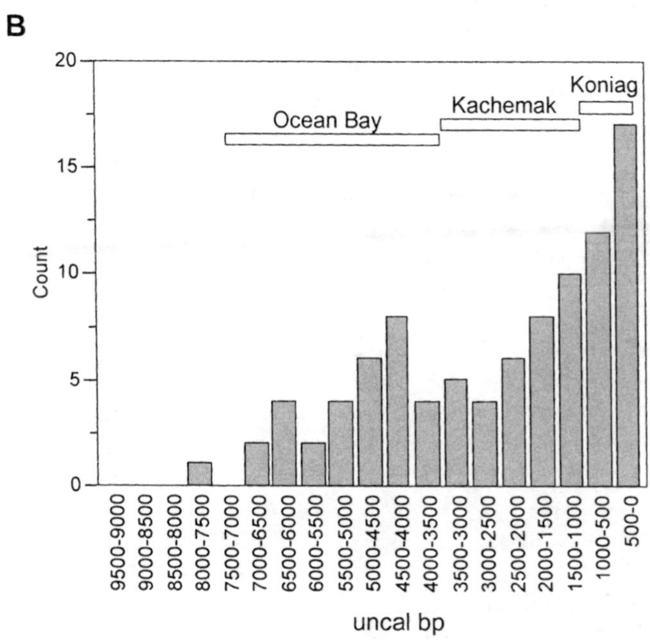

Figure 9.8. Population growth estimates (A) for the Sitkalidak region, based on ethnographic estimates of house populations, converted to per capita space allocation in the Historic Alutiiq house sample from the Sitkalidak Survey and projected into prehistoric contexts; and (B) for the Kodiak Archipelago as a whole, based on the frequency of radiocarbon dated sites per 500 year intervals. See text and Fitzhugh 1996 for detailed discussion.

product of density-driven social competition. Otherwise, we would expect these developments to have occurred much earlier in the Kodiak sequence.

This study demonstrates the utility of Binford's forager–collector model and the ecological theory that underlies it for tracking changes in settlement and subsistence through time. The model itself is not "evolutionary," but it can be used to structure observations on change through time, observations which can serve to elucidate evolutionary processes. One interesting conclusion here is that settlement pattern changes are relatively conservative during considerable time. Unlike the findings of Cannon for the British Columbia coast (this volume) or of Ames and Maschner (1999) for the broader northwest coast, the Kodiak data suggest that the earliest documented maritime hunter-gatherers were much less residentially sedentary and less logistically organized than those in later prehistoric times. Notwithstanding this conclusion, it is still probably reasonable to see these early groups as more logistically mobile than many terrestrial groups that have been the traditional focus of the forager–collector model. Interestingly, in a recent analysis of long-term settlement and land-use change in the Great Basin (Owen's Valley), Bettinger (1999) finds a very similar sequence of events in the evolution of a terrestrial hunting and gathering population. He also highlights the importance of a shift to a storage economy and the ways that this effects settlement patterns, land-use practices, and population densities. In the Kodiak case, the most significant differences are found in the later phases, when Kodiak population growth and social competition led to the development of greater social inequality.

Finally, these analyses support the view that hunter-gatherer evolutionary processes cannot be understood without attention to the combined ecological and social contexts in which people made decisions about ways to procure resources, where to position themselves across the landscape, and how to interact with the larger community in which they are embedded. The forager–collector model helps us to explain part of the changing dynamics in Kodiak settlement and land use, but not all of it. Particularly in later prehistoric times, it is the social environment that imposes the greatest proximate constraints of site locations (winter villages and defensive sites). But we cannot understand the social processes themselves without being sensitive to the dynamic constraints of population density and environmental variability.

ACKNOWLEDGMENTS

The research reported in this chapter was supported by grants from the Old Harbor Native Corporation, the National Science Foundation

(OPP-9311676), and the Wenner–Gren Foundation for Anthropological Research (Small Grant # 5774). Additional support was provided by the University of Michigan and the University of Washington. The chapter benefited from the generous advice of Donald K. Grayson. Laada Bilaniuk and Larisa Lumba provided valuable editorial assistance.

REFERENCES

Aigner, J. S., and Del Bene, T., 1982, Early Holocene Maritime Adaptation in the Aleutian Islands. In *Peopling of the New World*, edited by J. E. Ericson, R. E. Taylor, and R. Berger, Ballena Press, Los Altos.

Ames, K. M., and Maschner, H. D. G., 1999, *Peoples of the Northwest Coast: Their Archaeology and Prehistory*. Thames and Hudson, London.

Amorosi, T., 1987, *The Karluk and Uyak Archaeofaunas: An Approach Towards the Paleoeconomy of Kodiak Island, Alaska*. Paper Presented at the 52nd Annual Meeting of the Society for American Archaeology, Toronto.

Amorosi, T., 1988, *The Use of Faunal Remains in the Interpretation of Kodiak Island's Past.* Paper Presented at the First Kodiak Island Cultural Heritage Conference.

Barsch, R. L., 1985, Karluk River Study. Report submitted to the Kodiak Area Native Association. Kodiak, Alaska.

Bettinger, R. L., 1999, From Traveler to Processer: Regional Trajectories of Hunter-Gatherer Sedentism in the Inyo-Mono Region, California. In *Settlement Pattern Studies in the Americas: Fifty Years since Viru*, edited by B. R. Billman and G. M. Feinman, pp. 39–55. Smithsonian Institution Press, Washington, DC.

Binford, L. R., 1980, Willow Smoke and Dog's Tails: Hunter-Gatherer Settlement Systems and Archaeological Site Formation. *American Antiquity* 45(1):4–20.

Binford, L. R., 1982, The Archaeology of Place. *Journal of Anthropological Archaeology* 1:5–31.

Binford, L. R., 1990, Mobility, Housing, and Environment: A Comparative Study. *Journal of Anthropological Research* 46(2):119–152.

Black, L. T., 1992, The Russian Conquest of Kodiak. In *Contributions to the Anthropology of Southcentral and Southwestern Alaska*, edited by R. H. Jordan, F. de Laguna, and A. F. Steffian. *Anthropological Papers of the University of Alaska* 24(1–2):165–182.

Blurton, J. N. G., 1987, Tolerated Theft, Suggestions about the Ecology and Evolution of Sharing, Hoarding and Scrounging. *Social Science Information* 26:31–54.

Boehm, C., 1993, Egalitarian Behavior and Reverse Dominance Hierarchy. *Current Anthropology* 34:227–254.

Boone, J. L., 1992, Competition, Conflict, and Development of Social Hierarchies. In *Evolutionary Ecology and Human Behavior*, edited by E. A. Smith and B. Winterhalder, pp. 301–338. Aldine de Gruyter, New York.

Boone, J. L., and Smith, E. A., 1998, Is It Evolution Yet? A Critique of Evolutionary Archaeology, *Current Anthropology* 39 (Suppl):141–173.

Broughton, J. M., and O'Connell, J. F., 1999, On Evolutionary Ecology, Selectionist Archaeology, and Behavioral Archaeology. *American Antiquity* 64(1):153–165.

Carneiro, R. L., 1970, A Theory of the Origin of State. *Science* 169:733–738.

Clark, D. W., 1974, *Koniag Prehistory: Archaeological Investigations at Late Prehistoric Sites on Kodiak Island, Alaska*. Tubinger Monographien Zur Urgeschischte, Vol. 1, Tubingen.

Clark, D. W., 1979, *Ocean Bay: An Early North Pacific Maritime Culture*. National Museum of Man, Mercury Series, Archaeological Survey of Canada Paper No. 86. Ottawa.

Clark, D. W., 1982, An Example of Technological Change in Prehistory: The Origin of a Regional Ground Slate Industry in South-Central Coastal Alaska. *Arctic Anthropology* 19(1):103–126.

Clark, D. W., 1984, Prehistory of the Pacific Eskimo Region. In *Handbook of American Indians, vol. 5, Arctic*, edited by D. Damas, pp. 136–148. Smithsonian Institution Press, Washington, DC.

Clark, D. W., 1988, Pacific Eskimo Encoded Prehistoric Contact. In: *The Late Prehistoric Development of Alaska's Native People*, edited by R. D. Shaw, R. K. Harritt, and D. E. Dumond. *Aurora* 4:211–224, Alaska Anthropological Association, Anchorage.

Clark, D. W., 1992a, Archaeology on Kodiak: The Quest for Prehistory and Its Implications for North Pacific Prehistory. In *Contributions to the Anthropology of Southcentral and Southwestern Alaska*, edited by R. H. Jordan, F. de Laguna, and A. F. Steffian. *Anthropological Papers of the University of Alaska* 24(1–2):109–126.

Clark, D. W., 1992b, "Only a Skin Boatload or Two": The Role of Migration in Kodiak. *Arctic Anthropology* 29(1):2–17.

Clark, D. W., 1994, Still a Big Story: The Prehistory of Kodiak Island. In *Reckoning with the Dead: The Larsen Bay Repatriation and the Smithsonian Institution*, edited by T. L. Bray and T. W. Killion, pp. 137–149. Smithsonian Institution Press, Washington, DC.

Clark, D. W., 1996, The Old Kiavak Site, Kodiak Island, Alaska, and the Early Kachemak Phase. *Arctic Anthropology* 49(3):211–227.

Clark, D. W., 1997, *The Early Kachemak Phase on Kodiak Island at Old Kiavak*. Mercury Series, Archaeological Survey of Canada, Paper 155. Canadian Museum of Civilization, Hull, Quebec.

Clark, G. H., 1977, Archaeology on the Alaska Peninsula: The Coast of Shelikof Strait 1963–1965. *University of Oregon Anthropological Papers* No. 13, Eugene.

Crowell, A. C., 1986, An Archaeological Survey of Uyak Bay, Kodiak Island, Alaska, Final Report. Typescript. Department of Anthropology, National Museum of Natural History, Washington, DC.

Crowell, A. C., 1994, Koniag Eskimo Poison-Dart Whaling. In *Anthropology of the North Pacific Rim*, edited by W. W. Fitzhugh and V. Chaussonnet, pp. 217–242. Smithsonian Institution Press, Washington, DC.

Crowell, A. C., 1997, *Archaeology and the Capitalist World System: A Study from Russian America*. Plenum, New York.

Davydov, G. I., 1976, A Selection from G. I. Davydov: An Account of Two voyages to America. Translated by C. Bearne and edited by R. A. Pierce. *Arctic Anthropology* 8:1–30.

Dixon, E. J., 1993, *Quest for the Origins of the First Americans*. Univeristy of New Mexico Press, Albuquerque.

Dumond, D. E., 1987, *The Eskimos and Aleuts*, 2nd ed. Thames and Hudson, London.

Dumond, D. E., 1988a, The Alaska Peninsula as Super-Highway: A Comment. In *The Late Prehistoric Development of Alaska's Native People*, edited by R. D. Shaw, R. K. Harritt, and D. E. Dumond, pp. 379–388, *Aurora*. Alaska Anthropological Association Monograph Series 4, Anchorage.

Dumond, D. E., 1988b, Trends and Traditions in Alaskan Prehistory: A New Look at an Old View of the Neo-Eskimo. In *The Late Prehistoric Development of Alaska's Native People*, edited by R. D. Shaw, R. K. Harritt, and D. E. Dumond, pp. 17–26, *Aurora*. Alaska Anthropological Association Monograph Series 4, Anchorage.

Dumond, D. E., 1994, The Uyak Site in Prehistory. In *Reckoning with the Dead: The Larsen Bay Repatriation and the Smithsonian Institution*, edited by T. L. Bray and T. W. Killion, pp. 43–53. Smithsonian Institution Press, Washington, DC.

Dumond, D. E., 1998, Maritime Adaptation on the Northern Alaska Peninsula. *Arctic Anthropology* 35(1):187–203.

Dunnell, R. C., 1980, Evolutionary Theory and Archaeology. *Advances in Archaeological Method and Theory* 3:35–99.

Dyson-Hudson, R., and Smith, E. A., 1978, Human Territoriality: An Ecological Reassessment. *American Anthropologist* 80:21–41.

Easton, N. A., 1992, Mal de Mer above Terra Incognita, or "What Ails the Coastal Migration Theory?" *Arctic Anthropology* 29(2):28–42.

Ellison, P., 1994, Human Reproductive Ecology. *Annual Review of Anthropology* 23:255–275.

Erlandson, J. M., 1988, The Role of Shellfish in Coastal Economies: A Protein Perspective. *American Antiquity* 53:102–109.

Erlandson, J. M., Crowell, A., Wooley, C., and Haggarty, J. 1992, Spatial and Temporal Patterns in Alutiiq Paleodemography. *Arctic Anthropology* 29(2):42–62.

Fitzhugh, B., 1995, Clams and the Kachemak: Seasonal Shellfish Use on Kodiak Island, Alaska (1200–800 B.P.). In *Research in Economic Anthropology*, Vol. 16, edited by B. L. Isaac, pp. 129–176. JAI Press, Greenwich, CT.

Fitzhugh, J. B., 1996, *The Evolution of Complex Hunter-Gatherers in the North Pacific: An Archaeological Case Study from Kodiak Island, Alaska*. Unpublished Ph.D. Dissertation, University of Michigan, Ann Arbor.

Fitzhugh, B., 2001, Risk and Innovation in Human Technological Evolution. *Journal of Anthropological Archaeology* 20:125–167.

Fitzhugh, B., 2002, The Evolution of Complex Hunter-Gatherers on the Kodiak Archipelago. In *Hunter-Gatherers of the Pacific Rim*, edited by J. Habu, S. Koyama, J. Savelle, and H. Hongo. SENRI Ethnological Studies, Osaka, Japan (in press).

Fladmark, K. R., 1979, Routes: Alternative Migration Corridors for Early Man in North America. *American Antiquity* 44:55–69.

Goland, C., 1991, The Ecological Context of Hunter-Gatherer Storage: Environmental Predictability and Environmental Risk. In *Foragers in Context: Long Term, Regional, and Historical Perspectives in Hunter-Gatherer Studies*, edited by P. T. Miracle, L. E. Fisher, and J. Brown, Vol. 10, pp. 107–125. Michigan Discussions in Anthropology, Ann Arbor.

Grayson, D. K., 1984, *Quantitative Zooarchaeology: Topics in the Analysis of Archaeological Faunas*. Academic Press, New York.

Haggarty, J. C., Wooley, C. B., Erlandson, J. M., and Crowell, A., 1991, *The 1990 Exxon Cultural Resource Program: Site Protection and Maritime Cultural Ecology in Prince William Sound and the Gulf of Alaska*. Exxon Shipping Company and Exxon Company, Anchorage.

Halstead, P. and O'Shea, J. 1989, Introduction: Cultural Responses to Risk and Uncertainty. In *Bad Year Economics: Cultural Responses to Risk and Uncertainty*, edited by P. Halstead and J. O'Shea, pp. 1–7. Cambridge University Press, Cambridge.

Hawkes, K., 1992, Sharing and Collective Action. In *Evolutionary Ecology and Human Behavior*, edited by E. Smith and B. Winterhalder, pp. 269–300. Aldine de Gruyter, New York.

Hayden, B., 1994, Competition, Labor, and Complex Hunter-Gatherers. In *Key Issues in Hunter-Gatherer Research*, edited by E. S. Burch and L. J. Ellanna, pp. 223–239. Berg Press, Oxford.

Hayden, B., 1995, Pathways to Power: Principles for Creating Socioeconomic Inequalities. In *Foundations of Social Inequality*, edited by T. D. Price and G. M. Feinman, pp. 15–85. Plenum, New York.

Heizer, R. F., 1956, *Archaeology of the Uyak Site, Kodiak Island, Alaska*. University of California, Anthropological Record 17.

Hood, D. W., 1986, Physical Setting and Scientific History. In *The Gulf of Alaska: Physical Environment and Biological Resources*, edited by D. W. Hood and S. T. Zimmerman, pp. 5–27. National Oceanic and Atmospheric Administration (U. S. Dept. of Commerce) and Minerals Management Service (U. S. Dept. of Interior), Washington, DC.

Hrdliçka, A., 1944, The *Anthropology of Kodiak Island.* The Wistar Institute of Anatomy and Biology, Philadelphia.

Jewitt, J. R., 1987, *The Adventures and Sufferings of John R. Jewitt: Captive of Maquinna*, illustrated and annotated by H. Stewart. University of Washington Press, Seattle.

Jones, G. T., Grayson, D. K., and Beck, C., 1983, Artifact Class Richness and Sample Size in Archaeological Surface Assemblages. In *Anthropological Papers of the Museum of Anthropology*, No. 72. University of Michigan, Ann Arbor, pp. 55–74.

Jones, G. T., Beck, C., and Grayson, D. K., 1989, Measures of Diversity and Expedient Lithic Technologies. In *Quantifying Diversity in Archaeology*, edited by R. D. Leonard and G. T. Jones, pp. 69–78. Cambridge University Press, Cambridge.

Jordan, R. H., 1992, A Maritime Paleoarctic Assemblage from Crag Point, Kodiak Island, Alaska. In *Contributions to the Anthropology of Southcentral and Southwestern Alaska*, edited by R. H. Jordan, F. de Laguna, and A. F. Steffian. *Anthropological Papers of the University of Alaska* 24(1–2):127–140.

Jordan, R. H., and Knecht, R. A., 1988, Archaeological Research on Western Kodiak Island, Alaska: The Development of Koniag Culture. In *Late Prehistoric Development of Alaska's Native People*, edited by R. D. Shaw, R. K. Harritt, and D. E. Dumond. *Aurora IV*. Alaska Anthropological Association, Anchorage.

Kaplan, H., and Hill, K., 1992, The Evolutionary Ecology of Food Acquisition. In *Evolutionary Ecology and Human Behavior*, edited by E. A. Smith and B. Winterhalder, pp. 167–201. Aldine de Gruyter, New York.

Kelly, R. L., 1995, *The Foraging Spectrum*. Smithsonian Institution Press, Washington, DC.

Knecht, R. A., 1995, The Late Prehistory of the Alutiiq People: Culture Change on the Kodiak Archipelago from 1200–1750 A.D. Ph.D. Dissertation, Department of Anthropology, Bryn Mawr College, Bryn Mawr, PA.

Knecht, R. A., and Jordan, R. H., 1985, Nunakakhnak: An Historic Period Koniag village in Karluk, Kodiak Island, Alaska. *Arctic Anthropology* 22(2):17–35.

Lucier, C. V., and VanStone, J. W., 1995, *Traditional Beluga Drives of the Iñupiat of Kotzebue Sound, Alaska*. Fieldiana, Anthropology, New Series, No. 25, Publication 1468, Field Museum of Natural History, Chicago.

Mandryk, C. A. S., 1993, Hunter-Gatherer Social Costs and the Nonviability of Submarginal Environments. *Journal of Anthropological Research* 49(1):39–71.

Mann, D. H., Crowell, A. L., Hamilton, T. D., and Finney, B. P., 1998, Holocene Geologic and Climatic History Around the Gulf of Alaska. *Arctic Anthropology* 35(1):112–131.

Maschner, H. D. G., 1999, Settlement, Sedentism, and Village Organization on the Lower Alaska Peninsula: A Preliminary Assessment. In *Settlement Pattern Studies in the Americas: Fifty Years Since Viru*, edited by B. R. Billman and G. M. Feinman, pp. 56–76. Smithsonian Institution Press, Washington, DC.

Matson, R. G., and Coupland, G. 1995, *The Prehistory of the Northwest Coast*. Academic Press, New York.

Mills, R. O., 1994, Radiocarbon Calibration of Archaeological Dates from the Central Gulf of Alaska. *Arctic Anthropology* 31(1):126–149.

Mitchell, D. H., 1983, Seasonal Settlements, Village Aggregations, and Political Autonomy on the Central Northwest Coast. In *The Development of Political Organization in Native North America*, edited by E. Tooker, pp. 97–107. American Ethnological Society, Philadelphia.

Murra, J. V., 1980, *The Economic Organization of the Inka State*, Supplement 1 to Research in Economic Anthropology. JAI Press, Greenwich, CT.

O'Brien, M. J., (ed.), 1996, *Evolutionary Archaeology: Theory and Application*. University of Utah Press, Salt Lake City.

O'Connell, J., 1995, Ethnoarchaeology Needs a General Theory of Behavior. *Journal of Archaeological Research* 3(3):205–255.

Ogden, A., 1941, *The California Sea Otter Trade 1784–1848*. University of California Publications in History, Vol. 26, University of California Press, Berkley.

Ogden, A., 1991, Russian Sea Otter and Seal Hunting on the California Coast, 1803–1841. In *Fort Ross: California Outpost of Russian Alaska, 1812–1814*, edited by E. O. Essig, A. Ogden, and C. D. Kingston, pp. 35–59. The Limestone Press, Ontario.

Philibert, S., 1994, Ocre et le Traitement des Peaux: révision d'une Conception Traditionelle par l'Analyse Fonctionelle des Grattoirs Ocrés de la Balma Margineda (Andorre) [Ochre and the Treatment of Skins: Revision of a Traditional Concept for the Functional Analysis of Ochre Grinders in the Balma Margineda (Andorre) (in French)]. *Anthropologie* 98:447–453.

Price, T. D. and J. Brown (eds.), 1985, *Prehistoric Hunter-Gatherers: The Emergence of Cultural Complexity*. Academic Press, New York.

Shubin, V. O., 1994, Aleut in the Kurile Islands 1820–1870. In *Anthropology of the North Pacific Rim*, edited by W. W. Fitzhugh and V. Chaussonnet, pp. 337–345. Smithsonian Institution Press, Washington, DC.

Simon, J. J. K., and Steffian, A. F., 1994, Cannibalism or Complex Mortuary Behavior? An Analysis of Patterned Variability in the Treatment of Human Remains from the Kachemak Tradition of Kodiak Island. In *Reckoning with the Dead: The Larsen Bay Repatriation and the Smithsonian Institution*, edited by T. L. Bray and T. W. Killion, pp. 75–100. Smithsonian Institution Press, Washington, DC.

Spiess, A. E., 1979, *Reindeer and Caribou Hunters: An Archaeological Study*. Academic Press, San Francisco.

Steffian, A. F., 1992, Fifty years after Hrdlicka: Further Excavation of the Uyak Site, Kodiak Island, Alaska. In *Contributions to the Anthropology of Southcentral and Southwestern Alaska*, edited by R. H. Jordan, F. de Laguna, and A. F. Steffian. *Anthropological Papers of the University of Alaska* 24:141–164.

Steffian, A. F. (in prep), *Economic and Social Organization among the Kachemak Tradition Foragers of Kodiak Island, Alaska*. Ph.D. Dissertation, Department of Anthropology, University of Michigan, Ann Arbor.

Steffian, A. F., and Saltonstall, P. 1995, Markers of Identity: Labrets and Social Evolution on Kodiak Island, Alaska. Paper presented at the 60th Annual Meeting of the Society for American Archaeology, Minneapolis.

Steffian, A. F., Pontti, E., and Saltonstall, P. 1998, Archaeology of the Blisky Site: A Prehistoric Camp on Near Island, Kodiak Archipelago, Alaska. Manuscript of the Alutiiq Museum and Archaeological Repository.

Stephens, D. W., and Krebs, J. R. 1986, *Foraging Theory*. Princeton University Press, Princeton.

Testart, A., 1982, The Significance of Food Storage among Hunter-Gatherers: Residence Patterns, Population Densities, and Social Inequalities. *Current Anthropology* 23: 523–537.

Urcid, J., 1994, Cannibalism and Curated Skulls: Bone Ritualism on Kodiak Island. In *Reckoning with the Dead: The Larsen Bay Repatriation and the Smithsonian Institution*, edited by T. L. Bray and T. W. Killion, pp. 101–121. Smithsonian Institution Press, Washington, DC.

Wiessner, P., 1982, Beyond Willow Smoke and Dogs' Tails: A Comment on Binford's Analysis of Hunter-Gatherer Settlement Systems. *American Antiquity* 47(1):171–178.

Wilson, J. G., and Overland, J. E., 1986, Meteorology. In *The Gulf of Alaska: Physical Environment and Biological Resources*, edited by D. W. Hood and S. T. Zimmerman, pp. 31–54. National Oceanic and Atmospheric Administration (U. S. Dept. of Commerce) and Minerals Management Service (U. S. Dept. of Interior), Washington, DC.

Winterhalder, B. P., 1986, Diet Choice, Risk, and Food Sharing in a Stochastic Environment. *Journal of Anthropological Archaeology* 5:369–392.

Winterhalder, B. P., 1996, Social Foraging and the Behavioral Ecology of Intergroup Resource Transfers. *Evolutionary Anthropology* 5:46–57.

Winterhalder, B. P., 1997, Gifts Given, Gifts Taken: The Behavioral Ecology of Nonmarket, Intergroup Exchange. *Journal of Archaeological Research* 5:121–168.

Winterhalder, B. P. and Goland, C., 1997, Evolutionary Ecology Perspective on Diet Choice, Risk, and Plant Domestication. In *People, Plants, and Landscapes: Studies in Paleoethnobotany*. University of Alabama Press, Tuscaloosa.

Wobst, H. M., 1974, Boundary Conditions for Paleolithic Social Systems: A Simulation Approach. *American Antiquity* 39(2):147–178.

Woodburn, J., 1981, Hunters and Gatherers Today and Reconstruction of the Past. In *Soviet and Western Anthropology*, edited by E. Gellner. Duckworth, London.

Woodburn, J., 1982, Egalitarian Societies. *Man*. 17:431–451.

Workman, W. B., 1992, Life and Death in the First Millennium A.D. Gulf of Alaska: The Kachemak Tradition Ceremonial Complex. In *Ancient Images, Ancient Thought: The Archaeology of Ideology*, edited by A.1 S. Goldschmidt, S. Garvie, D. Seklin, and J. Smith, pp. 19–25. University of Calgary Archaeological Association, Calgary.

Workman, W. B., 1998, Archaeology of the Southern Kenai Peninsula Region. *Arctic Anthropology* 35(1):146–159.

Yesner, D. R., 1980, Maritime Hunter-Gatherers: Ecology and Prehistory. *Current Anthropology* 21(6):727–750.

Yesner, D. R., 1989, Osteological Remains from Larsen Bay, Kodiak Island, Alaska. *Arctic Anthropology* 26(2):96–106.

Yesner, D. R., 1998, Origins and Development of Maritime Adaptations in the Northwest Pacific Region of North America: A Zooarchaeological Perspective. *Arctic Anthropology* 35(1):204–222.

Part III

Beyond Ecological Approaches to Hunter-Gatherer Settlement Change

Introduction to Part III

Beyond Ecological Approaches to Hunter-Gatherer Settlement Change

Unlike the preceding ecologically oriented parts, the chapters in this part share an interest in exploring the roles that historical processes (idiosyncratic events), emergent cultural factors (e.g., ideology), and human agency and creativity play in long-term change of hunter-gatherer settlement and subsistence. Without contradicting the importance of ecological variables, each of these contributions argues in its own way that ecological factors are insufficient to explain hunter-gatherer settlement and subsistence change. This is an important perspective in a volume oriented toward ecologically grounded applications. Clearly, history matters, and cultural traditions and human actions have efficacy in directing the course of long-term cultural change.

It is notable that the challenge presented here to ecological approaches has had significant currency in biological and ecological debates, as well as in anthropology and archaeology. Historic ecology is emerging as a field that explicitly includes a role for historic contingency in shaping the development of particular environments or populations. Without abandoning the successful principles underlying ecological modes of explanation, it allows us to take the influence of unique, historic events into account. In the case of better understanding the evolution of hunter-gatherer settlement systems, we are inclined to view the different approaches here along a continuum from focus on explanatory generality (ecologically focused) to realism (historically and culturally focused). The chapters in this part are geared more toward the latter.

To some extent, these approaches take inspiration from the historical "revisionist" debate of hunter-gatherer social and economic organization (e.g., Schrire 1984; Wilmsen 1995). For example, Junker's chapter attempts to answer several key questions regarding the nature of the interaction between hunter-gatherers and food producers. The long time-depth that her database covers gives her a distinctive advantage in tackling this controversial issue. The two case studies by Cannon and Aldenderfer suggest possible ways in which ethnographic analogies may help us find alternative explanations to ecological or "rational" reasoning. Despite various potential problems, the use of ethnographic analogies will remain crucial in archaeological studies as an aid in constructing interpretive frameworks. This is particularly important when we attempt to approach history and ideology and also gender (see e.g., Cannon 1998). In this regard, case studies provided by the three chapters will positively contribute to the discussions on using ethnographic analogies and other nonarchaeological sources to structure models of interpretation.

The contributions by Cannon, Junker, and Aldenderfer to this volume are particularly significant when we think of the active debate between processual and postprocessual archaeologists since the mid-1980s. Because Binford was one of the original proponents of processual archaeology, the use of the forager/collector model is typically equated with adopting a processual approach. However, due to the growing interests in social and ideological issues in archaeology during the late 1980s and 1990s, the boundary between processual and postprocessual archaeologies is not nearly as clear-cut as it used to be (Trigger 1990, 1998). Instead of adopting a rigid ecological approach, the authors of these chapters use the forager/collector model as a heuristic device to identify places where ecological variables are insufficient to account for evolutionary sequences. It is worth noting that several archaeologists (e.g., Hodder 1999) have suggested that processual archaeology might prove more appropriate for studying small-scale hunter-gatherers but not necessarily for more complex societies. The fact that two of the three authors in Part III (Junker and Aldenderfer) deal with the interaction between hunter-gatherers and food producers (including herders) may support this general dichotomy.

Despite strong emphasis in these chapters on historical contingency and cultural logic as influential factors, none of them adopts an interpretive/narrative approach, which is increasingly gaining support among postprocessualists (e.g., Hodder 1999; Pluciennik 1999; see also Terell 1990). In other words, if we follow an epistemological criterion to draw a line between the two schools, these chapters may not fall into the category of postprocessual. Regardless of labeling, which we think is unimportant and to a certain extent inadequate, these chapters do demonstrate various ways in which long-term change in settlement practice can be explained with reference to the specific cultural setting and historical sequence of each case. (Eds.)

REFERENCES

Cannon, A., 1998, Contingency and Agency in the Growth of Northwest Coast Maritime Economies. *Arctic Anthropology* 35:57–67.

Hodder, I., 1999, *The Archaeological Process: An Introduction*. Blackwell, Oxford.

Pluciennik, M., 1999, Archaeological Narratives and Other Ways of Telling. *Current Anthropology* 40(5):653–678.

Schrire, C., 1984, Interactions of Past and Present in Arnhem Land, North Australia. In *Past and Present in Hunter Gatherer Studies*, edited by C. Schrire, pp. 67–93. Academic Press, Orlando.

Terrell, J., 1990, Storytelling and Prehistory, In *Archaeological Method and Theory*, Vol. 2, edited by M. B. Schiffer, pp. 1–29. University of Arizona Press, Tucson.

Trigger, B. G., 1990, The 1990s: North American Archaeology with a Human Face? *Antiquity* 64:778–787.

Trigger, B. G., 1998, Archaeology and Epistemology: Dialoguing Across the Darwinian Chasm. *American Journal of Archaeology* 102:1–34.

Wilmsen, E., 1995, Who were the Bushmen? Historical Process in the Creation of an Ethnic Construct. In *Articulating Hidden Histories: Exploring Influence of Eric R. Wolf*, edited by J. Schneider and R. Rapp, pp. 308–321. University of California Press, Berkeley.

Chapter **10**

Sacred Power and Seasonal Settlement on the Central Northwest Coast

AUBREY CANNON

The form of hunter-gatherer settlement in the vicinity of Namu on the central coast of British Columbia (Fig. 10.1) is consistent with the structure and constraint of environmental and social factors that Binford (1980) and Wiessner (1982) identified as important on the basis of their ethnographic observations. The long-term history of the settlement system, in contrast, appears to have developed according to a unique pattern of cultural perceptions and particular events. A system of winter villages, summer base camps, and specific purpose campsites structured around the seasonal gathering of resources is in evidence throughout the 10,000 years of archaeological history in the area. As the number and density of winter village settlements expanded at around 500 B.C., the number of site types increased to include possible summer village aggregations and a greater number and variety of resource extraction sites. The greater complexity of this pattern suggests an accommodation to the social opportunities and constraints caused by greater population density. Although these developments conform to expectations based on social and environmental considerations, their specific timing within the context of long-term stability cannot be

AUBREY CANNON • Department of Anthropology, McMaster University, Hamilton, Ontario, L8S 4L9 CANADA.

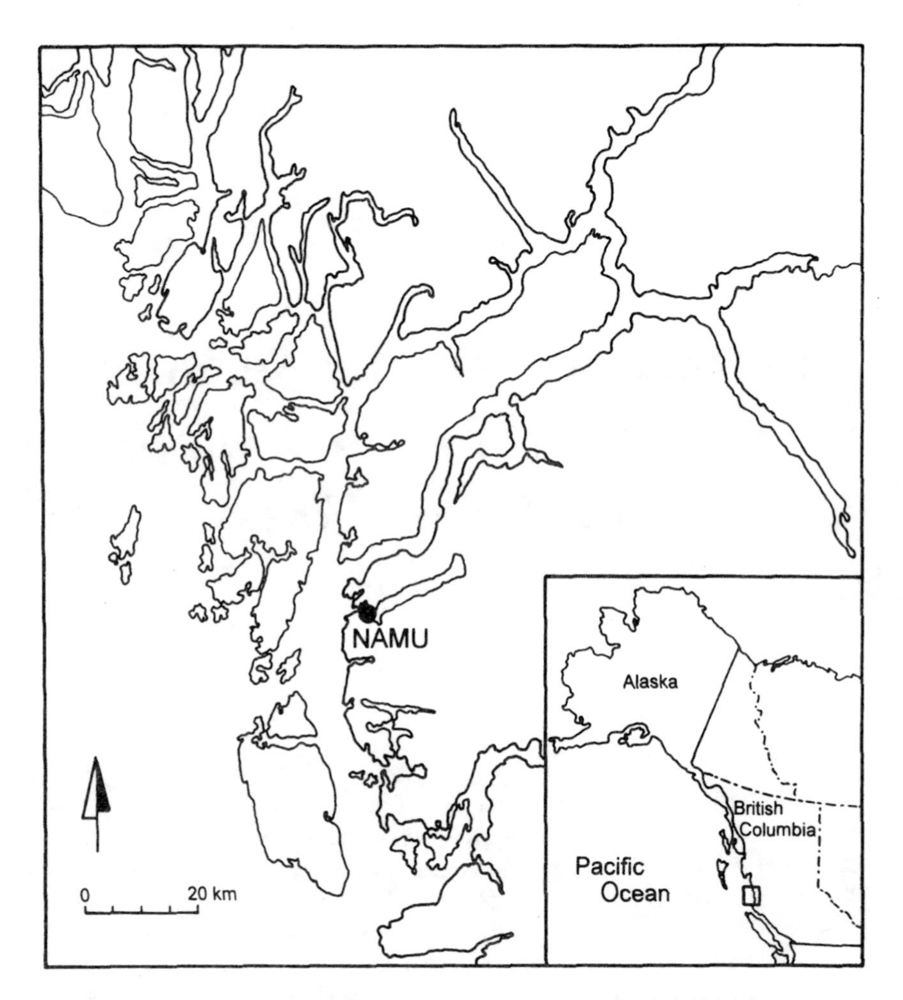

Figure 10.1. Location of Namu on the central coast of British Columbia.

explained with reference to those same considerations. Instead, expansion in the number of village settlements and long-term maintenance of Namu as a winter village appear related more to unique contingencies and cultural perceptions characteristic of ethnographically recorded Northwest Coast cultures. These perceptions relate to the means by which individuals acquired sacred power and the rights to food resources and to the ritual means by which power and rights were maintained and transferred over time.

Characterizing hunter-gatherer settlement systems as oriented more toward what Binford (1980) termed foraging or collecting strategies is a useful first step in organizing the complex array of settlement systems that might be encountered in any given context, but as an initial characterization, it takes into account only a limited range of relevant conditions. Binford outlined expectations of settlement systems based on spatial and seasonal availability of resources, modified by considerations of task group requirements and storage capabilities. Wiessner (1982), in commenting on Binford's framework, added considerations of social interaction, especially as related to strategies of risk reduction, to the set of conditions necessary to understand fully the form and distribution of hunter-gatherer settlements.

Neither Binford nor Wiessner was as concerned with long-term developments within systems as they were with characterizing settlement forms and relevant conditions that shaped systems at any given time. Binford (1980:19) acknowledged that his descriptive characterization of hunter-gatherer settlement concerned primarily short-term organizational strategies on the scale of the annual cycle rather than on the scale of long-term change. One obvious implication of his ecologically based model is that the characteristics of the settlement pattern should change along with any changes in the seasonal distribution or abundance of food resources. Wiessner's concerns with social interaction could be similarly addressed in considering long-term change by adding change in the abundance and distribution of people and the degree and frequency of risk as further factors that affect long-term developments. Essentially, the history and evolution of hunter-gatherer settlement would depend on change in the abundance and distribution of either resources or people. From this perspective, the long-term history of hunter-gatherers is a product of environmental change or population growth, whether based on natural growth rates, technologically based expansion of resource production, or migration.

Hunter-gatherer settlement systems in prehistory have shown changes predicated on changes in resource distribution or population. The general pattern of long-term trends in many regions is toward greater complexity of settlement types and reduced residential mobility, coupled with increased population density and a broader spectrum of resource utilization. This directional pattern of evolutionary change is variously attributed to environmental change (Binford 1968), technological developments (Hayden 1981), or the inevitable pressures of long-term population growth (Rosenberg 1998). The explanations vary for particular regions, but specific developments and long-term trends in most regions normally feature prominently one or a combination of environmental change, population growth, or technological innovation.

Binford explicitly avoided characterizing foraging and collecting strategies as opposing principles, but he also acknowledged that trends might be recognized. Archaeologists in many regions of the world have readily offered examples of settlement systems that become increasingly organized according to logistical, as opposed to residential mobility. Increasing seasonality of resource abundance, which was the critical factor identified by Binford that affects the degree of reliance on residential mobility, is rarely evident or cited as the cause of these trends. Changes in technological capacity, social organization, or population density are more commonly cited as factors that favor long-term reduction in residential mobility. Therefore, the evolution of settlement systems is viewed as a process of gradual adaptation to changing circumstances that results eventually in whatever pattern was recorded ethnographically. This is the way the long-term history of settlement on the Northwest Coast is generally viewed (Matson and Coupland 1995).

Ames and Maschner (1999:249) stress the dynamic and variable nature of Northwest Coast history and explicitly deny that it amounts simply to the gradual achievement of the ethnographic pattern. Nonetheless, they do chronicle what they see as a cumulative process of increasing sedentism and economic intensification based on greater reliance on storage. In contrast, the structure of resource distribution and seasonal availability and direct archaeological evidence of resource utilization, evident at least within a relatively small area of the central coast of British Columbia, suggest an alternative history that was characterized by long-term stability in the basic pattern of settlement types punctuated by specific and relatively short-term expansion in the number of winter villages. This pattern reinforces the basic validity of Binford's initial characterization and further suggests that apparent long-term trends in this and possibly other regions may be more the product of specific events and cultural perceptions than a function of ecological principles.

ETHNOGRAPHY, EVOLUTION, AND THE HISTORY OF NORTHWEST COAST SETTLEMENT

Northwest Coast archaeologists are well aware of the pervasive influence and potential dangers of ethnographically based inferences (Ford 1989; Moss and Erlandson 1995:28–29), yet archaeological interpretations of settlement systems rarely encompass even the full range of variability documented ethnographically. Mitchell (1983) provides one of the best summaries of ethnographic information on this topic. He notes a general acknowledgment that people on the coast moved around a lot to acquire various resources from various places during the year and that regular settlement moves were often accompanied by changes in settlement size and composition. Using a tripartite classification of settlement types as camps, villages, and village aggregates and the familiar four-season division of the year, Mitchell documented at least 16 distinct combinations of seasonal settlement arrangements among 69 local groups for which sufficient information was available. The most common pattern, consisting of village occupation during the winter and spring, a summer move to a multivillage aggregation, and then dispersion to fall camps, obtained in only 13 of the 69 groups (Mitchell and Donald 1988:309). From these observations, Mitchell (1983:99) reasonably concluded that no single pattern dominated the central Northwest Coast and that there was considerable variety in the seasonal sequence of concentration and dispersal.

Despite the availability of reasonably good seasonal indicators among faunal remains, archaeologists have rarely made seasonal variability in site use the specific focus of interpretation. Instead, they have concentrated on demonstrations of long-term trends in resource utilization (e.g., Croes and Hackenberger 1988; Matson 1992), even when seasonal variability in site use is an equally adequate explanation of observed variation in faunal assemblages. Another major focus of research has been the recovery of evidence of substantial structures that could be taken as an indication of the general pattern of semisedentary village settlement documented ethnographically. Unfortunately, due largely to the limited areal extent of excavations in deep, stratified midden deposits, recognition of substantial structures is limited to those rare sites that exhibit either surface indications of structures or clearly recognizable evidence of buried structures. As a result, the beginning of semisedentary village settlement in the region tends to be placed relatively late in time (Ames 1994; Ames and Maschner 1999; Matson and Coupland 1995).

In contrast to the long-term trends that have been the focus of major regional overviews, recent research projects in specific areas of the coast have begun to document a much greater range of settlement types and a variety of specific changes in these over time. Acheson's (1995) study of Kunghit Haida village settlement in southern Haida Gwaii (Queen Charlotte Islands), for example, documented a 1600-year precontact pattern of small, nucleated year-round settlements, which was replaced by the more familiar ethnographically described pattern of large multilineage villages, whose members shifted residence as part of a seasonal round. This shift is seen as a consequence of political and economic developments resulting from European contact and trade. In another example, Maschner (1997) documented a shift in settlement location in Tebenkof Bay in southeastern Alaska sometime between A.D. 300 and 500, which he attributes to an increase in conflict and a resulting concern for more defensible locations. He also notes an increase in village size, suggesting an amalgamation of lineages at this time. In a study of village sites in Prince Rupert harbor on the northern coast of British Columbia, Archer (2001) outlined a history of increasing social inequality based on an increase in the overall size and variability of house platforms. A transition toward villages with houses of larger and more variable size at around A.D. 100 is evident from dated samples taken from near the surface of visible house platforms. In all of these studies, the presence of identifiable surface features associated with residential structures is a key piece of evidence used to identify settlement forms and transitions. The late developments documented in these studies were readily identified and interpreted from data that could be obtained primarily from survey and minimal test excavation.

A tendency for extensive excavation projects to concentrate on larger village sites and the problem of establishing the time depth of settlement patterns from survey studies alone have made it difficult to develop an understanding of the long-term history of settlement systems in any particular region of the Northwest Coast. The small number of extensively excavated village sites has also made it very difficult to identify the long-term history of permanent village settlement. Even the few intensive investigations often do not provide the area extent of excavation needed to identify house structures. Despite this lack of direct evidence, indirect evidence of seasonality from the site of Namu, in the area of traditional Heiltsuk territory, does suggest a long and consistent history of winter village settlement, beginning from at least the 7000-year-old date of the earliest well-preserved vertebrate faunal remains.

SETTLEMENT HISTORY ON THE CENTRAL BRITISH COLUMBIA COAST

From as early as 5000 B.C., the Namu vertebrate fauna are dominated by very large quantities of salmon and herring (Cannon 1991, 2000a), which are available in greatest concentrations in the fall and late winter/early spring, respectively. The abundance of these resources suggests winter storage of salmon and residency at the site through the spring herring fishery. The consistent presence of neonatal harbor seal remains in all of the fauna-bearing deposits also indicates at least a consistent presence at the site during the late spring peak in the pupping season as well (Cannon 1991:59). A lack of major alternative food resources during the winter months would preclude the option of seasonal movement specifically to acquire resources between the time of the fall salmon and spring herring fisheries. Herring is the first resource to become available in great abundance following the winter period when stored foods were most relied upon. Despite a lack of evidence for substantial structures at Namu, the consistency of seasonal faunal indicators argues strongly in favor of the site's status as a winter village settlement. Given Binford's stress on seasonal resource availability as the key determinant in the development of logistically oriented settlement systems, the presence of permanent winter villages on the coast from at least as early as 7000 years ago is not unexpected. There have been no major environmental changes that affected the basic pattern of seasonal resource availability or abundance since that time. Therefore, the long-term evolution of settlement systems in the region, at least in relation to the establishment of permanent winter villages, is not an issue.

Despite continuity in the basic pattern of winter village settlement at Namu, there are indications of substantial changes in local resource availability and use over time that would have had implications for regional settlement. The analysis of vertebrate faunal remains recovered from excavations conducted by Roy Carlson of Simon Fraser University in 1977–78 clearly showed a substantial decline in the relative abundance of salmon beginning in the period after 1800 B.C., but especially after 200 B.C. (Cannon 1991). Subsequent analyses have established a variety of independent lines of evidence to substantiate the impression that the later subsistence economy was subject at least to periodic shortfalls in salmon production resulting in some measure of hardship for the resident population (Table 10.1). During this period there was also some contraction in the village size, and an area of the site adjacent to the mouth of the Namu River was abandoned sometime between 2000 and 2500 years ago (Carlson 1991). Although there is clear evidence, beginning around 2000 B.C., of at least periodic economic hardship predicated on intermittent salmon shortage, Namu remained a

Table 10.1. Evidence for Periodic Salmon Shortage and Dietary Stress at Namu after 2000 B.C.

Indicator	Evidence	Source
Decreased salmon production	Fewer salmon remains relative to other fish	Cannon (1991, 2000a)
Increased consumption of more marginal alternative foods	Greater numbers of ratfish and deer remains	Cannon (1995, 2000c)
	Lower ^{13}C values showing increase in terrestrial sources of protein in diet of dogs	Cannon et al. (1999)
	Higher ^{15}N values showing increase in lower trophic level protein sources (shellfish) in diet of dogs	Cannon et al. (1999)
	Thick layers and lenses of unbroken shell	Conover (1978)
	Increase in deer phalanges with medial–lateral breakage showing human processing for marrow	Zita (1997)
Greater dietary variability	More variable values of ^{13}C and ^{15}N in dog bones	Cannon et al. (1999)

major winter village for the next 3000 years. Evidence recovered from University of Colorado excavations in 1969–70 suggests that Namu may finally have been reduced to a seasonal campsite in the period from around A.D. 1000 to the time of European contact (Conover 1978:98).

A reduction in the size and later the seasonal duration of the Namu settlement is easily attributable to periodic shortfalls in local salmon production, but there are few clues from this one site to suggest how the establishment and decline of Namu as a major winter village related to any wider regional settlement system. Unfortunately, until recently, the remote location of the region and the logistical constraints of conducting even test excavations in deep shell-midden deposits limited the availability of information concerning the temporal depth and nature of activity at other locations in the vicinity. This limitation is now beginning to be overcome through a program of core and auger sampling of shell-midden sites (Cannon 2000a,b). The initial results show a great diversity of settlement types and a very particular history of expansion in site use that resulted over time in a regional system of much greater complexity, despite essential continuity in the basic form of settlement options.

Settlement Pattern

Fieldwork in 1996–97 obtained core and auger samples from 16 shell-midden sites in the vicinity of Namu (Fig. 10.2). Because this was an initial investigation to gain some insight into the range of site types, a variety of locations was selected for investigation. These included large middens that had evidence of structures or house platforms on the surface, as well as a number of moderate- and smaller sized middens. Locations ranged from the mainland and large islands immediately to the north and south of Namu and westward to include the more exposed outer island areas. The specific contexts of sites ranged from major mainland and large island rivers to smaller streams and coves and rocky islets. Not surprisingly, this variety of site sizes and locations yielded a variety of vertebrate and invertebrate faunal evidence indicating a range of seasonal occupations and activities.

The primary bases for inferring the nature and seasonality of site use are surface area and the overall density and variety of fish remains (Cannon 2000a). Based on these criteria, six different site types have been tentatively identified (Table 10.2). From their large size and very high density of salmon remains, two sites (ElSx-3, Kisameet Bay, and EkSx-12, Koeye River) have been identified as winter villages. Both exhibit a high density and variety of other fish remains, and the Koeye River site also exhibits evidence of large rectangular house platforms on the surface (Fig. 10.3). The density of salmon at each is comparable to the period of peak salmon production at Namu. The Kisameet Bay site also shows a comparable density of herring

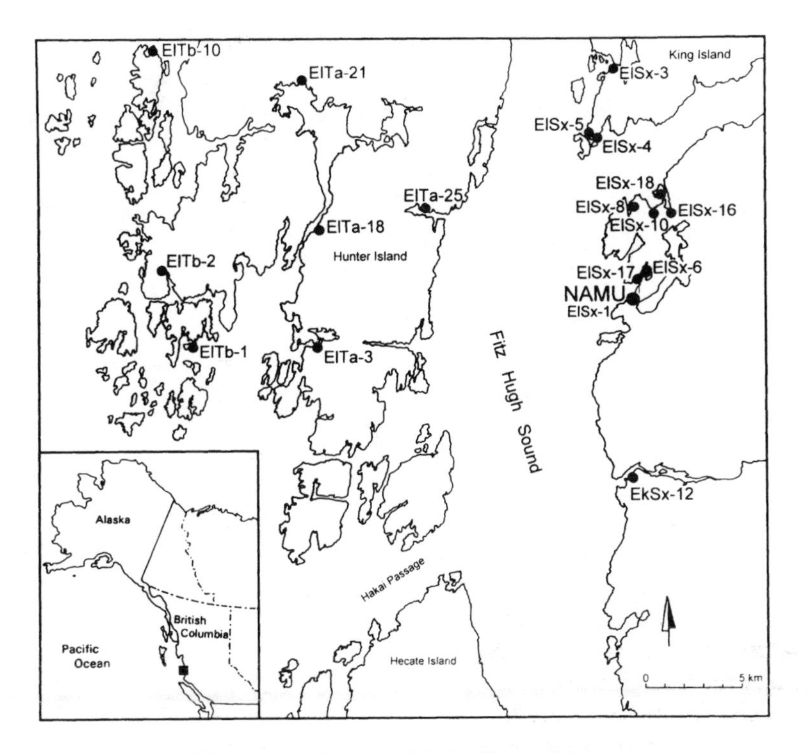

Figure 10.2. Sites tested in the Namu vicinity.

bones, indicating continued occupation through the late winter/early spring. The Koeye River site shows very little evidence of herring, suggesting that residents moved to another location for the late winter fishery. These two sites, together with Namu and the McNaughton Island site (ElTb-10), excavated in 1972 and 1974 (Carlson 1976; Pomeroy 1980), are the only likely winter villages so far identified in the region. Namu may also have served as a winter village in the period before 5000 B.C., for which faunal data are unavailable. The area of occupation and the size of the lithic artifact and debitage assemblages from these earlier deposits suggest that Namu was at least a major base camp if not a winter village at that time.

Another site (ElTb-1), located on Hurricane Island on the outer coast, has also been identified as a village on the basis of its size and high density of fish remains (Fig. 10.4). The fallen remains of relatively small contact-era, traditional post-and-beam cedar plank structures are evident on the surface (Fig. 10.5). The fish remains from this village show an unusually high density of herring, which is almost three times more abundant than at either Namu or Kisameet. The density of salmon is relatively low. The surviving remains of the seven house structures indicate an average

Table 10.2. Initial Dates and Characteristics of Site Types in the Namu Vicinity.

	Initial	Site	Sample	Density of Fish Remains (# per liter)				Identified	% Volume[b]
	Date[a]	Area (m^2)	Volume (L)	Salmon	Herring	Greenling	Other Fish	Fish Taxa	>2 mm
Winter village									
ElSx-1	5280–4720 B.C.	8100	62.7	51.4	37.8	0.7	3.3	15	17.9
ElTb-10[c]	830–400 B.C.	3700	NA	—	—	—	—	—	—
ElSx-3	770–50 B.C.	1630	30.7	50.7	52.4	3.2	2.3	14	25.4
EkSx-12	255 B.C.–A.D. 30	2500	26.2	61.1	7.6	1.9	5.2	18	12.2
Spring/summer village									
ElTb-1	805–410 B.C.	1110	29.8	14.2	139.0	3.2	4.6	13	18.4
Base camp									
ElSx-1	9600–8650 B.C.	?	NA	—	—	—	—	—	—
ElSx-5	4780–4510 B.C.	2190[d]	21.4	17.0	28.4	0.2	0.6	8	32.7
ElSx-10	4315–3960 B.C.	840	27.6	17.4	32.4	0.8	1.1	9	27.5
ElSx-18	1575–1310 B.C.	1310	13.2	20.8	20.5	0.4	1.1	9	35.6
Specific-purpose camp									
ElTa-25	2420–2025 B.C.	675	16.6	3.0	2.2	0.1	1.0	6	63.9
ElTb-2	A.D. 20–245	60	3.0	2.5	1.7	1.3	17.0	8	16.7
ElTa-21	A.D. 140–425	130	6.2	2.3	0.6	0.0	0.2	3	14.5
ElSx-8	A.D. 140–430	930	4.3	0.0	1.2	0.0	0.2	2	34.9
ElTa-3	A.D. 1160–1300	50	2.4	35.5	6.6	0.0	13.4	5	16.7
Small multipurpose camp									
ElTa-18	9605–9250 B.C.	1310[e]	12.1	2.2	0.1	0.2	0.0	3	11.6
ElSx-4	890–670 B.C.	140	6.0	4.5	2.8	0.0	0.2	3	25.0
ElSx-16	A.D. 660–940	300	4.1	3.9	37.1	0.0	0.5	3	39.0
Rocky Islet camp									
ElSx-6	A.D. 1550–1720	260	0.5	20.0	35.6	0.0	2.0	3	20.0
ElSx-17	A.D. 990–1165	320	0.8	1.2	0.0	0.0	0.0	1	12.5

[a]2 sigma calibrated date ranges, calibrations calculated using CALIB 4.1 from the University of Washington Quaternary Isotope Lab, using data sets from Stuiver et al. (1998), ElSx-1 dates from Carlson (1991), ElSx-3 date from Luebbers (1978), ElTb-10 date from Pomeroy (1980), and all other dates from Cannon (2000a,b).

[b]The volume of matrix larger than 2 mm generally includes quantities of rock, but it is largely composed of shell at most sites and therefore is a good indicator of relative shell density; the matrix from sites ElTb-2, ElTa-3, ElTa-18, ElTa-21, and ElSx-17 included only very small quantities of shell.

[c]Density data are not available, but excavation reports indicate a high density of salmon remains.

[d]Estimate based on partial mapping of the site.

[e]Area includes a steep slope in front of the site terrace as well as the 575 m^2 area of the terrace surface.

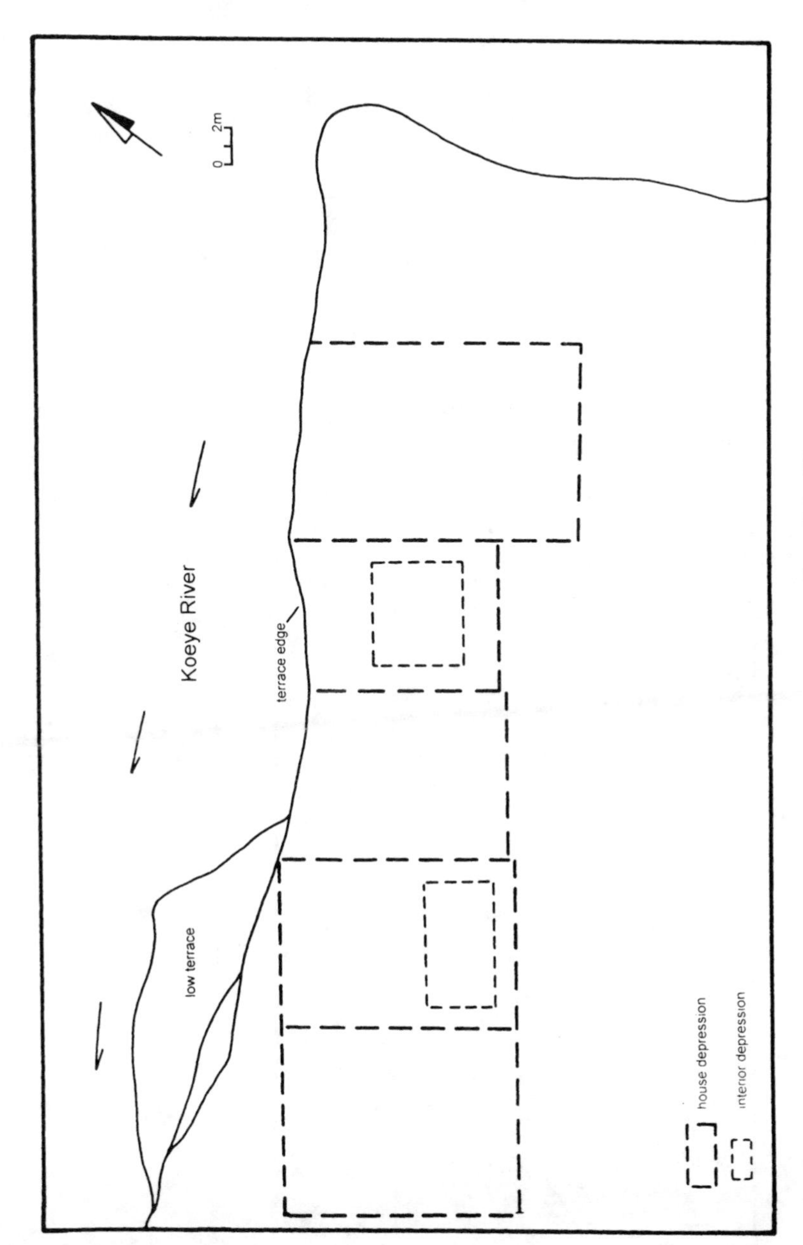

Figure 10.3. Koeye River village EkSx-12).

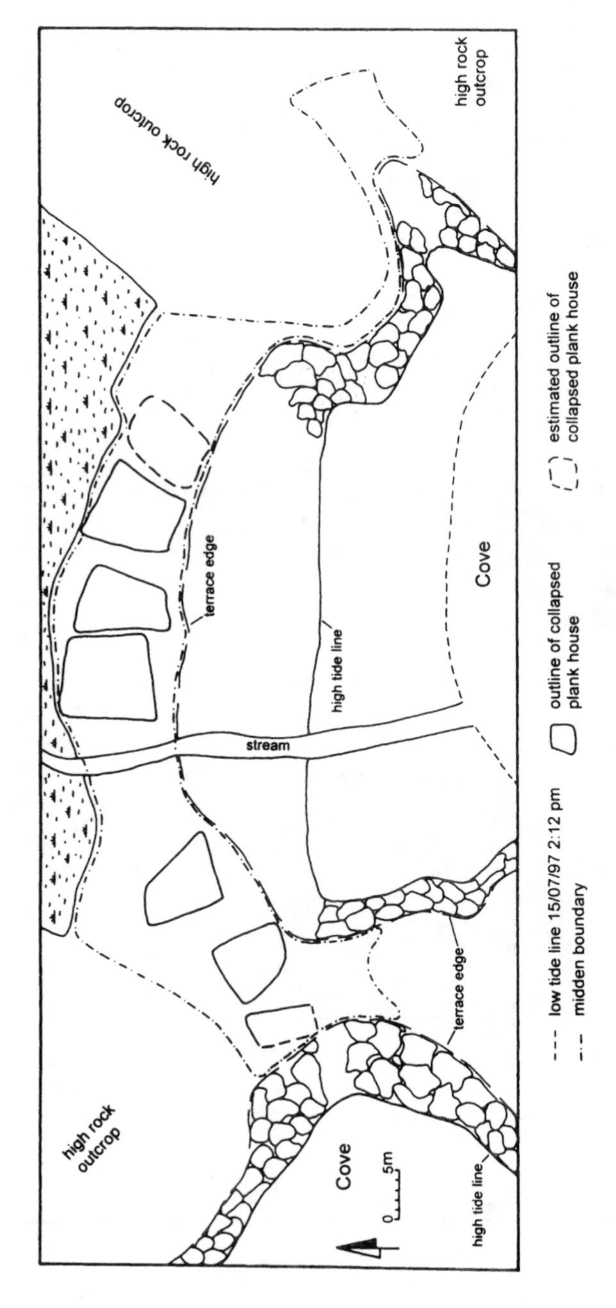

Figure 10.4. Hurricane Island village site (ElTb-1).

Figure 10.5. Fallen post-and-beam cedar plank houses at the Hurricane Island site (ElTb-1).

area dimension of $50\,m^2$, which is just half the average of the five house platforms at the Koeye River site. The fish remains and the small size of the structures suggest a spring village, established primarily to take advantage of the herring fishery, but which may have been used as a base camp for a variety of other activities throughout the spring and summer.

A third category of site (ElSx-5, ElSx-10, and ElSx-18) consists of moderate- to large-sized middens that show evidence of a range of subsistence activities. They lack the very high density or range of fish remains present in the village deposits but do not exhibit the low density evident at the more specialized shellfish gathering locations. Although clear evidence of the way these sites were used is not available, they are large enough and show sufficient evidence of a range of activities to qualify as multipurpose campsites, and possibly even as summer base camps. Task groups may have ranged from these locations to undertake specific subsistence pursuits at other sites in the immediate vicinity, which were less well suited for camping.

Vertebrate and invertebrate faunal evidence and the small size of a number of the other sites tested suggest a fourth category consisting of specific-purpose campsites. These include two sites that indicate a primary emphasis on shellfish gathering and processing (ElTa-25 and ElSx-8), and three probable fishing camps associated with stone fish traps at the mouths of small streams (ElTb-2, ElTa-3, and ElTa-21). One of these (ElTa-3) showed a very high density of salmon remains, but limited evidence of any other fishing or shellfish gathering that suggested a specialized salmon fishing station. A number of other smaller campsites (ElSx-4, ElSx-16, and ElTa-18) showed no clear indications of emphasis on any one specific type of activity but could have been used for a combination of fishing, hunting, shellfish gathering, or other subsistence-related pursuits. Two very small midden deposits on rocky islets in Namu harbor (ElSx-6 and ElSx-17) contained a range of fish and shellfish remains. Both sites are very small, and, given the lack of fresh water or other resources, neither is suited to any specific type of activity apart from short-term camping.

Given the still relatively modest extent of archaeological investigation, this range of site types is somewhat remarkable. At a minimum, it is clear evidence of a settlement system comparable in complexity to the ethnographically documented pattern for this and other regions of the coast. With the possible exception of Namu in its very latest period of occupation, all of the sites tested show overall consistency in their pattern of usage from the date of their establishment throughout subsequent occupation. The density and variety of faunal remains in the identified village sites is essentially the same throughout their deposits. Deposits at sites such as ElSx-5 and ElTa-25 are also consistent from their earliest dates (4700 and 2200 B.C., respectively).

Only the initial and terminal deposits of most of the sites tested in 1996–7 are dated, but those sites for which multiple dates are available show a continuous range of dates between the initiation and termination of settlement (Morlan 2001). The same is true for Namu, which has been radiocarbon dated extensively (Carlson 1991:85–95, 1996). There is no reason to think that any of the sites in the region were unused for appreciable periods from the time of their initial settlement. The terminal deposits at all but a few of the smallest sites also show that they were still in use into the European-contact era. On a site by site basis, therefore, the settlement system shows a remarkable degree of continuity, extending back to the time of the earliest fauna-bearing deposits at Namu. The basic pattern, consisting of permanent winter villages and seasonal camps for specific resource use, also appears to have been well established by this date. Despite this measure of long-term continuity, dates for the initial establishment of the various settlements indicate a history of development in the overall system that was a function of environmental, social, and cultural constraints and the result of very particular contingent circumstances. The effect over time was an increase in overall complexity based on an increase in the number of sites and variety of site types.

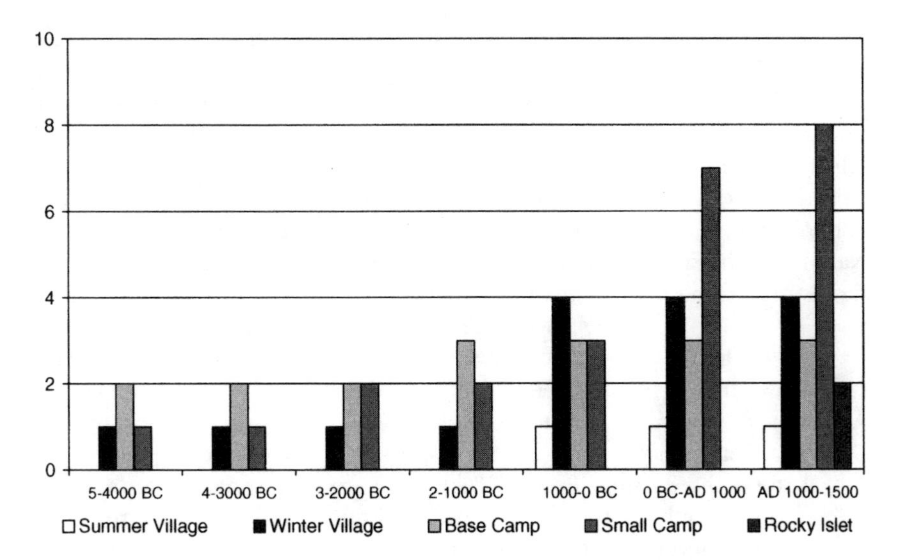

Figure 10.6. Number and variety of site types over time.

History

Figure 10.6 shows the settlement chronology of different site types in the vicinity of Namu. The two earliest dated sites, Namu and ElTa-18, provide little indication of the nature of their early use, but ElTa-18 is located on a small terrace and was unlikely to have been more than a small, possibly multipurpose campsite during this or any subsequent period. Namu is much larger at this time and has yielded a variety of artifactual evidence indicating more intensive occupation (Carlson 1996). There is no direct seasonal evidence for defining Namu as a winter village at this time, but it would certainly qualify as a generalized residential encampment and conceivably could have been a winter village even at this early date.

The earliest period for which good faunal data are available is between 5000 and 4000 B.C. During this time, there is good evidence of winter village settlement at Namu, as well as of the establishment of more generalized campsites at ElSx-5 and ElSx-10, which may have served as summer residential bases. Only two other sites among those tested were established during the next 3000 years. One was the generalized, possibly residential campsite at ElSx-18, which is located in the same small bay directly across from ElSx-10. The other was a specialized shellfish gathering and processing camp at ElTa-25. This period between 4000 and 1000 B.C. also encompasses the apparent peak in the Namu salmon-fishing economy and is the period to which the majority of the Namu burials date. Some of these are buried with grave goods indicative of variable wealth among individuals.

In contrast to the previous four millennia, the subsequent 1000-year period witnessed a dramatic expansion in the number of sites. All but one of the sites initially established in this period were villages. Although sampling and minor relative sea level decline over time may have had some effect on these results, the increase in the number of villages seems too clear to have been the result of sampling effects alone. The fact that this expansion corresponded to the period of increasing economic hardship at Namu and the time when evidence shows some contraction in the size of the Namu village also seems more than coincidental. The obvious conclusion is that an increase in the number of villages was somehow tied to Namu's declining prosperity. Subsequent to this brief period of village expansion, there is no evidence of the establishment of any new villages. All seven of the other sites tested, which were established during the last 2000 years, are small camps of either specific or unknown purpose.

If the eighteen sites listed in Table 10.2 are representative of trends in the region as a whole, then it is clear that although the basic pattern of settlement remained the same, there was an increase in the complexity of the overall system. Not surprisingly, there are more sites and a greater variety of site types over time, but what is surprising is the narrow time frame of village expansion and the subsequent increase in the number of specific-purpose and other small campsites. These latter include the two rocky islet middens, which date to within the last 1000 to 400 years.

The long-term stability in the number of villages before the period of expansion 2000–3000 years ago indicates that long-term gradual population growth was not a factor in the increase. Relative decline in sea levels may have been partly responsible for an increase in the number of suitable village locations, but the varied regional topography would have allowed growth in the number of villages before 1000 B.C. It is also likely that the productivity of salmon fishing at village locations such as Kisameet Bay and the Koeye River would have been as great before 1000 B.C. Village expansion appears to be related to a need or desire for new residential options at the time and possibly as a direct result of serious salmon shortfalls at Namu.

The available evidence, of course, does not rule out other possible scenarios, such as migration of peoples from adjacent areas. Mitchell (1990:357), for example, postulated a northward and southward expansion of Wakashan-speaking peoples from the northern part of the west coast of Vancouver Island at around 500 B.C. He suggests that speakers of Wakashan languages (which include Heiltsuk) displaced Salishan populations on the central coast, except for the remnant Nuxalk population at the mouth of the Bella Coola River. If such a migration occurred, it could explain the relatively sudden appearance of new villages. It would have been facilitated, perhaps, by economic decline at Namu, which would have reduced the capacity of the local population to resist migratory incursions. There is, however, no direct archaeological evidence of a late migration into the region. The continuing and consistent use of sites occupied before 500 B.C., such as Namu, ElTa-18, ElTa-25, ElSx-5, and ElSx-10, and the close proximity of new villages at the Koeye River and on Kisameet Bay also argue against the likelihood that migration of a distinct population was responsible for the increase in village density. The pattern of resource use at existing sites remained the same as it had been before the increase in villages. If the increased density was the result of an influx of migrant populations, greater disruption in the overall pattern of regional settlement might be expected.

Whatever the source of the new village populations, Namu continued to serve as a winter village, despite the continuing periodic hardships endured by its residents. Although there was the opportunity to take up residence at other locations, a sizable population continued to view Namu as its appropriate winter residence. The essential stability of the regional settlement pattern is also indicated by the fact that no new villages were established in the last 2000 years. Stability in the number of villages both before and after the relatively brief period of expansion also argues against a long-term process of continuous population growth. Population almost certainly grew over time, but that growth appears to have been the result of the increase in villages, rather than its cause.

One implication of the larger number of villages seems to have been the creation of new necessities and possibly new opportunities for an increase in the number of specific resource-gathering encampments and in the establishment of new site types, such as the spring village site on Hurricane Island. If the basic form of the settlement system was a function of environmental constraints, then these new developments may have been the result of the social implications of greater village density. The reasons for the establishment of new villages and for the limited duration of this period of expansion, however, appear to go beyond the material needs of subsistence and settlement. The lack of obvious material explanation suggests that the cause must be sought instead in an understanding of the relevant cultural perceptions and on actions based on those perceptions and contingent circumstances. The ritual activity conducted at winter villages to ensure the acquisition and transfer of sacred power and rights to gather food resources can explain the long-term stability and punctuated expansion of villages in the Namu area in ways that material conditions alone cannot.

Structural Constraints and Contingent Opportunities

Much of the basic form of the Namu vicinity settlement system can be explained adequately by the seasonal availability and abundance of subsistence resources, which Binford identified as critical determinants of hunter-gatherer settlement. The locations and forms of settlements established after the increase in winter villages can be explained with reference to these factors and to social factors that would have constrained the availability of locations for residence and resource extraction, but which may also have provided new opportunities for seasonal aggregations. The timing of village expansion and the apparent conservatism evident in the long-standing use of a restricted number of locations for winter settlement may be better explained by cultural conceptions particular to the region than by more generalized social or ecological considerations.

A logistically oriented settlement strategy has almost certainly been in place from the time that Northwest Coast subsistence economies became dependent on abundant but seasonally limited resources such as salmon and herring. Although the timing of this dependence may vary in different areas of the coast, the marked seasonality and scattered distribution of most other resources would always have favored some measure of residential stability. Between the time of the last salmon runs in late fall and the beginning of herring season in late winter/early spring, no other resources are available in sufficient quantity or concentration to warrant residential mobility. The available evidence does not show any greater concentration of land or sea mammal populations in earlier periods, and plant resources would always have been restricted seasonally in this northern climate. Therefore, although the timing of the development of permanent winter village settlements on the coast remains a subject of considerable debate, it is likely that there were few other equally viable options at any time in the past. Any advantages of greater residential mobility in the early period of settlement on the coast have yet to be articulated. The expectation of a logistical strategy based on longer term winter residence and a series of shorter term spring, summer, and fall residential campsites at locations of seasonally abundant resources is entirely consistent with Binford's model and our present understanding of the Northwest Coast environment. This pattern is also borne out by archaeological evidence from Namu and other sites in the vicinity from as early as 7000 years ago.

At any time during the past 7000 years, the settlement system of the central coast could be adequately described as consisting of winter residential villages and shorter term campsites for extracting specific subsistence resources. Although the evidence for summer residential campsites is not as clear, it is likely that these were in place throughout this period as well. As far as it goes, then, Binford's model is an accurate and adequate predictor of settlement organization in this, as well as any other region where there is marked seasonality in resource availability.

Despite this level of agreement with Binford's basic model, there are differences in the specific configuration of seasonal site use over time, especially during and following the period of village expansion. As the winter village density increases, a winter village without a spring herring fishing component is established on the Koeye River, and a late winter/early spring village site with a concentration on the herring fishery is established on Hurricane Island. This is a departure from the longer term seasonal occupation evident from very early times at Namu and later at Kisameet Bay. One possible explanation for these new forms of settlement is that they were the result of constraints imposed by greater village density, which prevented new villages from being established in locations where a full range of seasonal resources was available. In other words, because Namu and Kisameet were already established at better locations for gaining access to both salmon and herring, an additional village at the Koeye River site had to settle for access only to the fall salmon fishery at that location. New residential strategies were needed to access the spring herring fishery. A spring village, such as the Hurricane Island site, would represent just such an option.

Apart from being the result of constraints imposed by a more fully occupied landscape, the option for spring/summer villages like the one on Hurricane Island could also represent the development of more permanent social aggregations to bring together related family groups dispersed among the greater number of winter villages. This type of summer village aggregate, which was known ethnographically on the Northwest Coast (Mitchell 1983), could develop as the result of considerations such as the need to develop wider regional interaction networks of social integration and exchange. This type of social interaction to minimize the risk of resource shortage has been identified by Wiessner (1982) as a potentially important determinant of the form and distribution of hunter-gatherer settlements.

The greater density of villages also appears to have led to an increase in the number of specific resource extraction locations. From one perspective, this could be viewed as the result of population increase forcing use of a greater number and range of possibly less productive or less reliable resource locations. From another perspective, this could be seen as the opportunity to use a wider range of resource locations, enabled by an increase in village density and the resulting potential for pooling risk across a larger number of social groups. The risks associated with salmon fisheries at some of the very smallest streams, for example, would be offset by the opportunity to share in the returns of other small task groups that focused their efforts on a variety of similar locations. Although only three small fish trap locations have been dated thus far (ElTa-3, ElTa-21, and ElTb-2), their relatively late date is consistent with an increase in settlement variety resulting from greater population density and increased social interaction. Increased social interaction of a less cooperative nature might also be cited in the late addition of middens on rocky islets to the site inventory. These features have been identified in northern areas of the coast as possible refuges from raids (Maschner 1997; Moss and Erlandson 1992). If the increase in residential villages increased the potential for conflict in the region, then it would be reasonable to attribute their use to the more frequent threat of attack.

Therefore, the greater complexity evident in the settlement systems in later periods is directly attributable to the type of social considerations that, Wiessner suggested, were critical additions to Binford's environmentally based model. Although together the social and environmental variables account for the form of the settlement system as it appears at any time, they alone cannot account for the specific pattern of historical change in the system. The greater variety of site types following the increase in winter villages can be explained as a response to the opportunities and necessities this created, but the expansion itself cannot be explained on this basis. An explanation for the establishment of new villages simply as a response to resource shortages at Namu is also inadequate because it does not account for the lack of earlier expansion or for the continued use of Namu as a winter village. Although efforts to extend ethnographic observations into the archaeological past are often rightly criticized, the apparent conservatism evident in the settlement history of this region can be explained by aspects of recorded Northwest Coast cultures. These include beliefs concerning the acquisition of sacred power necessary to ensure resource abundance and the performance of rituals that legitimized status and rights to resources.

IDEOLOGY AND AGENCY

Among the Heiltsuk, as among neighboring peoples, the seasons of the year were divided into the summer secular, resource-gathering season and the sacred season when the Winter Ceremonial was performed (Hilton 1990:318). The former was marked by relaxed rules of social hierarchy and food sharing. Social groups were relatively small and group composition varied as individuals came and went to take advantage of local variation in resources (Harkin 1997:67). During the sacred winter season, society was much more structured, and food distribution became highly marked. According to Harkin (1997:68), the distribution of food reaffirmed that resources were owned by chiefs and that it was their supernatural power that made abundance possible. The basis of the ceremonies was reenactment of ancestral encounters with supernatural powers that enabled and legitimated the inherited power and rights of ranking individuals to control access to food resources. Therefore, the seasonal cycle of secular gathering and sacred performance was self-perpetuating because the accumulated food surplus was used by ranking individuals to sponsor ceremonies that legitimated their rights to those same food sources during the coming secular season.

The broader cosmology of the Heiltsuk and other Northwest Coast peoples linked people and animals, such as salmon, in a reciprocal relationship symbolized by consumption. This relationship was enacted in the ritual performances of the winter ceremonials. The associated feasts represented an investment of the previous season's harvest in the control of supernatural power and the legitimization of social power needed to ensure the next season's harvest. Therefore, the winter village was as much the site of ritual permanence (Bender 1985:26) as it was the site of residential permanence. The result would have been a sense of delayed return on investment in a location, which is more typical of farmers than of hunters and gatherers (Meillassoux 1972:99). The Northwest Coast consumption of stored foods has also been viewed conceptually as the conversion of hunting and gathering into farming (Ingold 1987:218). Strongly held beliefs that food production and the rights to food acquisition relied on ritual intercession with supernatural powers would create no less powerful a tie to the locality of ritual performance than the tie to the land that farmers create by tending and sowing their fields.

The temporal depth of religious beliefs and ritual practices is impossible to establish directly through archaeology, but ritual investment in a winter village to enact a relationship with a supernatural power to ensure food abundance would encourage maintenance of that settlement. A perception that individuals who have sacred power were responsible for resource abundance and a ritual practice among successive generations of enacting the acquisition and inheritance of that power would also restrain individuals from abrogating their responsibilities for investing in its continuing ritual legitimization. Validation of social status and contact with powerful superhuman forces took place in the winter village. The secular activities of resource gathering were, in large part, directed at accumulating food to be used in the feasts that accompanied all ritual activity (Harkin 1997:7). From this perspective, seasonal activities were inextricably linked. Ambitious individuals might seek to manipulate resource accumulation and feasting to bolster their own aspirations to status (Aldenderfer 1993), but the stability of the system on which status claims were based suggests that it was derived ultimately from strongly held beliefs and ritual practice, rather than mere competition for status.

From the presently available archaeological evidence, similar perceptions and ritual practices could be seen as constraining the location of winter village settlement at Namu during more than four millennia from at least 5000 B.C. up to the time of regional village expansion between 1000 and 500 B.C. The basis for this break with sacred tradition can also be explained with reference to the same cosmology and ritual practice. Aldenderfer (1993:13) has noted that successful ritual creates a body of conformists and a conformist tradition that has a strong basis for belief in the efficacy of the existing ritual system, but if the expected benefits fail to materialize, individuals may reject or seek to modify existing ritual practice. In the face of continuing periodic shortfall in the Namu salmon fishery, the expression of that rejection may have been the actions of some individuals to break with the ritual locus of Namu to establish new winter villages. There, the acquisition of their own power could be enacted and celebrated and form the basis for new ancestral traditions. According to Harkin (1997:164), a common theme in Heiltsuk mythology is that of the (potential) chief who goes out from a starving village to encounter a supernatural creature who gives him the power to obtain a wealth of food to feed his people. This theme can never be ascribed directly to the specific events evident in the archaeological history of settlement in the vicinity of Namu, but it suggests the means by which individual agents might break with ritual tradition in the face of its demonstrable failure. The contingency of periodic food shortage at one location would be legitimate basis for establishing new ritual traditions at alternative residential sites. A continuing and long-standing tradition of religious belief and ritual practice would also explain subsequent stability in the number and distribution of winter villages, as well as maintenance of Namu as a winter village in the face of continuing hardship and the availability of other residential options.

DISCUSSION AND CONCLUSIONS

On a broad scale, the history of settlement on the central coast embodies many of the themes commonly described in general evolutionary terms. The region witnessed an increase in the number of settlements and in the complexity of the settlement system over time. The seasonal availability and distribution of resources dictated the basic pattern, and social constraints and opportunities shaped later developments. Using only a rough chronology and limited evidence of seasonal activity, it would be easy to attribute this overall pattern to long-term population growth or regional circumscription that resulted in a gradual shift from small-scale, residentially mobile populations toward larger and more logistically organized populations. The actual history of settlement based on well-dated sites and the relative abundance of seasonally restricted resources is far more complex.

The common ethnographically described system of permanent winter villages and seasonal campsites appears to have been in place without any subsequent modification in form as early as 5000 B.C. From the available archaeological evidence, nothing in the physical or social environment of the vicinity of Namu changed sufficiently to encourage any change or growth in the settlement system during the next 4000 years. Because the social conflict that inevitably arises among sedentary populations should have been sufficient inducement for some growth in the number of regional settlements, it seems reasonable to conclude that some mechanism was in place to restrain individuals and groups from establishing alternative village locations. The constraint suggested here was one of religious belief and investment in the ritual, as well as the residential permanence of the Namu site. Aldenderfer (1993) described the increased role of ritual in the face of circumscription, but in this case ritual is more likely to have been the cause than the consequence of residential circumscription.

Growth in the size and complexity of the settlement system in this region also appears to have been the consequence of particular events. Demonstrated failure of the ritual power legitimated at the Namu winter ceremonials may have provided sufficient incentive and ritual license for some individuals and their families to establish new winter settlements. Continuing religious conservatism would then be sufficient to explain the subsequent stability in the number of winter villages. The emergence of the spring village and increasing use of smaller and more specialized resource extraction locales could be attributed to the opportunities made possible through increased social interaction among winter village populations and the advantages of pooling the risks of using smaller and more variable resource locations.

This interpretation of the history of the Namu vicinity settlement system is, of course, not the only one possible, and it depends for much of its content on sources that are not strictly archaeological. Specific ethnographic analogy to recorded practices and beliefs of the Heiltsuk is the basis for inferring beliefs regarding the supernatural origin of natural food resources and the ritual interaction of high ranking individuals with supernatural powers to ensure an abundance of food. Similar beliefs are not uncommon, however, and are recorded for such complex hunter-gatherers as the Calusa of southern Florida. There, the paramount chief acted as a broker with the spirit world, and his authority was legitimate because he intervened with the supernatural world to avert calamities and to ensure that the earth and the seas produced their riches (Marquardt 1992:120, 124). Raising the analogy to a more general level, however, does not necessarily increase its credibility, especially in the absence of direct archaeological evidence. The problem, of course, is that direct evidence of religious beliefs and of the designation of residential villages as sites of seasonal ritual performance will never be forthcoming in the absence of detailed iconography or enduring monumental construction.

The direct archaeological evidence is of a history of a settlement pattern that does not conform to any apparent corresponding change in material conditions, though the pattern itself is readily attributable to material and social conditions at any given point in its history. A continuous process of population growth is certainly not evident in the long-term paucity of major residential sites before 500 B.C. Nor is it evident in their punctuated expansion or subsequent stability. Whether or not migration was a factor in the relatively rapid increase in village settlements, the population resident at Namu clearly had not taken advantage of existing environmental opportunities to expand its numbers by establishing villages at suitable locations available in the vicinity. There is also no evidence of regional environmental changes that would have resulted in enhanced resource productivity at any of these alternative village sites. Local environmental changes at Namu that resulted in periodic failure of the salmon fishery also did not result in total abandonment of that location in favor of demonstrably more productive locations nearby. Defensive considerations have been cited in other regions of the coast as the cause of later village amalgamation (e.g., Maschner 1997). These could be considered a factor in maintaining a stable village residence at Namu, but there is no direct evidence to support such a claim for the earlier Namu occupation. It also seems incongruous to suggest greater concern for defense at a time when population density was so much lower than it was later when a greater variety of locations was used.

A more plausible alternative to this range of potential materialist explanations is that long-term stability in the settlement system, which is evident in the archaeological record, was structured by a system of belief and ritual practice. Failure of the salmon fishery at Namu, a periodic event more common after 500 B.C. and clearly evident in the archaeological evidence, contradicted the structural expectations of ritual and belief. This contingency would give some individual agents opportunity and reason to redefine and ultimately to restore the structure by establishing winter villages at new locations, an action also evident in the archaeological record. The enduring structure of belief and ritual then acted to maintain the stability of a new pattern of multiple winter village settlements in the region after 500 B.C., a pattern which again is evident archaeologically.

All aspects of this proposed scenario are evident archaeologically except for the structure and the agency, but some structure must be responsible for constraining winter villages before and after 500 B.C., and actions on the part of some agents were ultimately responsible for the establishment of new villages. Any alternative explanation would have to entail alternative structures and agency to account for the same pattern. That we cannot see these archaeologically does not mean they were not present or important as explanations. They are merely invisible and therefore must be extrapolated from what is visible, a practice that is often necessary in archaeology and in science more generally (Bailey 1983:176).

The explanation for the Namu region settlement history that I have proposed does take some of its content from ethnographic analogy, but it owes as much to its theoretical basis and to its fit with the archaeological data. Theoretically, resolution of the contradiction between structures of cultural perception, belief, and ritual action and the contingency of salmon fishery failure at Namu through the establishment of new winter villages would be an example of what Sahlins (1981:68, 1991:80–84) described as the structure of the conjuncture. Events that are contradictory to structures of cultural perception are precisely the occasions when individual action, change, and reestablishment of redefined structures are most favored.

I am suggesting that agents that acted in response to archaeologically documented events in Namu's history, shortfalls in salmon productivity, ultimately altered the history of settlement in the region but preserved the structural basis of the settlement system. The cultural perception that salmon shortages were contradictory to the ritual basis of the permanent winter village, which was designed to ensure the abundance of resources, resulted in resolving the contradiction by establishing new villages. The foundation of new villages as ritual and residential centers, however, was based on the same structures of cultural perception and belief that originally had been responsible for Namu's long history as a permanent winter village. These enduring structures that involved seasonal ritual reenactment of the supernatural encounters that ensured continuing resource abundance also subsequently ensured the stability and permanence of the new villages. The history of settlement expansion, in other words, was the product of the same structures that normally acted to constrain such expansion. The expansion itself was simply the resolution of enduring structures and contradictory events.

Events and their impact on existing structures of thought and action create the forms of change that we ultimately observe in the archaeological record. The advantage of taking such structures, events, and actions into consideration in the present case is the enhanced ability to account for the particular character of historical developments, while also accounting for the influence of a broad range of environmental, social, and ideological factors. Explanations of historical developments in hunter-gatherer settlement in most regions are likely to require similar considerations beyond what has so far been the focus of archaeological models. Archaeological evidence may not be particularly well suited to recognition of the conceptual schemes that influence settlement systems, but such schemes are not as inaccessible as some would suggest (e.g., Ingold 1987:218). Arguably, they become increasingly evident from the inability of simpler models to account fully for the richness of the archaeological data.

ACKNOWLEDGMENTS

Archaeological site investigations in the Namu vicinity in 1996–97 were undertaken with the financial support of a research grant from the Social Sciences and Humanities Research Council of Canada and were carried out in cooperation with the Heiltsuk Cultural Education Centre.

REFERENCES

Acheson, S. R., 1995, In the Wake of the Iron People: A Case for Changing Settlement Strategies among the Kunghit Haida. *Journal of the Royal Anthropological Institute* 1:273–299.

Aldenderfer, M., 1993, Ritual, Hierarchy and Change in Foraging Societies. *Journal of Anthropological Archaeology* 12:1–40.

Ames, K. M., 1994, The Northwest Coast: Complex Hunter-Gatherers, Ecology, and Social Evolution. *Annual Review of Anthropology* 23:209–229.

Ames, K. M., and Maschner, H. D. G., 1999, *Peoples of the Northwest Coast: Their Archaeology and Prehistory*. Thames and Hudson, London.

Archer, D. J. W., 2001, Village Patterns and the Emergence of Ranked Society in the Prince Rupert Area. In *North Coast Prehistory*, edited by J. S. Cybulski, pp. 203–222. Archaeological Survey of Canada, Mercury Series, Paper 160, Canadian Museum of Civilization, Hull.

Bailey, G. N., 1983, Concepts of Time in Quaternary Prehistory. *Annual Review of Anthropology* 12:165–192.

Bender, B., 1985, Prehistoric Developments in the American Midcontinent and in Brittany, Northwest France. In *Prehistoric Hunter-Gatherers: The Emergence of Cultural Complexity*, edited by T. D. Price and J. A. Brown, pp. 21–57. Academic Press, Orlando.

Binford, L. R., 1968, Post-Pleistocene Adaptations. In *New Perspectives in Archaeology*, edited by S. R. Binford and L. R. Binford, pp. 311–341. Aldine, Chicago.

Binford, L. R., 1980, Willow Smoke and Dog's Tails: Hunter-Gatherer Settlement Systems and Archaeological Site Formation. *American Antiquity* 45:4–20.

Cannon, A., 1991, *The Economic Prehistory of Namu: Patterns in Vertebrate Fauna*. Department of Archaeology, Simon Fraser University, Publication 19, Burnaby, British Columbia.

Cannon, A., 1995, The Ratfish and Marine Resource Deficiencies on the Northwest Coast. *Canadian Journal of Archaeology* 19:49–60.

Cannon, A., 2000a, Assessing Variability in Northwest Coast Salmon and Herring Fisheries: Bucket-Auger Sampling of Shell Midden Sites on the Central Coast of British Columbia. *Journal of Archaeological Science* 27:725–737.

Cannon, A., 2000b, Settlement and Sea-Levels on the Central Coast of British Columbia: Evidence from Shell Midden Cores. *American Antiquity* 65:67–77.

Cannon, A., 2000c, Faunal Remains as Economic Indicators on the Pacific Northwest Coast. In *Animal Bones, Human Societies*, edited by P. Rowley-Conwy, pp. 49–57. Oxbow, Oxford.

Cannon, A., Schwarcz, H. P., and Knyf, M., 1999, Marine-Based Subsistence Trends and the Stable Isotope Analysis of Dog Bones from Namu, British Columbia. *Journal of Archaeological Science* 26:399–407.

Carlson, R. L., 1976, The 1974 Excavations at McNaughton Island. In *Current Research Reports*, edited by R. L. Carlson, pp. 99–114. Department of Archaeology, Simon Fraser University, Publication No. 3, Burnaby, British Columbia.

Carlson, R. L., 1991, Appendix B: Namu Periodization and C-14 Chronology. In *The Economic Prehistory of Namu: Patterns in Vertebrate Fauna*, edited by A. Cannon, pp. 89–95. Department of Archaeology, Simon Fraser University, Publication 19, Burnaby, British Columbia.

Carlson, R. L., 1996 Early Namu. In *Early Human Occupation in British Columbia*, edited by R. L. Carlson and L. Dalla Bona, pp. 83–102. University of British Columbia Press, Vancouver.

Conover, K., 1978, Matrix Analyses. In *Studies in Bella Bella Prehistory*, edited by J. J. Hester and S. M. Nelson, pp. 67–99. Department of Archaeology, Simon Fraser University, Publication No. 5, Burnaby, British Columbia.

Croes, D. R., and Hackenberger S., 1988, Hoko River Archaeological Complex: Modeling Prehistoric Northwest Coast Economic Evolution. In *Prehistoric Economies of the Pacific Northwest Coast*, Research in Economic Anthropology, Supplement 3, edited by B. L. Isaac, pp. 19–85. JAI Press, Greenwich, CT.

Ford, P. J., 1989, Archaeological and Ethnographic Correlates of Seasonality: Problems and Solutions on the Northwest Coast. *Canadian Journal of Archaeology* 13:133–150.

Harkin, M. E., 1997, *The Heiltsuks: Dialogues of Culture and History on the Northwest Coast*. University of Nebraska Press, Lincoln, NE.

Hayden, B., 1981, Research and Development in the Stone Age: Technological Transitions among Hunter-Gatherers. *Current Anthropology* 22:519–548.

Hilton, S. F., 1990, Haihais, Bella Bella, and Oowekeeno. In *Northwest Coast*, edited by W. Suttles, pp. 312–322. Handbook of North American Indians, Vol. 7. Smithsonian Institution, Washington, DC.

Ingold, T., 1987, *The Appropriation of Nature: Essays on Human Ecology and Social Relations*. University of Iowa Press, Iowa City.

Luebbers, R., 1978, Excavations: Stratigraphy and Artifacts. In *Studies in Bella Bella Prehistory*, edited by J. J. Hester and S. M. Nelson, pp. 11–66. Department of Archaeology, Simon Fraser University, Publication No. 5. Burnaby, British Columbia.

Marquardt, W. H., 1992, Dialectical Archaeology. In *Archaeological Method and Theory*, Vol. 4, edited by M. B. Schiffer, pp. 101–104. University of Arizona Press, Tucson.

Maschner, H. D. G., 1997, The Evolution of Northwest Coast Warfare. In *Troubled Times: Violence and Warfare in the Past*, edited by D. L. Martin and D. W. Frayer, pp. 267–302. Gordon and Breach, Amsterdam.

Matson, R. G., 1992, The Evolution of Northwest Coast Subsistence. In *Long-Term Subsistence Change in Prehistoric North America*, Research in Economic Anthropology, Supplement 6, edited by D. R. Croes, R. A. Hawkins, and B. L. Isaac, pp. 367–428. JAI Press, Greenwich, CT.

Matson, R. G., and Coupland, G., 1995, *The Prehistory of the Northwest Coast*. Academic Press, San Diego.

Meillassoux, C., 1972, From Reproduction to Production: A Marxist Approach to Economic Anthropology. *Economy and Society* 1:93–105.

Mitchell, D. H., 1983, Seasonal Settlements, Village Aggregations, and Political Autonomy on the Central Northwest Coast. In *The Development of Political Organization in Native North America*, edited by E. Tooker, pp. 97–107. American Ethnological Society, Philadelphia.

Mitchell, D. H., 1990, Prehistory of the Coasts of Southern British Columbia and Northern Washington. In *Northwest Coast*, edited by Wayne Suttles, pp. 340–358. Handbook of North American Indians, Vol. 7. Smithsonian Institution, Washington, DC.

Mitchell, D., and Donald, L., 1988, Archaeology and the Study of Northwest Coast Economies. In *Prehistoric Economies of the Pacific Northwest Coast*, Research in Economic Anthropology, Supplement 3, edited by B. L. Isaac, pp. 293–351. JAI Press, Greenwich, CT.

Morlan, R. E., 2001, Canadian Archaeological Radiocarbon Database. (http://www.canadian-archaeology.com/radiocarbon/card/card.htm)

Moss, M., and Erlandson, J. M., 1992, Forts, Refuge Rocks, and Defensive Sites: The Antiquity of Warfare along the North Pacific Coast of North America. *Arctic Anthropology* 29:73–90.

Moss, M., and Erlandson, J. M., 1995, Reflections on North American Pacific Coast Prehistory. *Journal of World Prehistory* 9:1–45.

Pomeroy, J. A., 1980, *Bella Bella Settlement and Subsistence*, Unpublished Ph.D. Dissertation, Department of Archaeology, Simon Fraser University, Burnaby, British Columbia.

Rosenberg, M., 1998, Cheating at Musical Chairs: Territoriality and Sedentism in an Evolutionary Context. *Current Anthropology* 39:653–681.

Sahlins, M., 1981, *Historical Metaphors and Mythical Realities: Structure in the Early History of the Sandwich Islands Kingdom*. University of Michigan Press, Ann Arbor.

Sahlins, M., 1991, The Return of the Event, Again; With Reflections on the Beginnings of the Great Fijian War of 1843 to 1855 between the Kingdoms of Bau and Rewa. In *Clio in Oceania: Toward a Historical Anthropology*, edited by A. Biersack, pp. 37–99. Smithsonian Institution Press, Washington, DC.

Stuiver, M., Reimer, P. J., Bard, E., Beck, J. W., Burr, G. S., Hughen, K. A., Kromer, B., McCormac, G., van der Plight, J., and Spurk, M., 1998, INTCAL 98 Radiocarbon Age Calibration, 24,000–0 cal. B.P. *Radiocarbon* 40:1041–1084.

Wiessner, P., 1982, Beyond Willow Smoke and Dog's Tails: A Comment on Binford's Analysis of Hunter-Gatherer Settlement Systems. *American Antiquity* 47:171–178.

Zita, P., 1997, Hard Times on the Northwest Coast: Deer Phalanx Marrow Extraction at Namu, British Columbia. In *Drawing Our Own Conclusions, Proceedings of the 1997 McMaster Anthropology Society Students Research Forum*, edited by A. Dolphin and D. Strauss, pp. 62–71. McMaster Anthropology Society, Hamilton.

Chapter **11**

Long-term Change and Short-term Shifting in the Economy of Philippine Forager-Traders

LAURA LEE JUNKER

INTRODUCTION

When Europeans first made contact with mainland and island Southeast Asia populations, they found a complex amalgam of groups of extremely diverse economic orientations, levels of sociopolitical complexity, and linguistic and ethnic affiliations. Many researchers have stated that the considerable ecological diversity and geographic fragmentation of Southeast Asia contributed to the high degree of economic specialization and ubiquity of intensive interethnic exchange relations among various groups of tropical forest foragers, tribal swiddening populations, and complex chiefdoms and kingdoms focused on maritime trade and intensive rice farming (e.g., Dunn 1975; Hutterer 1974, 1976, 1983). The configurations of such interethnic trade systems in the historic period have been well documented by early texts associated with literate kingdoms of late first

LAURA LEE JUNKER • Department of Anthropology, University of Illinois at Chicago, Chicago, Illinois 60607.

millennium A.D. and early second millennium A.D. Southeast Asia, Chinese trade records, and later European histories (Andaya 1975; Hall 1985:1–20, 80–89; 1992:257–259; Junker 1999:239–259; Miksic 1984; Wheatley 1983; Wolters 1971:13–14). Hunter-gatherer populations that generally inhabited the interior uplands of Southeast Asia include the Semang (Orang Asli) of Malaysia; the Punan and Penan of Borneo; the Kudu of Sumatra; the Agta, Ata, and Batak of the Philippines; the Togucil of Maluku (the Moluccas or "Spice Islands"), the Nuaulu of Seram (in eastern Indonesia), the Andaman Islanders in the Indian Ocean, and various smaller and lesser known groups in Thailand and Vietnam (see Fig. 11.1).

Early historic sources suggest that many of these hunter-gatherer groups were well known to literate lowland farming populations and traders as collectors of tropical forest products (such as hunted meat, honey, rattan, resins, spices, and medicinal plants) that were much desired by lowland agriculturalists. Most historically known or ethnographically reported tropical forest foragers of the region received a significant portion of their carbohydrates, as well as various manufactured goods, from adjacent sedentary farming populations with whom they appeared to have long-standing economic and social interactions (Griffin 1984; Headland and Reid 1989). In addition, those foraging groups located relatively close to historically known Southeast Asian maritime trading kingdoms and chiefdoms such as Srivijaya (Sumatra), Majapahit (Java), Brunei (Borneo), Magindanao (Philippines), Sulu (Philippines), and Ternate (Maluku) gradually became enmeshed in more "commercialized" foraging for forest exports (particularly tropical hardwoods, spices, resins, animal pelts, and tropical bird feathers and nests). Control of these export products allowed the lowland elites of these various Southeast Asian complex societies to gain wealth and status through participation in South China Sea luxury good trade networks that linked them with China, India, and even African and Near Eastern polities to the west. Clearly, by the sixteenth century, most Southeast Asian foragers, even those of the remote interior, were tied economically into a larger "world system" through various kinds of symbiotic and exploitative exchange relations.

However, there is considerable controversy about the relative antiquity of these forager–farmer trade relations, their evolutionary dynamics, and their implications for ecologically based theory on the viability of "pure" hunter-gatherer adaptations in tropical forest environments. Some archaeologists favor a scenario in which hunter-gatherers spread at an early date (probably before the end of the Pleistocene) into the canopied, humid tropical forest interior of the Southeast Asian mainland and island archipelagos, gradually incorporating trade for grain with coastal farmers into an already flexible and diverse economic repertoire that may have included manipulation of plants

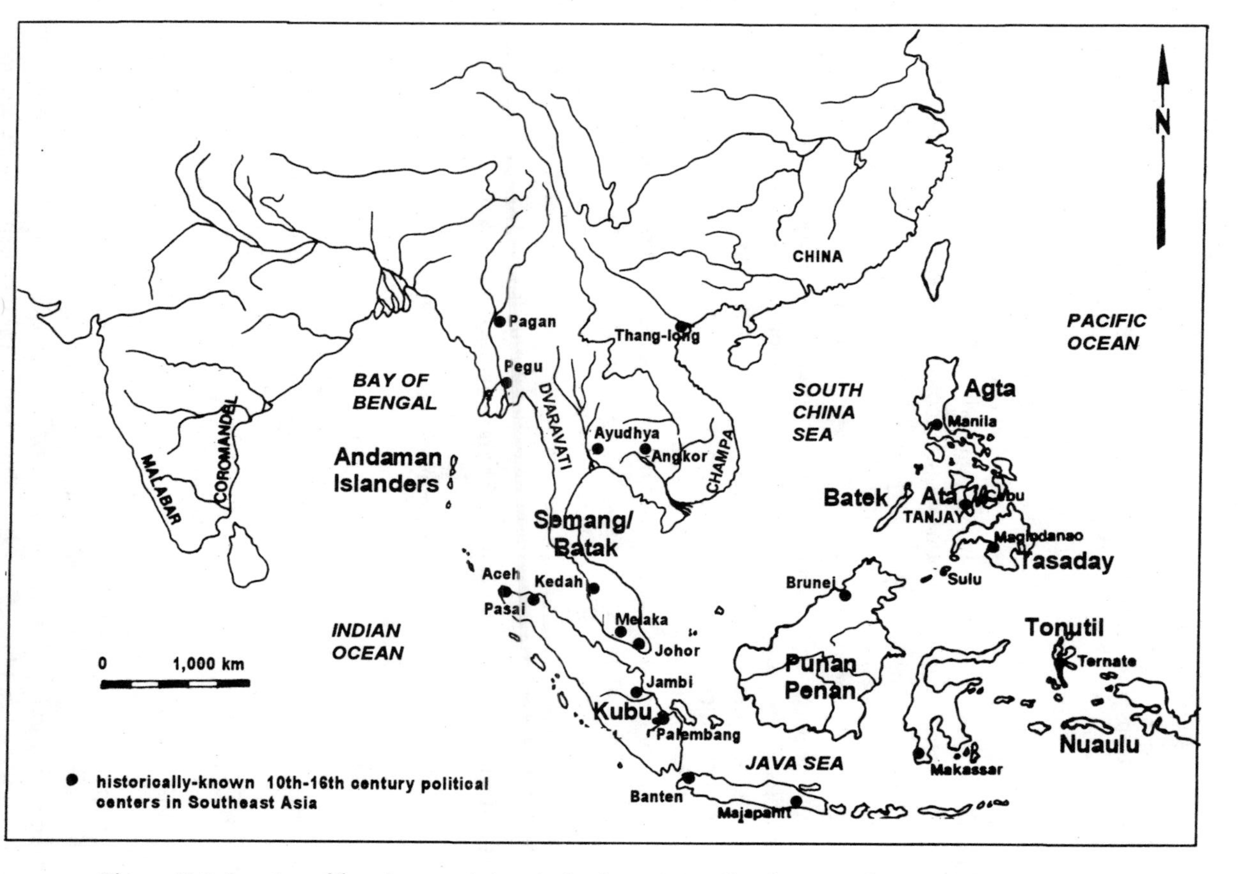

Figure 11.1. Location of foraging populations in Southeast Asia with reference to historically known states.

like wild sago and processing of certain plants for long-term storage to meet carbohydrate deficits. Other archaeologists, who cite ecological arguments and point to historically known trade-focused foragers such as the Punan of Borneo and Kubu of Sumatra, are skeptical of the presence, in any period, of "isolated" or "pristine" foragers in interior Southeast Asia. A few researchers have taken the extreme position that groups such as the Punan and the Kubu are actually "devolved" agriculturalists who became professional collectors in response to the demands of Chinese trade during the last thousand years. These questions require more than archaeological investigation and, in Southeast Asia at least, they have been the subject of intense debate involving archaeologists and also biological anthropologists, linguists, ethnographers, and historians. The first section of this chapter focuses on the way various lines of evidence have been brought to bear on the question of the long-term dynamics of forager–farmer relations and provides the wider theoretical context for my archaeological study of Philippine foragers.

Archaeological evidence presented here from a regional-scale project in the Bais-Tanjay Region of Negros Island in the Philippines addresses these key issues of long-term adaptations of foragers in the interior tropical forests of Southeast Asia. The Bais-Tanjay Region, comprised of a coastal river valley and adjacent interior mountains, was the center of an emerging lowland maritime trading chiefdom between the first and midsecond millennium A.D. (Junker 1999). The historically known Tanjay chiefdom was supported by an agriculturally based lowland population that had long-standing trade relations with both interior tribal swiddening populations and tropical forest foraging groups. Analysis of changing regional settlement organization, locational studies of archaeological sites identified as "forager camps" and "farming villages," and the material patterning of trade goods in the region materially document changing relationships over time between farmers incorporated into the lowland chiefdom and their forager trade partners. The archaeological evidence from the Bais-Tanjay Region shows that forager–farmer exchanges are of significant antiquity in the Philippines, as elsewhere in Southeast Asia, and that trade for carbohydrates and various technologies may have been one of several strategies used by Southeast Asian foragers to minimize risk in an environment that held some ecological hazards but that was by no means unusable without trade.

Changes in the intensity, volume, and content of this trade over time suggest that trade with farmers was not economically vital, that there was a great deal of flexibility in the use of imported resources, and that economic choices were situationally shifting. Management of particularly carbohydrate-rich forest resources, the incorporation of some swidden agriculture into the subsistence regime, and the maintenance of a generally high level of subsistence diversity were probably less archaeologically visible means of maintaining small foraging populations in these island interiors. The incorporation of foragers into foreign maritime trading markets introduced commercial aspects into forest collecting, resulted in more intensive exposure of foragers to lowland symbols of status and power, and created more asymmetrical relationships in this interior trade. The archaeological documentation from the Bais-Tanjay Region traces the penetration of exotic status goods into the island interior as foreign trade expanded, but suggests that it was primarily upland tribal swiddening populations that were socially transformed by accumulation of wealth and new concepts of competition for status.

DEBATE ON THE LONG-TERM DYNAMICS OF FORAGER–FARMER INTERACTIONS IN SOUTHEAST ASIA

Much of the recent literature on Southeast Asian foraging populations, including works by historians, cultural anthropologists, archaeologists, linguists, biological anthropologists, and tropical ecologists, has focused on several distinct but related issues. One of the long-standing controversies is whether these forager–farmer exchange relations are a recent phenomenon or a fundamental and long-term solution to the ecological limitations of tropical forest environments for both hunter-gatherers and agriculturalists (e.g., Eder 1988; Headland and Reid 1989, 1991; Hutterer 1974, 1976, 1983). Ethnohistoric, biological, linguistic, and archaeological investigations have thus sought to demonstrate the time depth of contacts between foragers and farmers by examining oral traditions of trade, biological and linguistic relatedness, and the time–space distributions of artifacts and sites believed to be associated with one or another of these groups (see Junker n.d.b).

Work on the historical linguistics of Southeast Asia has demonstrated that most extant hunter-gatherers of Southeast Asia speak Austronesian languages that are related to those of adjacent sedentary farming populations (Benjamin 1976; Blust 1976, 1989), who themselves acquired these languages as they migrated from (or had contacts with populations of) Southeast Asia's presumed "agricultural hearth" in South China. Glottochronology suggests that linguistic contacts between Southeast Asian foragers and Austronesian-speaking farmers go back at least 2000 years in many regions (Early and Headland 1998:16; Headland and Reid 1989). However, the linguistic situation is extremely complex. The Agta foragers and various other "negritos" of the Philippines speak Austronesian languages adopted from contacts with farmers as early as 3000 B.C., but their languages became mutually unintelligible with those of sedentary agriculturalists over time, suggesting significant long-term oscillation between intensive interaction and isolation from farming communities (Reid 1987).

In contrast, the fact that the Punan, Kubu, and Tasaday speak languages that are fully intelligible to the neighboring agricultural groups (the Kenyah, Malays, and Manobos, respectively) with which they trade, has led some anthropologists to suggest that these groups are not "genuine" foragers at all, but instead are ethnically and culturally indistinguishable from adjacent sedentary farmers, except for their recent specialized adaptation as commercial foragers (Blust 1989:54–55; Hoffman 1984, 1986). However, Spielmann and Eder (1994:307) note that hunter-gatherers who rely on frequent trade with food producers typically learn the language of their trade partners (as well as adopt important aspects of cultural communication, such as body language, concepts of fictive kinship, and rituals of social interaction), whereas the reverse is rarely true. Populations that rapidly acquire the languages of their economically and politically dominant trade partners during periods of particularly intense trade interaction may drift away from shared language patterns at other times either purposely (to preserve cultural identity when such distinctiveness is to their advantage) or as a result of greater isolation (due to a fall-off in trade), thwarting straightforward glottochronological analysis.

The distinct physical characteristics of "negrito" foragers such as the Ata, Semang, Andaman Islanders, and Batek have generally caused anthropologists to accept that these groups represent "aboriginal" inhabitants of insular Southeast Asia (more closely related to New Guinea populations) that lived in the region before the large-scale expansion of Austronesian-speaking, agricultural Mongoloid peoples into Southeast Asia around 3000–4000 B.C. (Bellwood 1997:132–135; Dentan et al. 1997:10; Omoto 1985:129–130; Solheim 1981). However, Fix's (1995, n.d.) genetic study of Semang Orang Asli foragers and their agricultural neighbors (the Senoi and Melayu Asli) shows that assumptions about the degree and meaning of biological variation in these populations may be flawed and that multiple scenarios for population histories can be derived from the same patterned biological differences. In fact, there is little evidence to support the traditional perception of sharp biological discontinuities between foragers and adjacent farming populations. This continuous variability could be explained by the idea of an "admixture" of once discrete aboriginal populations entering the peninsula in migratory "waves" (e.g. Bellwood 1992; Dentan et al. 1997:8–16), consistent with an "interactive" rather than "isolate" model. However, this pattern could also be explained by a common ancestral population thousands of years ago and subsequent divergence into distinct populations as each group developed distinct cultural adaptations to specific ecological niches (i.e., deep forest collecting, upland swidden farming, and coastal farming/fishing/trading), as argued by Benjamin (1976) and Rambo (1988). Thus, the biological evidence and linguistic evidence are equally ambiguous with regard to the antiquity and intensity of forager–farmer interactions.

Although archaeological investigations relevant to this issue are more limited, work at Philippine sites reveal that rice agriculturalists were living close to stone-tool-using hunter-gatherers by at least 3500 years ago (Hutterer 1974; Peterson 1974; Ronquillo 1981; Snow et al. 1986). Also of interest are the lithic industries known as "Hoabinhian," that consist of relatively amorphous assemblages dominated by unifacial tools manufactured on pebbles and flakes and generally associated with early Holocene to Late Holocene hunter-gatherers in Thailand, Vietnam, the Malay Peninsula, and other regions of both mainland and island Southeast Asia (Bellwood 1992:192–205; Bui Vinh 1998, Glover 1977; Reynolds 1990; Shoocondej 1996; Higham 1989:31–65). Hoabinhian sites are often interior cave sites (but with some coastal representations) where paleoethnobotanical and zooarchaeological studies have suggested broad based subsistence strategies focused on a diverse range of wild but sometimes "managed" plant and animal resources. What is significant is the fact that sites associated with Hoabinhian foragers date from the beginning of the Holocene around 10,000 B.C. and extend well into the period when we see the expansion of sedentary, food-producing peoples from around 4000–1000 B.C., suggesting a long period of interaction without assimilation of foraging groups.

A second, and more controversial, issue has been whether "pure" foraging adaptations were *ever* viable in the carbohydrate-poor, equatorial tropical forest interiors of Borneo and many of the islands of the Philippines, Malaysia, and Indonesia (e.g. Eder 1987:45–51; Griffin 1984). Many adopted the view that a hunting-and-collecting specialization in such areas was possible only from the advent of food production in adjacent lowland areas (e.g., Bailey et al. 1989; Dunn 1975; Headland 1987; Headland and Reid 1989). Linguistic, biological, and cultural evidence for a high degree of "relatedness" of foraging and farming populations, the absence of early archaeological evidence for colonization of many remote forest zones, and ecologically based arguments fuel the views of those who doubt the presence, in any period, of "isolated" or "pristine" foragers in interior Southeast Asia. Ecological arguments revolve around the general carbohydrate-poor environment of tropical forest ecosystems. The idea is that sparse preagricultural populations necessarily occupied richer near-coastal zones and were eventually pushed into trade-dependent specialization in the collection of forest products in what would have been "marginal" ecological zones. As noted above, some anthropologists have taken the extreme position that ethnographically known groups like the Punan of Borneo and the Kubu of Sumatra, as well as many other interior tropical forest foragers, are actually "devolved" agriculturalists who became professional collectors in response to the demands of Chinese trade during the last thousand years (e.g. Bellwood 1985; Hoffman 1984, 1986; Seitz 1981; for a contrasting perspective, see Brosius 1988; Rousseau 1984; Sellato 1988; Sandbukt 1988).

However, ecological studies of Southeast Asian tropical forests show the tremendous complexity and diversity of these environments (e.g., Flenley 1979; Whitmore 1975; Yamada 1997), suggesting that ecological generalizations for the region are problematic. A number of anthropologists have emphasized that too little is known about the ecological limitations of tropical forest habitats for human colonization to make an argument based purely on ecological grounds (e.g., Hayden 1981; Hutterer 1983). In addition, a negative view of the abilities of "pure" foragers to sustain themselves over the long term in these types of environments cannot adequately explain archaeological evidence for early Holocene occupation of the wet rain forests of the interior Malay Peninsula, northern Sumatra, southwestern Thailand, and Vietnam by stone-tool-using, clearly preagricultural "Hoabinhian" populations (for summaries, see Bellwood 1992:85–89, Higham 1989:31–65, and Reynolds 1990). Recent archaeological work by Matthew Spriggs (2000) in the rain forests of Melanesian island archipelagos such as the Solomons and Bismarcks supports the Southeast Asian archaeological evidence for early Holocene occupation of these supposed "depauperate" forests. Spriggs raises the intriguing possibility that the early island Melanesian foragers dealt with the resource limitations of the interior forests by importing high-yield plants that had storable products, such as canarium nut trees, from the more diverse and productive environments of mainland New Guinea, ultimately transplanting them far into the island interiors. Thus, the interior foragers, adept at plant manipulation themselves, had no need to wait for agriculturalist colonization of the island coasts to occupy the canopied rain forests of the interior successfully.

Spriggs' suggestion is consistent with the observations of a number of ethnographers who have studied recent deep forest foragers in Southeast Asia. Their research suggests that many groups reduce carbohydrate-related risks through such traditional practices as managing sago stands and intensively processing storable resources. Peter Brosius (1991) noted that the Penan of Borneo often "manage" wild sago stands by transplanting, allowing them sufficient carbohydrate calories even in the absence of trading opportunities with farmers. Roy Ellen (1988) similarly shows that the Nuaulu of Seram (eastern Indonesia) are not dependent on the swidden gardening and trading that are part of their current economic repertoire, but instead have traditionally relied on intensive processing of storable resources such as sago, canarium nuts, and smoked meat to sustain them in their tropical forest environment.

Other anthropologists have emphasized that strong linguistic, biological, and cultural similarities are exactly what one would expect in populations that have interacted regularly during many millennia (Benjamin 1985; Hutterer 1974, 1976, 1991) and that these similarities do not mean that these ethnic groups are not "real." At the same time, Griffin (1989) emphasizes that "situational shifting" among foraging, farming, maritime exploitation, trading, and other economic modes is a common phenomenon among both "hunter-gatherers" and "agriculturalists" in Southeast Asia and, in a sense, these dichotomous categories blur a great deal of variation over both time and space in the way various groups minimize risk through flexibility and diversification (also see Hutterer 1991). As in the controversy over the "pristineness" of hunter-gatherer groups in other supposed "marginal" environments (e.g., see Schrire 1980; Denbow 1984; and Wilmsen 1983 on the South African !Kung), this issue of independent tropical forest foragers in Southeast Asia will not soon be conclusively resolved through ecological reconstructions, ethnohistoric analysis, or archaeological data.

A third issue, which is more tightly bound to historical circumstances of the Southeast Asian case, is the impact of expanding maritime luxury good trade on the economic choices, mobility and settlement strategies, and the social dynamics of tropical forest foraging populations in Southeast Asia. Historical records and archaeological evidence document the rise of numerous maritime trading kingdoms and chiefdoms on the Malay Peninsula and in the island archipelagos of Southeast Asia at least by the beginning of the first millennium A.D. (Hall 1985, 1992). Because of aspects of geography, ecology, demography, and political structure, these growing complex societies had political economies heavily invested in foreign prestige goods trade, in which luxury goods from China and mainland Southeast Asia (primarily porcelains, silks, bronzes, and other precious metals) were imported for local elite status display and alliance-building (Junker 1999; Winzeler 1986; Wheatley 1983). The wealth-generating exports for these maritime trading polities were primarily interior forest products that could be obtained only by coercing intensified trade with foraging groups that occupied the island interiors. Historic sources indicate that lowland elites organized trade with interior upland populations through formalized and individually contracted long-term alliances, cemented through ceremonialism, gift exchange, and bequeathing impressive sounding royal titles on upland leaders (Junker 1999; Miksic 1984; Schlegel 1979). Lowland prestige goods were used for bridewealth payments, for status display, as payments for raiding reparations, and for ritual purposes particularly in the case of upland swidden farming groups with incipient concepts of social ranking and competition for status. In many cases, an escalating demand for these lowland symbols of power fueled an increasing economic emphasis on commercial procurement of forest products in these societies, institutionalizing political authority, and expanding social cleavages as more successful traders were able to accumulate more lowland political titles and wealth for social display (Benjamin 1985).

This raises the question of whether similar processes of wealth accumulation, the emergence of new political roles, and increased status differentiation might have occurred among foragers in periods of intensified participation in this type of nonsubsistence trade. One of the best examples of the effects of foreign trade on a foraging society are the Kubu of the Batang Hari in eastern Sumatra, if we discount the views of those who lump the Kubu with the Punan and others as peoples who have only recently adopted a foraging lifestyle. The ethnographer Sandbukt (1988) uses historical data to support the conclusion that the Kubu are long-term foragers who practiced some swidden agriculture and who collected forest resources for trade-oriented kingdoms such as Palembang and Jambi by the early second millennium A.D. Centuries of trade with an influx of politically manipulable wealth in cloth and metal tools and the bestowal of lowland Islamic titles on local "headmen" led to some degree of hierarchy in political authority and the emergence of various public forms of competition for status (particularly competition to accumulate cloth and metal implements for bride wealth). Similarly, Ellen (1988) reports that shell bangles, foreign porcelains, exotic textiles, and metal anklets obtained through trade were important in Nuaulu household ceremonies (associated with rites of passage, matrimonial rites, dispute settlement, and other events) and that inflationary competition and accumulation of these goods were tied to the emergence of corporate, property-owning political units not typical of foraging societies.

In contrast, Sandra Bowdler (n.d.) notes that for the Australian aborigines, contacts with the Macassans and other Southeast Asian complex societies, both on their own lands and on occasional voyages in which they accompanied Macassans to their homelands, failed to catalyze significant economic changes such as the adoption of agriculture or extreme specialization in procuring trade products. However, aspects of Southeast Asian ritual (such as Islamic style burials) and certain components of art along the northern coast of Australia were influenced by trade contacts with Southeast Asian peoples who came to collect trepang along the Arnhem Land and nearby coasts. Though it is difficult to support the conclusion that many contemporary Southeast Asia foragers are professional traders created out of the foreign trade demands of recent maritime trading kingdoms, ethnographic and historical analysis suggests that significant economic, social, and ideological transformations can occur among foragers who participate intensively in nonsubsistence trade systems controlled by socially stratified lowland societies.

ECONOMIC STRATEGIES IN PHILIPPINE FORAGING GROUPS

Tropical forest hunter-gatherers in general, faced with an environment of highly diverse species but relatively low edible biomass and patchy (and somewhat unpredictable) resources, generally have a high level of residential mobility, often depend very little on long-term storage of resources, and have few resources that can be targeted for intensive exploitation through logistical foraging (Hutterer 1983). With reference to Binford's (1980) dichotomy between highly mobile "foragers" and logistical "collectors," virtually all traditional Southeast Asian hunter-gatherers would be classified as "foragers." In a comparative study of residential mobility, Kelly (1983) found that the Punan of Borneo and the Semang of Malaysia both average more than 20 residential moves per year and camps are relocated on average about 5–15 kilometers from the previous base. Ethnographers who worked with Philippine groups report a similarly high level of mobility; Rai (1982:105–107) observed an average of 20 residential moves per year amongst the northeastern Luzon Agta group, and Eder (1987:32) recorded 17–26 residential moves annually among the Batak of Palawan. In both Philippine groups, the compact microenvironmental zones created by elevation meant that a variety of unexploited resources could be found in short moves of less than 10 kilometers (Rai reports an average distance of 5.3 kilometers per move). However, the length of time a camp is occupied and the distance covered in a subsequent residential move vary widely, and they depend on the "mix" of economic activities engaged in while in residence at a particular site (i.e., hunting, forest collecting, fishing, horticulture, labor for adjacent farmers, and trading activities), which in turn depends on seasonal factors of resource availability and the constraints of "fixed" resources (such as agricultural fields, lowland villages where trade partners reside, and certain stands of concentrated forest resources such as sago).

Among the Agta, Batek, and other ethnographically studied foraging groups of the Philippines (see Fig. 11.3), the coresidential unit generally consists of two to six nuclear families (typically 35–50 people) related through bilateral kin ties or various forms of fictive kinship (Eder 1987:30–31; Estioko-Griffin 1985:21; Griffin 1989:63). However, the sizes and compositions of groups vary widely both seasonally and from year to year according to the changing mix of economic opportunities (see discussion below) and according to social factors (e.g., social friction, exogamy and external affinal ties, social relations with sedentary populations) that favor residential group fluidity (Estioko-Griffin and Griffin 1975:243; see Plate 1). These extended family residential clusters tend to move upstream and downstream along specific river drainages which, though they do not constitute formalized "territories," are part of their long-term social and economic identities (Eder 1987:28–30; Griffin 1984:104–105, 1989:61; Rai 1982:61–63).

Most Philippine foraging groups recognize seasonal differences in the availability of resources, ease of transport, and possibilities of social interaction and cognitively divide their year minimally into "rainy" and "dry" seasons characterized by varying economic choices, mobility strategies, and settlement patterns (Allen 1985; Eder 1987:68–71; Griffin 1989).

For example, the Agta groups of northeastern Luzon (Estioko-Griffin 1985:21–23; Griffin 1984:105–113; 1989:64–69) nucleate in longer term camps along the upper or middle tributaries of the river system in the rainy season (approximately October to February), live off stored rice from the recent harvest, and hunt wild pig, deer, and monkey. The rainy season camps tend to be placed in the same or adjacent locales year after year (Griffin 1989:67), often close to reusable swidden fields cultivated in previous seasons. Although hunting is a year-round activity that dominates the Agta economy, the rainy season is considered the best time for hunting larger game such as pig and deer; these animals are fattest at this time and can be silently stalked in the wet forest. Large hunting parties of men and/or women radiate out in a "tethered" pattern along small streams near the sedentary rainy season camp (Griffin 1989:63).

The dry season (approximately May to September) is a time of greater mobility, settlement dispersion, and diversification of subsistence activities. In this season, smaller bands move upstream and downstream to target specific resources and build camps that consist of flimsily constructed lean-tos often occupied for only a few days (Estioko-Griffin and Griffin 1975:243). A wide variety of economic activities takes place from these small, short-term camps, including hunting (game drives that involve smaller animals are particularly favored), collecting (including wild roots, fruits, greens, and honey), spearfishing and mollusk collecting in rivers that were impassable in the rainy season, and engaging in periodic horticultural activities in nearby swidden clearings. Agta groups that rely more heavily on horticulture (such as the Nandukan Agta) tether their dry season camps to swidden fields to which they must return periodically for burning, planting, and harvesting activities that take place in the dry season. Although the inclusion of horticultural activities necessarily impacts time available for hunting and forest collecting, Griffin (1989:64–66) notes that it remains a relatively minor economic activity and does not exert a strong pull on settlement and mobility choices. The Agta who farm generally stagger their crop planting to fit with hunting and foraging priorities, they do not weed the plots between planting and harvesting, and they generally accord low priority to horticulture in subsistence (Estioko-Griffin and Griffin 1981:55; Headland 1975:249; Headland and Reid 1989:45). Agta who rely primarily on trade with farmers rather than their own horticultural activities for plant staples emphasize proximity to the villages of trade partners in locating these dry season camps. Of significance to the issue of upland-lowland trade are references to trading expeditions during the dry season when travel was less arduous, to meet lowland trade partners to exchange hunted meat and forest products for rice, lowland manufactured goods, and other lowland subsistence commodities (Griffin 1984:107–113). Allen (1985:23) notes that the Agta often made special collecting trips into the far interior forests at this time to procure the honey, rattan, orchids, and other tropical forest products desired by their lowland trade partners.

Seasonal movements and economic choices vary widely among Philippine foraging groups as a function of ecological differences among islands, microclimatic conditions that affect rainfall patterns in different areas of the archipelago, the proximity of agricultural groups as resource competitors and as trade magnets, varying access to coastal resources (determined by geography and also by the presence of other groups that impede access), and the proximity of maritime trade routes that linking foragers to larger scale trade systems (see Junker n.d.a). For example, because of their close proximity to major maritime trade routes and swidden farmers, the Batak foragers of Palawan in the southern Philippines rarely engaged in the horticultural activities common among the Agta; instead they emphasized trade for rice, tubers, and other carbohydrate-rich foods and devoted a larger portion of their economic energies to large-scale harvesting of forest products for export (Eder 1987).

Although we know less ethnographically about the Ata who traditionally occupied the upland portion of the Bais-Tanjay Region of Negros Island where my archaeological research was carried out, they appear to have practiced a similarly eclectic economic strategy that varied according to seasonal resource availability predicted by wet–dry monsoonal patterns, as well as opportunities for interaction with lowland trade partners (Figure 11.2). The Ata of Negros, more than the foragers of Palawan and Luzon, have been strongly affected by large-scale logging operations and encroachment by lowland farmers during the last 50 years. Today, they are almost fully sedentary and concentrated on a "reservation" at Mabini, north of the Bais-Tanjay Region. However, some early ethnographic work and historical reconstruction suggests that their economic activities included river fishing; pig and deer hunting; collecting tropical plants for subsistence; trading with agriculturalist lowlanders (for both farm produce and manufactured products such as pottery and metal); planting their own swiddens of rice, yams, taro, and papaya; and targeting upland resources such as rattan, resins, beeswax, and spices for external trade (Cadelina 1980; Oracion 1960; Rahmann and Maceda 1955). Deep forest hunting and

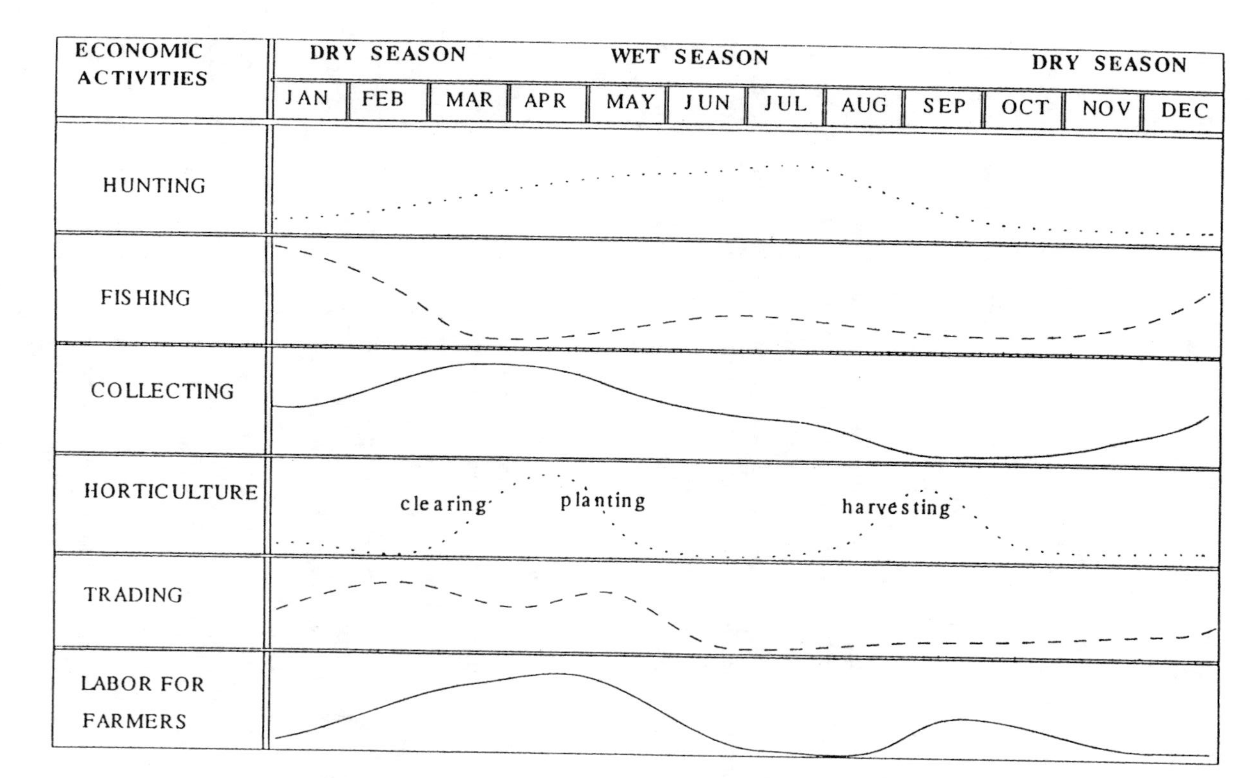

Figure 11.2. The seasonal intensity of various economic activities of the Negros Ata in the 1950s, reconstructed from the ethnographic accounts of Cadelina (1980), Oracion (1960), and Rahmann and Maceda (1955).

collecting, as well as occasional planting of swidden fields to supplement carbohydrate resources obtained through foraging, were the preferred activities in the rainy season. As the raging rivers of the wet season gave way to more navigable streams in the dry season, river fishing, large-scale collecting of exportable forest resources, and trips to the lowlands for trade became the primary economic emphasis (see Fig. 11.2). However, historical sources and archaeological evidence discussed below suggest that over the long-term, these foraging groups on Negros Island expanded or reduced their emphasis on trade according to changing circumstances. These changing conditions for external trade included the shifting economic and political fortunes of their lowland trade partners, the intensity of lowland slave raiding activities which heightened the risks of trade, year-to-year fluctuations in the productive potential of their own swidden gardens to meet carbohydrate shortfalls, and inconstant motivations to procure lowland-manufactured household products and prestige goods.

Elsewhere in Southeast Asia, we see the same flexibility and diversity in forager subsistence strategies. Some Punan and Penan groups of Borneo rely heavily on pig hunting and "management" of sago stands for carbohydrates and engage minimally in trade, whereas others apparently function as specialized and almost wholly commercial hunters and collectors dependent on farmers for basic foodstuffs (Brosius 1991; Hoffman 1984). Endicott (1988:112), in his ethnographic study of the Batek of Malaysia, emphasizes that it is very difficult to characterize a "typical" Batek economy because different Batek groups switch frequently among hunting, forest collecting for subsistence, planting swidden fields, managing concentrations of wild yams, and trading forest products (rattan, fragrant woods, hunted game) for rice and flour from lowland Malay traders. In his studies of Philippine foragers, Griffin (1989) makes the important point that the success of Southeast Asian foragers in tropical forest ecosystems has depended on this extreme flexibility in economic choices, both on a year-to-year basis and in the long term. Because Southeast Asian foragers live in complex and changing environments, they have long been part of larger cultural and political landscapes that appear to be almost constantly in flux, and their economic choices have long been affected by larger scale economic processes (such as trade policies of the Chinese and other foreigners), we must assume that most ethnographically or historically recorded economic patterns (and even archaeological sites isolated in time and space) represent behavior at a single moment and should not be projected to represent the whole range of hunter-gatherer adaptations.

Ethnographic accounts of the trade relations between Philippine foragers and adjacent agricultural populations suggest that groups like the Agta, Negros Ata, and Batak engaged in both direct trade with lowlanders who inhabited the "core" of maritime trading polities and indirect middleman trade that involved upland swidden groups as trade intermediaries. Early historic documents suggest that upland foragers came downstream far enough to make direct contacts with lowland farmers at riverbank trade centers in the early second millennium A.D. Because it is unlikely that Chinese maritime voyagers personally traveled beyond the coastal trading ports, Chau Ju-Kua's thirteenth century account of what are almost certainly members of some "negrito" group from either the Visayas or Palawan (Hirth and Rockhill 1911:162) suggests that these foragers ventured periodically into the lowlands near the coastal port to trade forest products directly with coastal agriculturalists. Chau Ju-Kua writes in his A.D. 1225 *Chu-Fan-Chi* (Zaide 1990, Vol. 1:7):

> In the remotest valleys there lives another tribe called the *Hai-tan*. They are small in stature and their eyes are round and yellow (brown), they have curly hair and their teeth show (between their lips). They nest in tree tops. Sometimes parties of three or five lurk in the jungle, from whence they shoot arrows on passers-by without being seen, and many have fallen victims to them. If thrown a porcelain bowl, they will stoop and pick it up and go away leaping and shouting for joy.

Chau Ju-Kua's highly accurate description of "negrito" physical characteristics suggests the strong possibility that this represents an actual eyewitness sighting at a coastal port, although ideas about their settlements and proclivities toward raiding appear to reflect the conveyed prejudices of lowland Filipinos who might have viewed their interior trade partners as inferior "savages." The reference to porcelains in association with the *Haitan* is significant in establishing that these interior foragers were indeed linked into coastal trade networks by at least the thirteenth century A.D.

Historic sources indicate that, in the prestige-goods-oriented political economy of lowland Philippine chiefdoms, even trade with tribal peoples and foragers of the interior was predicated on a series of formalized and individually contracted long-term alliances, cemented through ceremonialism and gift exchange. This ritual exchange often involved circulating "status goods" manufactured by specialists at coastal chiefly centers (e.g., fancy decorated or slipped earthenware, bronze weaponry, and gold jewelry) or obtained in foreign maritime trade (particularly Chinese porcelain and foreign beads) (Junker 1999). Most of the early ethnographic and historic accounts of upland–lowland trade focus on ritualized presentations of lowland political titles, ceremonial regalia, and status-conferring luxury goods to upland tribal leaders of swidden farming populations who already possessed concepts of social ranking and competition for status (e.g., the Tagbanua swiddeners with coastal Palawan chiefdoms, the tribal Manuvu and Tiruray with the Magindanao chiefdom, the tribal Hanunoo with the coastal Mindoro chiefdoms) (Conklin 1949; Schlegel 1979; Warren 1977). Interior tribal groups used these lowland prestige goods for bride wealth payments, for status display, as payments for raiding reparations, and for ritual purposes.

Ethnohistoric analysis suggests that, although interior foragers lacked the institutions of competition for status associated with these prestige goods, those groups that were particularly heavily involved in this trade eventually obtained small amounts of these luxury goods at least indirectly through exchange partnerships with other interior populations (see Junker 1999:241–246 for a summary). In fact, the passage quoted earlier from the Chinese Chau Ju-kua's description of "negritos" in the central Philippines suggests that the interior foragers occasionally procured Chinese porcelain bowls as early as the fourteenth century and that these exotics may have been further incentive for trade. This raises the question whether similar processes of wealth accumulation, the emergence of new political roles, and increased status differentiation might have occurred among foragers in periods of intensified participation in this type of nonsubsistence trade.

THE DYNAMICS OF A PRE-HISPANIC PHILIPPINE FORAGING POPULATION: THE TANJAY ARCHAEOLOGICAL PROJECT

Like many islands of the Philippine archipelago at the time of Spanish contact, the island of Negros was inhabited by lowland intensive agriculturalists associated with one or several maritime trading chiefdoms, who maintained complex interactions with upriver tribal swidden farmers and tropical forest foragers. In the roughly 315-km^2 river drainage of the Tanjay River on the eastern side of the island (an area designated as the Bais-Tanjay Region, shown in Figs. 11.3 and 11.4), ethnohistoric analysis and archaeological research indicate that a Visayan-speaking chiefdom-level society known to the Spaniards as *Tanay* grew in the vicinity of the Tanjay River mouth. It began as early as the late first millennium A.D. and peaked at the height of foreign trade around the fourteenth to early sixteenth centuries (Hutterer and Macdonald 1982; Junker 1999). In the uplands surrounding the 3–15 km wide coastal plain lived ethnically and linguistically distinct tribally organized swidden farmers known as the Bukidnon or Magahat and mobile hunter-gatherers known as Ata who were linked economically to the lowland farmers through extensive trade networks (see Fig. 11.3). The Negros Island Ata referred to in Spanish reports were probably related to the Ata foragers studied by ethnographers in the early to mid twentieth century and now settled in the vicinity of Mabini. What historic and ethnographic work has revealed about their economy and social

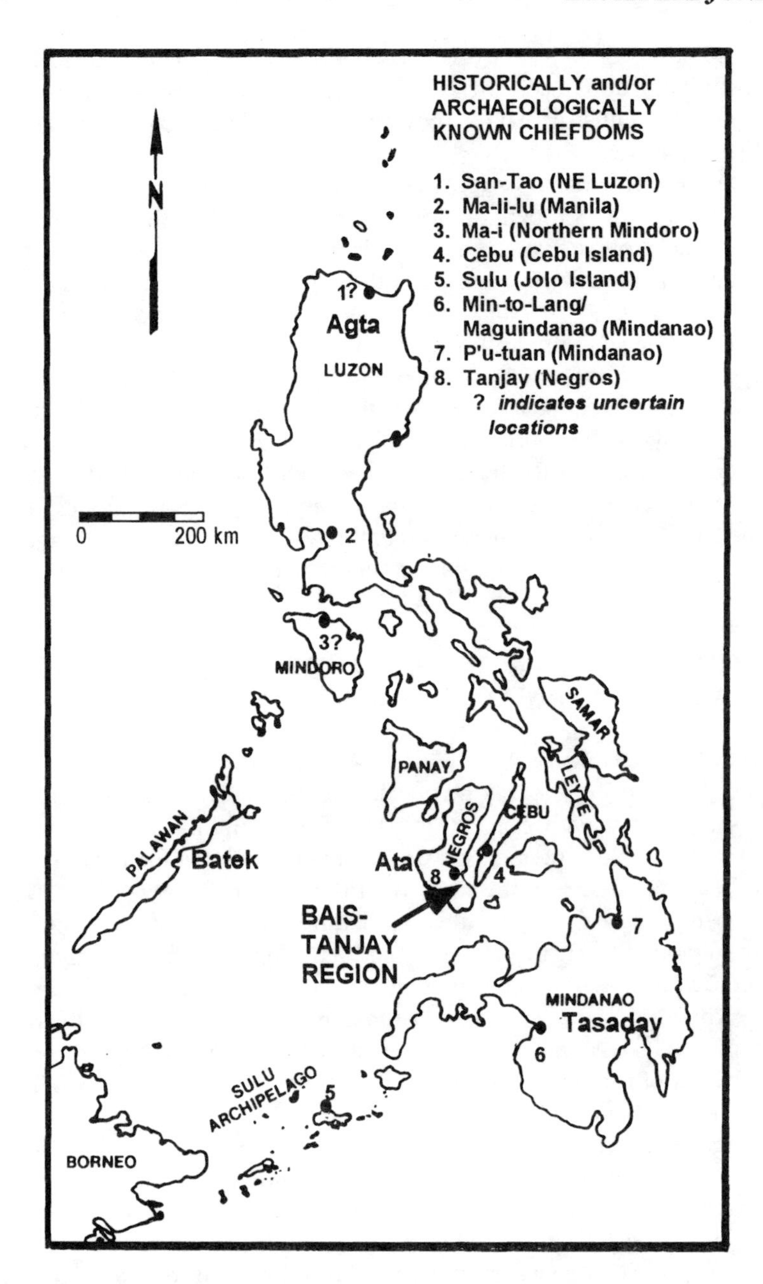

Figure 11.3. The Philippine islands and the locations of historically known and archaeologically known maritime trading chiefdoms, ethnographically known foragers, and the Bais-Tanjay region where the archaeological research discussed in this chapter was carried out.

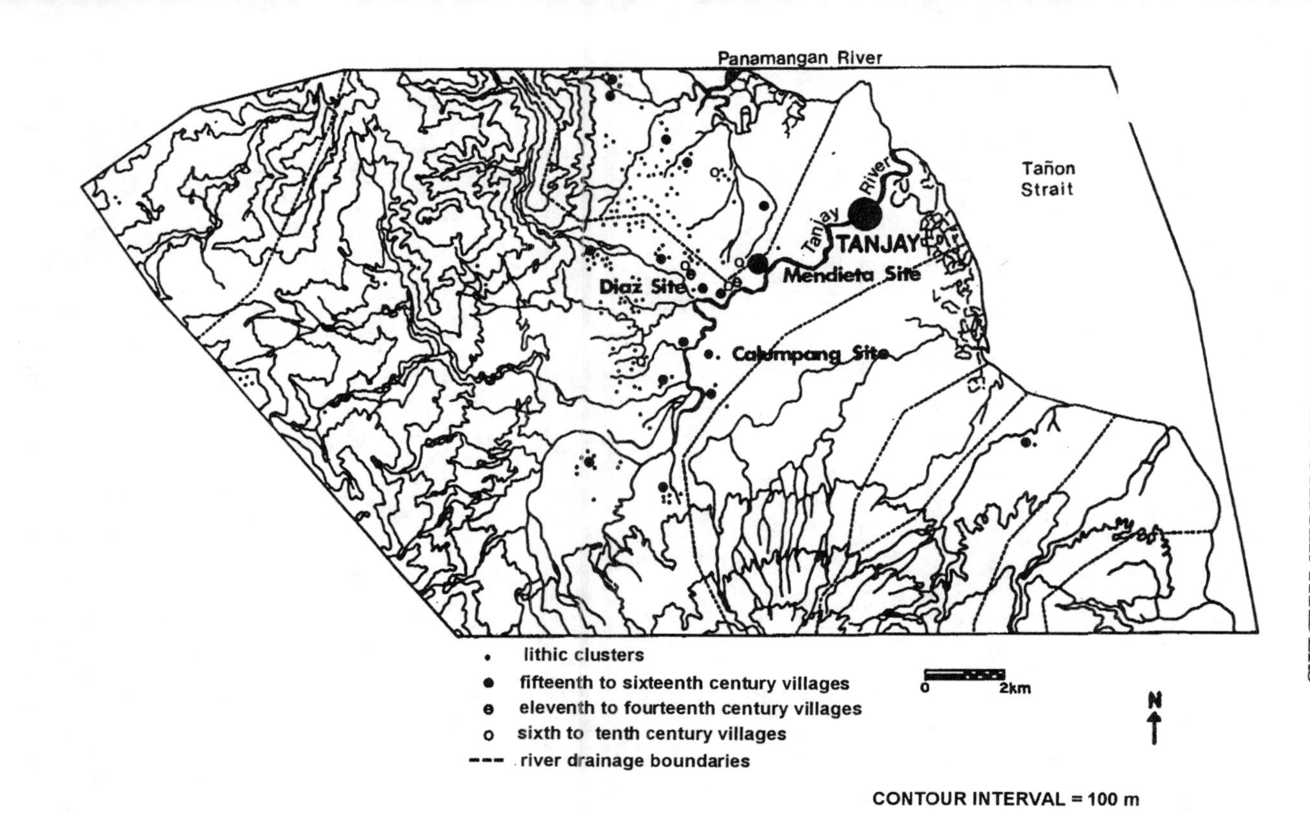

Figure 11.4. The location of chipped-stone sites identified as probable "forager camps" in the Bais-Tanjay region. Also shown are the locations of sedentary village sites of one or more hectares, dated to three pre-Hispanic phases of settlement in the region.

organization, as well as their connection to adjacent farming populations, has already been presented in some detail above.

Regional-scale archaeological survey and excavations have been carried out in the Bais-Tanjay Region during the last 20 years (Hutterer and Macdonald 1982; Hutterer 1981; Junker 1990, 1994, 1996, 1999; Macdonald 1982), and one of the long-standing research issues has been the changing nature of forager–farmer interactions during the past several thousand years. A combination of probability-based and contiguous-block (i.e., full coverage) surface surveys in 1979, 1982, 1994, and 1995 has allowed us to cover almost 30% of the Bais-Tanjay region and to record more than 400 sites that span roughly a 4000-year period before Spanish contact. Excavations were conducted at nine Tanjay region sites in 1979, 1981, 1985, 1986, 1994, and 1995, including the coastal chiefly center at the mouth of the Tanjay River (Tanjay), six villages at various distances upriver from Tanjay (including an iron production site and several "secondary" trade centers), a coastal fishing settlement, and an upland swidden farmstead. Six regional cultural phases have been identified based on local earthenware pottery, foreign porcelain, and radiocarbon determinations, dating from the second millennium B.C. to the Spanish period (see Junker 1999:45–53 for a detailed summary of chronology).

Only the most recent of the pre-Hispanic phases have yielded reliable regional settlement pattern data relevant to the issue of forager–farmer interactions on the lowland margins, due to problems of site visibility on the rapidly alluviating lowland plain and fewer chronologically diagnostic artifacts for periods before the midfirst millennium A.D. Thus, the research presented here concentrates on the following regional settlement phases: (1) the Aguilar Phase (approximately A.D. 500–1000), (2) the Santiago Phase (approximately A.D. 1100–1400), and (3) the Osmena Phase (approximately A.D. 1400–1600). Because of the failure to recover any pre second millennium B.C. deposits at excavated sites in the Bais-Tanjay region, no definitively "preagricultural" settlements have been identified. Therefore, the present archaeological database from the Bais-Tanjay region cannot contribute to the issue of "pristine" tropical forest foraging before food production. However, the archaeological evidence reveals that the Bais-Tanjay region foragers probably engaged in symbiotic trade for household goods with lowland farmers during a period of at least a thousand years before European contact. Further, the material remains from this roughly thousand-year period suggest that this trade affected their economic choices, as reflected in changing technological adaptations and settlement preferences. Finally, the archaeological record for the region indicates that some foraging groups, like those in the Bais-Tanjay region, began intensive collection of forest products for foreign markets during the late fourteenth to early sixteenth century peak of the Chinese porcelain trade. This new "commercial" component of foraging activities expanded lowland–upland interactions and introduced new "prestige" wealth into upland foraging communities.

One of the main research thrusts of the project has been the developing political economy of the historically known lowland chiefdom at Tanjay, and we are only beginning to focus specifically on the tropical forest forager-traders in the region (Junker 1996, 1999:239–259, n.d.a.), so there are limitations to the archaeological evidence presented in this study. Most of the full-coverage survey and settlement excavations carried out to date centered on the lowland alluvial plain surrounding the Tanjay River. We have surface-collected and mapped more than 130 probable camps associated with mobile foraging populations in lowland areas below 100 meters elevation (mostly along the interior branches of the Tanjay River), and we have test-cored (but not extensively excavated) 24 of these sites. However, in the upland areas above 100 meters elevation, we have a more limited sample of 11 surface-collected sites, none of which has been either cored or excavated below the surface. The small sample of upland sites is partly a result of research design (we presently have only an approximately 8% probability-based sample of 500 × 500 meter quadrats that have been systematically surface-collected in this upland zone) and partly a problem of poorer visibility in the rugged, forested interior (see Junker 1999 for a more detailed discussion of these methodological issues). Despite these limitations of the archaeological evidence, analyses of the sizes, locations, spatial relations, and artifact assemblages of probable "forager camps" have revealed some interesting patterns through time and over space that may relate to changing production choices and settlement strategies as Philippine foragers adapted to larger economic and social forces in their polyethnic world.

In the Bais-Tanjay region, probable "forager camps" are archaeologically distinct from what are likely to be permanent homesteads or villages of farming populations in site size and also in the content of their artifact assemblages. Although sites identified as farming settlements of various kinds can range from 0.1–7 hectares, they are generally characterized by relatively high densities of earthenware, shell, and animal bone and low to moderate densities of foreign trade porcelains (after the tenth century), foreign beads (after about A.D. 500), bronze and iron objects (after around 0 A.D.), and lithic material. In the lowlands, excavations at these settlements generally yield abundant evidence of permanent occupation (posthole patterns, hearths, pits, large midden areas, and sometimes craft production locales, ritual areas, and burials)(see Hutterer 1981; Junker 1993, 1994; Junker et al. 1996). They are interpreted as the varying sized villages of full sedentary, intensive agriculturalists incorporated into the lowland maritime trading chiefdom centered at Tanjay (see Junker 1990 and Junker 1999 for a discussion of lowland settlement hierarchies). In the upland zone, these sites are uniformly small (generally less than 1 hectare), less densely occupied, and have fewer lowland-derived prestige goods such as bronze and porcelain. Based on historic sources, the upland sites are most likely to represent the dispersed homesteads of less permanent upland tribal swiddening populations (the ancestors of the ethnographically known Bukidnon or Magahat) who imported some of their manufactured goods from the lowlands. Distinct from both types of sites that have substantial densities of pottery are both upland and lowland sites dominated by lithic material and animal bone; pottery, metal objects, and other manufactured goods are either absent or present in low densities. These sites are interpreted as the probable camps of mobile foragers and are generally compact (they average only around 0.25 hectares), and they tend to cluster in specific resource zones consistent with ethnographically reported seasonal camps (a point that will be elaborated later).

The stone technology from the Tanjay Region, like that of other forager-associated sites in the Philippines and elsewhere in Southeast Asia (Gorman 1971; Hutterer 1977), has little discernible "stylistic" elaboration that might allow us to define chronologically diagnostic forms. Therefore, the stone tools themselves provide no information on the relative period or periods of occupation, and we must rely on associated features and artifactual material. Sites that contain chronologically diagnostic earthenware, foreign porcelains, or glass beads in association with lithic artifact clusters have been assigned a tentative date. The largest number of lowland sites with stone tools have fifteenth to sixteenth century Osmena Phase artifacts; smaller numbers of sites contain twelfth to fourteenth century Santiago Phase artifacts or sixth to tenth century Aguilar Phase materials, and a significant portion of sites lack even tentative chronological assignments due to an absence of datable artifacts. In the preceding section, we already reviewed some of the problems of cross-dating based on surface-collected materials. To set out the problems of interpretation in simple terms, the presence of certain chronologically diagnostic lowland-manufactured goods at a site tells us that the foragers who occupied the site traded for goods A, B, and C at time X, but it does not rule out that the site is multicomponent, used at an earlier or later date by foragers who did not engage in trade for similarly archaeologically visible goods. Ethnographic work on Philippine hunter-gatherers supports the assumption that certain highly favorable locales (e.g., terraces above good water sources; locales near horticultural fields, trading partners, or other fixed resources) were repeatedly used by foraging groups such as the Agta and Batek as they moved up and down their river-defined "territories" in an annual round of seasonal activities (Eder 1987; Griffin 1989; Rai 1982). The presence of pottery dated to numerous regional phases at some of the surface-collected sites supports their interpretation as multicomponent, and subsurface coring at a sample of sites confirms these surface indications of chronologically complex stratigraphies. Cores at 7 out of the 24 sampled sites revealed lithic material in layers for which we have no surface indications of occupation (i.e., above or below assemblages that have datable pottery).

Because of the lack of extensive stratigraphic excavations at these sites, the large number of possible multicomponent sites and the provisional dating of most of these "forager camps" in the Bais-Tanjay region, the analyses of *settlement patterns* presented below (i.e., the location of forager camps relative to sedentary farming villages and various ecological and geographic features) will be largely synchronic. Given the current problems of dating, we are unable to say whether an increased emphasis on trade with lowland farmers is associated with changes in foragers settlement systems at *specific times*. However, the regional distribution of chronologically diagnostic ceramics, metal, and beads does allow us to track the *relative intensity of forager–farmer trade interactions over time* and to suggest, on a very general level, ways in which the increasing "lure" of foreign trade may have affected choices of resource exploitation (e.g., hunting for profit, development of new technologies to exploit marketable forest products), some of which are visible in the archaeological record.

Changing Forager Mobility Strategies and Settlement Systems

Ethnohistoric reconstructions of mobility patterns among extant Southeast Asian foragers suggest movement upstream and downstream for seasonal subsistence and exchange activities; the degree of dependence on trade was determined by whether and for how long foraging groups established camps near lowland farming communities. A regional archaeological survey in the Bais-Tanjay region yielded sites identified as probable forager camps that had high densities of chipped stone in three distinct resource zones: (1) along upstream tributaries of the Tanjay River (and adjacent rivers in the research region) in the mountainous interior above 100 meters elevation, (2) along the lower tributaries of the Tanjay River below 100 meters elevation in an area of gently rolling hills more than 5 kilometers upriver from Tanjay, and (3) on the flat alluvial plain surrounding Tanjay and other coastal sites (see Fig. 11.4).

As shown in Table 11.1, the sites in the three geographic and ecological zones differ considerably in density of sites, site location, site size, and densities of both lithic artifacts and imported pottery. The upland sites are

Table 11.1. Comparison of Lowland and Upland Sites that Have High Lithic Artifact Densities in the Bais-Tanjay Region and Are Identified as Possible "Forager Camps"

	Upland Sites (>100 m)	Lowland Sites (<100 m)	
		Upper Tanjay River (>5 km from Tanjay)	*Lower Tanjay River* (≤5 km from Tanjay)
Density Variables			
Number of sites analyzed[a]	11	120	22
Mean number of sites/km^2[b]	1.12	6.08[f]	1.08
Locational Variables			
Degree of site clustering (mean/variance ratio)[c]	5.1 Clustered	24.4 Highly clustered	2.3 Somewhat clustered
Average distance to river[d]	0.19	0.61[f]	1.1[f]
Site Sizes and Artifact Densities			
Mean site size (in hectares)	0.21	0.65[f]	0.12
Mean overall artifact densities[e] (in gms/1000 m^2)	512.4	265.6[f]	194.2[f]
Mean lithic densities (in gms/1000 m^2)	413.3	134.2[f]	124.7[f]
Mean pottery densities (in gms/1000 m^2)	28.4	74.5[f]	88.4[f]

[a]Twenty-two additional sites from the 1995 regional survey were added to the lowland total of 120 chipped-stone concentrations identified as probable "forager camps" from the 1982 regional survey which were analyzed in previous publications (Junker 1996, 1999, n.d.). All of these sites mapped in 1995 were located along the lower portion of the Tanjay River within 5 kilometers from Tanjay in an area of contiguous-block (i.e., total coverage) survey. The earlier analyzed lowland sites were recovered in the 1982 contiguous-block survey stretching from about 5 kilometers upriver from Tanjay to the edge of the upland zone about 15 kilometers from Tanjay. Like the 1982 survey sites, the 1995 survey sites were analyzed for size, artifact density, and locational variables. No additional work was carried out in the upland zone above 100 meters since the 1982 regional survey, but additional statistics on artifact densities and locational variables were calculated for the present analysis.
[b]The mean number of sites per square kilometer surveyed indicates the average number of lithic clusters recorded in each kilometer square survey block.
[c]We used the variance/mean ratio as a measure of the relative degree of clustering.
[d]The distance to the nearest river refers generally to the Tanjay River, Panamangan River, or one of their tributaries, as a measure of the degree to which river transportation routes and water sources affected settlement choices in the uplands and lowlands.
[e]Mean overall artifact densities refer to the average grams/1000 m^2 of artifactual material of any type recovered in systematic surface collections at sites, including earthenware pottery, porcelain, chipped stone, metal, animal bone, shell, burnt clay, beads, or other clearly human-manufactured or human-processed materials.
[f]Indicates statistically significant (at the 0.05 level) differences between lowland sites and upland sites, using the Student's t-test.

smaller and less numerous (a pattern that may be biased by sampling problems discussed above), but they generally have higher overall artifact densities and particularly high densities of lithic material within their compact surface areas. As shown in Figure 11.4, the upland camps are invariably found within 0.5 km of streams, tending to cluster on relatively flat terraces above the small tributaries of the Tanjay River that consistently provide water during the rainy season. Although the primary tropical forest had dwindled to about 5% of the upland vegetation by the 1980s (E. Hoffman 1982; Hutterer and Macdonald 1982), until the massive incursion of lowland loggers and colonizing farmers in the mid twentieth century, this area retained large stretches of primary tropical forest punctuated by occasional swidden fields and secondary growth created by upland tribal peoples and the Ata themselves (Wernstedt and Spencer 1967). The archaeological sites in this upland zone are consistent with settlement preferences of the Luzon Agta, the Negros Ata, and other historically known Philippine foragers, who generally established their interior base camps close to rivers for daily access to drinking water, fishing opportunities, and easy transport routes.

The more numerous, larger lowland sites along the ecotone between the flat alluvial plain of Tanjay and the upland margin are even more strongly clustered than those of the upland zone. The majority of sites in this area are tightly clustered in the upper reaches of the Tanjay River and Panamangan River to the north; most sites are within 3 kilometers of the abrupt transition to the mountainous interior (see Fig. 11.4). The site clustering index (variance/mean ratio) in Table 11.1, shows an extremely high degree of redundancy in site location along these small streams in the lowland margins. Furthermore, these highly clustered chipped-stone scatters are themselves clustered close to specific lowland agriculturalist settlements strung along the Tanjay and Panamangan rivers and their subsidiaries. A statistical test presented in Table 11.2 indicates that the surface-surveyed chipped-stone sites in the lowland are found near riverbank agricultural settlements more often than would be expected under conditions of purely random placement of the two site types.

If we separate out those lowland chipped-stone locales of more certain date (due to systematic subsurface auger coring at these sites) and compare their locations to lowland agricultural villages (larger than 1 hectare) of known date, the lowland "forager camps" are more likely to be located near sedentary villages in the fifteenth to sixteenth century Osmena Phase than in the late first-millennium A.D. Aguilar Phase (see Table 11.2). Too few chipped-stone sites were dated to the intervening Santiago Phase to carry out a comparable statistical analysis. The proximity to large riverbank settlements in the lowlands suggests that economic interactions with the lowland villagers were one of the magnets that drew upland hunter-gatherers

Table 11.2. Comparison of the Probability that an Individual Lowland Hunter-Gatherer Camp Has a Large Lowland Agricultural Village as Its "Nearest Neighbor"

	Surface Sites (dates uncertain)	*Aguilar Phase* Cored Sites	*Osmena Phase* Cored Sites
Total # of sites in analysis (*N*)	92	12	23
Lowland agricultural villages greater than one hectare (n_1)	24	5	12
Lowland forager camps (n_2)	68	7	11
Expected probability that nearest neighbor of forager camp was a large agricultural village (p_1)[a]	0.2637	0.4545	0.5454
Expected number of forager camps that have large agricultural villages as their nearest neighbors	17.93	3.18	6.00
Actual proportion of nearest neighbors that are large agricultural villages (p_2)	0.5735	0.5714	0.8182
Actual number of hunter-gatherer camps that have large agricultural villages as their nearest neighbors	39	4	9
Z statistic[b]	1.68[c]	0.47	1.10

[a]This is calculated as $p_1 = n_1/(N-1)$. $N-1$ is used as the denominator because a site cannot be a nearest neighbor to itself. Possible nearest neighbors are defined as sites within the transect and cannot be external sites of the transect boundaries.
[b]This is calculated as $Z = (x_1 - M)/S$ where $M = (p_1)(n_2)$, $x_1 = (p_2)(n_2)$, and $S = (n_2)(p_1)(1-p_1)$.
[c]Statistically significant at the 0.05 level.

periodically to these particular locales. Furthermore, there appears to be greater attraction to these lowland settlements at the midsecond millennium A.D. height of foreign trade, compared to the period before intensive commerce in tropical forest products for export. In the upland zone, mobile foragers probably had a long history of trading for staple resources (particularly carbohydrates) with adjacent tribal swidden farmers who occupied the larger pottery-rich sites, although some upland sedentary groups may have functioned as "middlemen" in forager interactions with lowland farmers for manufactured commodities and foreign prestige goods. We cannot presently sort out chronological phases for locational analyses on upland sites (due to poor dating and sample sizes), but upland chipped-stone sites exhibit no statistically significant "attraction" to upland pottery-yielding sites (Table 11.3), suggesting that upland camps are not specifically located near sedentary villages or homesteads for protracted trade activities.

Table 11.3. Comparison of the Probability that an Individual Upland Hunter-Gatherer Camp Has an Upland Agricultural Village as Its "Nearest Neighbor"

	Upland Sites
Total # of sites in analysis (N)	38
Upland agricultural villages larger than 1 hectare (n_1)	15
Upland forager camps (n_2)	23
Expected probability that nearest neighbor of forager camp was a large agricultural village (p_1)[a]	0.4054
Expected number of forager camps that have large agricultural villages as their nearest neighbors	9.32
Actual proportion of nearest neighbors that are large agricultural villages (p_2)	0.3913
Actual number of hunter-gatherer camps that have large agricultural villages as their nearest neighbors	9
Z statistic[b]	0.06

[a]This is calculated as $p_1 = n_1/(N-1)$. $N-1$ is used as the denominator because a site cannot be a nearest neighbor to itself. Possible nearest neighbors are defined as sites within the transect and cannot be external sites of the transect boundaries.
[b]This is calculated as $Z = (x_1 - M)/S$ where $M=(p_1)(n_2)$, $x_1=(p_2)(n_2)$, and $S=(n_2)(p_1)(1-p_1)$.
[c]Statistically significant at the 0.05 level.

Differences in site form and composition suggest that the upland and lowland sites represent different settlement components of a seasonal round of hunter-gatherer activities that involve movement along the river between the tropical forests of the mountainous interior and the lower elevation forests on the margins of the alluvial plain. The dense accumulation of artifactual material at the upland sites suggests longer term occupation, consistent with ethnographic accounts of rainy season camps established near swidden plots or productive forest zones when mobility is impeded by dangerously swelling rivers and constant downpours. In contrast, the lower elevation sites close to lowland agriculturalist villages in the upper Tanjay River plain appear to represent multiple, perhaps overlapping, occupations of relatively brief duration concentrated on exploiting lowland foraging opportunities and interactions with lowland farmers. These sites were probably used in the dry season because they contain freshwater shell that would have been exploitable then, and this portion of the river would have been impassable in the rainy season. Sites in the lower Tanjay River drainage are less numerous, less clustered, have lower artifact densities, and are considerably smaller than either those in the upper Tanjay River branches or those in the upland zone. These near-coastal sites are largely undated (except for four cored Osmena Phase sites) and are difficult to interpret, although the presence of significant quantities of marine shell in their artifact assemblages suggests a possible forager settlement component geared toward periodic exploitation of coastal resources. However, lacking good chronologies, we do not know whether these are early sites occupied before the extensive development of lowland farming and chiefdoms in the region, or if they represent some form of direct interaction with large coastal chiefly centers such as Tanjay.

Forager–Farmer Trade for Foodstuffs and Manufactured Products

We have argued purely on the basis of location that the chipped-stone sites clustered on the Tanjay lowland margins were loci of interaction between interior foragers and lowland agriculturalists. Analyses of artifact and ecofact assemblage composition tell us more about the content and intensity of this interaction and how it might have changed over time. In addition, we can attempt to infer some aspects of archaeologically "invisible" resources involved in interethnic trade by studying the technologies used to process them. Archaeological documentation of the diversity and volume of coastal-interior trade goods is considerably biased toward durable lowland-manufactured goods because of their archaeological visibility and long-term preservation compared to upland perishable products. Two of the most significant lowland-manufactured goods traded to interior groups, including Ata exchange partners, were ceramics and metal implements.

In the sixth to tenth century Aguilar Phase, lowland-manufactured earthenware (a diverse ware made probably at multiple production centers and known as "Aguilar spotted buff ware") is recovered at 21% of the chipped-stone clusters in the lowlands and at none of the upland chipped-stone clusters. The intermediate eleventh to fourteenth century Santiago Phase trade patterns are too poorly documented for quantitative analysis. However, by the fifteenth to sixteenth century Osmena Phase, more than 50% of the lowland foraging camps and 2 out of the 11 upland foraging camps contain pottery assemblages dominated by the highly standardized, Tanjay-produced earthenware ("Tanjay red ware"). As shown in Figure 11.5, both the number of sites containing earthenware and the relative densities of earthenware at the sites increase significantly in the fifteenth to sixteenth century Osmena Phase. A similar pattern is found for lowland-produced iron

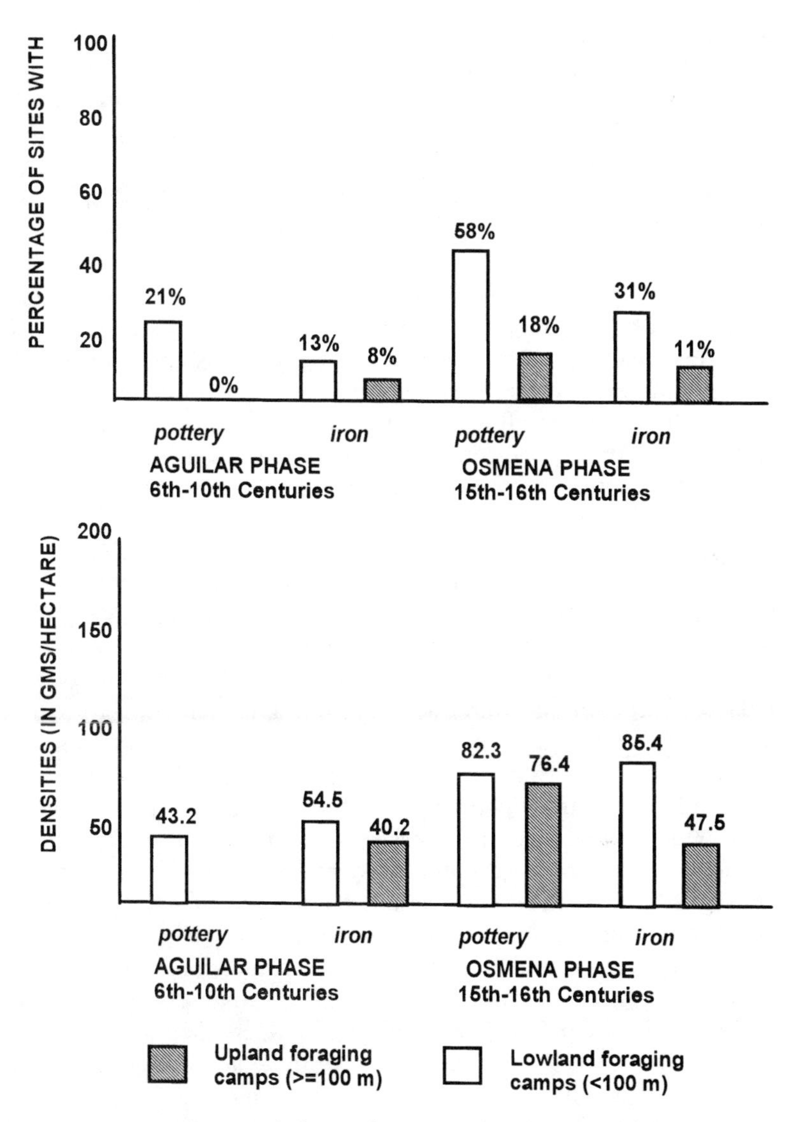

Figure 11.5. Average densities of plain earthenware and iron recovered from hunter-gatherer camps in lowland and upland zones in two pre-Hispanic phases of settlement in the Bais-Tanjay region (the sixth to tenth century Aguilar Phase and fifteenth to sixteenth century Osmena Phase). Sample sizes were too small for statistical analysis of the Santiago Phase (eleventh to fourteenth century) sites.

implements. Though iron objects are found at a few interior foraging camps in both the uplands and lowlands before the tenth century, the flow of iron to these hunter-gatherer populations increases significantly by the fifteenth to sixteenth centuries. Parry's (1982b) ethnographic study of tool manufacture and tool use among the contemporary Ata Negritos settled near Mabini reveals that iron blades obtained through trade or purchase are used as machetes (*bolos*) to clear swidden fields or they are reworked to manufacture iron arrowheads and spearheads for hunting.

Due to the larger sample of lowland earthenware and iron at fifteenth to sixteenth century hunter-gatherer camps, we can examine density–distance relationships in more detail for this period. As shown in Figure 11.6, there are significant correlations between the relative frequency of earthenware at these presumed hunter-gatherer camps and the distance of the camp from both the coastal center of Tanjay and upriver secondary centers. Access to iron correlates with distance from large sedentary villages along the Tanjay River, but unlike the earthenware pottery, it is not strongly dependent on the distance upriver from Tanjay. Archaeological studies of earthenware and metal manufacture in the Bais-Tanjay region suggest that earthenware pottery manufacture became more centralized in the vicinity of Tanjay in this later period, whereas iron manufacture was more regionally diffuse (Junker 1994, 1999:261–291; Junker et al. 1996). The discovery in 1994 of a probable iron production site at a settlement nearly 10 kilometers upriver from Tanjay suggests that the manufacture of iron implements, possibly specifically for trade into the interior, took place near ore sources and close to trade partners. This distance-dependent distribution of lowland trade potteries and iron tools suggests that the location of lowland hunter-gatherer camps adjacent to sedentary villages along the upper reaches of the Tanjay River was partially a function of regular, and perhaps seasonal, trade interactions between the two groups along the lowland–upland boundary. The volume of both pottery and iron objects increased with the emergence of a larger scale maritime trading chiefdom at Tanjay in the centuries just before European contact, suggesting a strong linkage between the expanding demands of foreign trade and the development of these forager–farmer trade systems.

Lowland-produced foods, including some tropical fruits, rice, fish, shellfish, and other marine resources also flowed into the interior. Though the interior movements of lowland rice and other perishable agricultural crops are difficult to document archaeologically, trade in marine products can be traced through an analysis of the content of shell middens at interior sites. Detailed quantitative data on shell species and environmental contexts have been collected for excavated shell assemblages from the coastal center of Tanjay and a secondary center about 5 kilometers upriver

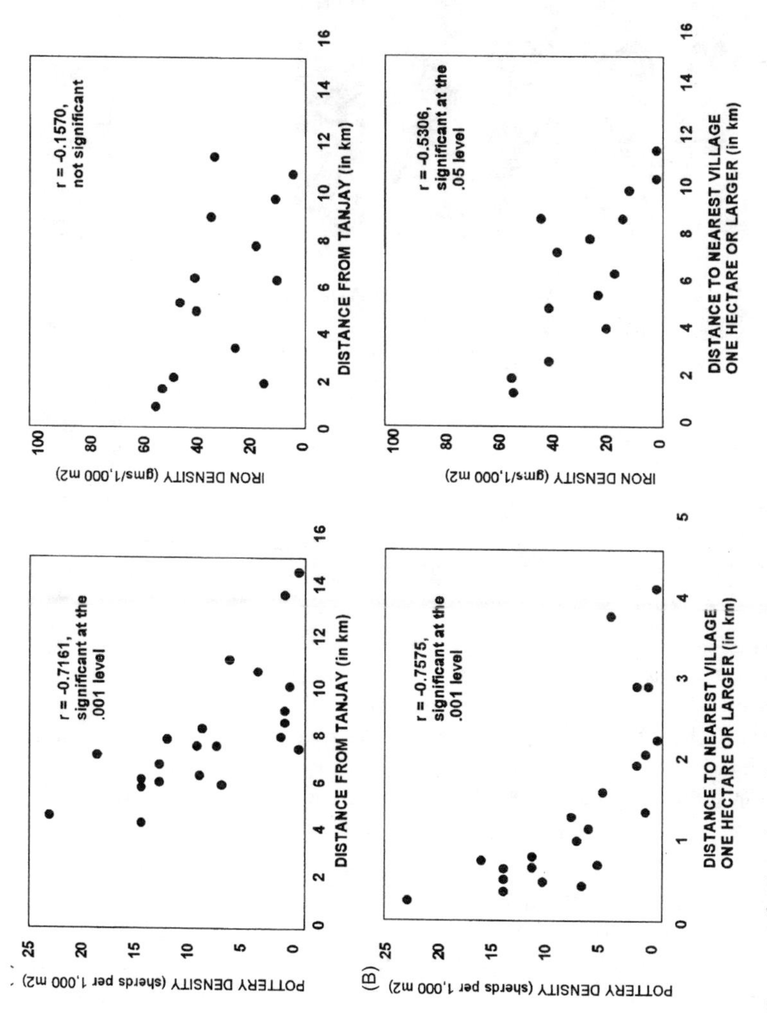

Figure 11.6. The relationship between densities of lowland-manufactured earthenware and iron recorded at sites with chipped stone dated to the Osmena Phase and the distance of these chipped-stone sites from (A) the coastal center of Tanjay and (B) the nearest upriver secondary center larger than 1 hectare.

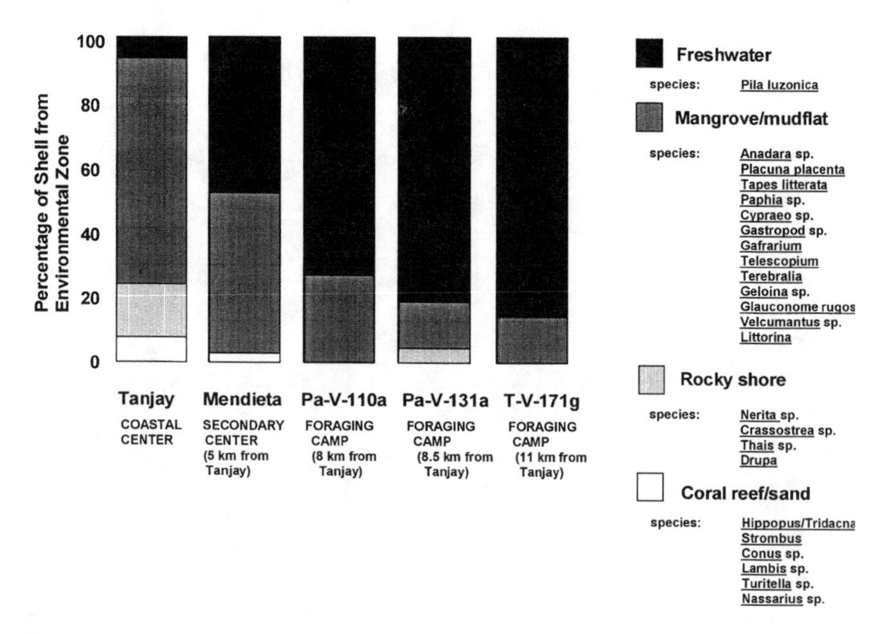

Figure 11.7. Comparison of Osmena Phase shell assemblages recovered from middens at Tanjay (coastal chiefly center), the Mendieta Site (an inland secondary center), and three surface-collected interior foraging camps.

(the Mendieta Site), as well as for surface-collected shell from three lithic clusters several kilometers upriver from the Mendieta Site (see Fig. 11.7). Because all of the sites, except Tanjay, yielded sufficient quantitative information on shell frequencies only for the most recent Osmena Phase, the discussion will be limited to this period. Not surprisingly, most of the shell species from the coastal center of Tanjay are marine species, whereas the majority of shells at inland Tanjay Region sites derives from riverine species. However, marine species represent a significant portion of shell assemblages at both the Mendieta Site and interior chipped-stone clusters, despite their considerable distance from the coast. The presence of substantial quantities of marine shells at upriver trading centers and at supposed adjacent hunter-gatherer camps suggests that marine products were traded into the interior as part of these symbiotic exchange relations.

It is considerably more difficult to assess archaeologically the range and volume of products amassed by interior hunter-gatherers and tribal agriculturalists for export to lowland exchange partners because many of these exports were perishable forest products (e.g., animal skins, hardwoods, resins, beeswax, medicinal plants, and possibly basketry or matting). With reference to interior hunted game as a potential hunter-gatherer export to agriculturalist exchange partners, it is relevant to note that most of the stone tool clusters concentrated along the upper Tanjay River yielded significant quantities of animal bone, the bulk of which are wild species (wild pig, deer, monkey, and a variety of small to medium mammals). Studies of the eleventh to sixteenth century faunal remains from midden deposits within the coastal chiefly center at Tanjay (Junker et al. 1994; Mudar 1997) show that more than one-third of the faunal material in both the Santiago and Osmena Phases derives from taxonomically wild species indigenous to the uplands. Recent analysis of charred and otherwise preserved plant macroremains from hearths and midden deposits at the coastal center of Tanjay (Gunn 1995, 1997) supports historic and ethnographic accounts of trading interior tropical forest plants and upland crops also to coastal populations. Through a detailed ecological analysis of environmental zones in the Tanjay region, Gunn (1997) suggested that the products of more agriculturally productive zones outside the lowland polity core may have been essential to support a burgeoning chiefdom that had an increasing economic emphasis on long-distance maritime trade. Rice varieties found in the Osmena Phase (fifteenth to sixteenth centuries) occupation levels at Tanjay include dry rice that grows best on well-drained hillsides. In addition, many of the nonfood plants recovered from this most recent pre-Hispanic occupation stratum are ornamental plants indigenous to higher elevation tropical forests of the interior. Thus, the botanical evidence suggests that a significant volume of both food and nonfood plants were imported from the Tanjay Region uplands in the centuries marking the height of foreign trade at Tanjay and other central Philippine ports. Because both the tribal swiddening populations, known as the Bukidnon and Magahat, and the Ata foragers collected forest products and engaged in horticultural activities, we cannot say how many of these resources were actually obtained from foraging groups.

A more indirect line of evidence for hunter-gatherer activities that might be related to trade is the analysis of variability in stone tool assemblages at Tanjay region sites. Philippine lithic artifacts of all periods consist primarily of relatively amorphous cores and flakes that have a low percentage of retouching and an even lower proportion of formalized "tools" such as knives and scrapers. This largely "expedient" industry appears to be primarily geared toward maintaining perishable tools rather than primary "extractive" activities. However, in a study of lithic material excavated and surface collected in the Tanjay region in 1979, Parry (1982) demonstrated that measurement of edge angles, overall tool or flake size and weight, and the incidence of edge damage (i.e., use) and retouch (i.e., purposeful modification) can provide insight into the types of activities taking place at sites that have chipped stone.

Table 11.4. Comparison of Lithic Assemblages at Lowland (<100 meters elevation) and Upland (100 meters or higher) Sites in the Tanjay Region Yielding Significant Lithic Components

	Lowland Sites (<100 m)	Upland Sites (≥100 m)
Number of sites analyzed	7	5
Total number of lithic fragments	529	352
Mean lithic densities	17.2 items/1000 m^2	78.9 items/1000 m^2
Mean pottery densities	20.2 shards/1000 m^2	5.2 shards/1000 m^2
Mean site sizes[a]	0.85 (0.43) ha	0.15 (0.32) ha
Percentage of artifacts that has one or more used edges	53%	25%
Percentage of artifacts with retouch	14%	5%
Mean edge angle[b]	57° (18°)	48° (12°)
Mean tool weight[c]	48 (105) gms	20 (32) gms

[a]Standard deviations in ( ).
[b]Used edges only.
[c]Retouched "tools" only.

A sample of seven lowland lithic clusters and five upland lithic clusters recorded in the 1982 Tanjay region survey were selected for detailed analyses of their lithic assemblages (Table 11.4), using some of the quantitative measures suggested by Parry (1982). Consistent with Parry's study, lowland lithic clusters were more spatially extensive, but yielded significantly lower stone artifact densities (and contrasting higher densities of animal bone and lowland products, such as pottery and marine shell). Differences in stone artifact densities appear to be at least partially a function of lithic assemblage differences. Upland sites primarily yielded cores and small, unretouched, unused flakes, whereas lowland sites contained a significant proportion of used flakes, as well as large flakes and pebbles that had been retouched into morphologically distinct "tools" such as scrapers, burins, notched tools, and knives (see Fig. 11.8). Heavy, steep-edged scrapers were unusually prominent in lithic clusters just west of the fifteenth to sixteenth century riverbank center known as the Diaz Site.

Interpretation of these differences in lithic assemblages is hampered by the lack of ethnographic and historic data on stone tool use among recent Philippine hunter-gatherers. Amorphous industries recovered from Philippine sites of various periods (comprised primarily of unretouched cores and flakes with little evidence for use-wear) have been interpreted as expedient tools and nonfunctionally specific tools used for a variety of both maintenance and primary extractive activities (e.g., Bevacqua 1972; Cherry

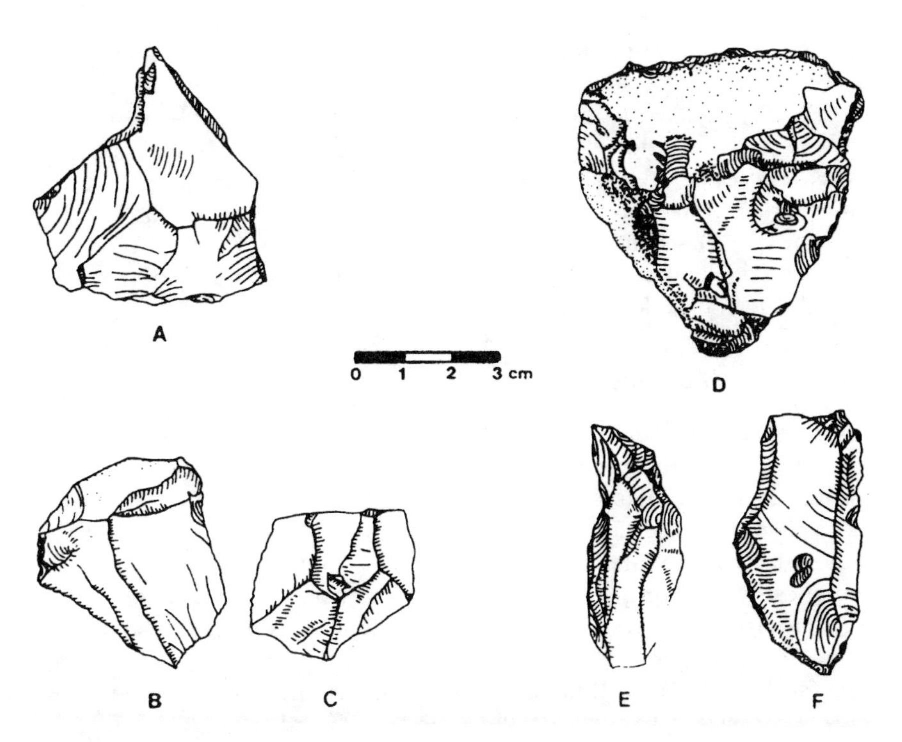

Figure 11.8. Some artifacts collected from upland chipped-stone clusters (A, B, C) and lowland chipped-stone clusters (D, E, F). A: core; B and C: unretouched flakes; D: bifacially flaked scraper; E: burin; F: retouched blade.

1978; Coutts 1984; Fox 1970; Ronquillo 1981). These activities might include butchering animals; processing plant materials; and producing arrow shafts, knives, and other tools from bamboo and wood. The lithic assemblages at the Tanjay region upland sites, consisting primarily of this "smash-and-grab" industry of unretouched flakes, blades, and cores, are likely to have been expediently used for a wide range of animal and plant processing and perishable tool manufacturing activities.

In contrast, the formalized tools and retouched components of the lowland lithic assemblages may represent more specialized emphasis on production activities associated with seasonal exchange relations with lowland agriculturalists. As noted in a previous study of Tanjay region lithic material by Parry (1982), tools that have steeply retouched edges are most likely to be intended as "maintenance tools" to work hard materials (i.e., to manufacture bone or wood tools) or as primary tools for scraping materials that cannot be processed effectively with softer wood or bamboo (i.e., for scraping large animal hides). Because hunted meat and animal hides were a significant exchange commodity for interior hunter-gatherers, we might speculate that the unusually large number of steep-edged scrapers reflects preparation of animal pelts intended for trade at the adjacent riverbank villages. In addition, intensive hunting and fishing along the lower river would have required the manufacture or repair of wooden bows, arrow shafts, and spears, which may have involved using the types of steep-edged scrapers, burins, and notched tools found in the lithic assemblages of the lowland camps.

The Impact of Foreign Trade

Scattered finds of Tang Period porcelains and pre-tenth century exotic bronzes and beads suggest that developing coastal chiefdoms in the Philippines established extra-archipelago trade contacts by at least the mid-first millennium A.D. However, Philippine chiefdoms such as Manila, Sulu, Cebu, Magindanao, and Tanjay grew in scale and complexity in the few centuries before European contact, supported by expanding trade for foreign prestige goods wealth (such as primarily porcelain, metals, and silks) with China and the developing maritime trading kingdoms of Southeast Asia (such as Champa, Brunei, Malacca, and Ayudhya) (Junker 1999: 183–220). As shown in Figure 11.9, Philippine chiefs had to amass large quantities of raw materials and forest products (including beeswax, raw cotton, pearls, tortoise shells, medicinal betelnuts, *abaca* or Southeast Asian hemp, coconut heart mats, various tropical hardwoods, and animal pelts) in exchange for the prestige goods offered by their foreign trade partners. Because most of these commodities derived from remote interior areas that the coastal chiefs did not directly control, systematic harvesting of them for export depended on expanding traditional lowland–upland trade systems. Because of the rugged terrain of many island interiors in Southeast Asia and limits on land-based militarism, even the most developed of the Southeast Asian kingdoms (such as Srivijaya and Malacca) found it impractical to obtain economic patrimony over these distant populations through conquest and enforced political hegemony (Andaya 1995:545–547; Gullick 1958:48–49; Miksic 1984). However, historic analysis has suggested some specific mechanisms whereby coastal chiefs may have expanded the volume and improved the reliability of lowland–upland trade systems, including (1) changes in river-based settlement systems and the emergence of large upriver secondary centers as energetically efficient collection points for interior resources; (2) centralized mass production of lowland products such as earthenware to stimulate greater volumes of interior trade; and (3) consolidation of interior exchange relations through the influx of

	San-Tao (northeast Luzon)	Ma-i (northern Mindoro)	Min-to-Lang (Mindanao)	Ma–li-lu (Maniila)	
CHINESE IMPORTS	*beeswax* textiles cotton	*kapok* *beeswax* tortoise shell *betelnut* textiles	*wuli wood* *musk* [1] *sandalwood* cotton *animal skins* [2]	tortoise shell *beeswax* *laka-wood* "jwu-bah" cloth [3] *kapok*	Upland forest products traditionally collected by foragers in the Philippines are shown in ***italics***.
CHINESE EXPORTS	*beads* *porcelain bowls* *iron*	iron cauldrons *other iron* red silk ivory silver coins	lacquer ware cauldrons (copper?) "Java" cloth red silk blue cloth tou [4] tin wine	ting bronzes blue cloth *porcelain jars* iron cauldrons "big pots" [5]	Chinese products that are specifically mentioned in historic documents as trade goods acquired by foragers in the Philippines are shown in ***italics***. Other products are assumed to have been restricted to lowland comple: in the Philippine:

Figure 11.9. Imports and exports associated with various fourteenth-century Philippine polities, as enumerated by the Yuan Period Chinese writer Wang Ta-yuan (from the *Tao i Chih Lueh*, A.D. 1349, translated in Zaide 1990, Vol. 1: 9–13).

Notes: 1. Musk, a greasy secretion produced in a glandular sac beneath the abdomen of certain mammals and often used in perfumes or scents, probably came from the civet cat in the Philippines. 2. Animal pelts from the interior tropical forests would have included civet cat, Philippine deer, and possibly wild boar. 3. "Jwu-bah" cloth, as a special form of textile, is unidentified. 4. The Chinese product known as "tou" is unidentified. 5. The Chinese product translated as "big pots" is unidentified.

lowland prestige goods such as porcelains and bronze, as well as status-conferring lowland titles and ceremonial regalia (Junker 1999:239–259).

In the preceding section, we already pointed to some of the changes in regional settlement organization and the increased volume of upland–lowland trade that may be related to expanding participation in foreign prestige goods trade by lowland chiefs in the Philippines. At the height of maritime luxury goods trade in the fifteenth and sixteenth centuries, large "secondary" centers emerge along the upper branches of the Tanjay River that are strategically located to control river-based trade between the interior and the coast. Analysis of the location of "forager camps" along the lowland margins shows that these sites are more numerous and tend to cluster more frequently near riverbank villages with substantial amounts of lowland-produced trade goods such as earthenware pottery and iron in these few centuries before European contact. In the fifteenth to sixteenth century Osmena Phase, these presumed forager camps are also larger than those of earlier periods. In addition, they are more likely than in earlier periods to contain lowland-manufactured goods in their artifact assemblages, and, where present, these goods are found in significantly larger quantities. The archaeological settlement data suggests that foraging populations in the Bais-Tanjay region, at the fifteenth to sixteenth century height of foreign trade, may have devoted more of their time and resources to lucrative trade interactions and more frequently visited or settled for longer periods near riverbank trade locales. Of course, an expanded emphasis on trade with lowland populations and the resulting "tethering" of dry season settlements to Tanjay River trade depots necessarily constricted other economic choices in other ecological zones, such as high elevation subsistence hunting and upland horticultural activities.

Archaeological evidence from the Bais-Tanjay region also suggests that lowland populations that were heavily invested in obtaining interior exports for foreign trade underwent changes in their production systems to expand coastal–interior trade and to bring it more directly under their control. Standardization studies on the earthenware ceramics from the Bais-Tanjay region suggest a pattern of dispersed household production of pottery in the periods before the fifteenth century; individual households manufactured and distributed pottery to their interior trade partners (see Junker 1994). By the fifteenth century, greater homogeneity of pottery forms in the region and evidence of substantial pottery-producing activities at Tanjay indicate a possible shift to more centralized ceramic production at the coastal chiefly center, perhaps oriented to intensified export into the interior (see Junker 1994, 1999:261–291). I have also suggested that the establishment of a substantial iron processing and manufacturing operation along the upper Tanjay River in the fifteenth to sixteenth centuries may also relate to efforts to intensify lowland–upland trade and, specifically, to bring it under more direct control by lowland chiefs (Junker et al. 1996). However, we have emphasized that our documentation of changes in trade goods and trade volumes is biased toward the trade initiatives of lowland groups. Though we can assume that certain forest products in strong demand by foreign traders (e.g., animal pelts, tropical wood, beeswax, and resins) might be targeted by lowland traders and foragers probably adjusted their collecting and production activities toward these resources, we cannot directly trace these changes due to the archaeological "invisibility" of most forest commodities.

Quantitative data on the regional distribution of lowland "prestige goods" indicate that Tanjay chiefs may have been intensifying their efforts to consolidate trade relations with interior tribal leaders in the fifteenth to sixteenth centuries through more frequent and more voluminous ceremonial gift exchange. As shown in Figure 11.10, less than 10% of the upland settlements of swidden agriculturalists and none of the interior hunter-gatherer camps yielded decorated or slipped earthenware in the sixth to tenth century Aguilar Phase, whereas only one upland farming village site yielded fragments of bronze bracelets. In the eleventh to fourteenth century Santiago Phase, prestige goods (including decorated earthenware, bronze objects, and Sung porcelain) are found at a slightly higher percentage of upland farmsteads, but forager camps still lack lowland prestige goods. By the fifteenth to sixteenth century Osmena Phase, around 40% of the highland settlements of swidden agriculturalists have archaeological evidence of access to lowland-manufactured fine earthenware, and many farmsteads have bronze objects, marine shell beads or glass beads, and Ming porcelain. However, the association of metals, porcelain, exotic beads, and decorated ceramics with hunter-gatherer camps is still rare in this period. The archaeological evidence from the Bais-Tanjay region suggests that interior tribal populations such as the Bukidnon and Magahat were the more strongly affected, in social terms, by an influx of lowland symbols of status and power. Lowland titles and status-imbuing wealth may have intensified the competitive interactions associated with marriage transactions, war negotiations, and ritual feasting, resulting in changing social dynamics in these upland farming societies. In contrast, there is no archaeological evidence of emerging competition for status and wealth accumulation among forager-traders in the region or in any other region of the Philippines where we have long-term forager histories (e.g., Eder 1987; Griffin and Estioko-Griffin 1985). Thus, the Kubu of interior Sumatra and the Nuaulu of central Seram are not widely representative of Southeast Asian foragers in general who engage heavily in foreign trade in terms of their unusual social complexity.

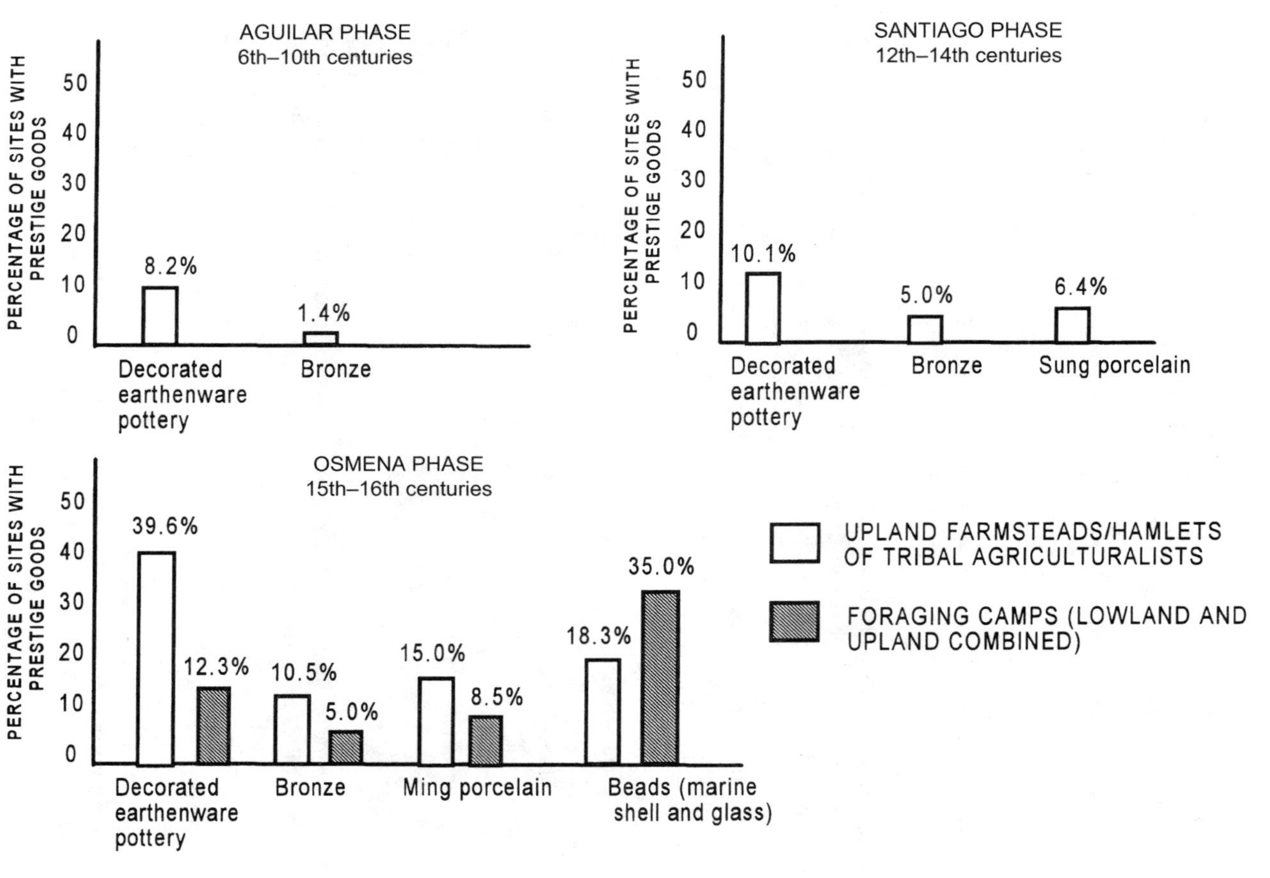

Figure 11.10. Percentage of interior forager camps that have various lowland prestige goods in their artifact assemblages.

SUMMARY AND CONCLUSIONS

For many of the lowland coastal chiefdoms in the Philippines, regular trade interaction with adjacent upland tribal swidden agriculturalists and mobile tropical forest hunter-gatherers who controlled resources not available in the lowlands, was a core element of their economies. Archaeological evidence from elsewhere in the Philippines suggests that this economic specialization and exchange between distinct ecological zones is significantly old, perhaps even beyond the period of complex society development in the Philippine lowlands. In the Bais-Tanjay region, the presence of lowland-manufactured pottery and iron goods at sites identified as probable "forager camps" by the late first millennium A.D. indicates that these types of forager–farmer exchanges considerably predate the advent of a coastal chiefly political economy focused on foreign prestige goods trade. Thus, although the present archaeological evidence fails to address the question of preagricultural foraging adaptations in the carbohydrate-limited interior tropical forest, it demonstrates that most contemporary Philippine hunter-gatherers are not "devolved" farmers whose forest specialization was catalyzed by relatively recent foreign demands for forest products accompanying the rise of the Chinese porcelain trade. The origins of the Punan of Borneo and the Tasaday of the Philippines as "genuine" foragers is still being debated in the absence of archaeological evidence, but the archaeological investigations in the Bais-Tanjay region leave little doubt that the ancestors of the Ata foragers were hunter-gatherers who had a complex set of economic options that allowed them to adapt successfully to the tropical rain forest. They had a long-term history of meeting shortfalls in carbohydrates through flexible strategies which, at various times, involved increasing exchanges of forest resources (particularly protein) for agricultural foods with adjacent farmers. Though these exchange relationships are visible in the archaeological record as settlement location preferences and long-distance movement of resources, we should also bear in mind that less archaeologically visible strategies, such as incorporating swidden farming into their own subsistence regime and managing (i.e., maintaining patches or transplanting) wild starchy plants such as sago, may have also been significant at various times.

Although the early to midsecond millennium A.D. maritime porcelain trade was unlikely to have been the initial impetus for creating a human niche of specialized tropical forest foragers in Southeast Asia, the commercialization of forest products had a visible impact on forager economies and social relations. Before foreign trade, local forager–farmer exchange networks probably involved individually contracted trade partnerships which circulated critical subsistence goods and household commodities without any control by regional political authorities and largely without the socially transforming emphasis on status goods and wealth accumulation. However, once interior trade systems became linked to an external market and enmeshed in lowland chiefly strategies for wealth accumulation, there was likely a shift in trade commodities (more luxury goods flowing upriver and more exportable forest products flowing downriver), a massive increase in trade volumes, and more coercive control by lowland chiefs. Archaeological evidence from the Bais-Tanjay region shows that, at the fifteenth to sixteenth century height of foreign trade, a number of large and strategically located "secondary centers" emerges upriver from the maritime trading center of Tanjay which yielded abundant quantities of lowland trade commodities and appear to have been magnets for seasonal encampments of hunter-gatherers. Although lowland-manufactured household goods are present at distant upland settlements from at least A.D. 500, the diversity and volume of lowland products at interior sites increases dramatically in the few centuries before European contact, attesting to heightened upland–lowland trade. Both locally manufactured and exotic prestige goods controlled by the lowland chiefs begin to penetrate more widely into interior settlements in the fifteenth to sixteenth centuries, suggesting that lowland chiefs became increasingly important in controlling and administering this interior trade.

ACKNOWLEDGMENTS

When this paper was first presented at the Society for American Archaeology symposium organized by Junko Habu and Ben Fitzhugh, it benefited tremendously from the comments of Douglas Price at the session and from later informal discussion with Junko, Ben, Sandra Bowdler, and many of the other authors of this volume. Much of my theoretical perspective on tropical hunter-gatherers comes from various past conversations with Karl Hutterer, Peter Brosius, Willie Ronquillo, and Kathy Morrison over several years, from an appreciation of the long-term ethnographic work of Bion Griffin and Tom Headland, and more recently from interactions with many colleagues at Brigham Young University, University of Oklahoma, and Western Michigan University who work with types of hunter-gatherers quite different from those in Southeast Asia. Archaeological fieldwork at the "forager camps" and other sites in the Bais-Tanjay region relevant to this research in 1985, 1986, 1992, 1994, and 1995 was supported by grants and fellowships from the Wenner–Gren Foundation, the Fulbright Foundation, the Social Science Research Council, the National Geographic Society, the Mellon Foundation, and the University

Research Council at Vanderbilt University. In the field, there were many individuals who contributed to my research, most notably Willie Ronquillo and Eusebio Dizon from the National Museum of the Philippines, Mary Gunn formerly from the University of Hawaii, Rolando Mascunana from Silliman University, and many, many fine student-archaeologists from Silliman University, the University of San Carlos, and the National Museum. The mayor of Tanjay (the Hon. Arturo Regalado) and numerous landowners in the Bais-Tanjay region (particularly the Mendieta, Diaz, and Calumpang families) were generous in giving permission for the 1994 and 1995 regional survey, auger coring program, and excavations in the Tanjay River drainage. The Department of Anthropology at University of Oklahoma, and particularly Patricia Gilman, generously provided logistical support for completing the laboratory and statistical portions of this research while I was a visiting scholar there.

REFERENCES

Allen, M., 1985, The Rain Forest of Northeastern Luzon and Agta Foragers. In *The Agta of Northeastern Luzon:Recent Studies*, edited by P. Bion Griffin and A. Estioko-Griffin, pp. 45–58. San Carlos, Cebu City, Philippines.

Andaya, B., 1975, The Nature of the State in Eighteenth Century Perak. In *Pre-Colonial State Systems in Southeast Asia*, edited by A. Reid and L. Castles, pp. 22–35. Monographs of the Malaysian Branch of the Royal Asiatic Society No. 6. Royal Asiatic Society, Hong Kong.

Andaya, B., 1995, Upstreams and Downstreams in Early Modern Sumatra. *The Historian* 57(3):537–552.

Bailey, R. C., Head, G., Jenike, M., Owen, B., Rechtman, R., and Zechenter, E., 1989, Hunting and Gathering in Tropical Rain Forest:Is it Possible? *American Anthropologist* 91:59–82.

Bevacqua, R. F., 1972, An Analysis of a Sample of Sohoton Cave Lithics. *Leyte-Samar Studies* 6(2):69–83.

Bellwood, P., 1992, Southeast Asia Before History. In *The Cambridge History of Southeast Asia, Vol. 1:From Early Times to c. 1800*, edited by N. Tarling. Cambridge University Press, Cambridge.

Bellwood, P., 1997, *Prehistory of the Indo-Malaysian Archipelago*, 2nd ed. University of Hawaii Press, Honolulu.

Benjamin, G., 1976, Austroasiatic Subgroupings and Prehistory in the Malay Peninsula. In *Austroasiatic Studies*, edited by P. Jenner, L. Thompson, and S. Starosta, pp. 37–128. University of Hawaii Press, Honolulu.

Benjamin, G., 1985, In the Long Term:Three Themes in Malayan Cultural Ecology. In *Cultural Values and Human Ecology in Southeast Asia*, edited by K. L. Hutterer, T. Rambo, and G. Lovelace, pp. 219–278. Center for South and Southeast Asian Studies, Ann Arbor.

Binford, L. R., 1980, Willow Smoke and Dog's Tails:Hunter-Gatherer Settlement Systems and Archaeological Site Formation. *American Antiquity* 45:4–20.

Blust, R., 1976, Austronesia Culture History:Some Linguistic Inferences and Their Relations to the Archaeological Record. *World Archaeology* 8:19–43.

Blust, R., 1989, Comment on Headland and Reid (1989). *Current Anthropology* 30(1):53–54.

Bowdler, S., In press, Hunters and Traders in Northern Australia. In *Forager-Traders in South and Southeast Asia*, edited by K. Morrison and L. Junker. Cambridge University Press, Cambridge.

Brosius, P., 1988, A Separate Reality:Comments on Hoffman's *The Punan:Hunter and Gatherers of Borneo*. *Borneo Research Bulletin* 20:81–106.

Brosius, P., 1991, Foraging in Tropical Rain Forests:The Case of the Penan of Sarawak, East Malaysia (Borneo). *Human Ecology* 19:123–150.

Bui Vinh, 1998, The Stone Age Archaeology in Viet Nam:Achievements and General Model. In *Southeast Asian Archaeology 1994*, edited by P.-Y. Manguin, pp. 5–12. Centre for Southeast Asian Studies, University of Hull, Hull.

Cadelina, R., 1980, Adaptive Strategies to Deforestation:The Case of the Ata of Negros Island, Philippines. *Silliman Journal* 27(2–3):93–112.

Cherry, R., 1978, An Analysis of the Lithic Industry of Buad Island, Samar. *Philippine Quarterly of Culture and Society* 6:3–80.

Conklin, H., 1949, Preliminary Report on Fieldwork on the Islands of Mindoro and Palawan, Philippines. *American Anthropologist* 51(2):268–273.

Coutts, P., 1984, A Hunter-Gatherer/Agriculturalist Interface on Panay Island. In *Southeast Asian Archaeology at the Fifteenth Pacific Science Congress*, edited by D. Bayard, pp. 254–271. University of Otago, New Zealand.

Denbow, J. R., 1984, Prehistoric Herders and Foragers of the Kalahari:The Evidence of 1500 Years of Interaction. In *Past and Present in Hunter Gatherer Studies*, edited by C. Schrire, pp. 175–193. Academic Press, Orlando.

Dentan, R., Endicott, K., Gomes, A., and Hooker, M., 1997, *Malaysia and the Original People*. Allyn and Bacon, Boston.

Dunn, F. L., 1975, *Rain-Forest Collectors and Traders:A Study of Resource Utilization in Modern and Ancient Malaya*. Malaysian Branch of the Royal Asiatic Society Monograph No. 5. Royal Asiatic Society, Kuala Lumpur.

Early, J. D., and Headland, T., 1998, *Population Dynamics of a Philippine Rain Forest People*. University of Florida Press, Gainesville.

Eder, J., 1987, *On the Road to Tribal Extinction:Depopulation, Deculturation, and Adaptive Well-Being Among the Batak of the Philippines*. University of California Press, Berkeley.

Eder, J., 1988, Hunter-gatherer/Farmer Exchange in the Philippines:Some Implications for Ethnic Identity and Adaptive Well-Being. In *Ethnic Diversity and the Control of Natural Resources in Southeast Asia*, edited by T. Rambo, K. Gillogly, and K. Hutterer, pp. 37–57. University of Michigan Center for South and Southeast Asian Studies, Ann Arbor.

Ellen, R., 1988, Foraging, Starch Extraction and the Sedentary Lifestyle in the Lowland Rainforest of Central Seram. In *Hunters and Gatherers, Vol. 1:History, Evolution and Social Change*, edited by T. Ingold, D. Riches, and J. Woodburn, pp. 117–134. Berg Press, Oxford, England.

Endicott, K., 1988, Property, Power and Conflict Among the Batek of Malaysia. In *Hunters and Gatherers 2:Property, Power and Ideology*, edited by T. Ingold, D. Riches, and J. Woodburn, pp. 110–128. St. Martin's Press, New York.

Estioko-Griffin, A., 1985, Women As Hunters:The Case of an Eastern Cagayan Agta Group. In *The Agta of Northeastern Luzon:Recent Studies*, edited by P. B. Griffin and A. Estioko-Griffin, pp. 18–32. University of San Carlos Press, Cebu City, Philippines.

Estioko-Griffin, A., and Bion P., Griffin, 1975, The Ebuked Agta of Northeastern Luzon. *Philippine Quarterly of Culture and Society* 3:237–244.

Fix, A., 1995, Malayan Paleosociology:Implications for Patterns of Genetic Variation among the Orang Asli. *American Anthropologist* 97:313–323.

Fix, A., In press, Foragers, Farmers, and Traders in the Malayan Peninsula:Origins of Cultural and Biological Diversity. In *Forager-Traders in South and Southeast Asia*, edited by K. Morrison and L. Junker. Cambridge University Press, Cambridge.

Flenley, J., 1979, *The Equatorial Rain Forest:A Geological History*, Butterworth, London.

Fox, R., 1970, *The Tabon Caves*. National Museum Publications, Manila.

Glover, I., 1977, The Hoabinhian Hunter-Gatherers or Early Agriculturalists of Southeast Asia. In *Hunters, Gatherers and First Farmers Beyond Europe*, edited by J. Megaw, pp. 145–166. Leicester University Press, Leicester.

Gorman, C., 1971, The Hoabinhian and After:Subsistence Patterns in Southeast Asia During the Latest Pleistocene and Early Recent Periods. *World Archaeology* 2:300–320.

Griffin, P. B., 1984, Forager Resource and Land Use in the Humid Tropics:The Agta of Northeastern Luzon, the Philippines. In *Past and Present in Hunter-Gatherer Studies*, edited by C. Schrire, pp. 175–193. Academic Press, Orlando.

Griffin, P. B., 1985, Problems and Prospects. In *The Agta of Northeastern Luzon:Recent Studies*, edited by P. Bion Griffin and A. Estioko-Griffin. San Carlos, Cebu City, Philippines.

Griffin, P. B., 1989, Hunting, Farming, and Sedentism in a Rain Forest Foraging Society, In *Farmers as Hunters*, edited by S. Kent, pp. 60–70. Cambridge University Press, Cambridge.

Gullick, J., 1958, *Indigenous Political Systems of Western Malaya*. Athlone Press, London.

Gunn, M. M., 1995, The Development of Pre-Hispanic Philippine Subsistence Exchange Networks:Preliminary Results From Flotation. *Convergence* 2(1):34–38.

Gunn, M. M., 1997, *The Devleopment of Social Networks:Subsistence Production and Exchange Between the Sixth and Sixteenth Centuries* A.D. in the Tanjay Region, Negros Oriental, Philippines, Ph. D. Dissertation, University of Hawaii, Honolulu.

Hall, K., 1985, *Maritime Trade and State Development in Early Southeast Asia*. University of Hawaii Press, Honolulu.

Hall, K., 1992, Economic History of Early Southeast Asia. In *The Cambridge History of Southeast Asia, Vol. 1:From Early Times to c. 1800*, edited by N. Tarling, pp. 183–275. Cambridge University Press, Cambridge.

Hayden, B., 1981, Subsistence and Ecological Adaptations of Modern Hunter-Gatherers. In *Omnivorous Primates*, edited by R. Harding and G. Teleki, pp. 344–421. Columbia University Press, New York.

Headland, T., 1975, The Casiguran Dumagats Today and in 1936. *Philippine Quarterly of Culture and Society* 3:245–257.

Headland, T., 1987, The Wild Yam Question:How Well Could Independent Hunter-Gatherers Live in a Tropical Rainforest Environment? *Human Ecology* 15(4):463–491.

Headland, T. and L. Reid, 1989, Hunter-Gatherers and Their Neighbors from Prehistory to the Present. *Current Anthropology* 30(1):43–66.

Headland, T. and L. Reid, 1991, Holocene Foragers and Interethnic Trade:A Critique of the Myth of Isolated Independent Hunter-Gatherers. In *Between Bands and States*, edited by S. Gregg, pp. 333–340, Occasional Paper No. 9. Center for Archaeological Investigations, Southern Illinois University, Carbondale.

Higham, C., 1989, *The Archaeology of Mainland Southeast Asia*. Cambridge University Press, Cambridge.

Hirth, F., and Rockhill W. R., 1911, *Chau Ju-kua:His Work on the Chinese and Arab Trade in the Twelfth and Thirteenth Centuries, Entitled Chu-fan-chih*. Imperial Academy of Sciences, St. Petersburg, Russia.

Hoffman, C., 1984, Punan Foragers in the Trading Networks of Southeast Asia. In *Past and Present in Hunter Gatherer Studies*, edited by C. Schrire, pp. 123–150. Academic Press, New York.

Hoffman, C., 1986, *The Punan:Hunters and Gatherers of Borneo*. UMI Research Press, Ann Arbor.

Hoffman, E., 1982, The Forest Frontier in the Tropics:Pioneer Settlement and Agricultural Intensification in Negros Oriental. In *Houses Built on Scattered Poles:Prehistory and Ecology in Negros Oriental, Philippines*, edited by K. Hutterer and W. Macdonald, pp. 53–84. University of San Carlos, Cebu City, Philippines.

Hutterer, K. L., 1974, The Evolution of Philippine Lowland Societies. *Mankind* 9(4):287–299.

Hutterer, K. L., 1976, An Evolutionary Approach to the Southeast Asian Cultural Sequence. *Current Anthropology* 17:221–242.

Hutterer, K. L., 1977, Reinterpreting the Southeast Asian Paleolithic. In *Sunda and Sahul*, edited by J. Allen, J. Golson, and R. Jones, pp. 31–71. Academic Press, London.

Hutterer, K. L., 1981, Bais Anthropological Project, Phase II:A First Preliminary Report. *Philippine Quarterly of Culture and Society* 9(4):333–341.

Hutterer, K. L., 1983, The Natural and Cultural History of Southeast Asian Agriculture:Ecological and Evolutionary Aspects. *Anthropos* 78:169–212.

Hutterer, K. L., 1991, Losing Track of the Tribes:Evolutionary Sequences in Southeast Asia. In *Profiles in Cultural Evolution*, edited by A. T. Rambo and K. Gillogly, pp. 219–234. University of Michigan Museum of Anthropology Paper No. 85. University of Michigan Museum of Anthropology, Ann Arbor.

Hutterer, K. L., and Macdonald W. K., (eds.), 1982, *Houses Built on Scattered Poles:Prehistory and Ecology in Negros Oriental, Philippines*. University of San Carlos, Cebu City, Philippines.

Junker, L. L., 1990, The Organization of Intra-Regional and Long-Distance Trade in Pre-Hispanic Philippine Complex Societies. *Asian Perspectives* 29(2):167–209.

Junker, L. L., 1993, Archaeological Excavations at the Late First Millennium and Early Second Millennium A.D. Settlement of Tanjay, Negros Oriental:Household Organization, Chiefly Production, and Social Ranking. *Philippine Quarterly of Culture and Society* 21(2):146–225.

Junker, L. L., 1994, The Development of Centralized Craft Production Systems in A.D. 500–1600 Philippine Chiefdoms. *Journal of Southeast Asian Studies* 25(1):1–30.

Junker, L. L., 1996, Hunter-Gatherer Landscapes and Lowland Trade in the Pre-Hispanic Philippines. *World Archaeology* 27(2):389–410.

Junker, L. L., 1999, *Raiding, Trading, and Feasting:The Political Economy of Philippine Chiefdoms*. University of Hawaii Press, Honolulu.

Junker, L. L., In press a, Economic Specialization and Inter-Ethnic Trade Between Foragers and Farmers in the Prehispanic Philippines. In *Forager-Traders in South and Southeast Asia*, edited by K. Morrison and L. Junker. Cambridge University Press, Cambridge.

Junker, L. L., In press b, Introduction:Southeast Asia. In *Forager-Traders in South and Southeast Asia*, edited by K. Morrison and L. Junker. Cambridge University Press, Cambridge.

Junker, L. L., Mudar, K., and Schwaller, M., 1994, Social Stratification, Household Wealth and Competitive Feasting in 15th–16th Century Philippine Chiefdoms. *Research in Economic Anthropology* 15:307–358.

Junker, L., Gunn, M., and Santos, M. J., 1996, The Tanjay Archaeological Project:A Preliminary Report on the 1994 and 1995 Field Seasons. *Convergence* 2(2):30–68.

Kelly, R., 1983, Hunter-Gatherer Mobility Strategies. *Journal of Anthropological Research* 39:277–306.

Macdonald, W. K., 1982, The Bais Anthropological Project, Phase III:A Preliminary Report With Some Initial Observations. *Philippine Quarterly of Culture and Society* 10:197–210.

Miksic, J., 1984, A Comparison Between Some Long-Distance Trading Institutions of the Malacca Straits Area and of the Western Pacific. In *Southeast Asian Archaeology at the*

Fifteenth Pacific Science Congress, edited by D. Bayard, pp. 235–253. University of Otago, New Zealand.

Mudar, K., 1997, Patterns of Animal Utilization in the Holocene of the Philippines:A Comparison of Faunal Samples from Four Archaeological Sites. *Asian Perspectives* 36(1):67–105.

Omoto, R. 1985, The Negritos:Genetic Origins and Microevolution. In *Out of Asia:Peopling the Americas and the Pacific*, edited by R. Kirk and E. Szathmary, pp. 123–131. Journal of Pacific Prehistory, Canberra.

Oracion, T., 1960, The Culture of the Negritos on Negros Island. *Silliman Journal* 7:201–218.

Parry, W., 1982a, Observations on the Arrow Technology of the Negritos. In *Houses Built on Scattered Poles:Prehistory and Ecology in Negros Oriental, Philippines*, edited by K. Hutterer and W. Macdonald, pp. 107–116. University of San Carlos, Cebu City, Philippines.

Parry, W., 1982b, Stone Tools From the Bais Area, Philippines:Technology, Function, and Distribution. In *Houses Built on Scattered Poles:Prehistory and Ecology in Negros Oriental, Philippines*, edited by K. Hutterer and W. Macdonald, pp. 303–330. University of San Carlos, Cebu City, Philippines.

Peterson, W., 1974, Summary Report of Two Archaeological Sites from North-Eastern Luzon. *Archaeology and Physical Anthropology in Oceania* 9:26–35.

Rahmann, R., and Maceda, M., 1955, Notes on the Negritos of Northern Negros. *Anthropos* 50:810–836.

Rai, N., 1982, *From Forest to Field:A Study of Philippine Negritos in Transition*, Ph.D. Dissertation, University of Hawaii, Honolulu.

Rambo, A. T., 1988, Why Are the Semang? Ecology and Ethnogenesis of Aboriginal Groups in Peninsular Malaysia. In *Ethnic Diversity and the Control of Natural Resources in Southeast Asia*, edited by A. T. Rambo, K. Gillogly, and K. Hutterer, pp. 303–330. University of Michigan Center for South and Southeast Asian Studies, Ann Arbor.

Reid, L., 1987, The Early Switch Hypothesis:Linguistic Evidence for Contact Between Negritos and Austronesians. *Man and Culture in Oceania* 3:41–59.

Reynolds, T. G., 1990, The Hoabinhian:A Review. In *Bibliographic Reviews of Far Eastern Archaeology*, edited by G. Barnes, pp. 1–30. Oxbow Books, Oxford.

Ronquillo, W., 1981, *The Technology and Functional Analysis of Lithic Flake Tools from Rabel Cave, Northern Luzon, Philippines*, National Museum of the Philippines Papers No. 13. National Museum of the Philippines, Manila.

Rousseau, J., 1984, Review Article:Four Theses on the Nomads of Central Borneo. *Borneo Research Bulletin* 16:85–95.

Sandbukt, O., 1988, Tributary Tradition and Relations of Affinity and Gender among the Sumatran Kubu. In *Hunters and Gatherers, Vol. 1:History, Evolution and Social Change*, edited by T. Ingold, D. Riches, and J. Woodburn, pp. 107–116. Berg Press, Oxford, England.

Schlegel, S., 1979, *Tiruray Subsistence*. Ateneo de Manila Press, Quezon City, Philippines.

Schrire, C., 1980, An Inquiry into the Evolutionary Status and Apparent Identity of San Hunter-Gatherers. *Human Ecology* 8:9–32.

Seitz, S., 1981, Die Penan in Sarawak und Brunei:Ihre Kulturhistorische Einordnung und Derzeitige Situation. *Paideuma* 27:275–311.

Sellato, B. J. L., 1988, The Nomads of Borneo:Hoffman and "Devolution." *Borneo Research Bulletin* 20:106–120.

Shoocondej, R., 1996, Working Towards an Anthropological Perspective on Thai Prehistory:Current Research on the Post-Pleistocene. In *Indo-Pacific Prehistory:The Chiang Mai Papers, Vol. 1*, edited by P. Bellwood, pp. 119–132. Indo-Pacific Prehistory Association, Australian National University, Canberra.

Snow, B., R. Shutler, D. Nelson, J. Vogel, and J. Southon, 1986, Evidence of Early Rice Cultivation in the Philippines. *Philippine Quarterly of Culture and Society* 14:3–11.

Solheim, W. G., 1981, Philippine Prehistory. In *The People and Art of the Philippines*, edited by G. Casals, pp. 17–83. Museum of Cultural History, University of California, Los Angeles.

Spielmann, K., and Eder, J., 1994, Hunters and Farmers:Then and Now. *Annual Review of Anthropology* 23:303–323.

Spriggs, M., 2000, Can Hunter-Gatherers Live in Tropical Rain Forests? The Pleistocene Island Melanesian Evidence. In *Hunters and Gatherers in the Modern World*, edited by P. Schweitzer, M. Biesele, and R. Hitchcock, pp. 287–304. Berghahn Books, New York.

Warren, C., 1977, Palawan. In *Insular Southeast Asia:Ethnographic Section 4:The Philippines*, edited by F. Lebar, pp. 229–290. HRAF, New Haven, CT.

Wernstedt, F., and Spencer, J. E., 1967, *The Philippine Island World:A Physical, Cultural and Regional Geography*. University of California Press, Berkeley.

Wheatley, P., 1983, *Nagara and Commandery:Origins of Southeast Asian Urban Traditions*. University of Chicago Dept. of Geography Research Paper No. 207–208, Chicago.

Whitmore, T. C., 1975, *Tropical Rainforest of the Far East*. Clarendon Press, Oxford.

Wilmsen, E., 1983, The Ecology of Illusion:Anthropological Foraging in the Kalahari. *Reviews in Anthropology* 10(1):9–20.

Winzeler, R., 1986, Ecology, Culture, Social Organization and State Formation in Southeast Asia. *Current Anthropology* 17:623–640.

Wolters, O. W., 1971, *The Fall of Srivijaya in Malay History*. Cornell University Press, Ithaca, NY.

Yamada, I., 1997, *Tropical Rain Forests of Southeast Asia:A Forest Ecologist's View*. University of Hawaii Press, Honolulu.

Zaide, G. (ed. and trans.), 1990, *Documentary Sources of Philippine History, Vol. 1*. National Bookstore, Manila.

Chapter 12

Explaining Changes in Settlement Dynamics across Transformations of Modes of Production

From Hunting to Herding in the South-Central Andes

MARK ALDENDERFER

INTRODUCTION

I think it no exaggeration to say that Lewis Binford's (1980) characterization of hunter-gatherer mobility strategies has become foundational to modern archaeological practice. In one form or another, Binford's formulation of the forager–collector continuum is taught in introductory archaeology classes, described in textbooks (see Fagan 1995: 88–90), used in countless professional discussions about hunter-gatherer adaptations, and debated at the level of theory by others (see Bettinger 1991; Kelly 1995). The model has been modified and refined through these discussions and examinations of case studies, both ethnographic as well as archaeological.

MARK ALDENDERFER • Department of Anthropology, University of California, Santa Barbara, California 93106.

This process of refinement and modification is important because it serves one of the central tenets of a scientific approach to archaeology, which is to define the explanatory adequacy of a model as it confronts empirical data and also its limits as to what it *should* be expected to explain. In other words, models and theories have domains of relevance.

My sense of the domain of the original formulation of the forager–collector continuum is the following: it deals primarily with terrestrial foragers who live in strongly seasonal environments that significantly affect resource predictability, availability, and abundance; these foragers have relatively simple social and political systems; both local and regional-scale population densities are low; and environment, within the parameters defined above, is relatively stable. That these conditions describe much of human history helps to explain why the forager–collector model has achieved such widespread currency. As long as these conditions are very generally met, the modeling exercise becomes one of identifying the ways in which local variation in resource configuration promotes one form of mobility strategy over another.

When we encounter empirical situations that begin to differ substantially from these conditions, the forager–collector model begins to run into trouble—that is, it begins to exceed its domain. But this is precisely what happens in many historical trajectories across the world: populations do grow in size and density, and social and political systems become considerably more complex. Further, these changes sometimes take place within a context of environmental change. Still other trajectories reflect changes in mode of production—plant cultivation, animal husbandry, or some combination of them replaces a foraging subsistence adaptation. These changes are not simply changes in model conditions but are changes in model character. Although structural elements of the model, such as definitions of residential or logistical mobility, may continue to be useful, the causal linkages that create a basis for a mobility strategy are substantially altered.

In fairness to the forager–collector model, it was never meant to be a universal, one-size-fits-all explanation of social change. It served best as a way to describe variability in a settlement system within the context of the foraging mode of production. In effect, it was a static model, not one that sought to explain changes across these modes. But as I noted earlier, it proved possible to modify the model to provide it with a capacity to explain the causal dynamics of change in locational choice while also expanding its domain and scope. The packing model, proposed by Price and Brown (1985) and modified by many others too numerous to mention here, is one such attempt to modify Binford's original model. The packing model defines aspects of locational choice and preference, and it creates a causal linkage between population growth, levels of mobility, and modifications to subsistence practice, such as widening niche breadth, focusing subsistence upon r-selected species, and storage. It also provides a context for examining social responses to these conditions, such as the concept of "social risk" by Brown (1985). Other synthetic models have been proposed, most derived from some variant of optimal foraging theory or evolutionary ecology, such as Bettinger's (1991) model of processors and travelers and my own model of montane, or high mountain, foraging (Aldenderfer 1998). Both models integrate factors of social choice within their structures.

But even these models have a hard time explaining transformations in modes of production. Although I was able to characterize settlement mobility in the Awati Phase (ca. 4400 to 3600 B.P.) in the Rio Asana valley in the western flanks of the Andes as primarily residential and fixed between base camps in two adjacent elevational zones, I was also hard pressed to explain how a herding economy came into being in what appeared to be a relatively short time frame (Aldenderfer 1998: 261–275).

In this chapter, then, I try to show how aspects of the original forager–collector model can be modified to account for changes in mode of production as observed through long-term historical trajectories. Following a brief definition of the mode of production concept, I then describe settlement and other data from three areas in the south-central Andes that demonstrate a transition from hunting to herding, Finally, I examine four "categories" of model structure: history, agency, contingency, and cultural logic, and explain how each of them can be used to build a more comprehensive model of settlement dynamics and locational choice within a context of increasing social complexity and the transformation of modes of production for each of these case studies.

MODES OF PRODUCTION: HUNTERS AND HERDERS

As with many terms in anthropology, the term "mode of production" has multiple, often conflicting definitions. The term originated with Karl Marx and has developed a strongly political interpretation consistent with the Marxian analysis of social relations. In this instance, I want to avoid its political connotations and instead focus on its key component parts, the *forces of production*, which refers to the resources, instruments, and labor committed to subsistence practice, and the *relations of production*, which describes the ways in which what is produced is distributed among the members of the group. As some authors have noted, these terms can be exquisitely vague and mean virtually anything. I intend to use them as heuristics, however, and will contrast what has been described as a foraging mode of production with a herding mode. I cannot stress too forcefully that these modes are not meant to serve as essentialist "stages" of cultural evolution or some kind of reified cultural type but are instead opposite locations on a continuum of variation in the ways in which foraging peoples differ from herders and in the ways in which land is perceived, labor allocated, and status defined. By contrasting these heuristics, we are in a better position to identify a full range of potential causal factors that may be implicated in a transition from hunting to herding and ultimately to expand the explanatory domain of the original forager–collector model.

The foraging mode of production, as defined by Leacock and Lee (1982: 8–9), has the following features:

1. Collective ownership of the means of production—that is, land and the resources on them are held in common by a social entity conveniently, but not satisfactorily, labeled the "band".
2. Rights of reciprocal access to the resources of others through marriage and other social ties.
3. Little emphasis on accumulation. Food storage may exist, but it tends to be limited to what is required by seasonal demands and is not perceived or tolerated as an end in itself.
4. Generalized reciprocity within the camp; this translates to widespread food sharing within the camp, but usually of a restricted range of resources.
5. Full equal access to the forces of production within a context of a sexual division of labor.
6. Individual ownership of the tools of production; note that the principle of ownership does not extend beyond the category of implements.

It should be apparent that this set of features has very clear overlap and compatibility with the explanatory domain of the forager–collector model as I outlined it earlier.

A mode of production for herders must be cobbled together from multiple sources (Ingold 1976; Paine 1971; Smith 1990). I focus my attention on carnivorous pastoralism [as opposed to milk pastoralism; see Ingold (1976: 192)].

1. Individuals, families (broadly defined), or some corporate group has ownership of the means of production. Land (i.e., pasturage) is held in common within this group, and individuals have use rights

to it. Animals are owned as property by families or individuals. These rights of ownership are exclusive within the group.

2. Rights of access to resources are highly limited and tend to be allocated through inheritance, sibling cooperation, and marriage. Thus cooperation tends to be channeled within these groups.
3. There is substantial emphasis on accumulation, which in this context is defined by the accumulation of more animals. Having a large herd is a cultural goal of most herders; prestige and status are often defined by having larger herds than one's fellows. The accumulation of good pasture or an expansion of one's right to use pasture is another form of accumulation.
4. Relationships beyond the family or corporate group tend to be based on balanced reciprocity, and animals are used in many pastoral groups as a basis for creating social debt or obligation but generally within the constraints described in (2). Food sharing beyond the domestic unit is uncommon—that is, rights of access to pastoral products lie within the domestic sphere and are not generally shared beyond the household.
5. A sexual division of labor is common in pastoral societies, but the role of women's labor stands in sharp contrast to that of women in foraging societies. In pastoral groups, women have limited rights to dispose of the products of pastoral production, which tend to be controlled by men. Though women's labor is important to societal reproduction, the status of women is lower than that in foraging groups.
6. The tools of production are owned by individuals, as are the forces of production—animals and land.

Clearly, there are significant differences in the ways in which land and the resources found on it are conceptualized when foragers are compared with herders. Herders own land, whereas foragers (adhering to the original domain of the forager–collector model) do not. Moreover, productive resources are owned by individuals and families in herding societies, whereas in foraging societies they are not. Accumulation is a further property, indeed, a necessity, of herding because even if the resource base is supplemented by exchange and hunting, it is predicated upon the destruction of herd numbers for consumption, and these must be replaced. This kind of accumulation is antithetical to the ethos of foraging peoples, who seek to level aggrandizers and hoarders through various mechanisms.

It is important to stress that any transition from foraging to herding in all likelihood does not happen through some sort of catastrophic replacement of the forces and relations of one mode of production with those of another. But it is also unclear how rapid or gradual these social transformations are and under what conditions—environmental, social, and ideological—the pace of change might be retarded or hastened.

THREE CASE STUDIES OF TRANSITIONS FROM FORAGING TO HERDING

The following case studies are drawn from my own research and that of my students in the south-central Andes: settlement along the Rio Asana in the western flanks of the Andes (Fig. 12.1; Aldenderfer 1998; Kuznar

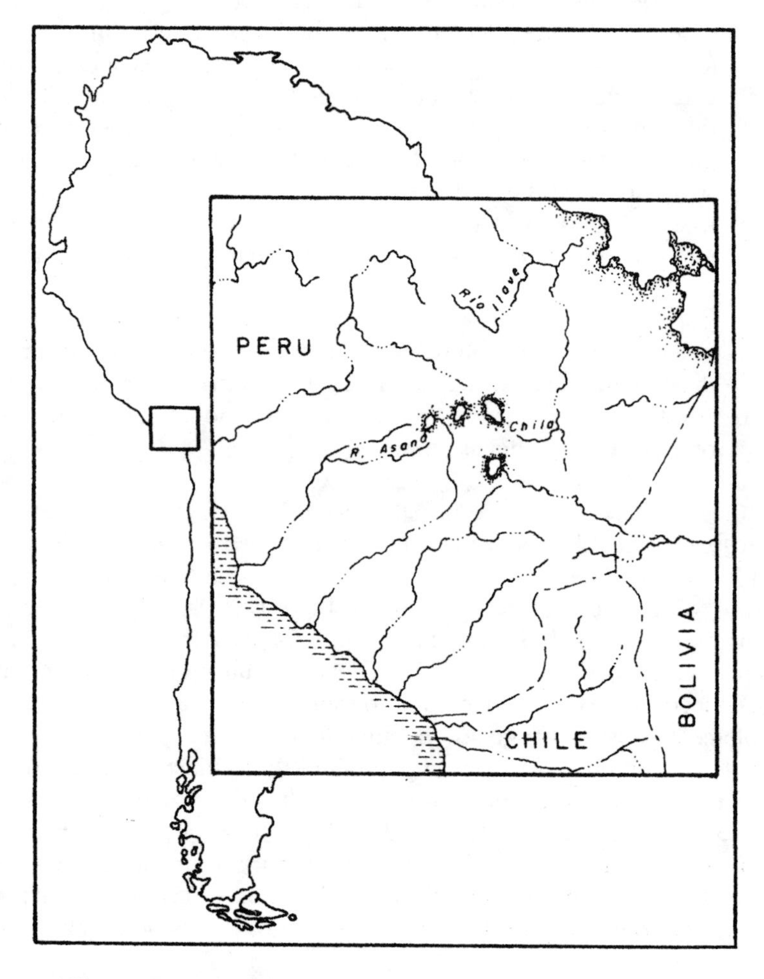

Figure 12.1. General location of research areas described in the text.

1990), the Rio Chila in the far southern fringes of the Lake Titicaca Basin (Aldenderfer in press; Klink 1999, in press; Kuznar 1989), and the Rio Ilave, one of the major tributary streams of Lake Titicaca (Aldenderfer 1997; Aldenderfer and Klink 1996).

In each of these three areas, there is a clear transition from a foraging lifeway to one based on carnivorous pastoralism (Asana and Chila) or some form of agropastoralism (Ilave). However, in each instance, there are apparently important and critical differences in the way in which this transformation takes place, and as I shall argue, though the forager–collector model is a useful descriptive tool, it offers little insight into the dynamics of each transition. Some of my arguments have been anticipated by Hesse (1982), especially those that deal with the social transformations implied by the shift from hunting to herding.

The Rio Asana Transition

At ca. 4600 B.P. (Qhuna Phase), the foragers who lived at Asana were residentially mobile and most likely were near sedentary. In contrast to previous phases, there is no evidence for the logistical (or residential) use of adjacent lower and high elevation ecological zones. The valley had no known logistical camps, and the stray finds of projectile points found up and down the valley suggest significant and intensive day use of a foraging radius of 15–20 km. Subsistence was based primarily upon hunting deer and camelids and intensive exploitation of a local form of chenopodium. There is no evidence of animal pens or corrals, and the faunal evidence reflects local consumption of animals and little exchange with adjacent groups. Residential structures were larger than in preceding phases and were grouped around a single public structure that was apparently used as either a dance floor, public space, or near the end of the occupation, as a probable focus of intensive, restricted worship.

By 4400 B.P. (Awati Phase), settlement in the valley was markedly different and was anchored by two residential bases—Asana and El Panteon, found some 17 km apart in distinct ecological zones. New logistical camps also appear on the landscape, as do field camps and stray finds. In contrast to the Qhuna Phase, plant use became far less important to subsistence, and it was replaced by a more intensive utilization of camelids—herding. Corrals are found at both Asana and El Panteon, and faunal indicators suggest a combination of local consumption of herds as well as trade and exchange with people living in ecological zones at lower elevations. No public structures are present at any site.

The transition to herding is quite abrupt as measured in archaeological time. The final cultural level of the Qhuna phase at Asana, CL V, is dated at 4580 B.P., whereas the first level of the Awati phase, CL IV, dates at

4340 B.P. The dates overlap substantially at two standard deviations, and are thus statistically indistinguishable. Putting this aside for the sake of discussion, the transition could have taken as little as 20 years and as much as 500. Importantly, there is no stratigraphic discontinuity between the two levels; no discernible sterile layer separates them, and neither layer is very thick. This situation, however, is complicated by the destruction of a portion of CL V by a massive landslide during or immediately after its use.

The Rio Chila Transition

Based on extensive excavation at the rock shelter called Quelcatani, a survey in its environs, and an additional survey in the nearby Rio Huenque drainage, from ca. 4500 to 4000 B.P. (the Late Archaic), the human use of the landscape of this harsh, high puna environment was characterized by a modified logistical mobility with an unknown frequency of residential moves. At this time, Quelcatani is most probably a short-term residential site or, more probably, a logistical camp. Mobility is confined to the high puna, but its scale is unknown. The faunal materials present indicate that hunting deer and camelids was important, although a wide variety of plant remains recovered in Terminal Archaic levels (ca. 4000–3600 B.P.) suggests a broad spectrum of plant use that supplemented the diet. There is no indication that camelids were herded; no animal control features like corrals or small pens for neonates or small camelids have been encountered in these levels at Quelcatani.

By at least 3700 B.P. and possibly earlier, Quelcatani becomes a residential base. Substantial structures are built within the rock shelter, ceramics appear for the first time in the cultural sequence, and plant use is more clearly focused on the use of probably domesticated chenopodium, which cannot be grown at the elevation of the shelter (4420 m), and therefore had to be obtained either through exchange or cultivated by moving part of the residential group to a lower elevation. However, the absence of high utility parts of camelids from the faunal assemblage suggests, that trade, not cultivation, was the most likely way in which the inhabitants of Quelcatani obtained these plants. Settlement patterns in the region also support the inference of decreased logistical mobility.

Once again, it appears as if the transition from hunting to herding is fairly abrupt, although not as clear-cut as the situation at Asana. Level WXXIV marks the appearance of the residential use of Quelcatani; in level WXXV, the pattern is clearly logistical. Although we have a secure radiocarbon assay of 3660 B.P. in WXXIV we have no date from WXXV. However, like Asana, there is no discernible sterile layer that separates them nor is there a stratigraphic discontinuity.

The Rio Ilave Transition

Although the excavation of key sites is still in progress, we have enough preliminary data available at least to outline the transition from hunting to herding in this area. During the Late Archaic (ca. 4500–4000 B.P.), land use is dominated by logistical mobility, although its frequency declines from the preceding Middle Archaic (ca. 8000–6000 B.P.). Moreover the proportion of large sites, here interpreted as residential bases, increases from the Middle to Late Archaic. Excavations at one of these, Jiskairumoko, shows a fairly large aggregation of small, circular residential pit-house structures and is probably a residential base. Subsistence reflects the hunting of deer and the use of camelids, although it is unclear if the latter are herded at this time. A large corral has been found at the margin of the site, but its date is currently unknown. Plant use is focused on chenopodium, and the preliminary analysis of charred seeds shows that both wild and transitional (i.e., thinning seed testa) plants are present. There is no known public architecture at the site that dates to the Late Archaic, but there is a small area whose ceremonial features are similar to those found at Qhuna Phase Asana.

In the following Terminal Archaic, which begins between 3800 and 4000 years ago, there is a significant change in settlement patterns in the region. The total number of sites decreases, but the proportion of large, presumably residential sites, increases compared to the Late Archaic. Moreover, the mean size of these large sites also increases, suggesting a process of population growth and aggregation. At Jiskairumoko, the aggregation of domestic structures continues, but the houses themselves are much larger, and now, rectangular in shape. The number of grinding tools associated with these structures increases as well compared to the Late Archaic, and the preliminary paleoethnobotanical data suggest that domesticated chenopods (i.e., those with thin testa seeds) are an important part of the diet. Faunal data, however, are unclear and show a continuation of hunting deer along with the use of camelids. Frequencies of faunal remains in Terminal Archaic levels are relatively low compared to the Late Archaic levels at the site or at other sites of similar date, such as Asana or Quelcatani. Overall, the impression is one of increasing reliance on plant foods. If herding is present, it is best characterized as an agropastoral, rather than a pastoral, adaptation (Aldenderfer 2001).

Just how rapid the transition was from a Late Archaic baseline to a Terminal Archaic agropastoralism is unknown. Once again, however, there are no obvious stratigraphic discontinuities at Jiskairumoko, where the final Late Archaic levels are immediately superior to and without intervening sterile levels below those of the Terminal Archaic. These data suggest at least the possibility of a relatively rapid transformation of modes of production.

Settlement patterns themselves, however, change little between the two periods.

In summary, solely the settlement data from the three areas show the following patterns as expressed in terms of the original forager–collector model:

> Rio Asana: Residential or near-sedentary mobility to mixed residential and logistical mobility; the time frame of the transition is relatively short.
> Rio Chila: Modified logistical mobility to a similar, but probably more residential mobility; time frame of the transition is short.
> Rio Ilave: Logistical mobility to a near-sedentary or residential mobility; the time frame of the transition is unknown.

Though the model helps to characterize the nature of mobility across the transition in a heuristic sense in these three cases, it cannot easily capture the variability present in each of the sequences. For instance, the Chila transition is described as one of a pattern of "modified" logistical mobility to a "more" residential mobility. "Modified" and "more" here are necessary weasel words. They are used to capture a sense of degree that is not present in the original model. In a sense, then, they are used in a crude, nonsystematic way to extend the precision of the original model.

But the terms, even in their modified forms, still lack something, and that is an explanation of the dynamic of the change itself. This is especially apparent when the possible duration of the change is considered. At least two of these changes take place over short time frames. Although the model certainly captures the "what" of the sites—a sense of their individual function in descriptive or mechanical terms, it fails rather badly to explain how the "place" (*sensu* Binford 1982) of each site is or is not modified across the transition. Put another way, the model does not easily deal with changes in the *state* of a settlement system, especially across modes of production, when the pace of change is relatively rapid or abrupt.

STRUCTURAL MODIFICATIONS OF THE ORIGINAL FORAGER–COLLECTOR MODEL

If we wish to improve and expand the explanatory domain of the forager–collector model, we need to consider adding to it the following four structural features: history, agency, contingency, and cultural logic. In one sense, they are similar to the "conditions" defined by Price and Brown (1985: 9) in their model of the emergence of complexity in foraging societies. In a real sense, however, they are a combination of conditions, potential causes, and model components. They are factors that must be considered when attempting to explain changes in settlement location across transformations of modes of production. They are also relatively open in content in that they are not derived from a single or specific body of theory. Although the choice of terms might be informed by a critique of processual archaeology, in no way are they incompatible with a scientific and empirically driven approach to archaeology. Two of the conditions—agency and cultural logic—are broadly compatible with the increasing emphasis upon methodological individualism, or agency, seen in a number of theoretical frameworks current in the field, whereas history and contingency have been long been part of archaeological thinking, albeit not strongly emphasized within some approaches to theory, such as systems thinking, neofunctionalism, or even a very doctrinaire selectionist perspective on the past. As an example of their theoretical flexibility, note their similarity to Ortner's (1989: 11–18 and 193–202) discussion of the four major features of modern practice theory: practice itself, structure, actors, and history.

The degree to which they are accessible archaeologically will vary, however, and it must be remembered that the logical status of each condition is based on a body of assumptions that must be questioned and evaluated within the context of a particular regional developmental trajectory. Even though some of these conditions might be relatively invisible from an archaeological standpoint, each may have been causal, and therefore their potential effects must be considered. If nothing else, a healthy awareness of the way they may have contributed to or conditioned a transition can help to stem a rush to premature closure in thinking about what and how things happened in the past. I will return to this problem later.

History

In each of the three case studies, there is a significant probability that each transformation from hunting to herding is not the outcome of an *in situ* evolution but instead a replacement of one mode of production with another or the adoption of the mode of production by local or indigenous inhabitants. The idea that camelid domestication evolved in a single region in the Andes and spread across the altiplano has had a long history (see Custred 1979; Jensen and Kautz 1974), but increasingly, empirical data from across the region suggest that at least some of the observed transitions from hunting to herding are likely to reflect *in situ* developments.

In these three case studies, replacement, rather than local development of herding, is an attractive hypothesis because if valid, it would be a parsimonious explanation of the relative abruptness of the transition at Asana and Quelcatani. And as I will discuss later, the idea of replacement, especially within the context of transformations across modes of production, would better explain the transformation of *values* implied by changes in mode of production. Here, I mean values to refer to the relations of production, especially the breaking of the ethos against accumulation, the cessation of widespread, extrahousehold sharing of meat resources, and the creation of private ownership of land and the resources on it. Such relations are highly conservative within society and do not change easily.

Problems abound with invoking replacement as a cause in these cases. Despite our need to explain a local trajectory, the social group that does the replacing has to come from somewhere, thus amplifying the scope of the problem to a regional scale. So far, unless we declare a central Andean diffusion locus as the "source," there are no obvious candidate populations known today. From a temporal perspective, the Asana transition could be a source for the other two, but Asana is one of the abrupt transitions that is unlikely to be explained adequately as an *in situ* process, especially considering the complex and preceding Qhuna Phase occupation of the valley.

Further, does replacement mean assimilation, outright displacement of existing landholders, or a kind of coexistence with local peoples? This question has certainly been one of signal importance in defining the transition from Mesolithic foragers to Neolithic farmers and herders all across Europe (see Gebauer and Price 1992; Price and Gebauer 1995; Zvelebil and Lillie 2000). These social relationships are extremely difficult to define empirically, and much depends on the local circumstances of indicators of group identity, the quality of chronological control, and the thoroughness of a regional-scale treatment of the problem. Not surprisingly, the issue has not arisen in the Andes aside from early discussions of "diffusion" precisely because few of these data-related conditions are met. Therefore, history in the broad sense, as discussed here, remains a viable alternative hypothesis to that of *in situ* development of herding in each of these case studies.

Agency

This term, like many in archaeology, has such a multiplicity of meanings that it borders on the useless. Here, I refer to agency as a more nuanced perspective on the way people make decisions or take courses of action within a primarily social context. In the original packing model, for instance, foragers, it was assumed, respond to an increasing number of neighbors through avoidance, territory formation, and boundary maintenance. This in turn promoted the formation of sedentary communities and thus a reduced scale and frequency of any kind of mobility. This model may capture aspects of the transition, but it is fairly coarse-grained, and it does not take into account alternative responses to these conditions. Further, there is an inevitability to the transition that may not apply in many local settings. In effect, then, the packing model asserts that decision-making under these conditions necessarily follows a single path. In part, this derives from the predominantly ecological domain of the model and the implicit optimization principles that structure it. But as I will discuss later, decisions are made in broader social contexts that may be at odds with principles of short-term economic utility.

In the Andes, the importance of maintaining strong social bonds with neighbors in adjacent ecological zones has long been recognized as one way in which these peoples manage their exposure to risk. Known as "complementarity" in the region, it takes on a number of forms that depend on the level of sociopolitical complexity of the peoples in question (Aldenderfer 1989, 1998; Salomon 1985) and range from the exchange of subsistence goods, the presentation of tokens of social ties, the establishment of marriage bonds, the ownership or rights of usufruct of land in adjacent zones, or in the most complex settings, the establishment of colonies or satellite communities in these places. Although these relationships can be described by using an ecological metaphor and modeling strategy (see Spielmann 1986; Aldenderfer 1998: 18–25), the social roles of these risk management features must not be discounted and may, in their own fashion, be causal, especially in circumstances of low overall population size and density (Aldenderfer 1998: 299–307). Admittedly, it is hard to identify intent in the past, and therefore, it may only be feasible to stick with the ecological narrative as our primary explanation of change while understanding fully that other bases for decision may have affected the direction of that change.

There is some suggestion that in at least one of these transitions—that of Asana—there may have been a set of so-called social responses directed at risk amelioration that was not fully "rational" in economic terms. Qhuna Phase settlement—preherding foragers—was characterized by near-sedentism in a relatively unproductive and very risky (in rainfall amount and frequency) ecological zone, the high sierra. There is strong evidence of a more intensive use of plants, a local chenopod, and complementarity is defined by the exchange of parts of hunted animals for other undefined goods because mobility does not appear to have been a means for ameliorating risk. However, neither of these strategies was sustainable unless fundamental transformations of the plants (through selective breeding or coevolution) and animals (through herding) were undertaken, and there is no empirical evidence (no change in seed sizes or morphology and no corrals or animal control features) for either. Instead, there is an intensification of ritual and ceremonial activity through time, and I have argued that this was a response to risk, albeit one that was not successful over the long term. Intensive ceremonialism and ritual are often used as responses to social stress (Aldenderfer 1993) and within a cultural context, may be seen as an appropriate way to deal with these crises.

This conception of agency within this context, then, supports a replacement explanation of the transition from foraging to herding at Asana because it provides grounds for a plausible "collapse" scenario of Qhuna Phase settlement dynamics.

Contingency

This condition resembles history, in that it refers to events or processes that occur during some developmental trajectory. These events, however, are not predictable, and in a sense, may be seen as "random." This does not mean that there is no material cause, but that the cause is unknown, or even unknowable, to those whom it affects. My sense of contingency here lies primarily within an environmental context and thus could refer to events such as an earthquake, extended drought, or some other similar process. They need not be abrupt, however, to be contingent. I would not include what might be considered as "normal" uncertainty or unpredictability of resource abundance or availability because these tend to be dealt with by anticipatory planning and management of risk.

Multiple contingent events appear to have affected at least two of the transitions to herding in these case studies. One of the most dramatic is the powerful landslide that occurred during the CL V occupation (Qhuna Phase) at Asana. This slide destroyed a large part of the site, including much of the ceremonial architecture that had become increasingly formalized at that time. Although it is impossible to tell if this slide "ended" the Qhuna Phase lifeway in the valley, the abrupt transition to a herding lifeway strongly suggests this possibility.

A different sort of contingent event affected the transition to herding in the Rio Ilave. Prior to 5000 B.P., conditions in the basin were characterized by extreme aridity, and sources of permanent, potable water, such as *bofedales*, springs, and the rivers themselves would have been significant resource pulls. Interior rivers, such as the Ilave, would have been deeply channeled, and their superior terraces (T3 and T4) would have formed before 8000 B.P. (Aldenderfer and Klink 1996: 15–17; Rigsby et al. 2000). Stream discharge rates would have been low, and seasonal inundation of lower terraces and floodplains would have been limited. Low rainfall, plus little floodplain transformation due to annual floods, would have led to limited, low-density and low-abundance plant growth, especially in the weedy species economically valuable to humans. Although animals would have been tethered to stream courses, their territories, especially those for the guanaco, would have been relatively large [because they are the modern sierra on the western flanks of the Andes (Aldenderfer 1998: 45)], and depending on the severity of aridity, may have led to some long-distance seasonal movement by this species. Encounter rates for both animals and plants, then, would have been lower than in the modern era.

Based on multiple lines of evidence, consensus is beginning to emerge that climate became wetter and probably more predictable some time around 3600 B.P. (Abbott et al. 1997a,b; Baker et al. 2001a,b; Seltzer et al. 1998). Water levels in both the northern and southern basins of Lake Titicaca begin to rise after this date, suggesting significantly increased rainfall. The lake level rise was apparently quite rapid; levels may have risen some 15 m in a span of 100 to 200 years. An alternative model (Wirrmann and Mourguiart 1995) suggests a slightly earlier timing to the lake level rise at 3800 B.P., and a slower increase in level. Whichever of these models is correct, however, it is clear that settlement mobility and sedentation, aggregation, and resource intensification in the Ilave basin occurred during an improving climatic context. It is likely that climatic improvement led to increases in animal numbers by creating more resource patches for them and also led to a decrease in territory size. Both of these outcomes would have improved the foraging efficiency of these species, especially for the guanaco, but to a lesser extent for the vicuña (sites of all periods on the valley floors are still distant from prime vicuña habitat) and the taruca (a cryptic species found on the high slopes surrounding the valley floor). Wetter conditions would have promoted increases in plant density, as well as changes in river bottom dynamics. The rise in lake level would have promoted stream aggradation, thus leading to the development of either meanders or braided stream segments (Aldenderfer and Klink 1996: 18). New terraces (the T2 series in the Ilave basin) may also have been created. Increased rainfall would have promoted seasonal inundation of these new terraces, and in turn, would have created ideal conditions for the spread of weedy species such as chenopods and tubers, much like the North American context of the floodplain weed theory described by Smith (1995: 194–96).

Although plant resource density may have increased over time, this would not in itself increase their foraging efficiency, especially for the chenopods and tubers. The wild ancestors of the chenopods are multistalk plants that are very difficult to harvest (Kuznar 1993; Aldenderfer 1998: Table 9.3), and given this, simply having more does not make the plant more efficient in resource ranking. The foraging efficiency of chenopods could have been improved by the appearance of single-stalk plants in combination with their cultivation near residential bases or their appearance in camelids corrals. Although similar studies on the foraging efficiency of tubers have not been done to my knowledge, the same situation obtains: their efficiency could not be increased until technologies were developed to detoxify these plants. Therefore, though there may have been "more" plants in larger patches as climate ameliorated, more did not automatically or necessarily translate to "more efficient." There is no question, however, that if this scenario is true, encounter rates for these and other weedy species would have increased. In fact, there is an apparent shift in site location between the Late and Terminal Archaic: in the former period, 14% of the known sites are located near the lower (T3 and T4) river terraces, whereas in the Terminal Archaic, this proportion increases to more than 25%. This seems consistent with the hypothesis that river valley resources increased in density and probably importance to human diet during the Late to Terminal Archaic transition. Further, if the ca. 3600 B.P. date of the rise of Lake Titicaca is accurate, we can hypothesize that plants became important to the diet very rapidly, and this further suggests that the plants themselves became more efficient through human intervention (selection for single-stalk chenopods, for example) or that new technologies (improved storage, processing, or food preparation techniques, for example; see Winterhalder and Goland 1997: Figure 7.4) improved their relative rankings.

In this case, a new habitat for both plants and animals was created across a relatively short period and thus offered the people in the region new opportunities for resource procurement. What brings this scenario into question, however, is the possibility that river valley plant densities improved significantly earlier than 3600–3900 B.P. Wirrmann and Mourguiart (1995) hypothesized a lake level increase driven by increased rainfall around 6000 B.P.; this could have led to river aggradation, seasonal flooding, and thus to increases in river valley plant densities, although most likely on a scale below that which occurred later. This process may also have created the lower (T2 series) of river terraces, which would have been the loci of plant growth. If correct, this would imply that although plant densities increased, encounter rates remained sufficiently low to slow or delay the transformation of their foraging efficiency. Thus, these plants might have entered the diet of pre-5000 B.P. foragers, but they did not have a significant short-term effect on settlement location and adaptive strategies. This in turn suggests that that the use of these plants intensified gradually during more than 2000 years, rather than across a span of just under 200 years. Though there may have been no short-term effect, the creation of these new resource patches would have formed the basis for long-term population growth which in turn would have led to the intensive use of chenopods by Late Archaic times.

To this point, I have discussed animal and plant intensification and use as separate processes when there is a significant possibility that the intensification toward domestication of chenopods was mutualistic with the herding toward domestication of camelids. Here, herding means the care and protection of camelids, most probably guanaco, and the selective management of these herds for meat. Whether or not these camelids were fully domesticated in these early herds is not likely to be directly measured, given the problematic nature of the skeletal morphology of Andean camelids, but if herding was present, the trend in faunal usage should be clear. Kuznar (1993:262–263) presents an interesting argument that suggests that the known timing of chenopod domestication across the Andes occurred between 4000 to 3000 years ago (Pearsall 1980: 198; 1989a:330), and that it is either contemporaneous with or follows, slightly in time, the domestication of the llama. The argument is simple: because wild chenopods are a preferred food of camelids, corralling these animals would have led to the appearance of higher than natural concentrations of chenopods in these corrals. This model, then, suggests that the first resources to be intensified in the basin were probably camelids, then very quickly chenopods. However, assigning temporal priority using standard dating methods to either camelids or chenopods will be almost impossible, and they will in effect appear to be simultaneous. Note, however, that Kuznar's model does not include consideration of other means by which the foraging efficiency of chenopods could be increased through environmental change, such as by creating extensive river bottom or T2 terrace patches. Thus if T2 terraces post date 6000 B.P., chenopod densities may have improved sufficiently to permit them to enter the diet, thus leading to their intensification, but across a longer period. In this case, camelids would enter slightly later, but should be herded by ca. 4500–4000 B.P. under the mutualism model.

Cultural Logic

This condition is perhaps the most problematic of those I have discussed in this chapter. It is certainly prone to misuse and misunderstanding. I mean cultural logic to refer to the ways in which people conceptualize their world and the way that conceptualization affects the ways in which they decide how to act. Though certainly about "mind" and "belief," it also is about the material consequences of action based on those beliefs. I make no brief here for those who think we can capture mind, at least in this sense, from the archeological record. Here, cultural logic refers to our informed assumptions about the way people conceptualize the world.

Archaeologists examine the past through a whole series of assumptions, and one of the most powerful has been that of humans as rational decision-makers. In both archaeology and the ethnographic analysis of many hunter-gatherers, this assumption has been implemented by using models derived from evolutionary ecology, which are often labeled broadly as optimal foraging theory. In these models, humans, it is assumed, act rationally and make choices about diet, settlement location, and mobility in a cost–benefit framework. There is no question that an optimality principle lies at the core of the original forager–collector model. In archaeology, these models have a powerful heuristic value because they can be used to create a set of expectations about what an "optimal" settlement dynamic would resemble.

Our problems begin when the archaeological cases fail to conform to model predictions, a not uncommon occurrence. We can blame the data (usually not enough of it for a compelling test), the model or some feature of it (use of an inappropriate currency), or the influence of extrinsic factors not directly modeled (model structure not up to the complexity of the situation). We seldom challenge one of the key theoretical assumptions behind the model; that of rationality defined in the purely economic sense of optimality.

To explore this idea more fully, I turn to a number of recent ethnographies of foraging peoples, most prominently Brightman (1993) and his description of Rock Cree human–animal relationships. Among the issues he discusses is the Cree cultural logic of hunger and scarcity. Fur traders and others who encountered the Cree during the eighteenth and early nineteenth centuries frequently remarked on what they considered as irrational behavior: minimal food storage for the harsh winter months, a propensity to eat all surplus available before going out to hunt in the winter, and the relative frequency of so-called sacred feasts despite winter food shortages. As the traders noted, the Cree were aware of the real possibility of starvation and death as consequences of their behavior but did not change their behavior accordingly (i.e., act frugally, store more, etc.).

Brightman explains this apparent lack of rationality by reference to the Cree perception of resources and how they become available to humans. Animals are given to hunters by the *ahcak* beings—the game rulers—when they need them, and further, animals are infinitely available as long as the hunter remains in a "right" relationship with game rulers (Brightman 1993: 286–291). Instead of rationally limiting kills or engaging in extensive food storage, the cultural logic of Cree hunting emphasized indiscriminate hunting and the likelihood of provisioning at any time of year. This kind of hunter–prey relationship was apparently widespread in many hunting groups (Paine 1971: 161–64).

If these examples are valid, it is clear that "nonrational" factors can condition even the most basic features of foraging life. But how does this help us to understand the hunting to herding transition, or more broadly, to improve the explanatory domain of the forager–collector model? At the least, it can help to critique models of the process that are *naively* constructed and that rely too much on a possibly untenable rationality assumption. Consider one of the earliest models of camelid domestication (Wheeler Pires-Ferreira et al. 1976), which outlines four broad stages of the domestication process:

specialized hunting of puna ungulates, including camelids;
specialized hunting of guanaco and vicuña;
semidomestication of an alpaca-like animal (derived from vicuña); and
fully domesticated camelids (alpacas and llama-like animals).

Although this model is best seen as a description of stages, its underlying dynamic is that as humans became more familiar with the puna environment over time, they focused their subsistence efforts on camelids because of their relatively great abundance and high spatial predictability—in other words, they are the basis of an optimal diet in this environment. Once the process begins, there is an inevitability to it. The model does not deal directly with settlement mobility, but it is compatible with a number of possible strategies, depending on our assumptions about their cultural logic. Hunters may have been residentially mobile if a Cree-like ethos existed, because these camelid species live in small, easily located family troops, and could thus have been quickly decimated. Wheeler and her colleagues imply that this kind of residential mobility is characteristic of the central Andean puna (Lavallee 1985). If, however, a rationality model is assumed, hunters may have been excellent conservationists and could have deliberately avoided overhunting. In this case, residential mobility could have been much lower. Rick (1980) has argued for something like this pattern on the Junin puna.

In my three south-central Andean case studies, the role of a cultural logic is equivocal. I have argued for the Asana transition that throughout the Archaic, hunters were likely to have been good managers and restricted the numbers of animals taken, especially during the Qhuna Phase, when they were essentially sedentary (Aldenderfer 1998: 286–287). This does not, however, help to explain the transition to herding in the subsequent Awati Phase. I think that a consideration of cultural logic has a important role to play in understanding transitions across modes of production. Reflection on cultural logic may help to develop more nuanced models that are less mechanistic by forcing us to consider alternative sets of assumptions about past behavior that had effects on behavior. A more complete understanding of cultural logic can also help to provide insight into the hows of the critical social transformations we see in these societies that move across modes of production. Cree hunters do not own animals, but Aymara pastoralists do. How does a model based on strict optimality principles deal with this transition except in the crudest manner? I offer no answers here, but we must learn how to take these processes into account in our reconstructions of the past.

DISCUSSION AND CONCLUSIONS

The expansion of the forager–collector model in the ways which I have outlined is both inevitable and necessary if it is have a prominent role in explaining change in the past. How this is to be accomplished archaeologically, however, is the salient issue, and although I offer no comprehensive solutions, I do outline some measures that can help to identify the effects of history, agency, contingency, and cultural logic in archaeological contexts.

History

Defining, then observing population replacement, the adoption of a mode of production, or the interaction of peoples with different modes of production is a challenging task for the archaeologist. Success is highly contingent on having copious quantities of high-quality, comparable data across a widespread area. An archaeological survey needs to be extensive, and it must be supplemented by extensive excavation. Moreover, chronology needs to be on a fine scale so that local developmental trajectories can be properly evaluated. These data in turn need to be synthesized into a coherent conception of archaeological cultures, traditions, or phases that have specific temporal and spatial domains. In this sense, history is consistent with a modern notion of defining group identity, ethnicity, or cultural affiliation in archaeological terms.

Relatively few regions satisfy these demands, but among them are Europe and the American Southwest. Across much of the European continent, indigenous populations of foragers have been defined. Source populations of agropastoralists who carry their suite of Neolithic cultural traits can be identified, and models of diffusion, migration, or interaction can thus be evaluated. Even here, however, there are still significant debates about varying interpretations of the archaeological evidence. Competing models are as simple as a dichotomy—colonization versus adoption

(Tringham 2000) or as complicated as this set of nonmutually exclusive possibilities: demic diffusion, folk migration, elite dominance, infiltration, enclave colonization, and individual frontier mobility (Zvelebil and Lillie 2000). Each model has its own set of expectations of the archaeological record, but even in this well-studied region, certain kinds of data, such as those on subsistence and settlement, are far too sparse to provide convincing tests of any of these alternatives. History, perhaps more than any other new condition to be added to the original forager–collector model, demands massive amounts of solid archaeological data.

Agency

Archaeologists are good at finding structure but are less adept at discovering agents or actors. Structure, of course, is the cultural framework in which agents act. Modes of production, as I have outlined them here, are structures within which individuals make choices, and among other things, the outcomes of those choices create the archaeological record. Archaeologists have become aware, however, that individuals do not simply respond blindly to their cultural training but can effect change. Practice theory, as outlined by Ortner (1989), demonstrates this from an ethnographic perspective and shows how individuals create structural change through manipulation and outright appropriation of existing structures for personal, but sometimes, group-oriented, ends.

In the transitions I have discussed, one profitable way in which to identify agents and agency is to consider the role of aggrandizers, or individuals prone to accumulation who attempt to dominate fellow community members, usually by economic means (Clark and Blake 1994). Accumulation is central to the pastoral mode of production because ownership of animals and pasturage are its key structural conditions. Archaeological indicators of aggrandizing behaviors are quite varied and depend substantially on social and political complexity (see Hayden 1995), as well as the relative success of the aggrandizer. In the case of pastoralists, we can expect status to be tied to larger herd sizes, which in archaeological terms may be observed by larger, more elaborate corrals (Aldenderfer 2001). The accumulation of valued goods may also be a part of this but should be subsidiary to herd size as a status marker. Indicators of feasting may also be present, especially those linked to alliance building and generating reciprocal obligations (Hayden 2001). Labor feasts are not likely in such contexts because the scarcity of labor does not tend to be a structural feature of pastoral society until herds become very large.

There should be no illusions about how easy it will be to see this kind of accumulation and feasting, but once it appears it is very suggestive of the erosion of an egalitarian ethos, further implying that a process of value transformation has begun and that a new mode of production may be emerging. Combined with insights obtained from historical analysis, a consideration of agency offers a powerful extension of the original forager–collector model.

Contingency

To approach contingency, as I have defined it in this chapter, archaeologists will have to work far more closely with specialists from allied fields than is often the case. In the Ilave transition, for example, the key contingent issue is the timing of the appearance of the T2 terrace system. Knowing if their formation is relatively early or relatively late provides a critical component in understanding how agents made decisions about diet and settlement mobility. Careful consultation with a geomorphologist and paleoclimatologist has led to the tentative conclusion that the terraces formed late. A similar issue confronted us in the Asana tradition, where detailed work by an engineering geologist helped us to understand the dynamics of landslides that radically transformed the environs of the site in a very short time in a number of instances. Understanding the role of contingency means identifying transformative events and also calculating their intensity and duration. The landslide at Asana was fairly local in almost every instance and probably did not affect resource availability across the valley. Other events, however, such as catastrophic mudslides that innundated whole villages on the southern Peruvian coast throughout much of prehistory, were far more serious and had much more severe, long-term effects.

Cultural Logic

As I have noted, the original forager–collector model is based in great part on an optimality assumption, that people make decisions about diet selection and mobility within a Western framework of rationality. As my example above shows, however, hunters and herders often make decisions in alternative frameworks that do not appear to be fully rational. It is not so much that these decisions are irrational, but instead that they are framed by very different sets of social and cultural structures and that they have an internal logic.

The problem, of course, is that we cannot directly observe a cultural logic because those who once carried it are long gone. We must, therefore, turn to ethnography and begin to build models that reflect alternative, but plausible logics. Although culture is highly labile, we are fortunate that the range of potential models is not infinite. We should not be seeking the meaning of the logics as much as what the archaeological signatures of those logics might be. I grant, though, that the problem of equifinality in this situation is substantial and that the models we construct may be no more than heuristics useful in guiding interpretation of the past.

I have argued in this chapter that although the forager–collector model has been a powerful heuristic tool within its domain, it requires substantial modification to expand that domain to deal with transitions of subsistence practice across modes of production as well as the emergence of more complex societies. These modifications include a more complete and nuanced consideration of history, agency, contingency, and cultural logic. I acknowledge that in many empirical settings it will be difficult to see the effects of these factors. I think it worth the effort to model these factors, however, because in the long run, I believe it will lead us to develop more plausible models of some of the most critical transformations that have characterized most of the history of our species.

ACKNOWLEDGMENTS

The assistance of the National Geographic Society (grant #5245–94) and two awards from the H. John Heinz III Charitable Trust and the National Science Foundation (SBR-9816313) are gratefully acknowledged.

REFERENCES

Abbott, M., Seltzer, G., Kelts, K., and Southon, J., 1997a, Holocene Paleohydrology of the Tropical Andes from Lake Records. *Quaternary Research* 47:70–80.

Abbott, M., Binford, M., Brenner, M., and Kelts, K., 1997b, A 3500 14C High-Resolution Record of Water-Level Changes in Lake Titicaca, Bolivia/Peru. *Quaternary Research* 47:169–180.

Aldenderfer, M., 1989, Archaic Period "Complementarity" in the Osmore Basin. In *Ecology, Settlement, and History in the Osmore Drainage, Peru*, edited by D. Rice, C. Stanish, and P. Scarr, pp. 101–128. British Archaeological Reports International Series 545, Oxford.

Aldenderfer, M., 1993, Ritual, Hierarchy, and Change in Foraging Societies. *Journal of Anthroplogical Archaeology* 12:1–40.

Aldenderfer, M., 1997, Informe Preliminar: Excavaciones Arqueologicas a Tres Sitios Arcaicos de la Cuenca del Rio Ilave, Sub-Region de Puno, Region "Jose Carlos Mariategui". Report submitted to Instituto Nacional de Cultura, Lima.

Aldenderfer, M., 1998, *Montane Foragers: Asana and the South-Central Andean Archaic*. University of Iowa Press, Iowa City.

Aldenderfer, M., 2001, Andean Pastoral Origins and Evolution: The Role of Ethnoarchaeology. In *The Ethnoarchaeology of Andean South America: Contributions to Archaeological Method and Theory*, edited by L. Kuznar, pp. 19–30. International Monographs in Prehistory, Ann Arbor.

Aldenderfer, M., in prep, *Quelcatani: The Prehistory of a Pastoral Lifeway on the High Puna*. Smithsonian Institution Press, Washington, DC.

Aldenderfer, M., and Klink, C., 1996, Archaic Period Settlement in the Lake Titicaca Basin: Results of a Recent Survey. Paper presented at the 36th Annual Meeting of the Institute for Andean Studies, Berkeley.

Baker, P., Seltzer, G., Fritz, S., Dunbar, R., Grove, M., Tapia, P., Cross, S., Rowe, H., and Broda, J., 2001a, The History of South American Precipitation for the Past 25,000 Years. *Science*. 291:640–643.

Baker, P., Rigsby, C., Seltzer, G., Fritz, S., Lowenstein, T., Bacher, N., and Veliz, C., 2001b, Tropical Climate Changes at Millennial and Orbital Timescales on the Bolivian Altiplano. *Nature* 409:698–701.

Bettinger, R., 1991, *Hunter-Gatherers: Archaeological and Evolutionary Theory*. Plenum, New York.

Binford, L., 1980, Willow Smoke and Dog's Tails: Hunter-Gatherer Settlement Systems and Archaeological Site Formation. *American Antiquity* 45:4–20.

Binford, L., 1982, The Archaeology of Place. *Journal of Anthropological Archaeology* 1:5–31.

Brown, J., 1985, Long-Term Trends to Sedentism and the Emergence of Complexity in the American Midwest. In *Prehistoric Hunter-Gatherers: The Emergence of Cultural Complexity*, edited by T.D. Price and J. Brown, pp. 201–231. Academic Press, Orlando.

Brightman, J., 1993, *Grateful Prey: Rock Cree Human-Animal Relationships*. University of California Press, Berkeley.

Clark, J., and Blake, M., 1994, The Power of Prestige: Competitive Generosity and the Emergence of Rank Societies in Lowland Mesoamerica. In *Factional Competition and Political Development in the New World*, edited by E. Brumfiel and J. Fox, pp. 17–30. Cambridge University Press, Cambridge.

Custred, G., 1979, Hunting Technologies in Andean Culture. *Journal de la Societe des Americanistes* 66:7–19.

Fagan, B., 1995, *Ancient North America*. Thames and Hudson: London.

Gebauer, A., and Price T. D., (eds.), 1992, *Transitions to Agriculture in Prehistory*. Prehistory Press, Madison, WI.

Hayden, B., 1995, Pathways to Power: Principles for Creating Socioeconomic Inequalities. In *Foundations of Social Inequality*, edited by T.D. Price and G. Feinman, pp. 15–86. Plenum, New York.

Hayden, B., 2001, Fabulous Feasts: A Prolegomenon to the Importance of Feasting. In *Feasts: Archaeological and Ethnographic Perspectives on Food, Politics, and Power*, edited by M. Dietler and B. Hayden, pp. 23–64. Smithsonian Institution Press, Washington, DC.

Hesse, B., 1985, Animal Domestication and Oscillating Environments. *Journal of Ethnobiology* 2:1–15.

Ingold, T., 1976, *Hunters, Pastoralists, and Ranchers: Reindeer Economies and their Transformations*. Cambrige University Press, Cambridge.

Jensen, P., and Kautz, R., 1974 Preceramic Transhumance and Andean Food Production. *Economic Botany* 28:43–55.

Kelly, R., 1995, *The Foraging Spectrum*. Smithsonian Institution Press, Washington, DC.

Klink, C., 1999, On the Edge: Prehistoric Trends of the Peruvian Altiplano Rim. Paper presented at the SAA Annual Meeting, Chicago.

Klink, C., In press, Archaic Period research in the Rio Huenque Valley, Peru. In *Advances in Andean Archaeology* 1, edited by C. Stanish, M. Aldenderfer, and A. Cohen. Cotsen Institute of Archaeology, University of California, Los Angeles.

Klink, C., and Aldenderfer, M., 1996, Archaic Period Settlement on the Altiplano: Comparison of Two Recent Surveys in the Southwestern Lake Titicaca Basin. Paper presented at the 24th Annual Midwest Conference of Andean and Amazonian Archaeology, Beloit, WI.

Kuznar, L., 1989, The Domestication of Camelids in Southern Peru: Models and Evidence. In *Ecology, Settlement, and History in the Osmore Drainage, Peru*, edited by D. Rice, C. Stanish, and P. Scarr, pp. 167–182. British Archaeological Reports International Series 545, Oxford.

Kuznar, L., 1990, Economic Models, Ethnoarchaeology, and Early Pastoralism in the High Sierra of the South-Central Andes, Ph.D. Dissertation, Department of Anthropology, Northwestern University.

Kuznar, L., 1993, Mutualism between Chenopodium, Herd Animals, and Herders in the South-central Andes. *Mountain Research and Development* 13:257–265.

Lavallee, D., 1985, *Chasseurs et Pasteurs Prehistoriques des Andes*, I. Editiones Recherches sur les Civilisations, Paris.

Leacock, E. and R. Lee, 1982, Introduction. In *Politics and History in Band Societies*, edited by E. Leacock and R. Lee, pp. 1–20. Cambridge University Press, Cambridge.

Ortner, S., 1989, *High Religion: A Cultural and Political History of Sherpa Buddhism*. Princeton University Press, Princeton, NJ.

Paine, R., 1971, Animals as Capital: Comparisons among Northern Nomadic Herders and Hunters. *Anthropological Quarterly* 44(3):157–172.

Pearsall, D., 1980, Pachamachay Ethnobotanical Report: Plant Utilization at a Hunting Base. In *Prehistoric Hunters of the High Andes*, edited by J. Rick, pp. 191–231. Academic Press, New York.

Pearsall, D., 1989a, Adaptation of Prehistoric Hunter-Gatherers to the High Andes: The Changing Role of Plant Resources. In *Foraging to Farming: The Evolution of Plant Exploitation*, edited by D. Harris and G. Hillman, pp. 318–332. Unwin Hyman, London.

Pearsall, D., 1989b *Paleoethnobotany*. Academic Press, Orlando, FL.

Pearsall, D., 1992 South America. In *The Origins of Agriculture: An International Perspective*, edited by C. Cowan and P. Watson, pp. 135–160. Smithsonian Institution Press, Washington, DC.

Pires-Ferreira, W., Pires-Ferreira, E., and Kaulicke, P., 1976, Preceramic Animal Utilization in the Central Peruvian Andes. *Science* 194:483–490.

Price, T. D., and Brown J., (eds.), 1985, *Prehistoric Hunter-Gatherers: The Emergence of Cultural Complexity*. Academic Press, Orlando, FL.

Price, T.D., and Gebauer A., (eds.), 1995, *Last Hunters, First Farmers: New Perspectives on the Prehistoric Transition to Agriculture*. School of American Research Press, Santa Fe.

Rick, J., 1980, *Prehistoric Hunters of the High Andes*. Academic Press, New York.

Rigsby, C. A., Baker, P. A., and Aldenderfer, M., 2000, Terrace Development along the Rio Ilave (Peru) in Response to Changes in Sediment Load and the Level of Lake Titicaca: *EOS Transactions, American Geophysical Union*, V. 81, Fall Meeting supplement, Abstract V21E-16.

Salomon, F. 1985, The Dynamic Potential of the Complementarity Concept. In *Andean Ecology and Civilization*, edited by S. Masuda, I. Shimada, and C. Morris, pp. 211–231. University of Tokyo Press, Tokyo.

Seltzer, G. P., Baker, P., Cross, S., and Dunbar, R., 1998, High-Resolution Seismic Reflection Profiles from Lake Titicaca, Peru-Bolivia: Evidence for Holocene Aridity in the Tropical Andes. *Geology* 26:167–170.

Smith, A., 1990, On Becoming Herders: Khoikhoi and San Ethnicity in Southern Africa. *African Studies* 49(2):51–74.

Smith, B. D., 1995, *The Emergence of Agriculture*. Scientific American Library, New York.

Spielmann, K., 1986, Interdependence among Egalitarian Societies. *Journal of Anthropological Archaeology* 5:279–312.

Tringham, R., 2000, Southeastern Europe in the Transition to Agriculture in Europe: Bridge, Buffer, or Mosaic. In *Europe's First Farmers*, edited by T. Douglas Price, pp. 19–56. Cambridge University Press, Cambridge.

Winterhalder, B., and Goland, C., 1997, An Evolutionary Ecology Perspective on Diet Choice, Risk, and Plant Domestication. In *People, Plants, and Landscapes: Studies in Paleoethnobotany*, edited by K. Gremillion, pp. 123–160. University of Alabama Press, Tuscaloosa, AL.

Wirrmann, D., and Mourguiart, P., 1995, Late Quaternary Spatio-Temporal Limnological Variations in the Altiplano of Bolivia and Peru. *Quaternary Research* 43:344–354.

Zvelebil, M., and Lillie, M., 2000, Transition to Agriculture in Eastern Europe. In *Europe's First Farmers*, edited by T. D. Price, pp. 57–92. Cambridge University Press, Cambridge.

Afterword

Beyond Foraging and Collecting: Retrospect and Prospect

T. DOUGLAS PRICE

In a summary of a volume such as this, it may be useful to look both backward and forward at the state of hunter-gatherer studies in archaeology. In the following pages then, I hope to provide that retrospect and prospect, along with comment on the chapters. I will begin with a very brief personal view of the history of the last 50 years or so of research. I will combine my remarks on the contents of the volume with some general observations on the current state of hunter-gatherer studies. I will conclude with a few thoughts on the future of such investigations.

RETROSPECT

Recent years have witnessed some stagnation in hunter-gatherer studies, following an extraordinarily productive period. The dispersal of anthropologists to remote parts of the planet after the Second World War fostered many new and exciting studies of foragers on five continents. The plethora of ideas and information that emerged culminated in the *Man the Hunter* conference held in Chicago in 1966.

In turn, that meeting and the subsequent volume (Lee and DeVore 1968a) spawned a multitude of new research and insights. Although the

T. DOUGLAS PRICE • Department of Anthropology, University of Wisconsin-Madison, Madison, Wisconsin 53706.

conference summary statement by the organizers emphasized small and mobile groups (Lee and DeVore 1968b), it was precisely those concepts that were challenged in the following years. Many significant hunter-gatherer studies appeared during the 1970s and 1980s as new ideas involving evolutionary ecology, affluence, storage, and other, more complex behaviors took hold.

Several major directions in these studies can be identified. One track has countered the small and mobile model of Richard Lee and Irven DeVore and emphasized larger, more sedentary, and more complex foragers. Marshall Sahlins' 1972 discourse on *Stone Age Economics* and his description of an original affluent society substantially revised thinking about the harshness and simplicity of foragers' lives (Sahlins 1972). Two seminal papers at the beginning of the 1980s emphasized the diversity of hunter-gatherer adaptations and initiated a new wave of archaeological investigations. Lewis Binford published "Willow Smoke and Dog's Tails: Hunter-Gatherer Settlement Systems and Archaeological Site Formation" (Binford 1980)—the focal point of the present volume—and James Woodburn wrote "Egalitarian Societies" (Woodburn 1982). These papers emphasized a dichotomy in hunter-gatherer lifestyles. Binford distinguished foraging versus collecting strategies in the structuring of settlement. Woodburn focused on economic behavior and defined immediate versus delayed returns as axes of hunter-gatherers activity.

Almost simultaneously, new volumes appeared that focused on affluence and greater complexity among hunter-gatherers. In 1981, Shuzo Koyama and David Thomas edited *Affluent Foragers*, a review of hunter-gatherers around the Pacific Rim (Koyama and Thomas 1981). The 1985 volume *Prehistoric Hunter-Gatherers: the Emergence of Cultural Complexity* that James Brown and I (Price and Brown 1985) compiled documented this distinction archaeologically among past groups on several continents.

A second major track in hunter-gatherer studies led from concerns with subsistence and the role of the environment in human adaptation toward the growth of ecological and mathematical models. In 1976 Michael Jochim published *Hunter-Gatherer Subsistence and Settlement* (Jochim 1976), a prelude to evolutionary ecological approaches. Brian Hayden's 1981 paper on "Subsistence and Ecological Adaptations of Modern Hunter-Gatherers" (Hayden 1981) continued this emphasis. A series of publications by Jim O'Connell (O'Connell and Hawkes 1981; O'Connell et al. 1988), Eric Alden Smith (Smith 1983), Bruce Winterhalder (Winterhalder 1986; Winterhalder and Smith 1981), and others moved optimal foraging theory from biology into archaeology.

The most comprehensive statement from this school of thought has been Robert Bettinger's 1991 volume *Hunter-Gatherers—Archaeological and Evolutionary Theory*. From this perspective, the behavior of foragers is determined largely by attempts to maximize their net rate of energy gain (Bettinger 1991:84). In that volume, Bettinger expounded the traveler-processor model for hunter-gatherers, along the lines of the earlier formulations of Binford and Woodburn. Travelers are characterized by low population densities, brief settlement stays, long distances between settlements, narrow spectrum resource use, and lower competitive fitness than processors who have the contrasting characteristics of high density, extended settlement duration, and broad spectrum resource exploitation. Food procurement costs among travelers come from movement and search time, whereas procurement and preparation place greater demands on processors.

A third direction in hunter-gatherer studies in the last 25 years has moved away from ecological approaches and emphasized the social and ideational aspects of forager behavior as primary. Much of this discussion revolved around change in hunter-gatherer adaptations, specifically the transition to farming. Barbara Bender principled this school arguing that the transition from foraging to food production was about change, not in food, but in production (Bender 1978). Kristina Jennbert expanded this concept in 1984 in *The Productive Gift* (my translation), arguing that the only obvious reason that hunter-gatherers in Mesolithic Sweden adopted farming was to generate surplus.

In 1982, Polly Wiessner pointed out that ecological models, that are focused toward environmental constraints on human populations ignore important components of behavior. Wiessner (1982) argued that societies construct social relationships to mediate differences in resource availability across time and space. Such relationships are as important as environmental constraints in structuring settlement. These studies presaged aspects of the postprocessual focus on social, ideational, and cognitive perspectives in archaeology.

Amid the emergence of these theoretical directions, the field of ethnoarchaeology was born. Extraordinary studies of the material culture of living hunter-gatherers were undertaken by archaeologists and others, for example, Wendell Oswalt and James VanStone (Oswalt and Stone 1967), Desmond Clark (Clark 1968), Richard Gould (Gould 1971), Wiessner (Wiessner 1977), John Yellen (Yellen 1977), Binford (Binford 1978), and O'Connell (O'Connell 1979). These investigations have profoundly impacted our thinking about foragers. Direct observation of the function and behavior of material culture in the hands of active, living peoples has provided a powerful means of searching for patterns in the past.

Compared to the previous 10 years, however, the last decade has been a dormant phase in the study of hunter-gatherers. Rather than new ideas and information, the 1990s witnessed restatements of the diversity of forager adaptations and reexaminations of fundamental assumptions and prior models.

Robert Kelly's *The Foraging Spectrum: Diversity in Hunter-Gatherer Lifeways* (Kelly 1995) was a bright spot in this otherwise pale decade, but nevertheless involved more surveying than pathbreaking. Susan Kent (Kent 1992) reiterated the prevalence of variability over pattern in a discussion of real versus ideal views of hunter-gatherers (e.g., Kent 1996). Such studies, along with our growing knowledge of hunter-gatherer adaptations, force us to admit that variability—rather than small and mobile, or dichotomies of simple versus complex—defines the world of foragers. Hunter-gatherers are fluid and flexible in their behaviors and difficult to categorize.

Ethnoarchaeological fieldwork also abated in the 1990s in spite of its utility and potential for theory building (O'Connell 1995). New ethnographies of hunter-gatherers have been few in number in recent years, as foraging groups have disappeared in the face of globalization. Moreover, the concerns of cultural anthropologists have shifted from descriptive studies to activist roles involved more with land tenure and native rights (e.g., Lee 1992). Ethnoarchaeological fieldwork has also declined as political and economic changes have reduced the options for such investigations.

Questions continue to be raised about some of the basic assumptions made regarding hunter-gatherers. Kaplan, for example, examined the "darker side" of affluence, specifically revisiting the original sources on affluent hunter-gatherers and concluding that these were flawed and misinterpreted (Kaplan 2000). Kaplan and others (e.g., Hawkes 1993; Hitchcock and Ebert 1984) questioned notions that hunter-gatherers work only 3–5 hours per adult per day in food collection, as Sahlins argued in his original depiction of affluent forager societies. As Kaplan noted, leisure sometimes has come at the expense of hunger.

THE PRESENT VOLUME

The splash of *Man the Hunter* is still felt in the disciplinary pool, but the ripples have almost disappeared. The trajectory of the 1990s suggests that the study of hunter-gatherers may be an endangered realm of scholarly investigation. Fortunately, several recent signs show that this is not the case. I anticipate that the coming years will see an efflorescence of hunter-gatherer studies in the wake of the publication of the *Cambridge Encyclopedia of Hunter-Gatherers*, edited by Richard Lee and Richard Daly (Lee and Daly 1999), and Lewis Binford's (2001) *Constructing Frames of Reference; Using Ethnographic and Environmental Data Sets.*

This volume also raises hopes and expectations for the future of hunter-gatherer studies and archaeology more generally. It has been a pleasure to read these chapters and to learn more of the intriguing investigations that are ongoing in the realm of prehistoric hunter-gatherers. My comments will address some of the themes that characterize these chapters and will try to highlight specific contributions. Although this volume is focused specifically on settlement dynamics and change in the context of Binford's forager/collector model, the chapters cover the gamut of hunter-gatherer adaptations.

1. Stasis versus Change

The title of the volume emphasized on evolutionary change in hunter-gatherer settlement systems. This is indeed an innovative aspect of the chapters. Most discussions of hunter-gatherers have assumed a condition of stasis or stability. Change is difficult to monitor and model, but a number of chapters deal directly with this question and provide valuable insight.

Fitzhugh examines the prehistory of the Kodiak archipelago during a 1500-year period. The basic settlement system of winter villages and seasonal campsites in this region was established as a consequence of the shift to a storage economy approximately 3500 years ago. Subsequent population growth and social competition led to the development of greater social inequality. Fitzhugh concludes that the emergence of inequality, extensive political alliance, warfare, and exchange during that period had relatively little effect on the underlying subsistence-settlement system. This study demonstrates both the utility of the forager/collector model and the ecological theory that underlies it for tracking changes in subsistence and settlement through time and also emphasizes the importance of the social environment. In one sense, Fitzhugh reports on stability in the context of change.

Junko Habu considers the role of long-term change among hunter-gatherer adaptations of Middle Jomon Japan. She notes that settlement pattern changes in one region during this period reflect a shift from collecting to foraging and that this shift influences Jomon adaptations in another region in Japan. Again, the point is made that factors beyond resource spacing and availability structure the organization of hunter-gatherers.

Laura Lee Junker also addresses the question of long-term change in hunter-gatherer settlement systems and contrasts such patterns with short-term shifts. Her fascinating study of the continuity of hunter-gatherer adaptation through short-term changes in strategies for obtaining foods in the limited environment of the tropical rain forest is a testament to the flexible and opportunistic nature of foragers. These studies document both the possibilities for examining long-term change among prehistoric hunter-gatherers and the variability that characterizes such adaptations.

2. Complexity and Inequality

One of the recurrent themes addressed in the volume is the issue of complexity and its corollary of status inequality among hunter-gatherers. More complex foragers are discussed from several areas including the Northwest Coast of North America, insular Alaska, and the Thule region. Both Ken Ames and Fitzhugh point to coasts as a special context and maritime hunter-gatherers as a special case of complexity characterized by social inequality, among other things. On the other hand, Junko Habu points out the absence of status differentiation in Jomon Japan and asks "Why?" Middle Jomon adaptations are defined by logistically organized collector systems, sophisticated material culture, major ceremonial features, long-distance trade, and large, sedentary communities. Nevertheless there is no indication of hereditary status differentiation in the archaeological record of this period.

Several points emerge. First, in spite of almost 20 years of discussion, the concept of complexity among foraging groups is still not well defined. There is general consensus that complexity means bigger groups, longer stays, more elaborate technology, intensified subsistence, broader resource utilization, and the like. But does complexity also mean status differentiation?

Status differentiation and social inequality are frequently attributed to more complex foragers. This certainly appears to be appropriate for certain ethnographically or historically known hunter-gatherers, especially for groups along the western coast of North America (e.g., Ames and Maschner 1999; Matson and Coupland 1995), among the Inuit whale hunters of northern Alaska (e.g., Sheehan 1985), and perhaps the Calusa on the southern coast of Florida (e.g., Marquardt and Payne 1992; Widmer 1988).

But the number of ethnographically known groups of hunter-gatherers that exhibit hereditary social inequality is strikingly small and limited in both recent time and geographic space. It is also clear that most ethnographically known groups of hunter-gatherers lack such hierarchical organization and can be viewed as largely egalitarian. This evidence suggests that we must define the meaning of complexity more rigorously and be very cautious in attributing status differentiation to hunter-gatherers in the past.

We must also be aware, as James Savelle clearly cautions, that shifts between more and less complex postures are another aspect of the flexibility of hunter-gatherer adaptations. Savelle, writing on the decline of Thule, emphasized that increasing complexity and the shift toward collecting strategies is not a one-way street for hunter-gatherers.

3. Diversity and Variability

Perhaps the most important lesson from this volume is a reiteration of the enormous diversity within and among hunter-gatherer adaptations. We are still learning how much variability exists in both the ethnographic and the archaeological record. In a sense, we are still placing points on the scatter plot. Until we more fully identify more of the variability that is present, it is going to be difficult to observe meaningful patterns.

Many of the chapters dealt with diversity and variability among prehistoric hunter-gatherers. David Zeanah's paper on pinyon harvesting is a document to hunter-gatherer diversity. Within the Great Basin, groups that share the same language, technology, and culture exhibit a wide range of adaptations from foragers to collectors. And within the activity class of pinyon harvesting alone, Zeanah notes a wide range of strategies to accomplish this task. Aubrey Cannon's comment that simple models are inadequate to account for the richness of hunter-gatherer data reflects this enormous diversity and emphasizes how little we know.

4. Unknowns and Missing Data

One of the important, yet unaddressed, aspects of any study is the role of the unknown and/or missing data. Often this is a situation in which we are aware that information is missing but conveniently ignore that fact in constructing models and offering interpretations. This is almost inevitable in the study of the past. We know very little about the function of most artifacts and features; we miss large components of a settlement system; subsistence and seasonality data are incomplete. Unknowns are also a major problem in terms of past environmental conditions, changes in sea level, and other factors that mask or modify the archaeological record and the conditions under which past foragers lived.

A brief example from northern Europe may convey the concern. During the early part of the Mesolithic period in this area, ca. 7000 B.C., sites are found only inland, they are small and seasonal, and have limited assemblages of tools and faunal remains. During the later Mesolithic, 2000 years later, most sites are coastal, large, sedentary, and have cemeteries for the dead and a wide range of resources. Do we interpret this as major shift from foraging to collecting, from simple to complex, as has often been suggested? The answer in this case is no. This shift is apparent, not real. Rising sea levels in the Holocene submerged the coastal facies of the earlier part of the Mesolithic, making it inaccessible to archaeology.

The urge to ignore missing data is strong. The absence of information in one sense makes solutions easier to find in archaeology. Nevertheless, it is essential to incorporate the range of possibilities that are absent, as well as present, in attempts to describe hunter-gatherer adaptations.

5. The Use of Models

A positive step in dealing with unknown variability and missing data is seen in the chapters in this volume. I particularly enjoyed Mark Aldenderfer's section in this regard. He is pretty smart for an enlightened processualist. Concerned with the shift from hunting to herding in the south-central Andes, Alderderfer notes how the forager–collector model fails to explain changes in site location over time. In the gap between model and data, he argues, are to be found the effects of history, agency, contingency, and cultural logic. His point, and a most important one, is that we must consider these issues as well if we are to understand settlement and subsistence in prehistory.

Many of the chapters in this volume point to missing variables in this gap between model and reality. Ken Ames, for example, highlights the role of transportation in models of hunter-gatherer adaptations and specifically the impact on the forager–collector model. Aubrey Cannon's discussion of sacred seasons among the hunter-gatherers of the central Northwest Coast of North America was particularly engaging in this context. Cannon reports on the disjunction between model and data that he explains in terms of social constraints and opportunities. The model settlement system based on resource availability and distribution offers a foundation for understanding the basic pattern of site location on the landscape, in place as early as 5000 B.C., but does not explain subsequent developments. The stability of this pattern is remarkable, given that population growth and social conflict should have produced a much more varied pattern of settlement. Cannon argues that religious belief and investment in both ritual and residential permanence prohibited the establishment of alternative village locations.

An important trend in the chapters in this volume is the use of models such as forager–collector as yardsticks, rather than as solutions. The model becomes a means to understand the past, rather than an end in itself. Focus is on the differences between the expected from the model and the observed in the record, rather than on the model itself. I would suspect that it is in this space between expected and observed that hunter-gatherer studies will advance in the coming years, pursuing some of the factors outlined by Aldenderfer and others.

6. Marginal Environments

In reading the chapters, I was struck by the active presence of hunter-gatherers in several parts of the world where no one should have lived during much of the Holocene. There are lots of interesting questions to be answered about such behavior, but a primary one is why did people live there in the first place? For example, Renato Kipnis describes Late Pleistocene/Early Holocene hunter-gatherers in an arid region of central Brazil where megafauna were extinct and large fauna mostly impoverished. Large stretches of the Great Basin are also less than optimal habitats for hunter-gatherers to occupy. It is surprising to find hunter-gatherers in such habitats at relatively early dates.

Were these groups forced into marginal environments by the expansion of farming populations? The deep prehistory of occupation suggests otherwise. Why did they go there? Why did they stay? Foraging theory does not answer these questions. If hunter-gatherer groups had taken a one-semester course in optimal foraging theory they would have never traveled to these places. On the other hand, the presence of these groups in some of the last places that the gods made suggests that we may have underestimated the environmental changes that have taken place in these areas or perhaps overestimated the human presence there. Marginal regions are clearly of great importance for enhancing our understanding of hunter-gatherers and should be the subject of further investigation.

7. Definition of Hunter-Gatherers

There is a final grand, and difficult, question that needs to be raised in this discussion. What do we really mean by "hunter-gatherers"? Does the term refer to an idealized concept of prefarming human adaptations, or do we intend the reality of the present and

> past world? How rigid is the definition of hunter-gatherers? Does this term define people who depend solely on the wild resources of the earth? Or do we allow that some horticulture, arboriculture, cultivation, or interaction with farmers may be part of this way of life?

If we accept that the Northwest Coast groups were planting roots and tubers (Ames and Maschner 1999), that European Mesolithic groups may have cultivated hazel groves (Pedersen 1997; Price et al. in press), that Jomon groups were involved with a variety of cultigens (Crawford 1992, Habu, this volume), or that Peruvian hunter-gatherers were interacting with pastoralists (Aldenderfer, this volume), where do we draw the line between the last hunters and the first farmers?

Can we discuss foraging adaptations outside the influence of farming and farming societies? We need to be source critical and distinguish the role of plant manipulation, and particularly cultivation, in the subsistence activities of hunter-gatherers. In the same vein, it is essential to consider potential relationships between hunter-gatherers and more hierarchical societies, particularly in the latter half of the Holocene. Junker's discussion of the exchange relationships between foragers and chiefdoms in the Philippines was very enlightening in this respect.

I am not suggesting that this question about the meaning of the term hunter-gatherer is an easy one to answer in a prehistoric context. But I do think that the role of plants and horticulture among hunter-gatherers and relations with farming populations needs a great deal more attention than it has previously received. Archaeologists and ethnographers have dismissed or ignored these relationships, confounding our understanding of what it means to be a hunter-gatherer. It is essential to decouple the impact of cultivation and external relationships in attempts to model hunter-gatherer behavior in deep time.

PROSPECT

Our understanding of the evolution of human society begins with hunter-gatherers as a fundamental baseline for comprehending ourselves. We cannot understand us without a view of our past. As this picture comes more clearly into view, most of the characteristics that distinguish preliterate states and even modern civilization can be seen to have their roots in our foraging past. But there are immediate obstacles to hunter-gatherers studies, including the demise of ethnography and the disappearance of both contemporary and prehistoric groups.

The beginning of the new millennium marks a sad milestone in the study of hunter-gatherers—the end of this way of life in most parts of the world. For years, anthropological textbooks have charted the geographic location of foraging groups on global maps and noted their declining numbers through time. Those maps are now blank.

In virtually every corner of the planet, hunter-gatherers have been incorporated into the global economy. There are charter excursions to the Andaman Islands. Kalahari Distillers Ltd. sells a glass calabash filled with "Kalahari Thirstland Liqueur." Skin drums have been replaced by electric guitars, seal liver by Carnation instant breakfast in the lives of the Netsilik. It is a poignant moment, yet a stark reminder of the great distance that remains to be covered in understanding the foraging way of life.

At the same time, archaeological sites are disappearing beneath the growing modern world. The expansion of cities, the growth of energy and transportation demands, agrobusiness, and acid rain daily reduce the quantity and quality of the archaeological record. There is urgent need to intensify our efforts to record, describe, and comprehend the hunter-gatherer way of life. This volume represents a major step in that direction.

ACKNOWLEDGMENTS

I thank the editors, Junko Habu and Ben Fitzhugh, for the invitation and opportunity to participate in their lively symposium at the 2000 meetings of the Society for American Archaeology and to prepare these comments as an afterword to this volume. I also sincerely thank Jim O'Connell for sharing parts of his knowledge and perspective on hunter-gatherer studies.

REFERENCES

Ames, K. M., and Maschner, H. D. G., 1999, *Peoples of the Northwest Coast: Their Archaeology and Prehistory*. Thames and Hudson, New York.

Bender, B., 1978, From Gatherer-Hunter to Farmer: A Social Perspective. *World Archaeology* 10:204–222.

Bettinger, R. L., 1991, *Hunter–Gatherers: Archaeological and Evolutionary Theory, Interdisciplinary Contributions to Archaeology*. Plenum, New York and London.

Binford, L. R., 1978, *Nunamiut Ethnoarchaeology*. Academic Press, New York.

Binford, L. R., 1980, Willow Smoke and Dog's Tails: Hunter-Gatherer Settlement Systems and Archaeological Site Formation. *American Antiquity* 45:4–20.

Binford, L. R., 2001, *Constructing Frames of Reference; Using Ethnographic and Environmental Data Sets*. University of California Press, Berkeley.

Clark, J. D., 1968, Studies of Hunter-Gatherers as an Aid to the Interpretation of Prehistoric Societies. In *Man the Hunter*, edited by R. Lee and I. DeVore, pp. 276–280. Aldine, Chicago.

Crawford, G. W., 1992, The Transitions to Agriculture in Japan. In *Transitions to Agriculture in Prehistory*, edited by A. B. Gebauer and T. D. Price, pp. 117–132. Prehistory Press, Madison, WI.

Gould, R. A., 1971, The Archaeologist as Ethnographer. *World Archaeology* 3:143–177.

Hawkes, K., 1993, Why Hunter-Gatherers Work: An Ancient Version of the Problem of Public Goods. *Current Anthropology* 34:341–361.

Hayden, B., 1981, Subsistence and Ecological Adaptations of Modern Hunter-Gatherers. In *Omnivorous Primates*, edited by R. S. O. Harding and G. Teleki, pp. 344–421. Columbia University Press, New York.

Hitchcock, R. K., and Ebert, J. I., 1984, Foraging and Food Production among Kalahari Hunter-Gatherers. In *From Hunters to Farmers: The Causes and Consequences of Food Production in Africa* edited by J. D. Clark and S. A. Brandt, pp. 328–348. University of California Press, Berkeley.

Jennbert, K., 1984, Den Produktiva Gåvan. Tradition och Innovation i Sydskandinavien för Omkring 5 300 År Sedan. *Acta Archaeologica Lundensia* 4:16.

Jochim, M., 1976, *Hunter-Gatherer Subsistence and Settlement, A Predictive Model*. Academic Press, New York.

Kaplan, D., 2000, The Darker Side of the "Original Affluent Society," *Journal of Anthropological Research* 56:301–324.

Kelly, R., 1995, *The Foraging Spectrum: Diversity in Hunter-Gatherer Lifeways*. Smithsonian Instution Press, Washington DC.

Kent, S., 1992, The Current Forager Controversy: Real Versus Ideal Views of Hunter-Gatherers. *Man* 27:45–70.

Kent, S., (ed.) 1996, *Cultural Diversity among Twentieth-Century Foragers*, Cambridge University Press, Cambridge.

Koyama, S., and Thomas, D. H., 1981, *Affluent Foragers*. Senri Ethnological Studies 9.

Lee, R. B., 1992, Art, Science, or Politics? The Crisis in Hunter-Gatherer studies. *American Anthropologist* 94:31–54.

Lee, R. B., and Daly, R., (eds.), 1999, *Cambridge Encyclopedia of Hunters and Gatherers*. Cambridge University Press, Cambridge.

Lee, R. B., and DeVore, I., 1968a, *Man the Hunter*. Aldine, Chicago.

Lee, R. B., and DeVore, I., 1968b, Problems in the Study of Hunter and Gatherers. In *Man the Hunter*, edited by R. B. Lee and I. DeVore, pp. 3–12. Aldine, Chicago.

Marquardt, W. H., and Payne, C., (eds.), 1992, *Culture and Environment in the Domain of the Calusa*. Institute of Archaeology and Paleoenvironmental Studies, University of Florida, Gainesville.

Matson, R. G., and Coupland, G., 1995, *Prehistory of the Northwest Coast*. Academic Press, San Diego.

O'Connell, J. F., 1979, Room to Move: Contemporary Alyawara Settlement Patterns and Their Implications for Aboriginal Housing Policy. In *A Black Reality: Aboriginal Camps and Housing in Remote Australia*, edited by M. Heppell. Australian Institute of Aboriginal Studies, Canberra.

O'Connell, J. F., 1995, Ethnoarchaeology Needs a General Theory of Behavior. *Journal of Archaeological Research* 3:205–255.

O'Connell, J. F., and Hawkes, K., 1981, Alywara Plant Use and Optimal Foraging Theory. In *Hunter-Gatherer Foraging Strategies: Ethnographic and Archaeological Analyses*. University of Chicago Press, Chicago.

O'Connell, J. F., Hawkes, K., and Blurton-Jones, N., 1988, Hadza Hunting, Butchering, and Bone Transport and Their Archaeological Implications. *Journal of Anthropological Research* 44:113–161.

Oswalt, W. H., and Stone, J. W. V., 1967, *The Ethnoarchaeology of Crow Village, Alaska*, Vol. 199. *Bureau of American Ethnology Bulletin*, Washington, DC.

Pedersen, L., 1997, They Put Fences in the Sea. In *The Danish Storebælt Since the Ice Age*, edited by A. F. L. Pedersen, and B. Aaby, pp. 124–144. A/S Storebælt Fixed Link, Copenhagen.

Price, T. D., Gebauer, A. B., Hede, S. U., Larsen, C. S., Mason, S., Nielsen, J., Noe-Nygaard, N., and Perry, D., In press, Excavations at Smakkerup Huse: Mesolithic Settlement in Northwest Zealand. *Journal of Field Archaeology*.

Price, T. D., and Brown, J. A., 1985, *Prehistoric Hunter-Gatherers: The Emergence of Cultural Complexity*. Academic Press, New York.

Sahlins, M., 1972. *Stone Age Economics*. Aldine, Chicago.

Sheehan, G. W., 1985, Whaling as an Organizing Focus in Northwestern Alaskan Eskimo Society. In *Prehistoric Hunter-Gatherers*, edited by T. D. Price and J. A. Brown, pp. 123–154. Academic Press, Orlando.

Smith, E. A., 1983, Anthropological Applications of Optimal Foraging Theory: A Critical Review. *Current Anthropology* 24:625–651.

Widmer, R. J., 1988, *The Evolution of the Calusa: A Nonagricultural Chiefdom on the Southwest Florida Coast*. University of Alabama Press, Tuscaloosa, AL.

Wiessner, P., 1982, Beyond Willow Smoke and Dog's Tails: A Comment on Binford's Analysis of Hunter-Gatherer Settlement Systems. *American Antiquity* 47:171–178.

Wiessner, P. W., 1977, Hxaro: A Regional System of Reciprocity for Reducing Risk among the !Kung San, Ph. D. Dissertation, University of Michigan, Ann Arbor.

Winterhalder, B., 1986, Diet Choice, Risk, and Food Sharing in a Stochastic Environment. *Journal of Anthropological Archaeology* 5:369–392.

Winterhalder, B. P., and Smith, E. A., 1981, *Hunter-Gatherer Foraging Strategies: Ethnographic and Archaeological Analyses*. University of Chicago Press, Chicago.

Woodburn, J., 1982, Egalitarian Societies. *Man* 17:431–451.

Yellen, J. E., 1977, *Archaeological Approaches to the Present: Models for Reconstructing the Past*. Academic Press, New York.

Index